VOLUME ONE HUNDRED AND TWENTY EIGHT

INTERNATIONAL REVIEW OF NEUROBIOLOGY

Big on BK: Current Insights into the Function of Large Conductance Voltage- and Ca^{2+}-Activated K^+ Channels at the Molecular, Cellular, and Systemic Levels

INTERNATIONAL REVIEW OF NEUROBIOLOGY

VOLUME 128

SERIES EDITORS

R. ADRON HARRIS

Waggoner Center for Alcohol and Drug Addiction Research
The University of Texas at Austin
Austin, Texas, USA

PETER JENNER

Division of Pharmacology and Therapeutics
GKT School of Biomedical Sciences
King's College, London, UK

VOLUME ONE HUNDRED AND TWENTY EIGHT

INTERNATIONAL REVIEW OF NEUROBIOLOGY

Big on BK: Current Insights into the Function of Large Conductance Voltage- and Ca^{2+}-Activated K^+ Channels at the Molecular, Cellular, and Systemic Levels

Edited by

CANDICE CONTET

Committee on the Neurobiology of Addictive Disorders,
The Scripps Research Institute,
La Jolla, CA, United States

ELSEVIER

AMSTERDAM • BOSTON • HEIDELBERG • LONDON
NEW YORK • OXFORD • PARIS • SAN DIEGO
SAN FRANCISCO • SINGAPORE • SYDNEY • TOKYO
Academic Press is an imprint of Elsevier

Academic Press is an imprint of Elsevier
50 Hampshire Street, 5th Floor, Cambridge, MA 02139, USA
525 B Street, Suite 1800, San Diego, CA 92101-4495, USA
The Boulevard, Langford Lane, Kidlington, Oxford OX5 1GB, UK
125 London Wall, London, EC2Y 5AS, UK

First edition 2016

Notices

Knowledge and best practice in this field are constantly changing. As new research and experience broaden our understanding, changes in research methods, professional practices, or medical treatment may become necessary.

Practitioners and researchers must always rely on their own experience and knowledge in evaluating and using any information, methods, compounds, or experiments described herein. In using such information or methods they should be mindful of their own safety and the safety of others, including parties for whom they have a professional responsibility.

To the fullest extent of the law, neither the Publisher nor the authors, contributors, or editors, assume any liability for any injury and/or damage to persons or property as a matter of products liability, negligence or otherwise, or from any use or operation of any methods, products, instructions, or ideas contained in the material herein.

ISBN: 978-0-12-803619-8
ISSN: 0074-7742

Publisher: Zoe Kruze
Acquisition Editor: Kirsten Shankland
Editorial Project Manager: Hannah Colford
Production Project Manager: Surya Narayanan Jayachandran
Cover Designer: Maria Ines Cruz

Typeset by SPi Global, India

CONTENTS

CONTRIBUTORS

A.L. Barth
Carnegie Mellon University, Pittsburgh, PA, United States

A.N. Bukiya
College of Medicine, The University of Tennessee Health Science Center, Memphis, TN, United States

C. Contet
Committee on the Neurobiology of Addictive Disorders, The Scripps Research Institute, La Jolla, CA, United States

A.M. Dopico
College of Medicine, The University of Tennessee Health Science Center, Memphis, TN, United States

P.J. Duncan
Centre for Integrative Physiology, College of Medicine & Veterinary Medicine, University of Edinburgh, Edinburgh, United Kingdom

R.K. Duncan
Kresge Hearing Research Institute, University of Michigan, Ann Arbor, MI, United States

T.-M. Gao
State Key Laboratory of Organ Failure Research, Key Laboratory of Psychiatric Disorders of Guangdong Province, School of Basic Medical Sciences, Southern Medical University, Guangzhou, China

M.L. Garcia
Kanalis Consulting, L.L.C., Edison, NJ, United States

S.P. Goulding
Committee on the Neurobiology of Addictive Disorders, The Scripps Research Institute, La Jolla, CA, United States

S.H. Heinemann
Friedrich Schiller University Jena & Jena University Hospital, Jena, Germany

T. Hoshi
University of Pennsylvania, Philadelphia, PA, United States

G.J. Kaczorowski
Kanalis Consulting, L.L.C., Edison, NJ, United States

H. Kim
Chicago Medical School, Rosalind Franklin University of Medicine and Science, North Chicago, IL, United States

M. Koide
University of Vermont, Burlington, VT, United States

G. Krishnamoorthy-Natarajan
Nutriflour International Inc. LLC, Norwalk, CT, United States

D.A. Kuljis
Carnegie Mellon University, Pittsburgh, PA, United States

G. Kuntamallappanavar
College of Medicine, The University of Tennessee Health Science Center, Memphis, TN, United States

B. Li
State Key Laboratory of Organ Failure Research, Key Laboratory of Psychiatric Disorders of Guangdong Province, School of Basic Medical Sciences, Southern Medical University, Guangzhou, China

Q. Li
The University of Texas MD Anderson Cancer Center, Houston, TX, United States

J. Liu
College of Medicine, The University of Tennessee Health Science Center, Memphis, TN, United States

K.H. Oh
Chicago Medical School, Rosalind Franklin University of Medicine and Science, North Chicago, IL, United States

R. Olcese
David Geffen School of Medicine, University of California at Los Angeles, Los Angeles, CA, United States

A. Pantazis
David Geffen School of Medicine, University of California at Los Angeles, Los Angeles, CA, United States

S.J. Pyott
University Medical Center Groningen, Groningen, The Netherlands

M.J. Shipston
Centre for Integrative Physiology, College of Medicine & Veterinary Medicine, University of Edinburgh, Edinburgh, United Kingdom

L. Tian
Centre for Integrative Physiology, College of Medicine & Veterinary Medicine, University of Edinburgh, Edinburgh, United Kingdom

J. Yan
The University of Texas MD Anderson Cancer Center, Houston, TX, United States

PREFACE

In 1981, the ion channel community marveled at single-channel recordings of calcium-dependent potassium currents showing an exceptionally high unitary conductance, which later earned them the designation of "Big Potassium," or BK, channels. This discovery was made possible by the recently developed patch-clamp method and was achieved quasi-simultaneously in two laboratories on each side of the Atlantic: in chromaffin cells by Dr. Alain Marty at the Max Planck Institute for Biophysical Chemistry and in muscle cells by Drs. Barry Pallotta, Karl Magleby, and John Barrett at the University of Florida School of Medicine. The following 35 years of research has continued to mesmerize BK channel aficionados by revealing the structural details that underlie the unique dual gating of BK channels by voltage and calcium, as well as the diversity and complexity of mechanisms exploited by BK channels to adjust their activity to a variety of cellular contexts. Not surprisingly, these superlative features have turned out to be critical determinants of tissue function throughout the body, most notably in smooth muscle and in the nervous system. This volume of *International Review of Neurobiology* is meant to provide an integrated perspective on our current understanding of the structure, function, and regulation of BK channels, with levels of analysis ranging from the submolecular conformational changes that drive channel pore opening to the systemic implications of BK channel dysfunction in diseases such as epilepsy.

In Chapter 1, Drs. Antonios Pantazis and Riccardo Olcese discuss the structural principles that govern BK channel activation by membrane potential and calcium, and they explain how BK channels represent both a model and a rebel in the ion channel realm. The following three chapters provide an overview of the mechanisms regulating BK channel biophysical properties, subcellular localization, and pharmacological sensitivity. Chapter 2 by Drs. Qin Li and Jiusheng Yan addresses the influence of auxiliary β and γ subunits, Chapter 3 by Drs. Michael Shipston and Lijun Tian details the role of alternative splicing and posttranslational modifications, and Chapter 4 by Drs. Hongkyun Kim and Kelly Oh reviews the functional impact of BK channels' interaction with a dense network of proteins. While BK channels are primarily located at the plasma membrane, they are also present in intracellular organelles and Chapter 5 by Drs. Boxing Li and Tian-Ming Gao describes the properties and function of mitochondrial and nuclear BK channels.

The activity of BK channels can be modulated by a multitude of small molecules. In Chapter 6, Drs. Toshinori Hoshi and Stefan Heinemann propose a classification of BK channel modulators according to their molecular mode of action, and they analyze the effects of several endogenous messengers and synthetic compounds in this framework. Chapter 7 by Drs. Alex Dopico, Anna Bukiya, Guruprasad Kuntamallappanavar, and Jianxi Liu focuses on the effects of ethanol on BK channel activity and their relevance for alcohol-driven behaviors.

The following four chapters analyze the role of BK channels in the physiology of different types of excitable cells located in the central nervous system, pituitary gland, inner ear, and blood vessels, respectively. In Chapter 8, Drs. Candice Contet, Scott Goulding, Dika Kuljis, and Alison Barth describe the complex influence of BK channels on neuronal excitability, along with their role in suppressing neurotransmitter release. They also review the numerous neurological disorders that are associated with an imbalance in BK channel activity. In Chapter 9, Drs. Peter Duncan and Michael Shipston contrast the facilitatory influence of BK channels on the bursting activity of anterior pituitary cells with their inhibitory influence on hormone secretion by posterior lobe neuroendocrine terminals. In Chapter 10, Drs. Sonja Pyott and Keith Duncan explain the evolutionary dichotomy that differentiates the role of BK channels in electrical tuning of nonmammalian vs mammalian hair cells. In Chapter 11, Drs. Gayathri Krishnamoorthy-Natarajan and Masayo Koide review the critical contribution of BK channels to vasodilation and neurovascular coupling. Finally, Chapter 12 by Drs. Gregory Kaczorowski and Maria Garcia discusses the therapeutic potential of BK channel modulators and provides a unique insider's perspective on the failures and successes of medicinal chemistry efforts targeted at BK channels.

Collectively, these chapters offer a fascinating journey along the varied and interconnected trails of BK channel research, where every reader will hopefully find vantage points of interest.

I would like to extend my sincere gratitude to Drs. Andrew Braun, Dan Cox, Yolima del Pilar Torres Ruiz, Jonathan Jaggar, Christopher Lingle, David McCobb, Dhasakumar Navaratnam, Jonathan Pierce-Shimomura, Brad Rothberg, Harpreet Singh, Bernd Sokolowski, and Stanko Stojilkovic for reviewing these chapters and providing constructive comments.

Candice Contet
Committee on the Neurobiology of Addictive Disorders,
The Scripps Research Institute, La Jolla, CA, United States

CHAPTER ONE

Biophysics of BK Channel Gating

A. Pantazis, R. Olcese[1]

David Geffen School of Medicine, University of California at Los Angeles, Los Angeles, CA, United States

[1]Corresponding author: e-mail address: rolcese@ucla.edu

Contents

Abstract

BK channels are universal regulators of cell excitability, given their exceptional unitary conductance selective for K^+, joint activation mechanism by membrane depolarization and intracellular [Ca^{2+}] elevation, and broad expression pattern. In this chapter, we discuss the structural basis and operational principles of their activation, or gating, by membrane potential and calcium. We also discuss how the two activation mechanisms interact to culminate in channel opening. As members of the voltage-gated potassium channel superfamily, BK channels are discussed in the context of archetypal family members, in terms of similarities that help us understand their function, but also seminal structural and biophysical differences that confer unique functional properties.

International Review of Neurobiology, Volume 128

ISSN 0074-7742

http://dx.doi.org/10.1016/bs.irn.2016.03.013

1. INTRODUCTION

The large conductance voltage- and Ca^{2+}-activated K^+ channel (BK, also known as MaxiK or Slo1), first cloned from *Drosophila* (Adelman et al., 1992), is the universal regulator of cellular excitability (Contreras et al., 2013; Hoshi, Pantazis, & Olcese, 2013; Salkoff, Butler, Ferreira, Santi, & Wei, 2006). BK pore-forming α subunits (encoded by *KCNMA1*) are near-ubiquitously expressed, endowing cells with a voltage- and intracellular $[Ca^{2+}]$-sensitive K^+ conductance (Berkefeld, Fakler, & Schulte, 2010; Contreras et al., 2013; Cui, Yang, & Lee, 2009; Hoshi, Pantazis, et al., 2013; Latorre & Brauchi, 2006; Latorre, Morera, & Zaelzer, 2010; Lee & Cui, 2010; Salkoff et al., 2006; Yang, Zhang, & Cui, 2015). BK channel roles include setting the vascular tone and blood pressure (Brayden & Nelson, 1992; Dunn & Nelson, 2010; Eichhorn & Dobrev, 2007; Ledoux, Werner, Brayden, & Nelson, 2006) (see Chapter "BK Channels in the Vascular System" by Krishnamoorthy-Natarajan and Koide), and regulating neuronal excitability and neurotransmitter release (Faber & Sah, 2003; Meredith et al., 2006) (see Chapter "BK Channels in Neurons" by Barth and Contet). In addition, intracellular BK channels have been detected in the inner mitochondrial membrane and other organelles (Balderas, Zhang, Stefani, & Toro, 2015; Singh, Stefani, & Toro, 2012; Szabo & Zoratti, 2014). Opening of mitochondrial BK channels in the heart confers protection from ischemia–reperfusion injury (Cordeiro, Terentyev, & Clements, 2015; Singh et al., 2013; Testai et al., 2013; Xu et al., 2002) (also see Chapter "Functional Role of Mitochondrial and Nuclear BK Channels" by Gao and Li).

BK channels are tetramers of channel-forming α subunits (Shen et al., 1994). The α polypeptide (Fig. 1) is characterized by the modular organization of its domains. Starting with an extracellular N-terminus (Meera, Wallner, Song, & Toro, 1997; Wallner, Meera, & Toro, 1996), the α subunit crosses the cell membrane seven times before folding into globular intracellular regulator of conductance of K^+ domains: RCK1 and RCK2 (Jiang et al., 2002; Kim, Lim, Rho, Eom, & Park, 2006; Wu, Yang, Ye, & Jiang, 2010; Yuan, Leonetti, Pico, Hsiung, & MacKinnon, 2010; Yusifov, Javaherian, Pantazis, Gandhi, & Olcese, 2010; Yusifov, Savalli, Gandhi, Ottolia, & Olcese, 2008). The first five transmembrane helices (S0–S4) organize into a bundle forming the BK channel voltage sensor (Diaz et al., 1998; Liu et al., 2010; Ma, Lou, & Horrigan, 2006; Pantazis & Olcese, 2012; Savalli, Kondratiev, Toro, & Olcese, 2006;

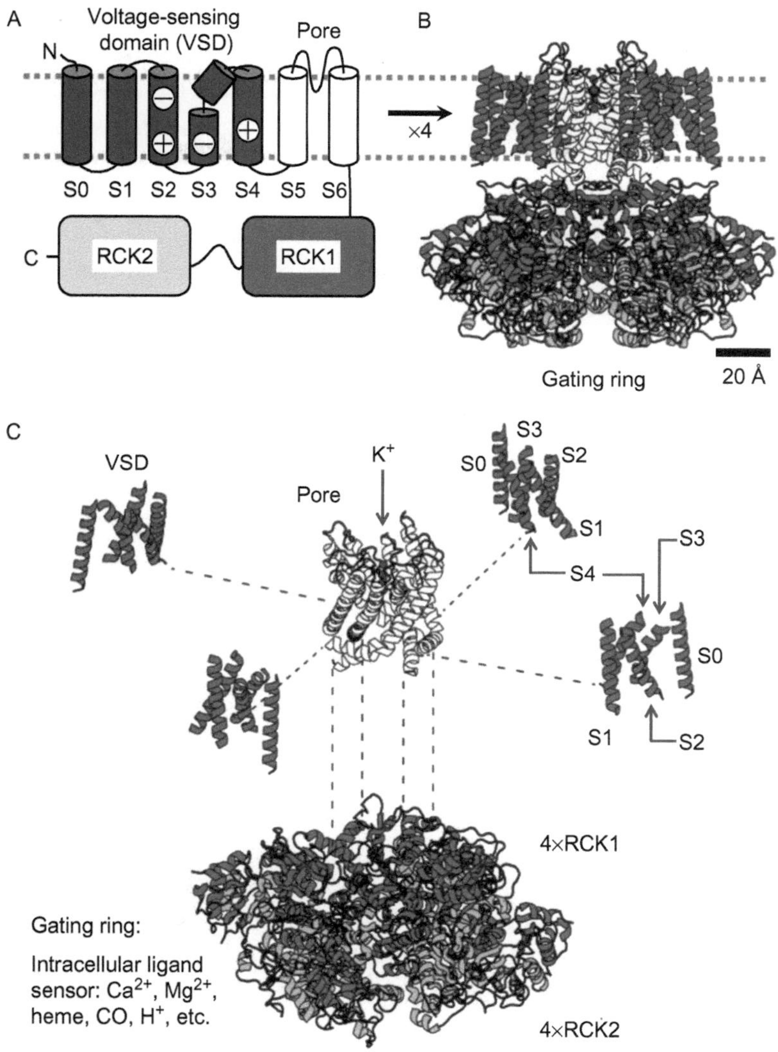

Fig. 1 BK channel topology, putative structure, and modular assembly. (A) Membrane topology of a single BK α subunit (Meera et al., 1997; Wallner et al., 1996). The cell membrane is indicated by the *dotted lines*. The N-terminal one-third of the channel crosses the cell membrane seven times. The voltage-sensing domain (VSD, *maroon*, gray in the print version) comprises transmembrane helices S0–S4. S2, S3, and S4 bear voltage-sensing residues, indicated by their charge (Ma et al., 2006). Transmembrane helices S5 and S6 (*white*) contribute to the central, K^+-selective pore in the tetrameric channel. The intracellular C-terminal domain encodes two structurally homologous regulator of conductance for K^+ (RCK) domains (Yusifov et al., 2008), which possess BK intracellular ligand-binding sites. (B) Side view of the putative structure of the BK channel composed of four α subunits (Hoshi, Pantazis, & Olcese, 2013). The transmembrane portion is the

(Continued)

Wang & Sigworth, 2009). S5 and S6 from each α subunit contribute to the channel central, K^+-selective pore (Brelidze, Niu, & Magleby, 2003; Haug et al., 2004; Nimigean, Chappie, & Miller, 2003). RCK1 and RCK2 assemble into a heterooctameric superstructure, the gating ring (Javaherian et al., 2011; Wang & Sigworth, 2009; Wu, Yang, et al., 2010; Yuan, Leonetti, Hsiung, & MacKinnon, 2012; Yuan et al., 2010), a sensor for multiple intracellular signaling molecules (Hou, Heinemann, & Hoshi, 2009). The broad expression pattern of BK channels means that they operate within a wide spectrum of signaling milieux: from spiking electrical activity in neurons to slower-repolarizing arterial smooth muscle and even inexcitable cells. In vivo, the activity of BK channels is tailored to their environment by modulatory β and γ subunits, which modify their biophysical properties (Contreras et al., 2013; Orio, Rojas, Ferreira, & Latorre, 2002; Sun, Zaydman, & Cui, 2012; Wu & Marx, 2010; Zhang & Yan, 2014). Since the expression pattern of auxiliary subunit subtypes is highly tissue-specific, they tune BK channel properties to their local signaling environment. β subunits localize between α subunit voltage sensor helices (Liu et al., 2010); indeed, BK channels can exist as complexes with up to four β subunits, exhibiting incremental effects of subunit modification (Wang, Ding, Xia, & Lingle, 2002). In contrast, γ subunits elicit an all-or-nothing response (Gonzalez-Perez, Xia, & Lingle, 2014). The same BK channel can accommodate both β and γ subunit types, greatly broadening the spectrum of physiological BK channel regulation (Gonzalez-Perez, Xia, & Lingle, 2015).

A key aspect of BK function in vivo is their close spatial and functional association and voltage-gated Ca^{2+} (Ca_V) channels, to integrate Ca^{2+} influx and cell excitability (Berkefeld et al., 2010, 2006; Cox, 2014; Loane, Lima, & Marrion, 2007; Milescu, 2015; Nelson & Quayle, 1995; Rehak et al., 2013; Roberts, Jacobs, & Hudspeth, 1990; Sun, Yazejian, & Grinnell, 2004).

Fig. 1—Cont'd structure of a related K_V1.2–2.1 chimeric channel structure (PDB #: 2R9R) (Long, Tao, Campbell, & MacKinnon, 2007), with the addition of S0 modeled as ideal alpha helices. Four RCK1–RCK2 assemble into the gating ring superstructure (PDB #: 3NAF) (Wu, Yang, et al., 2010). (C) Exploded diagram of the core BK channel functional modules. Four VSDs (*maroon*, gray in the print version) surround and control the central K^+-selective pore (*white*). Resolved K^+ is shown as *purple* (light gray in the print version) *spheres* (Long et al., 2007). S0 has been positioned according to the Karlin–Marx model (Liu et al., 2010). The pore is also coupled to the intracellular gating ring, which confers sensitivity to intracellular ligands. All structure images were generated with PyMOL (Schrodinger, 2015).

Eminent biophysicist David T. Yue often referred to the Ca_V channel as "the queen of ion channels" (Yue, 2004). Considering the BK/Ca_V cell signaling paradigm and the exceptionally high BK unitary conductance, likely afforded by "crowns" of negatively charged residues at the channel pore (Brelidze et al., 2003; Haug et al., 2004; Nimigean et al., 2003), BK channels may be considered the "king of ion channels." In addition to understanding cell function in health and disease, the motivation behind a basic scientific study of the BK channel must also have been a fascination evoked by this intricate molecular puzzle of ~1000 amino acids and all the energetic interactions underlying its structure and function. Studying BK channels has yielded invaluable information on fundamental macromolecule properties such as protein structure and conformational change, voltage and ligand sensitivity, allostery, signal transduction, and functional modulation. In this chapter, we present the structural and biophysical basis for BK channel activation by membrane depolarization and intracellular [Ca^{2+}] elevation, as well as how these processes are integrated to govern channel activity.

2. VOLTAGE-DEPENDENT GATING IN BK CHANNELS

BK channels respond to changes in the membrane potential by virtue of specialized voltage-sensing domains (VSDs) (Diaz et al., 1998; Liu et al., 2010; Ma et al., 2006; Pantazis & Olcese, 2012; Savalli et al., 2006; Wang & Sigworth, 2009). As each BK α subunit encodes for one VSD, four identical VSDs surround, and confer depolarization-dependent gating to, the central K^+-selective pore in the whole, homotetrameric BK channel (Fig. 1C). VSDs are conserved protein domains that are shared among voltage-gated K^+, Na^+, Ca^{2+}, and H^+ channels, as well as voltage-sensitive phosphatases (Chanda & Bezanilla, 2008; Nelson, Kuan, Saier, & Montal, 1999; Palovcak, Delemotte, Klein, & Carnevale, 2014; Yu & Catterall, 2004). What follows is a brief introduction on our current understanding of VSD function. While BK channels largely conform to the basic principles of voltage-dependent activation outlined later, they also exhibit seminal differences, discussed in the following Section 2.2.

2.1 Our Current Understanding of Voltage-Dependent Channel Gating

Our current understanding of the voltage dependence of membrane ionic conductances is based on over six decades' worth of electrophysiological experiments and mathematical modeling, as well as more recent contributions from fluorescence spectroscopy, VSD-specific toxins, molecular and structural biology, bioinformatics and in silico simulations, primarily in the archetypal fruit fly Shaker K^+ channel, but also in other members of the "VGL-Chanome" (Yu & Catterall, 2004). VSD structure and operation is an extensively reviewed topic (Bezanilla, 2008; Borjesson & Elinder, 2008; Catterall, 2010; Chanda & Bezanilla, 2008; Swartz, 2008; Tombola, Pathak, & Isacoff, 2006); the objective of this section is to provide a context for the properties of the BK VSD, which are discussed in the next section.

Canonical VSDs consist of four transmembrane helices (S1–S4), which arrange counter-clockwise into a bundle (Long, Campbell, & MacKinnon, 2005a; Long, Tao, Campbell, & MacKinnon, 2007). The widely accepted view is that the principal effector of voltage sensing is helix S4, which exhibits a motif of repeated positively charged residues (Arg or Lys) at every third position (Diaz et al., 1998; Logothetis, Movahedi, Satler, Lindpaintner, & Nadal-Ginard, 1992; Papazian, Timpe, Jan, & Jan, 1991; Perozo, Santacruz-Toloza, Stefani, Bezanilla, & Papazian, 1994; Stuhmer et al., 1989). Some of these voltage-sensing charges are immersed in the membrane electric field; upon membrane depolarization, these charges experience an electromotive force, compelling the outward movement of S4 and the adoption of the active VSD state. The movement of S4 generates most "gating charge" displacement in the VSDs of archetypal K^+ Shaker channels (Aggarwal & MacKinnon, 1996; Seoh, Sigg, Papazian, & Bezanilla, 1996) and other proteins (Ishida, Rangel-Yescas, Carrasco-Zanini, & Islas, 2015; Villalba-Galea, Frezza, Sandtner, & Bezanilla, 2013; Zhang, Liu, & Tseng, 2004): as the S4 transitions between the active and resting states, the resulting charge displacement of Arg and Lys residues through the membrane field generates "gating" or "sensing" current, recorded by electrophysiological experiments in the absence of ionic current (Armstrong & Bezanilla, 1973). Shaker VSD activation results in the displacement of 12–14*e* per channel (Aggarwal & MacKinnon, 1996; Islas & Sigworth, 1999; Noceti et al., 1996; Schoppa, McCormack, Tanouye, & Sigworth, 1992; Zagotta, Hoshi, Dittman, & Aldrich, 1994). This conformational rearrangement is electromechanically transduced to the channel pore, stabilizing the open state. The active and resting states of the VSD are stabilized by salt bridges formed by the S4 positively charged residues and negative counter-charges in S2 and S3 (DeCaen, Yarov-Yarovoy, Sharp, Scheuer, & Catterall, 2009; DeCaen, Yarov-Yarovoy, Zhao, Scheuer, & Catterall, 2008; Papazian et al., 1995; Tiwari-Woodruff, Lin, Schulteis, & Papazian, 2000; Wu, Delaloye, et al., 2010).

While the active state of the S4 has been captured in high atomic resolution in numerous crystallographic studies (Long et al., 2005a, 2007; Payandeh, Gamal El-Din, Scheuer, Zheng, & Catterall, 2012; Payandeh, Scheuer, Zheng, & Catterall, 2011; Tao, Lee, Limapichat, Dougherty, & MacKinnon, 2010), the resting state remains largely elusive. The movement of S4 can be ascertained by many different experimental approaches. The formation of metal or disulfide bridges can pinpoint state-dependent interactions within the VSD (Henrion et al., 2012; Lin, Abramson, & Papazian, 2010). Certain animal toxins known as "gating modifier toxins" have evolved to preferentially bind to, and stabilize, the VSD active or resting states, providing yet another experimental tool (Bosmans & Swartz, 2010; Catterall et al., 2007; Kalia et al., 2015)—particularly engineered chemoselective, fluorescent derivatives thereof (Tilley et al., 2014). Another approach to study S4 movement is to make the VSD permeable by mutagenesis, generating proton, or other ionic "ω" currents at the resting or active state (Capes, Arcisio-Miranda, Jarecki, French, & Chanda, 2012; Gamal El-Din, Scheuer, & Catterall, 2014; Sokolov, Scheuer, & Catterall, 2005; Starace & Bezanilla, 2004; Tombola, Pathak, Gorostiza, & Isacoff, 2007; Tombola, Pathak, & Isacoff, 2005). Structural biological studies are beginning to capture S4 in the elusive resting state (Li et al., 2014, 2015), while the S4 transitions have also been simulated in silico, to provide atomistic interpretations for experimental data (Jensen et al., 2012; Pathak et al., 2007; Schwaiger, Bjelkmar, Hess, & Lindahl, 2011; Schwaiger et al., 2012; Vargas, Bezanilla, & Roux, 2011; Yarov-Yarovoy, Baker, & Catterall, 2006). Of note, S4 movements can be directly optically tracked by the fluorescence deflections reported by site-directed fluorescent probes: the voltage clamp fluorometry (VCF) technique (Cha & Bezanilla, 1997; Claydon & Fedida, 2007; Gandhi & Olcese, 2008; Horne & Fedida, 2009; Mannuzzu, Moronne, & Isacoff, 1996; Priest & Bezanilla, 2015; Talwar & Lynch, 2015; Zhu, Varga, & Silva, 2015).

The VSD helix bundle allows water-filled crevices to form deep into the bilayer, "focusing" the electric field in a very narrow dielectric (Ahern & Horn, 2005; Asamoah, Wuskell, Loew, & Bezanilla, 2003; Starace & Bezanilla, 2004). In fact, recent electrophysiological, crystallographic, and computational studies have shown that the electric field is limited to the side chain of a single residue: a highly conserved Phe in helix S2 (Lacroix & Bezanilla, 2011; Schwaiger et al., 2012; Schwaiger, Liin, Elinder, & Lindahl, 2013; Tao et al., 2010). Finally, another structural feature of the VSD is a conserved Pro that splits the S3 helix into S3a (N-terminal) and

S3b (C-terminal): a structural feature important for VSD toxin association (Li-Smerin & Swartz, 2001).

2.2 Operation of the Noncanonical BK Channel VSD

The principles of voltage-dependent ion channel gating described earlier are thought to also apply to BK channels. The structural organization of the BK channel VSD largely conforms to the consensus VSD model: four homologous transmembrane helices (S1–S4); three conserved Arg residues in S4 following the RxxR motif (R207, R210, R213) corresponding to R1, R2, and R3 in Shaker; negative counter-charges in S2 and S3; two consecutive Pro residues in S3 (P191, P192) which likely split the helix into S3a and S3b; and a conserved Phe in S2 (F160) which could act as the hydrophobic charge transfer center (Fig. 2A). A cryo-EM structure of the whole BK channel (Wang & Sigworth, 2009) revealed the similarity of the BK transmembrane core with an atomic-resolution structure of $K_V1.2$ transmembrane regions, including the four VSDs (Long et al., 2005a).

In addition to these similarities, the BK channel VSD exhibits a few important deviations from the canonical VSD model: first, the BK channel VSD possesses additional transmembrane helix S0 (Meera et al., 1997; Wallner et al., 1996). S0 is required for functional BK channel expression (Meera et al., 1997; Morrow et al., 2006; Wallner et al., 1996) and localizes at the periphery of the VSD, proximal to voltage-sensing helix S4 (Koval et al., 2007; Liu et al., 2010; Liu, Zakharov, Yao, Marx, & Karlin, 2015; Niu et al., 2013; Pantazis, Kohanteb, & Olcese, 2010) (Fig. 2B). Trp substitutions at specific S0 positions mapping to a face of its hypothetical α-helical structure can perturb voltage-dependent gating, potentially defining a region in S0 which interacts with S4 (Koval et al., 2007). The extracellular flanks of S0 and S4 are proximal at rest, but diverge upon VSD activation (Pantazis, Kohanteb, et al., 2010) (Fig. 2C). The lack of observable functional effects in S0/S4 disulfide cross-linking experiments suggest that S0 does not directly affect VSD operation (Liu, Zakharov, Yang, Deng, et al., 2008; Niu et al., 2013). However, S0 serves as a site of auxiliary β subunit association (Liu, Zakharov, Yang, Wu, et al., 2008; Liu et al., 2010, 2015; Meera et al., 1997; Morrow et al., 2006; Wallner et al., 1996; Wu, Chudasama, et al., 2009; Wu et al., 2013), which results in the direct modification of VSD biophysical properties in a tissue-specific manner (Contreras, Neely, Alvarez, Gonzalez, & Latorre, 2012; Savalli, Kondratiev, de Quintana, Toro, & Olcese, 2007); more on this topic in

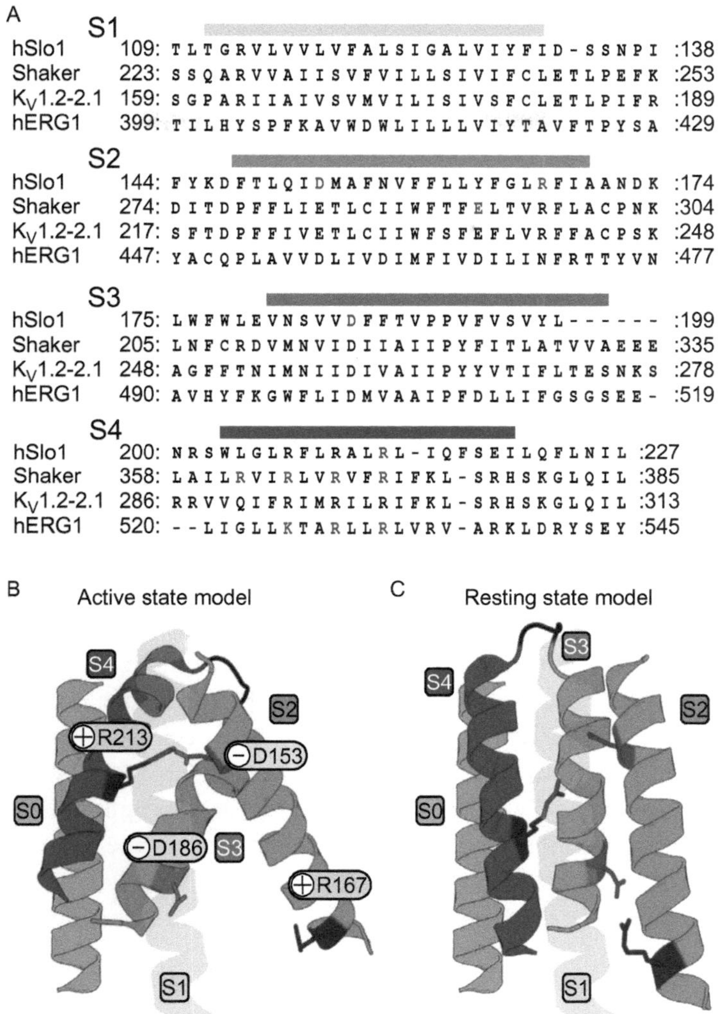

Fig. 2 The BK voltage sensor: homology and structural model. (A) Alignment of the VSDs of human BK channel (hSlo1), archetypal *Drosophila* Shaker K^+ channel, the Kv1.2–2.1 paddle chimera used in a VSD crystal structure (Long et al., 2007) and the human HERG1 channel. Charged residues that contribute to the gating charge are shown in *red* (dark gray in the print version) (Aggarwal & MacKinnon, 1996; Ma et al., 2006; Seoh et al., 1996; Zhang et al., 2004). The helical regions corresponding to S1, S2, S3, and S4 in the crystal structure (Long et al., 2007) are shown as *bars*. Note that only BK channels have an S0 helix in this group, not shown in the alignment. Alignment from Pantazis, Gudzenko, Savalli, Sigg, and Olcese (2010). (B) The crystal structure of a related K_V channel construct (K_V1.2–2.1 chimera, PDB #: 2R9R) (Long et al., 2007) voltage sensor in the active state was used as a template for the BK channel VSD. S0 (*olive*, light gray in the print version) was modeled as an ideal α helix and positioned

(Continued)

Chapter "Modulation of BK Channel Function by Auxiliary Beta and Gamma Subunits" by Yan and Li.

Second, BK VSD activation displaces ~fivefold less charge compared with Shaker: 2.6*e* per channel (Horrigan & Aldrich, 1999). Moreover, the BK VSD exhibits a decentralized distribution of voltage-sensing charges, as shown by a thorough mutagenesis scan of conserved charged residues (Ma et al., 2006) (Fig. 2A). While S4 is clearly an important segment for voltage-dependent BK channel activation (Diaz et al., 1998), only one of its three Arg contributes to voltage sensing: R213, which corresponds to the Shaker R3 (Ma et al., 2006). In fact, most voltage-sensing charge is contributed by residues outside S4: D153 and R167 (in S2) and D186 (in S3) (Ma et al., 2006) (Fig. 2A and B). Since voltage-sensing residues are thought to compel the voltage-dependent rearrangement of VSD transmembrane helices, the BK VSD structure likely responds differently to depolarization than other VSDs, where S2 and S3 are practically insensitive to depolarization (Li et al., 2014; Vargas et al., 2011). Given the short, helix–loop–helix structure of the extracellular S3–S4 linker (Semenova et al., 2009; Unnerstale, Lind, Papadopoulos, & Maler, 2009), S3 and S4 may move as a single voltage-sensing unit. On the other hand, the voltage-dependent transitions of S2 have a subtly different voltage dependence from those of S3/S4 (Pantazis, Gudzenko, et al., 2010), so this segment likely activates independently of S3/S4: as the opposite charges of S2 voltage-sensing residues D153 and R167 would compel them to move in opposite directions upon membrane depolarization, S2 may experience torque and undergo a tilting motion. The activation transition of the BK VSD helix bundle was resolved by a special implementation of VCF using site-directed fluorescent

Fig. 2—Cont'd in the VSD periphery next to S4, from electrophysiological, disulfide cross-linking and VCF data (Koval, Fan, & Rothberg, 2007; Liu et al., 2010; Pantazis, Kohanteb, & Olcese, 2010). S1 is in semitransparent gray for clarity. S2 (*green*, gray in the print version) bears voltage-sensing residues D153 (*red*, dark gray in the print version) and R167 (*blue*, gray in the print version), while S3 (*light blue*, gray in the print version) bears D186 (*red*, dark gray in the print version) (Ma et al., 2006; Pantazis, Gudzenko, et al., 2010). The S3–S4 (*black*) linker, which spans only three residues, is shown as a simple helix–loop–helix motif, as deduced by fluorescence quenching (Semenova, Abarca-Heidemann, Loranc, & Rothberg, 2009). In S4 (*red*, dark gray in the print version), only one residue contributes to the total gating charge (R213, *blue*, gray in the print version) (Ma et al., 2006; Pantazis, Gudzenko, et al., 2010). (C) A model of the BK VSD in the resting state, which is consistent with VCF data: upon deactivation, the extracellular flank of S4 approaches S0, S1, and S2, while S2 moves away from S1 (Pantazis & Olcese, 2012).

labeling and quenching (Pantazis, Kohanteb, et al., 2010; Pantazis & Olcese, 2012): upon membrane depolarization, the extracellular portion of S3/S4 diverges from S0, S1, and S2, while S2 approaches S1 (Fig. 2C).

While the movement of S3/S4 is thought to confer voltage sensitivity to the channel gate via electromechanical coupling (see Sections 4.2 and 4.3), what function do S2 movements serve in the BK VSD? Neutralization of S2 charged residues not only decreased gating charge and impaired voltage-dependent BK channel gating (Ma et al., 2006), but it also perturbed the voltage-dependent movements of S3/S4 (Pantazis, Gudzenko, et al., 2010). Likewise, neutralization of S4 voltage-sensing residue R213 impaired voltage-dependent channel opening (Ma et al., 2006) and the activation of S2 (Pantazis, Gudzenko, et al., 2010). This is evidence for a form of cooperativity underlying the operation of the BK VSD, whereby activation of S2 or S3/S4 facilitates the ability of the juxtaposed segment to efficiently respond to changes in the membrane potential, for example, by focusing the electric field (Pantazis, Gudzenko, et al., 2010), as previously proposed to occur in Shaker (Cha & Bezanilla, 1997; Chanda & Bezanilla, 2008). This voltage-sensing mechanism offers multiple opportunities for modulation, which are discussed later.

2.3 Modulation of BK Voltage-Dependent Gating

A major modulator of the BK voltage-sensing properties is intracellular [Ca^{2+}], elevation of which shifts the voltage dependence of ionic conductance to more negative potentials (Barrett, Magleby, & Pallotta, 1982; Cox, Cui, & Aldrich, 1997; Cui, Cox, & Aldrich, 1997; Magleby, 2003; Moczydlowski & Latorre, 1983). Horrigan and Aldrich revealed that Ca^{2+} binding allosterically modulates the voltage-sensing process of the BK channel, noting from gating current measurements that the VSDs deactivate more slowly in the presence of Ca^{2+} (Horrigan & Aldrich, 2002). Savalli et al. showed that Ca^{2+} binding to RCK1 contributes ~fourfold more energy (per subunit) toward VSD activation than RCK2 (Savalli, Pantazis, Yusifov, Sigg, & Olcese, 2012). Ca^{2+}-sensing by BK channels is discussed in more detail in Section 3.1, while the interplay of the voltage- and Ca^{2+}-dependent gating mechanisms of the BK channel is discussed in Section 4.6.

Ca^{2+} mediates its effects on the VSD allosterically, by binding to the gating ring (Budelli, Geng, Butler, Magleby, & Salkoff, 2013; Wei, Solaro, Lingle, & Salkoff, 1994); the BK VSD is also modulated by the direct association of other divalent cations. Inhibition of BK channels by extracellular Cu^{2+} (Morera, Wolff, & Vergara, 2003) is due to Cu^{2+} coordination by extracellular-facing VSD residues D133 (in S1), Q151 and D153 (S2), and R207 (S4) with micromolar affinity, which opposes VSD activation (Ma, Wong, & Horrigan, 2008).

On the intracellular side of the VSD, residues D99 (S0–S1 loop) and N172 (S2–S3 loop) contribute to a millimolar-affinity Mg^{2+}-binding site formed along with residues from the intracellular RCK1 domain of a neighboring subunit (Horrigan & Hoshi, 2008; Yang et al., 2008). The occupancy of this Mg^{2+} site is thought to stabilize open channels with an active VSD, in effect facilitating VSD activation and channel opening (Chen, Geng, & Magleby, 2011; Horrigan & Ma, 2008; Yang et al., 2007). The role of Mg^{2+} on the biophysics of BK channel gating is discussed more in Section 3.2.

The BK β subunits (four homologs, numbered β1–β4) can profoundly affect the voltage dependence of BK channels, both in terms of VSD voltage dependence and kinetics, as well as the transduction of VSD activation to pore opening (Bao & Cox, 2005; Contreras et al., 2012; Cox & Aldrich, 2000; McManus et al., 1995; Nimigean & Magleby, 2000; Orio & Latorre, 2005; Orio et al., 2002; Savalli et al., 2007; Sun et al., 2012). An exception is β3, which appears not to perturb voltage-dependent BK channel activation (Contreras et al., 2012), even though its transmembrane helices localize near those of the VSD (Wu et al., 2013), very similarly to the other β subunits (Liu, Zakharov, Yang, Wu, et al., 2008; Wu, Chudasama, et al., 2009; Wu et al., 2013). This suggests that merely localizing near the VSD does not suffice for modulation, excluding nonspecific, steric interactions; rather, specific protein–protein interactions must take place between the channel-forming α subunit and the β subunits that modulate voltage sensing. In fact, key residues in the extracellular loop of β1 (Gruslova, Semenov, & Wang, 2012) and its intracellular N-terminus (Castillo et al., 2015) have been implicated in modifying voltage-dependent opening. The modulation of BK channels by auxiliary β and γ subunits is presented in Chapter "Modulation of BK Channel Function by Auxiliary Beta and Gamma Subunits" by Yan and Li.

The voltage-dependent activation of BK channels can also be modified by pharmacological compounds; for example, polyphenol compound mallotoxin (rottlerin) potently augments the voltage-dependent activation of BK channels (Zakharov, Morrow, Liu, Yang, & Marx, 2005).

3. LIGAND-DEPENDENT GATING IN BK CHANNELS

BK channels can be gated or modified by a wide range of intracellular ligands, including Ca^{2+}, Mg^{2+}, fatty acids, phospholipids, cholesterol, CO, NO, H_2S, protons, heme, and ethanol (Bukiya, Belani, Rychnovsky, & Dopico, 2011; Bukiya et al., 2014; Dopico & Bukiya, 2014; Hoshi, Pantazis, et al., 2013; Hoshi, Tian, Xu, Heinemann, & Hou, 2013; Hoshi, Wissuwa, et al., 2013; Hoshi, Xu, Hou, Heinemann, & Tian, 2013; Hou et al., 2009; Latorre & Contreras, 2013; Singh, McMillan, et al., 2012; Tang, Zhang, Meng, Cui, & Logothetis, 2014). Most of these molecules mediate their effects by binding to the intracellular gating ring, which acts as a multiligand biological sensor for BK channels. In this section, we limit our discussion to gating ring structure, as well as the binding of Ca^{2+} and Mg^{2+}, as these divalent cations are central to the biophysics of BK channel gating. The action of other signaling molecules is presented in other chapters of this book, particularly Chapters "Modulation of BK Channels by Small Endogenous Molecules and Pharmaceutical Channel Openers" by Hoshi and Heinemann and "Modulation of BK Channels by Ethanol" by Dopico.

3.1 Ca^{2+}-Dependent Gating

In addition to membrane depolarization, BK channels can also be gated by elevation of the internal Ca^{2+} concentration (Latorre, Vergara, & Hidalgo, 1982; Marty, 1981; Pallotta, Magleby, & Barrett, 1981). In fact, Ca^{2+}-dependent gating is considered rate limiting for BK activity in neurons (Berkefeld & Fakler, 2013). In contrast with most Ca^{2+}-sensitive proteins, which gain Ca^{2+} sensitivity by their association with Ca^{2+}-binding calmodulin proteins (Kursula, 2014), the BK channel polypeptide encodes intrinsic Ca^{2+} sensors, localized in a distinct intracellular domain (Wei et al., 1994). The C-terminal tail of the BK α subunit comprises two tandem, homologous but nonidentical, regulator of conductance for potassium (RCK) domains: RCK1 and RCK2 (Jiang et al., 2002; Kim et al., 2006; Yusifov et al., 2008) (Fig. 1A). In bacterial K^+ channels, such as MthK and TvoK, RCK domains form octameric ligand-sensing gating ring structures (Dong, Shi, Berke, Chen, & Jiang, 2005; Jiang et al., 2002; Parfenova, Abarca-Heidemann, Crane, & Rothberg, 2007). In the whole, tetrameric BK channel, four RCK1–RCK2 pairs assemble to form a heterooctameric gating ring (Javaherian et al., 2011; Wu, Yang, et al., 2010; Yuan et al., 2010, 2012) (Figs. 1B, C and 3). Highlighting RCK domain modularity and functional independence, purified RCK1 and RCK2 proteins are each capable of forming octameric structures which, in solution, undergo Ca^{2+}-dependent conformational changes (Yusifov et al., 2008, 2010). In another

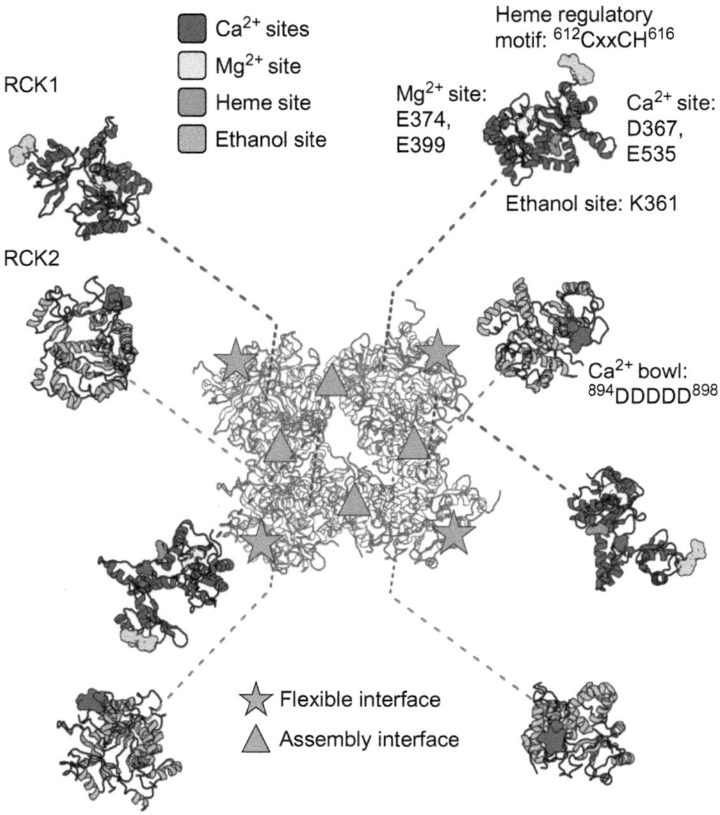

Fig. 3 Anatomy of the BK gating ring. In the whole tetrameric BK channel, the gating ring is a heterooctameric structure consisting of four RCK1 (*blue*, gray in the print version) and four RCK2 (*gold*, light gray in the print version) domains, which contain sites for binding small intracellular signaling molecules such as Ca^{2+} (*red*, dark gray in the print version), ethanol (*violet*, light gray in the print version), and heme (*green*, gray in the print version) (Bukiya et al., 2014; Schreiber & Salkoff, 1997; Tang et al., 2003; Xia, Zeng, & Lingle, 2002; Zhang et al., 2010). Mg^{2+} coordination sites are formed by residues contributed by RCK1 (*yellow*, light gray in the print version) (Yang et al., 2007) and the VSD (not shown) (Yang et al., 2008). Flexible (intrasubunit) and assembly (intersubunit) RCK1/RCK2 interfaces are indicated with *light green* (light gray in the print version) *stars* and *triangles*, respectively. Structure contributed by the Jiang laboratory (PDB #: 3NAF) (Wu, Yang, et al., 2010).

demonstration of BK channel protein modularity, expressed BK channel constructs without the C-terminal tail retain voltage-dependent gating, but exhibit no Ca^{2+} sensitivity (Budelli et al., 2013).

BK RCK domains possess specialized Ca^{2+}-binding sites (Fig. 3). Several RCK1 residues have been implicated in Ca^{2+} sensing, including D362/D367 (Xia et al., 2002) and E535 (Zhang et al., 2010), which likely coordinate Ca^{2+} by electronegative oxygen atoms (Zhang et al., 2010). RCK2 binds Ca^{2+} at the "Ca^{2+} bowl" site, which comprises five consecutive Asp residues (^{894}DDDDD^{898}) (Bian, Favre, & Moczydlowski, 2001; Schreiber & Salkoff, 1997); particularly D895 and D897 (Bao, Kaldany, Holmstrand, & Cox, 2004): a property captured in recent crystallographic studies (Yuan et al., 2010, 2012).

How tightly do the intrinsic gating ring sites bind Ca^{2+}? Given their distinct molecular identity, it is not surprising that the two sites exhibit different affinities for Ca^{2+}: while the precise Ca^{2+} dissociation constants differ slightly between studies, RCK2 has been consistently found to bind Ca^{2+} more tightly than RCK1 (Bao, Rapin, Holmstrand, & Cox, 2002; Savalli et al., 2012; Sweet & Cox, 2008; Xia et al., 2002). Since Ca^{2+} binding stabilizes the conducting state of the channel, it follows that open channels have a higher affinity for Ca^{2+}. By testing the response of unitary BK channel currents over a wide $[Ca^{2+}]$ range at −80 and 0 mV, Sweet and Cox found that the RCK1 affinity for Ca^{2+} is 13.7 ± 2.3 μM (closed channels) or 2.8 ± 0.5 μM (open channels). RCK2 binds Ca^{2+} more tightly, at 3.13 ± 0.22 μM (closed channels) or 0.88 ± 0.06 μM (open channels) (Sweet & Cox, 2008). Another study reported intrinsic Ca^{2+} affinities of 10.03 and 2.98 μM, for RCK1 and RCK2, respectively (Savalli et al., 2012). Savalli et al. estimated these values by the simultaneous fitting of channel opening and VSD activation sampled from BK channels with intact or neutralized Ca^{2+}-binding sites at different membrane potentials and $[Ca^{2+}]$ to an allosteric model of BK channel activation; since the allosteric influences of the VSDs and the pore were accommodated in the model, these values best correspond to affinity measurements from closed channels with deactivated VSDs. Finally, a study of purified gating rings in solution showed that $[Ca^{2+}]$ elevation elicits a spectroscopic response (quenching of native Trp residue fluorescence intensity) with two Hill components: $EC_{50} = 0.29 \pm 0.0043$ and 3.5 ± 0.16 μM (Javaherian et al., 2011). These responses, which reflect Ca^{2+}-induced conformational rearrangements of the gating ring, likely correspond to Ca^{2+} binding to RCK2 and RCK1, respectively, since neutralization of the Ca^{2+} bowl resulted in a single

Trp quenching component with an EC_{50} of 2.0 ± 0.19 μM (Javaherian et al., 2011). Since the isolated gating ring can bind Ca^{2+} more tightly than the whole channel, it appears that the transmembrane core may exert steric effects on the Ca^{2+} sites or alter overall gating ring structure. By comparison, small-conductance K^{+} channels (SK), which derive their Ca^{2+} sensitivity from their association with calmodulin, exhibit a higher sensitivity to the intracellular second messenger, eg, 0.35 μM in the case of SK2 (Xia et al., 1998).

The different structural identities of the two BK Ca^{2+}-binding sites are also evidenced by their different selectivity for divalent cations: RCK1 Ca^{2+}-sensing residues can also coordinate Sr^{2+} and Cd^{2+} (Zeng, Xia, & Lingle, 2005); while the Ca^{2+} bowl (RCK2) responds to Sr^{2+} (Zeng et al., 2005), Mg^{2+} (Javaherian et al., 2011), and Ba^{2+} (Zhou, Zeng, & Lingle, 2012). Ba^{2+}-dependent BK channel activation has been challenging to resolve, as Ba^{2+} also blocks BK channel conduction (Miller, 1987; Neyton & Miller, 1988); the Ca^{2+} bowl affinity for this divalent cation was reported to be 16.6 μM (closed channels) or 6.6 μM (open channels) (Zhou et al., 2012). Isolated gating rings with an intact Ca^{2+} bowl respond to Mg^{2+} with an apparent affinity of 154 ± 20.7 μM (Javaherian et al., 2011). Together, the earlier studies (Javaherian et al., 2011; Zeng et al., 2005; Zhou et al., 2012) indicate that the RCK1 site can accommodate divalent metal ions within a more restricted range of ionic radii (Cd^{2+}: 0.97 Å—Sr^{2+}: 1.13 Å) than the Ca^{2+} bowl (Mg^{2+}: 0.65 Å—Ba^{2+}: 1.35 Å). The ionic radius of Ca^{2+} is 0.99 Å (Pauling, 1960).

A lot of information on the structure, assembly, and operation of RCK domains and gating rings has come from crystallographic studies in prokaryotic channel domains (Dong et al., 2005; Jiang et al., 2002; Jiang, Pico, Cadene, Chait, & MacKinnon, 2001; Pau et al., 2011; Smith, Pau, Cingolani, & Rothberg, 2012, 2013; Ye, Li, Chen, & Jiang, 2006). A leap in our understanding of the BK gating ring structure and Ca^{2+}-dependent operation came about from the atomic-level resolution of human and zebrafish BK channel gating ring structures in the Ca^{2+}-free/closed (Wu, Yang, et al., 2010) (Figs. 1, 3, 4), Ca^{2+}-bound/closed (Yuan et al., 2010) and Ca^{2+}-bound/open (Yuan et al., 2012) states. According to these structures, Ca^{2+} binding results in the widening of the central gating ring aperture, while most Ca^{2+}-induced structural differences occur in the pore-facing N-terminal structures of the gating ring (Hoshi, Pantazis, et al., 2013)—particularly the structures between the βA and αC elements, called the AC region (Fig. 4). This may seem surprising since the pore/RCK1 interface and the Ca^{2+} bowl are separated by ~500 residues. However,

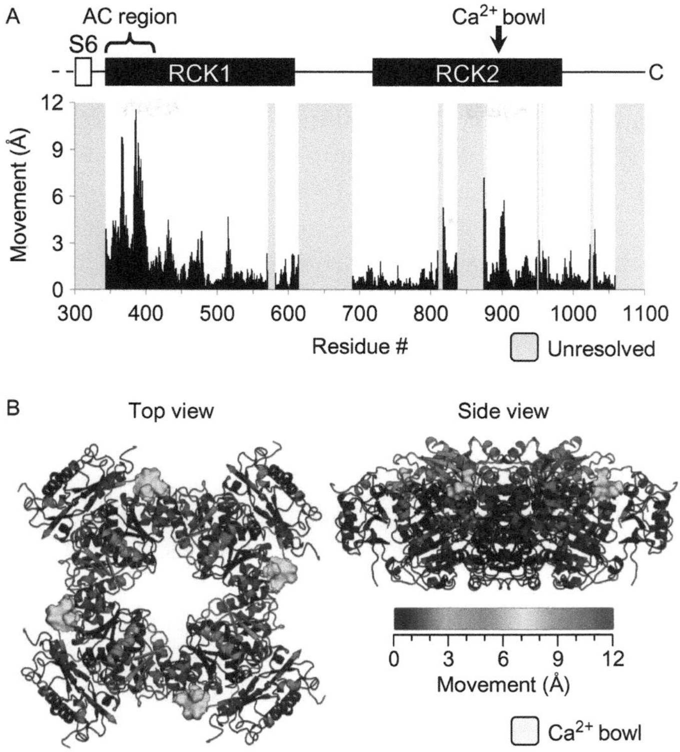

Fig. 4 Gating ring structural rearrangements evoked by Ca^{2+} binding to the Ca^{2+} bowl. (A) Movement of common, resolved Cα atoms in a human, Ca^{2+}-free gating ring structure (PDB #: 3ANF) (Wu, Yang, et al., 2010) and a zebrafish gating ring with Ca^{2+} resolved in the Ca^{2+} bowl (PDB #: 3U6N) (Yuan et al., 2012). Areas in *pink* (*light gray* in the print version) were unresolved (no distance information). A layout of the relative positions of the S6 transmembrane helix, RCK1, RCK2, the Ca^{2+} bowl, and the carboxy-terminus (C) is shown above. The gating ring undergoes a modest overall rearrangement in response to Ca^{2+} binding to the Ca^{2+} bond, with most movement (up to 12 Å) in the RCK1 N-terminal structures (AC region) and some near the Ca^{2+} bowl. (B) Structure of the Ca^{2+}-free gating ring (Wu, Yang, et al., 2010). The residues are colored according to their movement in response to Ca^{2+} binding to the Ca^{2+} bowl (*light gray*). Note that the Ca^{2+} bowl is located near the moving structures of RCK1, which are on the membrane-facing part of the gating ring; these movements likely cause Ca^{2+}-dependent gating. The structures were aligned in PyMOL (Schrodinger, 2015). *Figure recreated after Hoshi, T., Pantazis, A., & Olcese, R. (2013). Transduction of voltage and Ca^{2+} signals by Slo1 BK channels.* Physiology (Bethesda), *28(3), 172–189.*

the gating ring structures resolved by the MacKinnon and Jiang laboratories show that the Ca^{2+} bowl localizes in the "northern hemisphere" of the gating ring, proximal to the pore (Figs. 3 and 4). Thus the gating ring structure transduces ligand-binding free energy from distal sites (Ca^{2+} bowl) to the pore of the channel. Intriguingly, the pore-facing area of the gating ring (AC region) encompasses the site of a naturally occurring mutation (D434G) which causes increased Ca^{2+} sensitivity and has been associated with generalized epilepsy and paroxysmal dyskinesia (Du et al., 2005; Yang et al., 2010). The AC region is also thought to mediate Ca^{2+}-dependent channel gating from Ca^{2+} binding to RCK1 (Yang et al., 2010).

To date, there are no available BK gating ring structures showing Ca^{2+} associated with RCK1. Does Ca^{2+} binding to the RCK1 site stabilize the same open gating ring conformation captured in the RCK2-occupied structure (Yuan et al., 2012)? Analysis of single-channel activity in saturating $[Ca^{2+}]$ showed that the BK channel stays open for a greater proportion of time when both RCK1 and RCK2 Ca^{2+}-binding sites are occupied (Qian, Niu, & Magleby, 2006). Spectroscopic analysis of purified gating rings in solution shows more pronounced evidence of conformational changes (more Trp fluorescence quenching) when both high-affinity sites are intact (Javaherian et al., 2011). Ca^{2+} binding to RCK1 facilitates channel opening with different efficiency than RCK2 (Savalli et al., 2012; Sweet & Cox, 2008; Xia et al., 2002). The differential interpretations for these data are that Ca^{2+} binding to RCK1 or RCK2 stabilizes (i) the same gating ring structure (each sensor with low efficacy and different Ca^{2+} affinity), or (ii) different structures, each with their own efficiency of channel opening. There are two compelling pieces of evidence for the latter case: (i) Ca^{2+} photouncaging facilitates BK voltage sensor activation only in channels with an intact Ca^{2+} bowl (Savalli et al., 2012) and (ii) Ca^{2+} binding produces gating ring movements expected to be larger than those observed in the crystal structures, as measured by FRET between fluorescent protein gating ring inserts (Miranda et al., 2013). These results suggest that there is yet more to know about the mechanism of Ca^{2+}-dependent BK channel activation at the atomic level. From a macroscopic perspective, dynamic light scattering measurements showed that free $[Ca^{2+}]$ elevation produces an overall reduction of the surface-area-to-volume ratio of the gating ring (Javaherian et al., 2011), which may arise from a rotation of the RCK domains (Miranda et al., 2013), as suggested in gating ring-endowed prokaryotic MthK channels (Jiang et al., 2002). The AC region, at the N-terminus of RCK1, is connected to the channel gate via the S6–RCK1 linker and could therefore be a point of convergence for the conformational changes evoked by Ca^{2+} binding to either RCK1 (Yang et al., 2010; Zhang et al., 2010) or RCK2 (Yuan et al., 2012).

Membrane depolarization can modulate Ca^{2+}-binding properties, altering the apparent affinity of the channel for Ca^{2+} (Moczydlowski & Latorre, 1983). The interplay between Ca^{2+}- and voltage-dependent activation in BK channels is discussed further in Section 4.6.

In tissues, Ca^{2+} sensitivity is modified by the association of modulatory β subunits. These effects are discussed in Section 4.4.

3.2 Mg^{2+}-Dependent Gating

Very soon following their discovery, BK channels were found to respond to millimolar concentrations of another divalent cation, Mg^{2+} (Golowasch, Kirkwood, & Miller, 1986; Oberhauser, Alvarez, & Latorre, 1988). Where in BK is the Mg^{2+}-binding site? Shi and Cui reported that Mg^{2+} does not activate BK channels through the high-affinity Ca^{2+}-binding sites (Shi & Cui, 2001). Moreover, mutations that abolish Ca^{2+} sensitivity did not abolish Mg^{2+} sensitivity to channel opening and vice versa (Shi et al., 2002).

Comprehensive electrophysiological investigations revealed that, in BK channels, Mg^{2+} is coordinated by residues D99 (in the intracellular S0–S1 loop), D172 (S2–S3 loop) (Horrigan & Hoshi, 2008; Yang et al., 2008), E374 and E399 (RCK1) (Shi et al., 2002; Yang, Hu, Shi, & Cui, 2006) (Fig. 3). That is, the BK Mg^{2+}-selective site consists of residues contributed from both the transmembrane core of the BK channel and the gating ring. Accordingly, BK channels without the gating ring exhibit no Mg^{2+}-dependent gating (Budelli et al., 2013), while purified gating rings in solution (with a neutralized Ca^{2+} bowl) do not spectroscopically exhibit Mg^{2+}-dependent conformational changes (Javaherian et al., 2011). According to BK channel topology (Meera et al., 1997; Wallner et al., 1996), D99 and D172 are within short intracellular loops connecting transmembrane VSD helices, whereas the crystallographic studies of the C-terminal domains (Wu, Yang, et al., 2010; Yuan et al., 2010, 2012) place E374 and E399 on the exterior of the gating ring structure, facing the membrane (Fig. 3), and therefore the cytoplasm-facing portion of the VSD; thus, the four oxygen-bearing residues are ideally located near each other to coordinate a single Mg^{2+} atom. Mg^{2+} is typically coordinated by six electronegative oxygen atoms (Dudev & Lim, 2003); in BK, two water molecules are thought to complete Mg^{2+} coordination (Yang et al., 2008). In an example of how electrophysiological investigations can provide deep structural insights, the core and gating ring residues comprising the Mg^{2+} site cannot be contributed from the same subunit (Yang et al., 2008). Therefore, the

C-terminal tail of one subunit is structurally coupled (via Mg^{2+}) to the VSD of a neighboring subunit (Horrigan & Hoshi, 2008; Yang et al., 2008).

How does Mg^{2+} binding increase BK open probability? In the case of voltage- or Ca^{2+}-dependent activation, a structural transition at the sensor domain (VSD activation or Ca^{2+}-induced conformational changes) is communicated to the channel gate, stabilizing its open, conducting state. It is not known whether Mg^{2+} binding also induces conformational changes to the BK channel. However, following work showing that VSD-perturbing interventions also affect Mg^{2+} sensing (Hu et al., 2003), it was proposed that Mg^{2+} interacts electrostatically with R213 (Yang et al., 2007), the voltage-sensing residue of the S4 helix in the VSD (Ma et al., 2006). Electrostatic repulsion between Mg^{2+} and R213 would favor the outward movement of the residue, acting as membrane depolarization, stabilizing the active VSD state, and facilitating channel opening (Yang et al., 2007). Subsequent findings agree that Mg^{2+} binding occurs when the VSD is active, favoring VSD activation and BK channel opening (Chen et al., 2011; Horrigan & Ma, 2008).

The principal Mg^{2+}-binding site does not account for all Mg^{2+} sensitivity: a modest effect on channel activation remains after neutralization of key Mg^{2+}-binding residues, supporting the existence of a low-affinity site elsewhere in the channel (Yang et al., 2008). $[Mg^{2+}]$ elevation from 10 to 100 mM acts similarly to Ca^{2+} elevation, increasing open probability even at very negative potentials, ie, under resting voltage sensors (Horrigan & Ma, 2008).

4. THE BK ALLOSTERIC GATING MECHANISM

Membrane depolarization or cytosolic Ca^{2+} elevation is each sufficient to activate BK channels. The two gating mechanisms act in parallel to bring about channel activation in complex signaling environments: the interplay, or synergy, of the voltage- and Ca^{2+}-dependent gating was recognized early in their discovery, whereby membrane potential was found to affect Ca^{2+}-dependent gating (Barrett et al., 1982; Moczydlowski & Latorre, 1983), and has resulted in complex gating mechanisms with multiple kinetic states (Cox et al., 1997; Cui & Aldrich, 2000; Horrigan & Aldrich, 2002; Horrigan, Cui, & Aldrich, 1999; Magleby, 2003; Pantazis, Gudzenko, et al., 2010; Rothberg & Magleby, 1999, 2000; Savalli et al., 2012), while establishing BK channels as model system for the study of allostery, or conformational coupling, in proteins (Horrigan, 2012; Latorre et al., 2010; Sigg, 2014). The following section discusses two seminal models that characterize BK channel gating: the Horrigan–Cui–Aldrich (HCA) and Horrigan–Aldrich (HA) models; and reviews the physical connections between the major BK-sensing domains and the pore, as well as how they are used by signaling molecules and auxiliary subunits to modulate BK channel activity.

4.1 The HCA and HA Models of Allosteric BK Channel Activation

Despite the complex response of BK channels to voltage and Ca^{2+}, Horrigan and Aldrich constructed a relatively simple scheme to account for their activation mechanism, the HA model (Horrigan & Aldrich, 2002), by extending a previous model by Horrigan, Cui and Aldrich (the HCA model), used to account for voltage-dependent BK channel activation (Horrigan et al., 1999) (Fig. 5A). The HCA model could be described by the transitions of the pore and four VSDs, reflecting the BK channel's structural organization. The HA model added four Ca^{2+}-sensing domains (Fig. 5B), likely reflecting the Ca^{2+} bowl: Horrigan and Aldrich potentially missed, by a few months, the discovery of high-affinity sites in RCK1, bringing the total number of high-affinity Ca^{2+} sensors per BK channel to eight (Bao et al., 2002; Xia et al., 2002).

Each of these domains was allowed to exist in two states: closed/open (pore), resting/active (VSDs), Ca^{2+}-free/Ca^{2+}-bound (Ca^{2+} sensors). Thus there are three equilibrium constants (VSDs: *J*; Ca^{2+} sensors: *K*; pore: *L*) and three allosteric factors quantifying the domain interactions (Ca^{2+} sensor/pore: *C*; VSD/pore, electromechanical coupling: *D*; Ca^{2+} sensors/VSD: *E*). In Ca^{2+}-free conditions and very negative membrane potentials, in the absence of VSD activation or Ca^{2+}-binding, the BK channel can open, albeit highly reluctantly ($P_O \approx 10^{-7}$) (Horrigan & Aldrich, 2002). Activation of each sensor increases pore opening equilibrium constant by *C*- or *D*-fold, (Ca^{2+} and voltage sensors, respectively), tuning BK channel opening under physiologically relevant conditions.

By excluding intersubunit interactions as an initial simplifying assumption, the HA model consisted of 70 states. The importance of this scheme is that it sought to define the simplest physically plausible mechanism, rather than simply the least number of kinetic states. Experimental determination of model parameters was achieved by studying gating and ionic currents over a wide range of potentials and [Ca^{2+}], to capture the vast dynamic range of BK channel open probability, over seven orders of magnitude; most

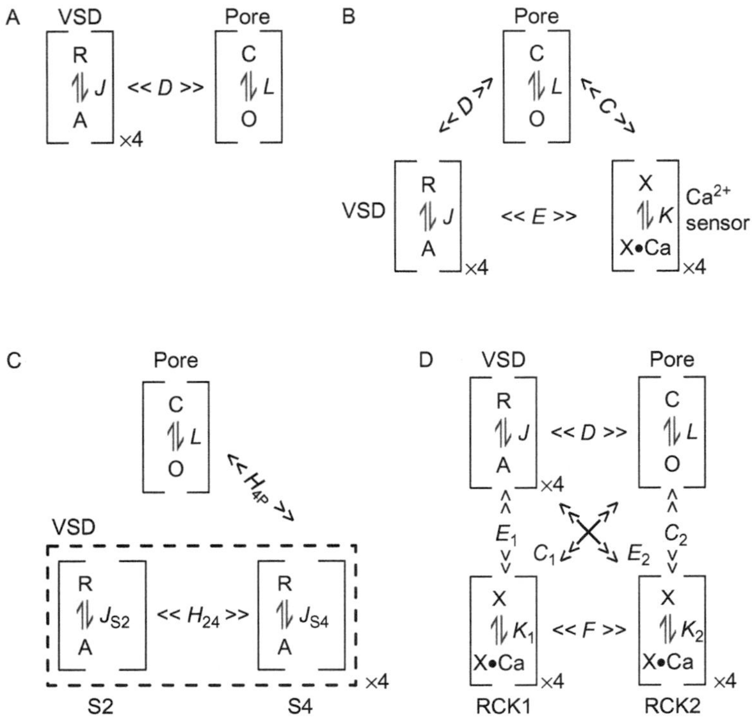

Fig. 5 Models of allosteric BK channel gating. (A) The Horrigan–Cui–Aldrich (HCA) model used to account for voltage-dependent BK channel activation in the absence of Ca^{2+} (Horrigan et al., 1999). Only the pore and VSD transitions are modeled. The pore exists in two states (closed/open) governed by equilibrium constant *L*. The VSD exists in resting/active states, with equilibrium constant *J*. The pore can open independently of the VSD state; VSD activation facilitates pore opening by allosteric factor *D*, and vice versa. (B) The Horrigan–Aldrich (HA) model (Horrigan & Aldrich, 2002) is an extension of the HCA model to include four Ca^{2+}-binding sites per channel. The Ca^{2+} sensor can be Ca^{2+}-free (X) or Ca^{2+}-bound (X•Ca) states, with equilibrium (dissociation) constant *K*. Ca^{2+} binding facilitates pore opening by allosteric factor *C*. The authors also found a weak cooperative interaction between the VSD and the Ca^{2+} sensors, quantified by allosteric factor *E*. (C) An extension of the HCA model to account for the different voltage dependency of S2 and S4 helices in the BK VSD, and the effect of charge neutralization in one segment on the voltage sensitivity of its neighbor (Pantazis, Gudzenko, et al., 2010). Each VSD contains two voltage-sensing subdomains representing S2 and S4, with their own states and equilibrium constants (J_{S2}, J_{S4}). S2 and S4 facilitate the activation of each other via energetic coupling H_{24}. S4 activation also facilitates pore opening via energetic coupling H_{4P}, equivalent to the allosteric factor *D* of the HCA model. (D) An extension of the HA model to include two types of Ca^{2+} sensors, representing RCK1 and RCK2, together accounting for all high-affinity Ca^{2+}-sensing sites, and the distinct effects of RCK1 or RCK2 neutralization on VSD activation (Savalli et al., 2012). Four of each type bring the total number of Ca^{2+}-binding sites per channel to eight. Each Ca^{2+} sensor has its own Ca^{2+} dissociation constant (K_1, K_2), and can interact with the pore (allosteric factors C_1, C_2) and the VSD (E_1, E_2). The two RCK domains can also interact with each other, via allosteric factor *F*.

importantly, at extremes of membrane potential and [Ca^{2+}], where the voltage or Ca^{2+} activation mechanism could be studied in almost complete isolation (Horrigan & Aldrich, 2002; Horrigan et al., 1999).

The usefulness of the HCA and HA models exceeded the thorough biophysical characterization of BK channel gating, because they set a framework for the mechanistic characterization of BK channel modulation used by the authors and other investigators. The HA model was used to dissect the effects of charge-neutralizing mutations in the VSD and identify voltage-sensing residues (Ma et al., 2006). An extension of the HCA model, where is VSD was split into two subdomains representing S2 and S4 (Fig. 5C) was used to quantify the cooperative interaction of these two BK VSD helices (Pantazis, Gudzenko, et al., 2010), described in Section 2.2. An extension of the HA model to include two Ca^{2+} sensor types, representing RCK1 and RCK2 (Fig. 5D), was used to interpret VCF and Ca^{2+} uncaging data, to dissect the effect of each Ca^{2+}-binding site on VSD activation and pore opening (Savalli et al., 2012), described in Section 3.1. More recently, an allosteric model of voltage-dependent gating in $Ca_V1.2$ channel, where the activation of up to four homologous, but nonidentical, voltage sensor domains is coupled to pore opening, was constructed inspired by the structural relevance and allosteric principles of the HA model (Pantazis, Savalli, Sigg, Neely, & Olcese, 2014). In the following sections, we discuss the mechanisms by which the VSDs and the Ca^{2+} sensors facilitate pore opening in the context of the HA model.

4.2 Electromechanical Coupling

Each VSD activation increases the pore opening equilibrium by ~18- to 25-fold (Horrigan & Aldrich, 2002; Shelley, Niu, Geng, & Magleby, 2010); from a statistical-mechanical point of view, this amounts to the reduction of the open state free energy by up to 1.93 kcal/mol, 3.2 kT, or 80 meV. How does VSD activation lead to BK channel opening?

The mechanism by which the voltage sensor activation causes conduction pore opening is referred to as electromechanical coupling (Blunck & Batulan, 2012; Vardanyan & Pongs, 2012). In related K_V channels, the ionic permeation pathway begins with a conserved ionic selectivity filter near the extracellular portion of the pore. As its name suggests, the selectivity filter only allows a certain ion to go through based on its charge, atomic or hydration radius, and/or desolvation energetics (Jensen et al., 2012; Roux et al., 2011). At the intracellular portion of the permeation pathway, the selectivity filter is followed by the S6 helix bundle crossing, which is widely considered the channel gate in voltage-gated ion channels (del Camino & Yellen, 2001; Long et al., 2005a; Long, Campbell, & MacKinnon, 2005b). At resting conditions, the selectivity filter is considered to be open and the S6 bundle crossing shut, preventing conductance. Upon membrane depolarization, activation of the VSD (outward movement of S4) pulls on the covalently attached S4–S5 intracellular linker, which in turn interacts noncovalently with the S6 helix bundle (Chowdhury & Chanda, 2012; Long et al., 2005b; Lu, Klem, & Ramu, 2002; Tristani-Firouzi, Chen, & Sanguinetti, 2002). This is thought to cause an "opening of the trapdoor" and ionic permeation. Opening of the S6 bundle gate triggers a collapse at the selectivity filter region, which causes C-type inactivation (Blunck, Cordero-Morales, Cuello, Perozo, & Bezanilla, 2006; Cuello, Jogini, Cortes, & Pan, et al., 2010, Cuello, Jogini, Cortes, & Perozo, 2010; Wu, Yang, et al., 2010). The fact that the selectivity region can act as a gate is also encountered in K_V-related cyclic nucleotide-gated (CNG) channels, where is it thought to be the main ionic permeation gate (Contreras & Holmgren, 2006; Contreras, Srikumar, & Holmgren, 2008).

BK channels, which do not undergo C-type inactivation, may also be gated at the selectivity filter region (Cox & Hoshi, 2011); while conformational changes do occur in the S6 bundle crossing of BK channels upon opening and closing (Brelidze & Magleby, 2005; Li & Aldrich, 2006; Wu, Xiong, et al., 2009), there is evidence that the inner vestibule of BK channels is wider than that of K_V (Brelidze & Magleby, 2005; Geng, Niu, & Magleby, 2011; Li & Aldrich, 2004; Moldenhauer, Diaz-Franulic, Gonzalez-Nilo, & Naranjo, 2016; Zhou, Xia, & Lingle, 2011, 2015), allowing relatively large blockers through while the channel is closed (Wilkens & Aldrich, 2006). Moreover, deep-pore residues seem important for channel gating (Chen & Aldrich, 2011; Chen, Yan, & Aldrich, 2014; Lippiat, Standen, & Davies, 2000; Zhou et al., 2011), while the action of small blockers depends on the permeant ion, ie, the state of the selectivity filter (Thompson & Begenisich, 2012). The role of residue F380, in the inner vestibule of BK channels, in ionic conductance (Lippiat et al., 2000) and voltage-dependent gating (Carrasquel-Ursulaez et al., 2015) identify this region as critical for electromechanical coupling and gating; particularly, the hydrophobic interaction of F380 with residue L377 of a neighboring subunit (Carrasquel-Ursulaez et al., 2015). Thus, gating in BK channels shows important differences from canonical K_V channel gating and seems to resemble the CNG channel mechanism more so than that of

K_V. Regardless, selectivity filter gating does not exclude interactions between the S4–S5 linker and S6 in BK channels.

In Shaker-type channels, electromechanical coupling is very tightly implemented, ie, the channel gate will only open after all four VSDs have activated (Bezanilla, Perozo, & Stefani, 1994; Schoppa et al., 1992; Zagotta, Hoshi, & Aldrich, 1994), acting as the tumblers in a lock mechanism. Voltage-independent gating probability was measured to be as low as 10^{-9} in Shaker channels (Islas & Sigworth, 1999). However, the ability of BK channels to exhibit voltage-independent opening at very negative potentials, where the voltage sensors are resting ($P_O = 10^{-7}$ to 10^{-3}, depending on [Ca^{2+}] (Horrigan & Aldrich, 2002)), brought forth the need for a different type of mechanism, whereby opening of the channel gate can occur independently of the state of the four VSDs, but each VSD activation progressively favors channel opening (Cox et al., 1997; Cui et al., 1997; Horrigan, 2012; Horrigan & Aldrich, 1999; Horrigan et al., 1999; Rothberg, 2004; Rothberg & Magleby, 2000). This mechanism is referred to as allosteric, since a conformational change in one domain (VSD activation) has an effect in another, independent protein domain (pore opening). Allosteric voltage-dependent gating in BK channels is also evidenced by the crossing over of the voltage dependences of ionic conductance and charge displacement upon [Ca^{2+}] elevation (Stefani et al., 1997). KCNQ1 ($K_V7.1$) channels also exhibit allosteric voltage-dependent opening (Liin, Barro-Soria, & Larsson, 2015; Meisel et al., 2012; Osteen et al., 2012), while this type of conformational coupling mechanism is compatible with the VSD activation and gating properties of L-type voltage-gated Ca^{2+} channels ($Ca_V1.2$) (Pantazis et al., 2014). More recently, an allosteric mechanism was proposed for the cyclic nucleotide-dependent activation of HCN2 channels (DeBerg, Brzovic, Flynn, Zagotta, & Stoll, 2016).

Upon Shaker K^+ channel opening, the channel's N-terminal tail, the "ball" peptide, binds to the open channel gate, blocking conduction: the "N-type," fast inactivation mechanism (Hoshi, Zagotta, & Aldrich, 1990). Upon membrane repolarization, the lingering ball delays gate closure and, since the active VSDs are tightly coupled to the gate, their deactivation is also delayed. This phenomenon, termed voltage sensor immobilization, can be observed in both gating current (Bezanilla, Perozo, Papazian, & Stefani, 1991) and optical (Savalli et al., 2007) measurements of Shaker VSD activity. While BK channel α subunits do not have an intracellular ball peptide, they achieve a similar type of inactivation by their association with β2 or β3 subunits, which do possess N-terminal ball-type peptides (Behrens et al., 2000; Bentrop, Beyermann, Wissmann, & Fakler, 2001; Brenner, Jegla, Wickenden, Liu, & Aldrich, 2000; Lingle, Zeng, Ding, & Xia, 2001; Wallner, Meera, & Toro, 1999). Interestingly, upon membrane repolarization in the presence of β2 subunits, BK channel voltage sensors return to their resting state while most of the channels are still inactivated: there is no immobilization (Savalli et al., 2007), which adds weight to the premise that BK channels do not gate as canonical K_V-type channels.

Even though they were originally studied in the context of obligatory gated channels, interactions between the S4–S5 linker and S6 could also mediate allosteric gating, where the VSD and the pore can activate/open independently: (i) S4–S5 and S6 could interact at all times, albeit with a flexible interface; (ii) S4–S5 and S6 could interact when the VSDs are active and the pore is open, stabilizing these states; or (iii) S4–S5 and S6 could interact when the pore is open but the VSDs are resting, destabilizing the open pore through steric hindrance (Horrigan, 2012).

4.3 Modulation of Electromechanical Coupling

The BK channel allosteric mode of electromechanical coupling can be modulated to tune BK channel activity. Mg^{2+} was found to increase electromechanical coupling, practically doubling the *D* factor and facilitating VSD activation, increasing its activation equilibrium constant by ~1.5-fold (Horrigan & Ma, 2008).

Another signaling molecule, heme, binds to a conserved Heme Regulatory Motif at the C-terminus of the RCK1 domain in the gating ring, at the beginning of the flexible linker connecting RCK1 and RCK2 (Tang et al., 2003) (Fig. 3). Heme binding favors BK channel opening at negative potentials, but inhibits BK channels at positive potentials (Tang et al., 2003). According to HCA model fits, heme binding facilitates BK pore gating by 10-fold but also decreased electromechanical coupling (*D*) by 75% (Horrigan, Heinemann, & Hoshi, 2005). By contrast, Ca^{2+} binding facilitates channel opening (Section 3.1) and favors VSD activation (Section 4.6) (Horrigan & Aldrich, 2002; Savalli et al., 2012). Even though both heme and Ca^{2+} ligands bind to the gating ring, their distinct effects on BK channel gating likely means that they induce distinct conformational changes, highlighting the versatility of the gating ring in sensing and discriminating between different intracellular signals. Experimental evidence of this hypothesis will greatly add to our understanding of ligand sensing and molecular recognition in proteins.

Electromechanical coupling is also subject to modulation by auxiliary subunits: β1 subunits reduce *D* by 24% (Bao & Cox, 2005), which may be due to reduced Mg^{2+} binding in the presence of β1 (Qian & Magleby, 2003). In contrast, γ1 subunits (LRRC26) markedly increase electromechanical coupling by ~20-fold, facilitating BK channel opening in nonexcitable cells (Yan & Aldrich, 2010). Identification of the γ1 site of action in the BK α subunit is of high interest, to understand the all-or-nothing modification of BK channels (Gonzalez-Perez et al., 2014), distinct from the progressive modification resulting from the association of up to four β subunits in the same BK channel (Wang et al., 2002). See Chapter "Modulation of BK Channel Function by Auxiliary Beta and Gamma Subunits" by Yan and Li for more information on auxiliary subunit modification of BK channels.

4.4 Coupling Between the Ca^{2+} Sensors and the Gate

Using their model, Horrigan and Aldrich determined that an occupancy of a single Ca^{2+} site increases the pore opening equilibrium by 7.4-fold (allosteric factor *C*), amounting to 1.2 kcal/mol, 2 kT, or 51 meV (Horrigan & Aldrich, 2002). However, the HA model includes four high-affinity Ca^{2+}-binding sites; subsequent work taking into account the eight high-affinity sites in BK (Fig. 3) reported that the RCK1–pore allosteric factor is 3.7 (Xia et al., 2002), 4.2 (Bao et al., 2002), or 4.57/7.52 at −80 or 0 mV, respectively (Sweet & Cox, 2008). RCK2 is generally less efficient at opening the pore, with allosteric factors of 2.3 (Xia et al., 2002), 4.4 (Bao et al., 2004), or 3.8 (Sweet & Cox, 2008). A reason for discrepancies among these figures could be the different models of voltage-dependent activation used by the investigating groups. Since VSDs can also have an effect on RCK Ca^{2+}-binding (Horrigan & Aldrich, 2002; Sweet & Cox, 2008), Savalli et al. used an extended HA model with two high-affinity Ca^{2+}-sensing sites representing RCK1 and RCK2, each coupled to both the pore and the VSD (Fig. 5D); they found that the RCK1 sensor is in fact less efficient at opening the channel pore than RCK2 ($C_1 = 2.84$; $C_2 = 10.41$), but RCK1 is more efficient at activating the VSD ($E_1 = 10$; $E_2 = 2.28$), which would in turn mediate channel opening via electromechanical coupling (Savalli et al., 2012).

Ca^{2+} binding to the Ca^{2+} bowl results in a widening of the gating ring aperture and a movement of the portions nearest the S6–RCK1 linker (Hoshi, Pantazis, et al., 2013; Wu, Yang, et al., 2010; Yuan et al., 2012) (Fig. 4). The S6–RCK1 linker, which connects the channel pore and the gating ring, has been described as the critical transducer the prokaryotic MthK Ca^{2+}-activated K^+ channel, which also has gating rings composed of RCK domains (Jiang et al., 2002). The S6–RCK1 linker in BK channels exhibits the mechanical properties of a passive spring, with a Hooke's spring constant of 15.4 pN/nm at 0 Ca^{2+} (Niu, Qian, & Magleby, 2004). In fact, BK channel openers NS1619 and Cym04 bind to this linker, effectively shortening the "spring" and facilitating channel activation (Gessner et al., 2012). The spring model is supported by the finding that the flexibility of the RCK1 AC region is important for Ca^{2+}-dependent gating (Yang et al., 2010). Given the evidence that BK channels gate by conformational changes deep in their pore region, perhaps their selectivity filter (Chen et al., 2014; Cox & Hoshi, 2011; Thompson & Begenisich, 2012; Zhou et al., 2015) (see Section 4.2), it is likely that the conformational change is communicated from the S6–RCK1 linker, along the length of S6, to regions deep within the permeation pathway, toward the selectivity filter near the extracellular portion of the channel.

4.5 Modulation of BK Ca^{2+} Sensitivity

We previously mentioned that a naturally occurring mutation (D434G) linked with paroxysmal epilepsy and dyskinesia (Du et al., 2005) occurs in the RCK1 "AC region," resulting in increased response to Ca^{2+} (Yang et al., 2010). This mutation is thought to lie on the pathway between the RCK1 Ca^{2+}-binding site and the channel gate and causes increased coupling of Ca^{2+} binding with channel gating, perhaps by increasing the rigidity of the AC region (Yang et al., 2010).

The Ca^{2+} response of the BK channels, both in terms of affinity for Ca^{2+} and the coupling between the Ca^{2+} sensors and the channel gate, are subject to modulation in vivo. β1 subunits apparently potentiate the BK channel response to Ca^{2+} (McManus et al., 1995). However, since β1 subunits also inhibit basal channel activation (Wang & Brenner, 2006) and facilitate voltage-dependent gating (Contreras et al., 2012), producing an overall Ca^{2+}-independent increase of channel activity (Nimigean & Magleby, 2000), it is difficult to isolate its effects on specific channel domains. By using multiple experimental Ca^{2+} concentrations, Sweet and Cox reported that β1 association increases the affinity of RCK1 for Ca^{2+} when the channels are open (K_d from 2.1 to 0.52 μM), while decreasing the affinity of RCK2 (Ca^{2+} bowl) for Ca^{2+} when the channels are closed (K_d from 2.1 to 5.9 μM); the result of both modifications is that the two sensors are practically more efficient at stabilizing the open state (Sweet & Cox, 2009). Double mutant cycle analysis revealed that the structural basis for this modification is both electrostatic and hydrophobic interactions between the sites in RCK1 and RCK2, and the β1 intracellular N-terminal tail (Liu et al., 2014).

4.6 Coupling Between Voltage and Ca^{2+} Sensors

In BK channels, VSD activation is effectively coupled to Ca^{2+} binding, and vice versa, since both sensors jointly facilitate pore opening. This is readily evidenced in the leftward shift of the voltage dependence of ionic conductance brought about by Ca^{2+} (Cox et al., 1997); the effect of membrane potential on Ca^{2+} binding (Moczydlowski & Latorre, 1983) and measurements showing that open channels have a higher affinity for Ca^{2+} than closed channels (Bao et al., 2004; Sweet & Cox, 2008; Xia et al., 2002). Intriguingly, the work of Horrigan and Aldrich revealed that a single Ca^{2+} binding facilitates VSD activation equilibrium, and vice versa, by 2.4-fold (0.5 kcal/mol, 0.9 kT, or 22 meV) (Horrigan & Aldrich, 2002). More recently, Sweet and Cox found that Ca^{2+} binding in RCK1 is voltage-dependent (Sweet & Cox, 2008), while Savalli and colleagues optically demonstrated that $[Ca^{2+}]$ elevation facilitates VSD activation and reported that RCK1 is more efficient at facilitating VSD activation than RCK2 ($E=10$ and 2.28, respectively) (Savalli et al., 2012). A FRET-based study of gating ring conformational rearrangements in whole BK channels showed that the gating ring undergoes pronounced voltage-dependent rearrangements (Miranda et al., 2013).

A candidate for the physical connection between the gating ring and the voltage sensor is Mg^{2+}, which is coordinated by residues contributed from both domains (Yang et al., 2008) (see Section 3.2). Moreover, since the two domains come near enough each other to coordinate Mg^{2+}, it is not unlikely that direct interaction also occurs between them. The AC region in RCK1, which faces the membrane, undergoes conformational changes in response to Ca^{2+} binding to RCK1 (Yang et al., 2010), while portions of the gating ring nearer the S6–RCK1 linker show most conformational change in response to Ca^{2+} binding to RCK2 (Fig. 4) (Hoshi, Pantazis, et al., 2013; Wu, Yang, et al., 2010; Yuan et al., 2012): it is not unlikely that these movements, in addition to increasing open probability, facilitate activation of the VSD.

4.7 Coupling Between RCK1 and RCK2 Ca^{2+} Sensors

In proteins with gating rings, two types of RCK/RCK interface arise from the symmetry of octameric RCK oligomerization: flexible and assembly

(Jiang et al., 2002). In the BK channel, the flexible interface exists between RCK1 and RCK2 of the same subunit, while the assembly interface connects the RCK1 and RCK2 domains of adjacent subunits (Fig. 3) (Yuan et al., 2010). The Park group identified conserved hydrophobic residues in the assembly (intersubunit) RCK1–RCK2 interface (Fig. 3) which are involved in interactions critical for BK function (Kim et al., 2006). However, RCK1 and RCK2 sensors across different subunits were found not to influence Ca^{2+} binding in each other directly (Niu & Magleby, 2002; Qian et al., 2006). This suggests that the hydrophobic interactions in the assembly interface reported by Kim et al. may be important for the structural integrity of the gating ring, rather than functional RCK1/RCK2 conformational coupling, consistent with the operation of a prokaryotic channel gating ring, where the assembly interface is thought to be rigid, while the flexible interface mediates Ca^{2+}-dependent conformational changes (Jiang et al., 2002).

On the other hand, the RCK1 and RCK2 domains of the same subunit have been reported to influence each other positively (ie, Ca^{2+} binding in RCK1 increases the effective affinity of RCK2, and vice versa) (Qian et al., 2006), or negatively (Savalli et al., 2012; Sweet & Cox, 2008). Interestingly, Savalli et al. (2012) also reported that in BK channels with a neutralized Ca^{2+} bowl, the RCK1–RCK2 interaction appeared to be cooperative; this suggests that neutralization of the five negatively charged Ca^{2+} bowl residues caused a structural perturbation that altered RCK1–RCK2 interaction and may account for the finding of a previous study reporting positive RCK1–RCK2 interaction (Qian et al., 2006). Consistent with the findings that RCK1 and RCK2 of the same subunit interact when binding Ca^{2+}, mutations of a residue in the RCK1/RCK2 flexible (intrasubunit) interface (G803 in the rat BK α, homologous to G733 in the human subunit), perturb channel activation (Kim et al., 2008). While the position of G733 is practically the same in the closed (Wu, Yang, et al., 2010) and open (Yuan et al., 2012) gating ring crystal structures ($C\alpha$–$C\alpha = 0.86$ Å; Fig. 4A), its backbone flexibility may be critical for transducing conformational coupling between RCK1 and RCK2.

5. THE EXTRAORDINARY CONDUCTANCE OF BK CHANNELS

The parallel BK gating mechanisms by voltage and $[Ca^{2+}]$, and the subsequent opening of their pore, culminate to no less but the highest known selective conductance for K^+, up to ~300 pS (Latorre & Miller, 1983; Latorre et al., 1982; Marty, 1981; Pallotta et al., 1981). For comparison, the related Shaker K^+ channel exhibits a smaller conductance by over an order of magnitude (~12 pS) (Diaz-Franulic, Sepulveda, Navarro-Quezada, Gonzalez-Nilo, & Naranjo, 2015). Despite the "Big K^+" conductance of BK channels literally being their defining feature, its precise mechanism remains elusive. A major determinant of ionic conductance is the local concentration of the permeant ion near the pore. In ion channels, charged residues near the pore can generate a local electrostatic potential to attract ions and locally raise their concentration above that of the bulk solution. In BK channels, this is thought to be mediated by rings of negatively charged Asp residues in both the internal (Brelidze et al., 2003; Nimigean et al., 2003) and external (Haug et al., 2004) cavities of the channel pore. Neutralization of these residues dramatically reduces their conductance, implicating a role for electrostatics (Carvacho et al., 2008). The unique geometry of the BK pore inner cavity (Brelidze & Magleby, 2005; Geng et al., 2011; Moldenhauer et al., 2016; Zhou, Tang, Xia, & Lingle, 2010; Zhou et al., 2011, 2015) may also be pertinent to their superlative conductive properties, as its occlusion also diminishes BK conductance (Brelidze & Magleby, 2005).

Not only do BK channels exhibit a uniquely large K^+ conductance, but also a highly selective one: BK channels preferentially conduct K^+ over Na^+ by more than 300-fold (Eisenman, Latorre, & Miller, 1986; Latorre & Miller, 1983; Yellen, 1984). The BK channel selectivity filter exhibits the signature G–Y–G sequence at the membrane-reentrant P loop between helices S5 and S6 that defines K^+ conductance (Doyle et al., 1998; Heginbotham, Abramson, & MacKinnon, 1992; Heginbotham, Lu, Abramson, & MacKinnon, 1994). K^+ ion permeation through this filter is implemented by the sequential, thermodynamically favorable coordination of dehydrated K^+ ions by backbone carbonyl oxygen atoms (Doyle et al., 1998; Morais-Cabral, Zhou, & MacKinnon, 2001; Zhou, Morais-Cabral, Mann, & MacKinnon, 2001). In contrast, Na^+ ion coordination in the selectivity filter of K^+ channels is thought to be less thermodynamically favorable (Luzhkov & Aqvist, 2001; Noskov, Berneche, & Roux, 2004). While Na^+ ions can occupy sites within the selectivity filter (Thompson et al., 2009), their complete permeation is inhibited by steep free energy barriers, resulting in K^+ selectivity over Na^+ and other cations (Egwolf & Roux, 2010). While these insights have been gained from structural, electrophysiological, and in silico studies primarily in the KcsA bacterial K^+ channel model, similar mechanisms are thought to apply to BK channels. A study attempting to bestow high conductance to Shaker channels proposed that the BK selectivity filter may also be more efficient at K^+ translocation than that of Shaker-type channels, contributing to large conductance (Diaz-Franulic et al., 2015).

Thus, BK channels are notoriously efficient at exclusively transporting K^+ ions. Their spectacular conductance seems to justify its tight regulation by multiple signals, including membrane depolarization, intracellular Ca^{2+}, Mg^{2+}, and other cytosolic messengers, as well as splice variation, posttranslational modifications, and regulation by auxiliary subunits: BK channel complexes endow their cells with a potent repolarizing K^+ efflux capability, as well as the means to control it by a wide variety of signals.

6. CONCLUDING REMARKS: THE BK CHANNEL AS A REBEL AND A MODEL

Model systems are often used in science to explain complex, or difficult-to-resolve, phenomena. Biology is no exception: for example, model organisms have provided valuable paradigms in genetics and physiology. The same stands for ion channel biophysics: priceless insights on ion channel structure and function have been gained from seminal work on ion channels that have attracted, and continue to attract, intense study. This chapter has provided ample examples of how our understanding of BK channels has greatly benefited from the principles of voltage dependence, ionic permeation, and ligand sensitivity even from bacterial or insect homologs that serve as model ion channels, such as the MthK, KcsA, and Shaker K^+ channels, to name just a few.

However, BK channels do not merely conform to paradigms of ion channel function. In this chapter, we have also seen how BK channels break the mold. BK channel VSDs are homologous to those of Shaker, but they also exhibit distinct functional properties (Sections 2.1 and 2.2). Their pore possesses the overall structure and signature sequence of other K^+-selective channels; yet permeability and structural studies reveal important differences in BK gating (Section 4.2) and conductance (Section 5). The BK gating ring does not merely sense Ca^{2+} but, in addition, it integrates the signals of a wide gamut of cytosolic messenger molecules, from divalent cations to lipids and gasotransmitters, to fine-tune BK activity in any signaling milieu this almost ubiquitous ion channel operates (see Chapters "Modulation of BK Channels by Small Endogenous Molecules and Pharmaceutical Channel Openers" by

Hoshi and Heinemann and "Modulation of BK Channels by Ethanol" by Dopico). Perhaps most famously, BK channels employ allostery, or conformational coupling, to concurrently respond to electrical and Ca^{2+} signaling (Section 4). While these deviations from the K_V canon make BK channels the "rebels" of the voltage-gated K^+ channel superfamily, they are also becoming model ion channels themselves. Evidence of VSD/pore allosteric coupling was recently shown for the KCNQ1 channel. We derived inspiration from the HA mechanism to develop allosteric models for $Ca_V1.2$ channels (Pantazis et al., 2014).

Still, our understanding of the BK activation mechanism remains far from complete. We anticipate that with innovative experimental methods, new insights are within reach. We also anticipate that the physiological roles of BK channels will expand following the development of BK subunit knock-out animals (Brenner, Perez, et al., 2000; Martinez-Espinosa, Yang, Gonzalez-Perez, Xia, & Lingle, 2014; Meredith, Thorneloe, Werner, Nelson, & Aldrich, 2004; Zemen et al., 2015), investigations of intracellular BK channels (Balderas et al., 2015; Singh, Stefani, et al., 2012), and the increasing number of BK molecular partners (Toro et al., 2014; Zhang & Yan, 2014). Will these studies eventually establish the BK channel as a rebel, or a model?

ACKNOWLEDGMENTS

We thank Enrique Balderas, Taleh Yusifov, and other members of the Olcese laboratory, for useful feedback on the manuscript. We are also grateful to the National Institutes of Health/National Heart Lung and Blood Institute (5R01HL107418 to R.O.) and the American Heart Association, National Center (14SDG20300018 to A.P.) for financial support.

REFERENCES

Adelman, J. P., Shen, K. Z., Kavanaugh, M. P., Warren, R. A., Wu, Y. N., Lagrutta, A., et al. (1992). Calcium-activated potassium channels expressed from cloned complementary DNAs. *Neuron, 9*(2), 209–216.

Aggarwal, S. K., & MacKinnon, R. (1996). Contribution of the S4 segment to gating charge in the shaker K^+ channel. *Neuron, 16*(6), 1169–1177.

Ahern, C. A., & Horn, R. (2005). Focused electric field across the voltage sensor of potassium channels. *Neuron, 48*(1), 25–29.

Armstrong, C. M., & Bezanilla, F. (1973). Currents related to movement of the gating particles of the sodium channels. *Nature, 242*(5398), 459–461.

Asamoah, O. K., Wuskell, J. P., Loew, L. M., & Bezanilla, F. (2003). A fluorometric approach to local electric field measurements in a voltage-gated ion channel. *Neuron, 37*(1), 85–97.

Balderas, E., Zhang, J., Stefani, E., & Toro, L. (2015). Mitochondrial BKCa channel. *Frontiers in Physiology, 6*, 104.

Bao, L., & Cox, D. H. (2005). Gating and ionic currents reveal how the BKCa channel's Ca^{2+} sensitivity is enhanced by its beta1 subunit. *The Journal of General Physiology*, *126*(4), 393–412.
Bao, L., Kaldany, C., Holmstrand, E. C., & Cox, D. H. (2004). Mapping the BKCa channel's "Ca^{2+} bowl": Side-chains essential for Ca^{2+} sensing. *The Journal of General Physiology*, *123*(5), 475–489.
Bao, L., Rapin, A. M., Holmstrand, E. C., & Cox, D. H. (2002). Elimination of the BK(Ca) channel's high-affinity Ca(2+) sensitivity. *The Journal of General Physiology*, *120*(2), 173–189.
Barrett, J. N., Magleby, K. L., & Pallotta, B. S. (1982). Properties of single calcium-activated potassium channels in cultured rat muscle. *The Journal of Physiology*, *331*, 211–230.
Behrens, R., Nolting, A., Reimann, F., Schwarz, M., Waldschutz, R., & Pongs, O. (2000). hKCNMB3 and hKCNMB4, cloning and characterization of two members of the large-conductance calcium-activated potassium channel beta subunit family. *FEBS Letters*, *474*(1), 99–106.
Bentrop, D., Beyermann, M., Wissmann, R., & Fakler, B. (2001). NMR structure of the "ball-and-chain" domain of KCNMB2, the beta 2-subunit of large conductance Ca^{2+}- and voltage-activated potassium channels. *The Journal of Biological Chemistry*, *276*(45), 42116–42121.
Berkefeld, H., & Fakler, B. (2013). Ligand-gating by Ca^{2+} is rate limiting for physiological operation of BK(Ca) channels. *The Journal of Neuroscience*, *33*(17), 7358–7367.
Berkefeld, H., Fakler, B., & Schulte, U. (2010). Ca^{2+}-activated K^+ channels: From protein complexes to function. *Physiological Reviews*, *90*(4), 1437–1459.
Berkefeld, H., Sailer, C. A., Bildl, W., Rohde, V., Thumfart, J. O., Eble, S., et al. (2006). BKCa-Cav channel complexes mediate rapid and localized Ca^{2+}-activated K^+ signaling. *Science*, *314*(5799), 615–620.
Bezanilla, F. (2008). How membrane proteins sense voltage. *Nature Reviews. Molecular Cell Biology*, *9*(4), 323–332.
Bezanilla, F., Perozo, E., Papazian, D. M., & Stefani, E. (1991). Molecular basis of gating charge immobilization in Shaker potassium channels. *Science*, *254*(5032), 679–683.
Bezanilla, F., Perozo, E., & Stefani, E. (1994). Gating of Shaker K^+ channels: II. The components of gating currents and a model of channel activation. *Biophysical Journal*, *66*(4), 1011–1021.
Bian, S., Favre, I., & Moczydlowski, E. (2001). Ca^{2+}-binding activity of a COOH-terminal fragment of the Drosophila BK channel involved in Ca^{2+}-dependent activation. *Proceedings of the National Academy of Sciences of the United States of America*, *98*(8), 4776–4781.
Blunck, R., & Batulan, Z. (2012). Mechanism of electromechanical coupling in voltage-gated potassium channels. *Frontiers in Pharmacology*, *3*, 166.
Blunck, R., Cordero-Morales, J. F., Cuello, L. G., Perozo, E., & Bezanilla, F. (2006). Detection of the opening of the bundle crossing in KcsA with fluorescence lifetime spectroscopy reveals the existence of two gates for ion conduction. *The Journal of General Physiology*, *128*(5), 569–581.
Borjesson, S. I., & Elinder, F. (2008). Structure, function, and modification of the voltage sensor in voltage-gated ion channels. *Cell Biochemistry and Biophysics*, *52*(3), 149–174.
Bosmans, F., & Swartz, K. J. (2010). Targeting voltage sensors in sodium channels with spider toxins. *Trends in Pharmacological Sciences*, *31*(4), 175–182.
Brayden, J. E., & Nelson, M. T. (1992). Regulation of arterial tone by activation of calcium-dependent potassium channels. *Science*, *256*(5056), 532–535.
Brelidze, T. I., & Magleby, K. L. (2005). Probing the geometry of the inner vestibule of BK channels with sugars. *The Journal of General Physiology*, *126*(2), 105–121.
Brelidze, T. I., Niu, X., & Magleby, K. L. (2003). A ring of eight conserved negatively charged amino acids doubles the conductance of BK channels and prevents inward

rectification. *Proceedings of the National Academy of Sciences of the United States of America, 100*(15), 9017–9022.

Brenner, R., Jegla, T. J., Wickenden, A., Liu, Y., & Aldrich, R. W. (2000). Cloning and functional characterization of novel large conductance calcium-activated potassium channel beta subunits, hKCNMB3 and hKCNMB4. *The Journal of Biological Chemistry, 275*(9), 6453–6461.

Brenner, R., Perez, G. J., Bonev, A. D., Eckman, D. M., Kosek, J. C., Wiler, S. W., et al. (2000). Vasoregulation by the beta1 subunit of the calcium-activated potassium channel. *Nature, 407*(6806), 870–876.

Budelli, G., Geng, Y., Butler, A., Magleby, K. L., & Salkoff, L. (2013). Properties of Slo1 K^+ channels with and without the gating ring. *Proceedings of the National Academy of Sciences of the United States of America, 110*(41), 16657–16662.

Bukiya, A. N., Belani, J. D., Rychnovsky, S., & Dopico, A. M. (2011). Specificity of cholesterol and analogs to modulate BK channels points to direct sterol-channel protein interactions. *The Journal of General Physiology, 137*(1), 93–110.

Bukiya, A. N., Kuntamallappanavar, G., Edwards, J., Singh, A. K., Shivakumar, B., & Dopico, A. M. (2014). An alcohol-sensing site in the calcium- and voltage-gated, large conductance potassium (BK) channel. *Proceedings of the National Academy of Sciences of the United States of America, 111*(25), 9313–9318.

Capes, D. L., Arcisio-Miranda, M., Jarecki, B. W., French, R. J., & Chanda, B. (2012). Gating transitions in the selectivity filter region of a sodium channel are coupled to the domain IV voltage sensor. *Proceedings of the National Academy of Sciences of the United States of America, 109*(7), 2648–2653.

Carrasquel-Ursulaez, W., Contreras, G. F., Sepulveda, R. V., Aguayo, D., Gonzalez-Nilo, F., Gonzalez, C., et al. (2015). Hydrophobic interaction between contiguous residues in the S6 transmembrane segment acts as a stimuli integration node in the BK channel. *The Journal of General Physiology, 145*(1), 61–74.

Carvacho, I., Gonzalez, W., Torres, Y. P., Brauchi, S., Alvarez, O., Gonzalez-Nilo, F. D., et al. (2008). Intrinsic electrostatic potential in the BK channel pore: Role in determining single channel conductance and block. *The Journal of General Physiology, 131*(2), 147–161.

Castillo, K., Contreras, G. F., Pupo, A., Torres, Y. P., Neely, A., Gonzalez, C., et al. (2015). Molecular mechanism underlying beta1 regulation in voltage- and calcium-activated potassium (BK) channels. *Proceedings of the National Academy of Sciences of the United States of America, 112*(15), 4809–4814.

Catterall, W. A. (2010). Ion channel voltage sensors: Structure, function, and pathophysiology. *Neuron, 67*(6), 915–928.

Catterall, W. A., Cestele, S., Yarov-Yarovoy, V., Yu, F. H., Konoki, K., & Scheuer, T. (2007). Voltage-gated ion channels and gating modifier toxins. *Toxicon, 49*(2), 124–141.

Cha, A., & Bezanilla, F. (1997). Characterizing voltage-dependent conformational changes in the Shaker K^+ channel with fluorescence. *Neuron, 19*(5), 1127–1140.

Chanda, B., & Bezanilla, F. (2008). A common pathway for charge transport through voltage-sensing domains. *Neuron, 57*(3), 345–351.

Chen, X., & Aldrich, R. W. (2011). Charge substitution for a deep-pore residue reveals structural dynamics during BK channel gating. *The Journal of General Physiology, 138*(2), 137–154.

Chen, R. S., Geng, Y., & Magleby, K. L. (2011). Mg(2+) binding to open and closed states can activate BK channels provided that the voltage sensors are elevated. *The Journal of General Physiology, 138*(6), 593–607.

Chen, X., Yan, J., & Aldrich, R. W. (2014). BK channel opening involves side-chain reorientation of multiple deep-pore residues. *Proceedings of the National Academy of Sciences of the United States of America, 111*(1), E79–E88.

Chowdhury, S., & Chanda, B. (2012). Perspectives on: Conformational coupling in ion channels: Thermodynamics of electromechanical coupling in voltage-gated ion channels. *The Journal of General Physiology, 140*(6), 613–623.
Claydon, T. W., & Fedida, D. (2007). Voltage clamp fluorimetry studies of mammalian voltage-gated K(+) channel gating. *Biochemical Society Transactions, 35*(Pt. 5), 1080–1082.
Contreras, G. F., Castillo, K., Enrique, N., Carrasquel-Ursulaez, W., Castillo, J. P., Milesi, V., et al. (2013). A BK (Slo1) channel journey from molecule to physiology. *Channels (Austin, Tex.), 7*(6), 442–458.
Contreras, J. E., & Holmgren, M. (2006). Access of quaternary ammonium blockers to the internal pore of cyclic nucleotide-gated channels: Implications for the location of the gate. *The Journal of General Physiology, 127*(5), 481–494.
Contreras, G. F., Neely, A., Alvarez, O., Gonzalez, C., & Latorre, R. (2012). Modulation of BK channel voltage gating by different auxiliary beta subunits. *Proceedings of the National Academy of Sciences of the United States of America, 109*(46), 18991–18996.
Contreras, J. E., Srikumar, D., & Holmgren, M. (2008). Gating at the selectivity filter in cyclic nucleotide-gated channels. *Proceedings of the National Academy of Sciences of the United States of America, 105*(9), 3310–3314.
Cordeiro, B., Terentyev, D., & Clements, R. T. (2015). BKCa channel activation increases cardiac contractile recovery following hypothermic ischemia/reperfusion. *American Journal of Physiology. Heart and Circulatory Physiology, 309*(4), H625–H633.
Cox, D. H. (2014). Modeling a Ca(2+) channel/BKCa channel complex at the single-complex level. *Biophysical Journal, 107*(12), 2797–2814.
Cox, D. H., & Aldrich, R. W. (2000). Role of the beta1 subunit in large-conductance Ca(2+)-activated K(+) channel gating energetics. Mechanisms of enhanced Ca(2+) sensitivity. *The Journal of General Physiology, 116*(3), 411–432.
Cox, D. H., Cui, J., & Aldrich, R. W. (1997). Allosteric gating of a large conductance Ca-activated K^+ channel. *The Journal of General Physiology, 110*(3), 257–281.
Cox, D. H., & Hoshi, T. (2011). Where's the gate? Gating in the deep pore of the BK(Ca) channel. *The Journal of General Physiology, 138*(2), 133–136.
Cuello, L. G., Jogini, V., Cortes, D. M., Pan, A. C., Gagnon, D. G., Dalmas, O., et al. (2010a). Structural basis for the coupling between activation and inactivation gates in K(+) channels. *Nature, 466*(7303), 272–275.
Cuello, L. G., Jogini, V., Cortes, D. M., & Perozo, E. (2010b). Structural mechanism of C-type inactivation in K(+) channels. *Nature, 466*(7303), 203–208.
Cui, J., & Aldrich, R. W. (2000). Allosteric linkage between voltage and Ca(2+)-dependent activation of BK-type mslo1 K(+) channels. *Biochemistry, 39*(50), 15612–15619.
Cui, J., Cox, D. H., & Aldrich, R. W. (1997). Intrinsic voltage dependence and Ca^{2+} regulation of mslo large conductance Ca-activated K^+ channels. *The Journal of General Physiology, 109*(5), 647–673.
Cui, J., Yang, H., & Lee, U. S. (2009). Molecular mechanisms of BK channel activation. *Cellular and Molecular Life Sciences, 66*(5), 852–875.
DeBerg, H. A., Brzovic, P. S., Flynn, G. E., Zagotta, W. N., & Stoll, S. (2016). Structure and energetics of allosteric regulation of HCN2 ion channels by cyclic nucleotides. *The Journal of Biological Chemistry, 291*(1), 371–381.
DeCaen, P. G., Yarov-Yarovoy, V., Sharp, E. M., Scheuer, T., & Catterall, W. A. (2009). Sequential formation of ion pairs during activation of a sodium channel voltage sensor. *Proceedings of the National Academy of Sciences of the United States of America, 106*(52), 22498–22503.
DeCaen, P. G., Yarov-Yarovoy, V., Zhao, Y., Scheuer, T., & Catterall, W. A. (2008). Disulfide locking a sodium channel voltage sensor reveals ion pair formation during activation. *Proceedings of the National Academy of Sciences of the United States of America, 105*(39), 15142–15147.

del Camino, D., & Yellen, G. (2001). Tight steric closure at the intracellular activation gate of a voltage-gated K(+) channel. *Neuron, 32*(4), 649–656.

Diaz, L., Meera, P., Amigo, J., Stefani, E., Alvarez, O., Toro, L., et al. (1998). Role of the S4 segment in a voltage-dependent calcium-sensitive potassium (hSlo) channel. *The Journal of Biological Chemistry, 273*(49), 32430–32436.

Diaz-Franulic, I., Sepulveda, R. V., Navarro-Quezada, N., Gonzalez-Nilo, F., & Naranjo, D. (2015). Pore dimensions and the role of occupancy in unitary conductance of Shaker K channels. *The Journal of General Physiology, 146*(2), 133–146.

Dong, J., Shi, N., Berke, I., Chen, L., & Jiang, Y. (2005). Structures of the MthK RCK domain and the effect of Ca^{2+} on gating ring stability. *The Journal of Biological Chemistry, 280*(50), 41716–41724.

Dopico, A. M., & Bukiya, A. N. (2014). Lipid regulation of BK channel function. *Frontiers in Physiology, 5*, 312.

Doyle, D. A., Morais, C. J., Pfuetzner, R. A., Kuo, A., Gulbis, J. M., Cohen, S. L., et al. (1998). The structure of the potassium channel: Molecular basis of K^{+} conduction and selectivity. *Science, 280*(5360), 69–77.

Du, W., Bautista, J. F., Yang, H., Diez-Sampedro, A., You, S. A., Wang, L., et al. (2005). Calcium-sensitive potassium channelopathy in human epilepsy and paroxysmal movement disorder. *Nature Genetics, 37*(7), 733–738.

Dudev, T., & Lim, C. (2003). Principles governing Mg, Ca, and Zn binding and selectivity in proteins. *Chemical Reviews, 103*(3), 773–788.

Dunn, K. M., & Nelson, M. T. (2010). Potassium channels and neurovascular coupling. *Circulation Journal, 74*(4), 608–616.

Egwolf, B., & Roux, B. (2010). Ion selectivity of the KcsA channel: A perspective from multi-ion free energy landscapes. *Journal of Molecular Biology, 401*(5), 831–842.

Eichhorn, B., & Dobrev, D. (2007). Vascular large conductance calcium-activated potassium channels: Functional role and therapeutic potential. *Naunyn-Schmiedeberg's Archives of Pharmacology, 376*(3), 145–155.

Eisenman, G., Latorre, R., & Miller, C. (1986). Multi-ion conduction and selectivity in the high-conductance Ca++-activated K^{+} channel from skeletal muscle. *Biophysical Journal, 50*(6), 1025–1034.

Faber, E. S., & Sah, P. (2003). Calcium-activated potassium channels: Multiple contributions to neuronal function. *The Neuroscientist, 9*(3), 181–194.

Gamal El-Din, T. M., Scheuer, T., & Catterall, W. A. (2014). Tracking S4 movement by gating pore currents in the bacterial sodium channel NaChBac. *The Journal of General Physiology, 144*(2), 147–157.

Gandhi, C. S., & Olcese, R. (2008). The voltage-clamp fluorometry technique. *Methods in Molecular Biology, 491*, 213–231.

Geng, Y., Niu, X., & Magleby, K. L. (2011). Low resistance, large dimension entrance to the inner cavity of BK channels determined by changing side-chain volume. *The Journal of General Physiology, 137*(6), 533–548.

Gessner, G., Cui, Y. M., Otani, Y., Ohwada, T., Soom, M., Hoshi, T., et al. (2012). Molecular mechanism of pharmacological activation of BK channels. *Proceedings of the National Academy of Sciences of the United States of America, 109*(9), 3552–3557.

Golowasch, J., Kirkwood, A., & Miller, C. (1986). Allosteric effects of Mg^{2+} on the gating of Ca^{2+}-activated K^{+} channels from mammalian skeletal muscle. *The Journal of Experimental Biology, 124*, 5–13.

Gonzalez-Perez, V., Xia, X. M., & Lingle, C. J. (2014). Functional regulation of BK potassium channels by gamma1 auxiliary subunits. *Proceedings of the National Academy of Sciences of the United States of America, 111*(13), 4868–4873.

Gonzalez-Perez, V., Xia, X. M., & Lingle, C. J. (2015). Two classes of regulatory subunits coassemble in the same BK channel and independently regulate gating. *Nature Communications, 6*, 8341.

Gruslova, A., Semenov, I., & Wang, B. (2012). An extracellular domain of the accessory beta1 subunit is required for modulating BK channel voltage sensor and gate. *The Journal of General Physiology, 139*(1), 57–67.

Haug, T., Sigg, D., Ciani, S., Toro, L., Stefani, E., & Olcese, R. (2004). Regulation of K^+ flow by a ring of negative charges in the outer pore of BKCa channels. Part I: Aspartate 292 modulates K^+ conduction by external surface charge effect. *The Journal of General Physiology, 124*(2), 173–184.

Heginbotham, L., Abramson, T., & MacKinnon, R. (1992). A functional connection between the pores of distantly related ion channels as revealed by mutant K^+ channels. *Science, 258*(5085), 1152–1155.

Heginbotham, L., Lu, Z., Abramson, T., & MacKinnon, R. (1994). Mutations in the K^+ channel signature sequence. *Biophysical Journal, 66*(4), 1061–1067.

Henrion, U., Renhorn, J., Borjesson, S. I., Nelson, E. M., Schwaiger, C. S., Bjelkmar, P., et al. (2012). Tracking a complete voltage-sensor cycle with metal-ion bridges. *Proceedings of the National Academy of Sciences of the United States of America, 109*(22), 8552–8557.

Horne, A. J., & Fedida, D. (2009). Use of voltage clamp fluorimetry in understanding potassium channel gating: A review of Shaker fluorescence data. *Canadian Journal of Physiology and Pharmacology, 87*(6), 411–418.

Horrigan, F. T. (2012). Perspectives on: Conformational coupling in ion channels: Conformational coupling in BK potassium channels. *The Journal of General Physiology, 140*(6), 625–634.

Horrigan, F. T., & Aldrich, R. W. (1999). Allosteric voltage gating of potassium channels II. Mslo channel gating charge movement in the absence of Ca(2+). *The Journal of General Physiology, 114*(2), 305–336.

Horrigan, F. T., & Aldrich, R. W. (2002). Coupling between voltage sensor activation, Ca^{2+} binding and channel opening in large conductance (BK) potassium channels. *The Journal of General Physiology, 120*(3), 267–305.

Horrigan, F. T., Cui, J., & Aldrich, R. W. (1999). Allosteric voltage gating of potassium channels I. Mslo ionic currents in the absence of Ca(2+). *The Journal of General Physiology, 114*(2), 277–304.

Horrigan, F. T., Heinemann, S. H., & Hoshi, T. (2005). Heme regulates allosteric activation of the Slo1 BK channel. *The Journal of General Physiology, 126*(1), 7–21.

Horrigan, F. T., & Hoshi, T. (2008). Integration of an electric-metal sensory experience in the Slo1 BK channel. *Nature Structural & Molecular Biology, 15*(11), 1130–1132.

Horrigan, F. T., & Ma, Z. (2008). Mg^{2+} enhances voltage sensor/gate coupling in BK channels. *The Journal of General Physiology, 131*(1), 13–32.

Hoshi, T., Pantazis, A., & Olcese, R. (2013). Transduction of voltage and Ca^{2+} signals by Slo1 BK channels. *Physiology (Bethesda), 28*(3), 172–189.

Hoshi, T., Tian, Y., Xu, R., Heinemann, S. H., & Hou, S. (2013). Mechanism of the modulation of BK potassium channel complexes with different auxiliary subunit compositions by the omega-3 fatty acid DHA. *Proceedings of the National Academy of Sciences of the United States of America, 110*(12), 4822–4827.

Hoshi, T., Wissuwa, B., Tian, Y., Tajima, N., Xu, R., Bauer, M., et al. (2013). Omega-3 fatty acids lower blood pressure by directly activating large-conductance Ca(2)(+)-dependent K(+) channels. *Proceedings of the National Academy of Sciences of the United States of America, 110*(12), 4816–4821.

Hoshi, T., Xu, R., Hou, S., Heinemann, S. H., & Tian, Y. (2013). A point mutation in the human Slo1 channel that impairs its sensitivity to omega-3 docosahexaenoic acid. *The Journal of General Physiology, 142*(5), 507–522.

Hoshi, T., Zagotta, W. N., & Aldrich, R. W. (1990). Biophysical and molecular mechanisms of Shaker potassium channel inactivation. *Science, 250*(4980), 533–538.

Hou, S., Heinemann, S. H., & Hoshi, T. (2009). Modulation of BKCa channel gating by endogenous signaling molecules. *Physiology (Bethesda)*, *24*, 26–35.

Hu, L., Shi, J., Ma, Z., Krishnamoorthy, G., Sieling, F., Zhang, G., et al. (2003). Participation of the S4 voltage sensor in the Mg^{2+}-dependent activation of large conductance (BK) K^+ channels. *Proceedings of the National Academy of Sciences of the United States of America*, *100*(18), 10488–10493.

Ishida, I. G., Rangel-Yescas, G. E., Carrasco-Zanini, J., & Islas, L. D. (2015). Voltage-dependent gating and gating charge measurements in the Kv1.2 potassium channel. *The Journal of General Physiology*, *145*(4), 345–358.

Islas, L. D., & Sigworth, F. J. (1999). Voltage sensitivity and gating charge in Shaker and Shab family potassium channels. *The Journal of General Physiology*, *114*(5), 723–742.

Javaherian, A. D., Yusifov, T., Pantazis, A., Franklin, S., Gandhi, C. S., & Olcese, R. (2011). Metal-driven operation of the human large-conductance voltage- and Ca^{2+}-dependent potassium channel (BK) gating ring apparatus. *The Journal of Biological Chemistry*, *286*(23), 20701–20709.

Jensen, M. O., Jogini, V., Borhani, D. W., Leffler, A. E., Dror, R. O., & Shaw, D. E. (2012). Mechanism of voltage gating in potassium channels. *Science*, *336*(6078), 229–233.

Jiang, Y., Lee, A., Chen, J., Cadene, M., Chait, B. T., & MacKinnon, R. (2002). Crystal structure and mechanism of a calcium-gated potassium channel. *Nature*, *417*(6888), 515–522.

Jiang, Y., Pico, A., Cadene, M., Chait, B. T., & MacKinnon, R. (2001). Structure of the RCK domain from the E. coli K^+ channel and demonstration of its presence in the human BK channel. *Neuron*, *29*(3), 593–601.

Kalia, J., Milescu, M., Salvatierra, J., Wagner, J., Klint, J. K., King, G. F., et al. (2015). From foe to friend: Using animal toxins to investigate ion channel function. *Journal of Molecular Biology*, *427*(1), 158–175.

Kim, H. J., Lim, H. H., Rho, S. H., Bao, L., Lee, J. H., Cox, D. H., et al. (2008). Modulation of the conductance-voltage relationship of the BK Ca channel by mutations at the putative flexible interface between two RCK domains. *Biophysical Journal*, *94*(2), 446–456.

Kim, H. J., Lim, H. H., Rho, S. H., Eom, S. H., & Park, C. S. (2006). Hydrophobic interface between two regulators of K^+ conductance domains critical for calcium-dependent activation of large conductance Ca^{2+}-activated K^+ channels. *The Journal of Biological Chemistry*, *281*(50), 38573–38581.

Koval, O. M., Fan, Y., & Rothberg, B. S. (2007). A role for the S0 transmembrane segment in voltage-dependent gating of BK channels. *The Journal of General Physiology*, *129*(3), 209–220.

Kursula, P. (2014). The many structural faces of calmodulin: A multitasking molecular jackknife. *Amino Acids*, *46*(10), 2295–2304.

Lacroix, J. J., & Bezanilla, F. (2011). Control of a final gating charge transition by a hydrophobic residue in the S2 segment of a K^+ channel voltage sensor. *Proceedings of the National Academy of Sciences of the United States of America*, *108*(16), 6444–6449.

Latorre, R., & Brauchi, S. (2006). Large conductance Ca^{2+}-activated K^+ (BK) channel: Activation by Ca2+ and voltage. *Biological Research*, *39*(3), 385–401.

Latorre, R., & Contreras, G. (2013). Keeping you healthy: BK channel activation by omega-3 fatty acids. *The Journal of General Physiology*, *142*(5), 487–491.

Latorre, R., & Miller, C. (1983). Conduction and selectivity in potassium channels. *The Journal of Membrane Biology*, *71*(1-2), 11–30.

Latorre, R., Morera, F. J., & Zaelzer, C. (2010). Allosteric interactions and the modular nature of the voltage- and Ca^{2+}-activated (BK) channel. *The Journal of Physiology*, *588*(Pt. 17), 3141–3148.

Latorre, R., Vergara, C., & Hidalgo, C. (1982). Reconstitution in planar lipid bilayers of a Ca^{2+}-dependent K^+ channel from transverse tubule membranes isolated from rabbit skeletal muscle. *Proceedings of the National Academy of Sciences of the United States of America, 79*(3), 805–809.

Ledoux, J., Werner, M. E., Brayden, J. E., & Nelson, M. T. (2006). Calcium-activated potassium channels and the regulation of vascular tone. *Physiology (Bethesda), 21*, 69–78.

Lee, U. S., & Cui, J. (2010). BK channel activation: Structural and functional insights. *Trends in Neurosciences, 33*(9), 415–423.

Li, W., & Aldrich, R. W. (2004). Unique inner pore properties of BK channels revealed by quaternary ammonium block. *The Journal of General Physiology, 124*(1), 43–57.

Li, W., & Aldrich, R. W. (2006). State-dependent block of BK channels by synthesized shaker ball peptides. *The Journal of General Physiology, 128*(4), 423–441.

Li, Q., Shen, R., Treger, J. S., Wanderling, S. S., Milewski, W., Siwowska, K., et al. (2015). Resting state of the human proton channel dimer in a lipid bilayer. *Proceedings of the National Academy of Sciences of the United States of America, 112*(44), E5926–E5935.

Li, Q., Wanderling, S., Paduch, M., Medovoy, D., Singharoy, A., McGreevy, R., et al. (2014). Structural mechanism of voltage-dependent gating in an isolated voltage-sensing domain. *Nature Structural & Molecular Biology, 21*(3), 244–252.

Liin, S. I., Barro-Soria, R., & Larsson, H. P. (2015). The KCNQ1 channel—Remarkable flexibility in gating allows for functional versatility. *The Journal of Physiology, 593*(12), 2605–2615.

Lin, M. C., Abramson, J., & Papazian, D. M. (2010). Transfer of ion binding site from ether-a-go-go to Shaker: Mg^{2+} binds to resting state to modulate channel opening. *The Journal of General Physiology, 135*(5), 415–431.

Lingle, C. J., Zeng, X. H., Ding, J. P., & Xia, X. M. (2001). Inactivation of BK channels mediated by the NH(2) terminus of the beta3b auxiliary subunit involves a two-step mechanism: Possible separation of binding and blockade. *The Journal of General Physiology, 117*(6), 583–606.

Lippiat, J. D., Standen, N. B., & Davies, N. W. (2000). A residue in the intracellular vestibule of the pore is critical for gating and permeation in Ca^{2+}-activated K^{+} (BKCa) channels. *The Journal of Physiology, 529*(Pt. 1), 131–138.

Li-Smerin, Y., & Swartz, K. J. (2001). Helical structure of the COOH terminus of S3 and its contribution to the gating modifier toxin receptor in voltage-gated ion channels. *The Journal of General Physiology, 117*(3), 205–218.

Liu, H. W., Hou, P. P., Guo, X. Y., Zhao, Z. W., Hu, B., Li, X., et al. (2014). Structural basis for calcium and magnesium regulation of a large conductance calcium-activated potassium channel with beta1 subunits. *The Journal of Biological Chemistry, 289*(24), 16914–16923.

Liu, G., Niu, X., Wu, R. S., Chudasama, N., Yao, Y., Jin, X., et al. (2010). Location of modulatory beta subunits in BK potassium channels. *The Journal of General Physiology, 135*(5), 449–459.

Liu, G., Zakharov, S. I., Yang, L., Deng, S. X., Landry, D. W., Karlin, A., et al. (2008a). Position and role of the BK channel alpha subunit S0 helix inferred from disulfide crosslinking. *The Journal of General Physiology, 131*(6), 537–548.

Liu, G., Zakharov, S. I., Yang, L., Wu, R. S., Deng, S. X., Landry, D. W., et al. (2008b). Locations of the beta1 transmembrane helices in the BK potassium channel. *Proceedings of the National Academy of Sciences of the United States of America, 105*(31), 10727–10732.

Liu, G., Zakharov, S. I., Yao, Y., Marx, S. O., & Karlin, A. (2015). Positions of the cytoplasmic end of BK alpha S0 helix relative to S1-S6 and of beta1 TM1 and TM2 relative to S0-S6. *The Journal of General Physiology, 145*(3), 185–199.

Loane, D. J., Lima, P. A., & Marrion, N. V. (2007). Co-assembly of N-type Ca^{2+} and BK channels underlies functional coupling in rat brain. *Journal of Cell Science, 120*(Pt. 6), 985–995.

Logothetis, D. E., Movahedi, S., Satler, C., Lindpaintner, K., & Nadal-Ginard, B. (1992). Incremental reductions of positive charge within the S4 region of a voltage-gated K^+ channel result in corresponding decreases in gating charge. *Neuron, 8*(3), 531–540.

Long, S. B., Campbell, E. B., & MacKinnon, R. (2005a). Crystal structure of a mammalian voltage-dependent Shaker family K^+ channel. *Science, 309*(5736), 897–903.

Long, S. B., Campbell, E. B., & MacKinnon, R. (2005b). Voltage sensor of Kv1.2: Structural basis of electromechanical coupling. *Science, 309*(5736), 903–908.

Long, S. B., Tao, X., Campbell, E. B., & MacKinnon, R. (2007). Atomic structure of a voltage-dependent K^+ channel in a lipid membrane-like environment. *Nature, 450*(7168), 376–382.

Lu, Z., Klem, A. M., & Ramu, Y. (2002). Coupling between voltage sensors and activation gate in voltage-gated K^+ channels. *The Journal of General Physiology, 120*(5), 663–676.

Luzhkov, V. B., & Aqvist, J. (2001). K(+)/Na(+) selectivity of the KcsA potassium channel from microscopic free energy perturbation calculations. *Biochimica et Biophysica Acta, 1548*(2), 194–202.

Ma, Z., Lou, X. J., & Horrigan, F. T. (2006). Role of charged residues in the S1-S4 voltage sensor of BK channels. *The Journal of General Physiology, 127*(3), 309–328.

Ma, Z., Wong, K. Y., & Horrigan, F. T. (2008). An extracellular Cu^{2+} binding site in the voltage sensor of BK and Shaker potassium channels. *The Journal of General Physiology, 131*(5), 483–502.

Magleby, K. L. (2003). Gating mechanism of BK (Slo1) channels: So near, yet so far. *The Journal of General Physiology, 121*(2), 81–96.

Mannuzzu, L. M., Moronne, M. M., & Isacoff, E. Y. (1996). Direct physical measure of conformational rearrangement underlying potassium channel gating. *Science, 271*(5246), 213–216.

Martinez-Espinosa, P. L., Yang, C., Gonzalez-Perez, V., Xia, X. M., & Lingle, C. J. (2014). Knockout of the BK beta2 subunit abolishes inactivation of BK currents in mouse adrenal chromaffin cells and results in slow-wave burst activity. *The Journal of General Physiology, 144*(4), 275–295.

Marty, A. (1981). Ca-dependent K channels with large unitary conductance in chromaffin cell membranes. *Nature, 291*(5815), 497–500.

McManus, O. B., Helms, L. M., Pallanck, L., Ganetzky, B., Swanson, R., & Leonard, R. J. (1995). Functional role of the beta subunit of high conductance calcium-activated potassium channels. *Neuron, 14*(3), 645–650.

Meera, P., Wallner, M., Song, M., & Toro, L. (1997). Large conductance voltage- and calcium-dependent K^+ channel, a distinct member of voltage-dependent ion channels with seven N-terminal transmembrane segments (S0-S6), an extracellular N terminus, and an intracellular (S9-S10) C terminus. *Proceedings of the National Academy of Sciences of the United States of America, 94*(25), 14066–14071.

Meisel, E., Dvir, M., Haitin, Y., Giladi, M., Peretz, A., & Attali, B. (2012). KCNQ1 channels do not undergo concerted but sequential gating transitions in both the absence and the presence of KCNE1 protein. *The Journal of Biological Chemistry, 287*(41), 34212–34224.

Meredith, A. L., Thorneloe, K. S., Werner, M. E., Nelson, M. T., & Aldrich, R. W. (2004). Overactive bladder and incontinence in the absence of the BK large conductance Ca^{2+}-activated K^+ channel. *The Journal of Biological Chemistry, 279*(35), 36746–36752.

Meredith, A. L., Wiler, S. W., Miller, B. H., Takahashi, J. S., Fodor, A. A., Ruby, N. F., et al. (2006). BK calcium-activated potassium channels regulate circadian behavioral rhythms and pacemaker output. *Nature Neuroscience, 9*(8), 1041–1049.

Milescu, L. S. (2015). A stochastic relationship. *Biophysical Journal, 108*(2), 225–227.

Miller, C. (1987). Trapping single ions inside single ion channels. *Biophysical Journal, 52*(1), 123–126.

Miranda, P., Contreras, J. E., Plested, A. J., Sigworth, F. J., Holmgren, M., & Giraldez, T. (2013). State-dependent FRET reports calcium- and voltage-dependent gating-ring motions in BK channels. *Proceedings of the National Academy of Sciences of the United States of America, 110*(13), 5217–5222.
Moczydlowski, E., & Latorre, R. (1983). Gating kinetics of Ca^{2+}-activated K^{+} channels from rat muscle incorporated into planar lipid bilayers. Evidence for two voltage-dependent Ca^{2+} binding reactions. *The Journal of General Physiology, 82*(4), 511–542.
Moldenhauer, H., Diaz-Franulic, I., Gonzalez-Nilo, F., & Naranjo, D. (2016). Effective pore size and radius of capture for K(+) ions in K-channels. *Scientific Reports, 6*, 19893.
Morais-Cabral, J. H., Zhou, Y., & MacKinnon, R. (2001). Energetic optimization of ion conduction rate by the K^{+} selectivity filter. *Nature, 414*(6859), 37–42.
Morera, F. J., Wolff, D., & Vergara, C. (2003). External copper inhibits the activity of the large-conductance calcium- and voltage-sensitive potassium channel from skeletal muscle. *The Journal of Membrane Biology, 192*(1), 65–72.
Morrow, J. P., Zakharov, S. I., Liu, G., Yang, L., Sok, A. J., & Marx, S. O. (2006). Defining the BK channel domains required for beta1-subunit modulation. *Proceedings of the National Academy of Sciences of the United States of America, 103*(13), 5096–5101.
Nelson, R. D., Kuan, G., Saier, M. H., Jr., & Montal, M. (1999). Modular assembly of voltage-gated channel proteins: A sequence analysis and phylogenetic study. *Journal of Molecular Microbiology and Biotechnology, 1*(2), 281–287.
Nelson, M. T., & Quayle, J. M. (1995). Physiological roles and properties of potassium channels in arterial smooth muscle. *The American Journal of Physiology, 268*(4 Pt. 1), C799–C822.
Neyton, J., & Miller, C. (1988). Potassium blocks barium permeation through a calcium-activated potassium channel. *The Journal of General Physiology, 92*(5), 549–567.
Nimigean, C. M., Chappie, J. S., & Miller, C. (2003). Electrostatic tuning of ion conductance in potassium channels. *Biochemistry, 42*(31), 9263–9268.
Nimigean, C. M., & Magleby, K. L. (2000). Functional coupling of the beta(1) subunit to the large conductance Ca(2+)-activated K(+) channel in the absence of Ca(2+). Increased Ca(2+) sensitivity from a Ca(2+)-independent mechanism. *The Journal of General Physiology, 115*(6), 719–736.
Niu, X., Liu, G., Wu, R. S., Chudasama, N., Zakharov, S. I., Karlin, A., et al. (2013). Orientations and proximities of the extracellular ends of transmembrane helices S0 and S4 in open and closed BK potassium channels. *PLoS One, 8*(3), e58335.
Niu, X., & Magleby, K. L. (2002). Stepwise contribution of each subunit to the cooperative activation of BK channels by Ca2+. *Proceedings of the National Academy of Sciences of the United States of America, 99*(17), 11441–11446.
Niu, X., Qian, X., & Magleby, K. L. (2004). Linker-gating ring complex as passive spring and Ca(2+)-dependent machine for a voltage- and Ca(2+)-activated potassium channel. *Neuron, 42*(5), 745–756.
Noceti, F., Baldelli, P., Wei, X., Qin, N., Toro, L., Birnbaumer, L., et al. (1996). Effective gating charges per channel in voltage-dependent K^{+} and Ca^{2+} channels. *The Journal of General Physiology, 108*(3), 143–155.
Noskov, S. Y., Berneche, S., & Roux, B. (2004). Control of ion selectivity in potassium channels by electrostatic and dynamic properties of carbonyl ligands. *Nature, 431*(7010), 830–834.
Oberhauser, A., Alvarez, O., & Latorre, R. (1988). Activation by divalent cations of a Ca^{2+}-activated K^{+} channel from skeletal muscle membrane. *The Journal of General Physiology, 92*(1), 67–86.
Orio, P., & Latorre, R. (2005). Differential effects of beta 1 and beta 2 subunits on BK channel activity. *The Journal of General Physiology, 125*(4), 395–411.

Orio, P., Rojas, P., Ferreira, G., & Latorre, R. (2002). New disguises for an old channel: MaxiK channel beta-subunits. *News in Physiological Sciences*, *17*, 156–161.

Osteen, J. D., Barro-Soria, R., Robey, S., Sampson, K. J., Kass, R. S., & Larsson, H. P. (2012). Allosteric gating mechanism underlies the flexible gating of KCNQ1 potassium channels. *Proceedings of the National Academy of Sciences of the United States of America*, *109*(18), 7103–7108.

Pallotta, B. S., Magleby, K. L., & Barrett, J. N. (1981). Single channel recordings of Ca^{2+}-activated K^+ currents in rat muscle cell culture. *Nature*, *293*(5832), 471–474.

Palovcak, E., Delemotte, L., Klein, M. L., & Carnevale, V. (2014). Evolutionary imprint of activation: The design principles of VSDs. *The Journal of General Physiology*, *143*(2), 145–156.

Pantazis, A., Gudzenko, V., Savalli, N., Sigg, D., & Olcese, R. (2010). Operation of the voltage sensor of a human voltage- and Ca^{2+}-activated K^+ channel. *Proceedings of the National Academy of Sciences of the United States of America*, *107*(9), 4459–4464.

Pantazis, A., Kohanteb, A. P., & Olcese, R. (2010). Relative motion of transmembrane segments S0 and S4 during voltage sensor activation in the human BK(Ca) channel. *The Journal of General Physiology*, *136*(6), 645–657.

Pantazis, A., & Olcese, R. (2012). Relative transmembrane segment rearrangements during BK channel activation resolved by structurally assigned fluorophore-quencher pairing. *The Journal of General Physiology*, *140*(2), 207–218.

Pantazis, A., Savalli, N., Sigg, D., Neely, A., & Olcese, R. (2014). Functional heterogeneity of the four voltage sensors of a human L-type calcium channel. *Proceedings of the National Academy of Sciences of the United States of America*, *111*(51), 18381–18386.

Papazian, D. M., Shao, X. M., Seoh, S. A., Mock, A. F., Huang, Y., & Wainstock, D. H. (1995). Electrostatic interactions of S4 voltage sensor in Shaker K^+ channel. *Neuron*, *14*(6), 1293–1301.

Papazian, D. M., Timpe, L. C., Jan, Y. N., & Jan, L. Y. (1991). Alteration of voltage-dependence of Shaker potassium channel by mutations in the S4 sequence. *Nature*, *349*(6307), 305–310.

Parfenova, L. V., Abarca-Heidemann, K., Crane, B. M., & Rothberg, B. S. (2007). Molecular architecture and divalent cation activation of TvoK, a prokaryotic potassium channel. *The Journal of Biological Chemistry*, *282*(33), 24302–24309.

Pathak, M. M., Yarov-Yarovoy, V., Agarwal, G., Roux, B., Barth, P., Kohout, S., et al. (2007). Closing in on the resting state of the Shaker K(+) channel. *Neuron*, *56*(1), 124–140.

Pau, V. P., Smith, F. J., Taylor, A. B., Parfenova, L. V., Samakai, E., Callaghan, M. M., et al. (2011). Structure and function of multiple Ca^{2+}-binding sites in a K^+ channel regulator of K^+ conductance (RCK) domain. *Proceedings of the National Academy of Sciences of the United States of America*, *108*(43), 17684–17689.

Pauling, L. (1960). *The nature of the chemical bond and the structure of molecules and crystals: An introduction to modern structural chemistry*. Ithaca, NY: Cornell University Press.

Payandeh, J., Gamal El-Din, T. M., Scheuer, T., Zheng, N., & Catterall, W. A. (2012). Crystal structure of a voltage-gated sodium channel in two potentially inactivated states. *Nature*, *486*(7401), 135–139.

Payandeh, J., Scheuer, T., Zheng, N., & Catterall, W. A. (2011). The crystal structure of a voltage-gated sodium channel. *Nature*, *475*(7356), 353–358.

Perozo, E., Santacruz-Toloza, L., Stefani, E., Bezanilla, F., & Papazian, D. M. (1994). S4 mutations alter gating currents of Shaker K channels. *Biophysical Journal*, *66*(2 Pt. 1), 345–354.

Priest, M., & Bezanilla, F. (2015). Functional site-directed fluorometry. *Advances in Experimental Medicine and Biology*, *869*, 55–76.

Qian, X., & Magleby, K. L. (2003). Beta1 subunits facilitate gating of BK channels by acting through the Ca^{2+}, but not the Mg^{2+}, activating mechanisms. *Proceedings of the National Academy of Sciences of the United States of America, 100*(17), 10061–10066.
Qian, X., Niu, X., & Magleby, K. L. (2006). Intra- and intersubunit cooperativity in activation of BK channels by Ca^{2+}. *The Journal of General Physiology, 128*(4), 389–404.
Rehak, R., Bartoletti, T. M., Engbers, J. D., Berecki, G., Turner, R. W., & Zamponi, G. W. (2013). Low voltage activation of KCa1.1 current by Cav3-KCa1.1 complexes. *PLoS One, 8*(4), e61844.
Roberts, W. M., Jacobs, R. A., & Hudspeth, A. J. (1990). Colocalization of ion channels involved in frequency selectivity and synaptic transmission at presynaptic active zones of hair cells. *The Journal of Neuroscience, 10*(11), 3664–3684.
Rothberg, B. S. (2004). Allosteric modulation of ion channels: The case of maxi-K. *Science's STKE: Signal Transduction Knowledge Environment, 2004*(227), pe16.
Rothberg, B. S., & Magleby, K. L. (1999). Gating kinetics of single large-conductance Ca^{2+}-activated K^{+} channels in high Ca^{2+} suggest a two-tiered allosteric gating mechanism. *The Journal of General Physiology, 114*(1), 93–124.
Rothberg, B. S., & Magleby, K. L. (2000). Voltage and Ca^{2+} activation of single large-conductance Ca^{2+}-activated K^{+} channels described by a two-tiered allosteric gating mechanism. *The Journal of General Physiology, 116*(1), 75–99.
Roux, B., Berneche, S., Egwolf, B., Lev, B., Noskov, S. Y., Rowley, C. N., et al. (2011). Ion selectivity in channels and transporters. *The Journal of General Physiology, 137*(5), 415–426.
Salkoff, L., Butler, A., Ferreira, G., Santi, C., & Wei, A. (2006). High-conductance potassium channels of the SLO family. *Nature Reviews. Neuroscience*, 7(12), 921–931.
Savalli, N., Kondratiev, A., de Quintana, S. B., Toro, L., & Olcese, R. (2007). Modes of operation of the BKCa channel beta2 subunit. *The Journal of General Physiology, 130*(1), 117–131.
Savalli, N., Kondratiev, A., Toro, L., & Olcese, R. (2006). Voltage-dependent conformational changes in human Ca(2+)- and voltage-activated K(+) channel, revealed by voltage-clamp fluorometry. *Proceedings of the National Academy of Sciences of the United States of America, 103*(33), 12619–12624.
Savalli, N., Pantazis, A., Yusifov, T., Sigg, D., & Olcese, R. (2012). The contribution of RCK domains to human BK channel allosteric activation. *The Journal of Biological Chemistry, 287*(26), 21741–21750.
Schoppa, N. E., McCormack, K., Tanouye, M. A., & Sigworth, F. J. (1992). The size of gating charge in wild-type and mutant Shaker potassium channels. *Science, 255*(5052), 1712–1715.
Schreiber, M., & Salkoff, L. (1997). A novel calcium-sensing domain in the BK channel. *Biophysical Journal, 73*(3), 1355–1363.
Schrodinger, L. L. C. (2015). *The PyMOL molecular graphics system*. Version 1.8. https://www.pymol.org/citing.
Schwaiger, C. S., Bjelkmar, P., Hess, B., & Lindahl, E. (2011). 3(1)(0)-helix conformation facilitates the transition of a voltage sensor S4 segment toward the down state. *Biophysical Journal, 100*(6), 1446–1454.
Schwaiger, C. S., Borjesson, S. I., Hess, B., Wallner, B., Elinder, F., & Lindahl, E. (2012). The free energy barrier for arginine gating charge translation is altered by mutations in the voltage sensor domain. *PLoS One*, 7(10), e45880.
Schwaiger, C. S., Liin, S. I., Elinder, F., & Lindahl, E. (2013). The conserved phenylalanine in the K^{+} channel voltage-sensor domain creates a barrier with unidirectional effects. *Biophysical Journal, 104*(1), 75–84.
Semenova, N. P., Abarca-Heidemann, K., Loranc, E., & Rothberg, B. S. (2009). Bimane fluorescence scanning suggests secondary structure near the S3-S4 linker of BK channels. *The Journal of Biological Chemistry, 284*(16), 10684–10693.

Seoh, S. A., Sigg, D., Papazian, D. M., & Bezanilla, F. (1996). Voltage-sensing residues in the S2 and S4 segments of the Shaker K^+ channel. *Neuron*, *16*(6), 1159–1167.

Shelley, C., Niu, X., Geng, Y., & Magleby, K. L. (2010). Coupling and cooperativity in voltage activation of a limited-state BK channel gating in saturating Ca^{2+}. *The Journal of General Physiology*, *135*(5), 461–480.

Shen, K. Z., Lagrutta, A., Davies, N. W., Standen, N. B., Adelman, J. P., & North, R. A. (1994). Tetraethylammonium block of Slowpoke calcium-activated potassium channels expressed in Xenopus oocytes: Evidence for tetrameric channel formation. *Pflügers Archiv*, *426*(5), 440–445.

Shi, J., & Cui, J. (2001). Intracellular Mg(2+) enhances the function of BK-type Ca(2+)-activated K(+) channels. *The Journal of General Physiology*, *118*(5), 589–606.

Shi, J., Krishnamoorthy, G., Yang, Y., Hu, L., Chaturvedi, N., Harilal, D., et al. (2002). Mechanism of magnesium activation of calcium-activated potassium channels. *Nature*, *418*(6900), 876–880.

Sigg, D. (2014). Modeling ion channels: Past, present, and future. *The Journal of General Physiology*, *144*(1), 7–26.

Singh, H., Lu, R., Bopassa, J. C., Meredith, A. L., Stefani, E., & Toro, L. (2013). MitoBK (Ca) is encoded by the Kcnma1 gene, and a splicing sequence defines its mitochondrial location. *Proceedings of the National Academy of Sciences of the United States of America*, *110*(26), 10836–10841.

Singh, A. K., McMillan, J., Bukiya, A. N., Burton, B., Parrill, A. L., & Dopico, A. M. (2012). Multiple cholesterol recognition/interaction amino acid consensus (CRAC) motifs in cytosolic C tail of Slo1 subunit determine cholesterol sensitivity of Ca^{2+}- and voltage-gated K^+ (BK) channels. *The Journal of Biological Chemistry*, *287*(24), 20509–20521.

Singh, H., Stefani, E., & Toro, L. (2012). Intracellular BK(Ca) (iBK(Ca)) channels. *The Journal of Physiology*, *590*(Pt. 23), 5937–5947.

Smith, F. J., Pau, V. P., Cingolani, G., & Rothberg, B. S. (2012). Crystal structure of a Ba(2+)-bound gating ring reveals elementary steps in RCK domain activation. *Structure*, *20*(12), 2038–2047.

Smith, F. J., Pau, V. P., Cingolani, G., & Rothberg, B. S. (2013). Structural basis of allosteric interactions among Ca^{2+}-binding sites in a K^+ channel RCK domain. *Nature Communications*, *4*, 2621.

Sokolov, S., Scheuer, T., & Catterall, W. A. (2005). Ion permeation through a voltage-sensitive gating pore in brain sodium channels having voltage sensor mutations. *Neuron*, *47*(2), 183–189.

Starace, D. M., & Bezanilla, F. (2004). A proton pore in a potassium channel voltage sensor reveals a focused electric field. *Nature*, *427*(6974), 548–553.

Stefani, E., Ottolia, M., Noceti, F., Olcese, R., Wallner, M., Latorre, R., et al. (1997). Voltage-controlled gating in a large conductance Ca^{2+}-sensitive K^+ channel (hslo). *Proceedings of the National Academy of Sciences of the United States of America*, *94*(10), 5427–5431.

Stuhmer, W., Conti, F., Suzuki, H., Wang, X. D., Noda, M., Yahagi, N., et al. (1989). Structural parts involved in activation and inactivation of the sodium channel. *Nature*, *339*(6226), 597–603.

Sun, X. P., Yazejian, B., & Grinnell, A. D. (2004). Electrophysiological properties of BK channels in Xenopus motor nerve terminals. *The Journal of Physiology*, *557*(Pt. 1), 207–228.

Sun, X., Zaydman, M. A., & Cui, J. (2012). Regulation of voltage-activated K(+) channel gating by transmembrane beta subunits. *Frontiers in Pharmacology*, *3*, 63.

Swartz, K. J. (2008). Sensing voltage across lipid membranes. *Nature*, *456*(7224), 891–897.

Sweet, T. B., & Cox, D. H. (2008). Measurements of the BKCa channel's high-affinity Ca^{2+} binding constants: Effects of membrane voltage. *The Journal of General Physiology*, *132*(5), 491–505.

Sweet, T. B., & Cox, D. H. (2009). Measuring the influence of the BKCa {beta}1 subunit on Ca^{2+} binding to the BKCa channel. *The Journal of General Physiology*, *133*(2), 139–150.

Szabo, I., & Zoratti, M. (2014). Mitochondrial channels: Ion fluxes and more. *Physiological Reviews*, *94*(2), 519–608.

Talwar, S., & Lynch, J. W. (2015). Investigating ion channel conformational changes using voltage clamp fluorometry. *Neuropharmacology*, *98*, 3–12.

Tang, X. D., Xu, R., Reynolds, M. F., Garcia, M. L., Heinemann, S. H., & Hoshi, T. (2003). Haem can bind to and inhibit mammalian calcium-dependent Slo1 BK channels. *Nature*, *425*(6957), 531–535.

Tang, Q. Y., Zhang, Z., Meng, X. Y., Cui, M., & Logothetis, D. E. (2014). Structural determinants of phosphatidylinositol 4,5-bisphosphate (PIP2) regulation of BK channel activity through the RCK1 Ca^{2+} coordination site. *The Journal of Biological Chemistry*, *289*(27), 18860–18872.

Tao, X., Lee, A., Limapichat, W., Dougherty, D. A., & MacKinnon, R. (2010). A gating charge transfer center in voltage sensors. *Science*, *328*(5974), 67–73.

Testai, L., Martelli, A., Marino, A., D'Antongiovanni, V., Ciregia, F., Giusti, L., et al. (2013). The activation of mitochondrial BK potassium channels contributes to the protective effects of naringenin against myocardial ischemia/reperfusion injury. *Biochemical Pharmacology*, *85*(11), 1634–1643.

Thompson, J., & Begenisich, T. (2012). Selectivity filter gating in large-conductance Ca(2+)-activated K^+ channels. *The Journal of General Physiology*, *139*(3), 235–244.

Thompson, A. N., Kim, I., Panosian, T. D., Iverson, T. M., Allen, T. W., & Nimigean, C. M. (2009). Mechanism of potassium-channel selectivity revealed by Na(+) and Li(+) binding sites within the KcsA pore. *Nature Structural & Molecular Biology*, *16*(12), 1317–1324.

Tilley, D. C., Eum, K. S., Fletcher-Taylor, S., Austin, D. C., Dupre, C., Patron, L. A., et al. (2014). Chemoselective tarantula toxins report voltage activation of wild-type ion channels in live cells. *Proceedings of the National Academy of Sciences of the United States of America*, *111*(44), E4789–E4796.

Tiwari-Woodruff, S. K., Lin, M. A., Schulteis, C. T., & Papazian, D. M. (2000). Voltage-dependent structural interactions in the Shaker K(+) channel. *The Journal of General Physiology*, *115*(2), 123–138.

Tombola, F., Pathak, M. M., Gorostiza, P., & Isacoff, E. Y. (2007). The twisted ion-permeation pathway of a resting voltage-sensing domain. *Nature*, *445*(7127), 546–549.

Tombola, F., Pathak, M. M., & Isacoff, E. Y. (2005). Voltage-sensing arginines in a potassium channel permeate and occlude cation-selective pores. *Neuron*, *45*(3), 379–388.

Tombola, F., Pathak, M. M., & Isacoff, E. Y. (2006). How does voltage open an ion channel? *Annual Review of Cell and Developmental Biology*, *22*, 23–52.

Toro, L., Li, M., Zhang, Z., Singh, H., Wu, Y., & Stefani, E. (2014). MaxiK channel and cell signalling. *Pflügers Archiv*, *466*(5), 875–886.

Tristani-Firouzi, M., Chen, J., & Sanguinetti, M. C. (2002). Interactions between S4-S5 linker and S6 transmembrane domain modulate gating of HERG K^+ channels. *The Journal of Biological Chemistry*, *277*(21), 18994–19000.

Unnerstale, S., Lind, J., Papadopoulos, E., & Maler, L. (2009). Solution structure of the HsapBK K^+ channel voltage-sensor paddle sequence. *Biochemistry*, *48*(25), 5813–5821.

Vardanyan, V., & Pongs, O. (2012). Coupling of voltage-sensors to the channel pore: A comparative view. *Frontiers in Pharmacology*, *3*, 145.

Vargas, E., Bezanilla, F., & Roux, B. (2011). In search of a consensus model of the resting state of a voltage-sensing domain. *Neuron*, *72*(5), 713–720.

Villalba-Galea, C. A., Frezza, L., Sandtner, W., & Bezanilla, F. (2013). Sensing charges of the Ciona intestinalis voltage-sensing phosphatase. *The Journal of General Physiology, 142*(5), 543–555.

Wallner, M., Meera, P., & Toro, L. (1996). Determinant for beta-subunit regulation in high-conductance voltage-activated and Ca(2+)-sensitive K^+ channels: An additional transmembrane region at the N terminus. *Proceedings of the National Academy of Sciences of the United States of America, 93*(25), 14922–14927.

Wallner, M., Meera, P., & Toro, L. (1999). Molecular basis of fast inactivation in voltage and Ca^{2+}-activated K^+ channels: A transmembrane beta-subunit homolog. *Proceedings of the National Academy of Sciences of the United States of America, 96*(7), 4137–4142.

Wang, B., & Brenner, R. (2006). An S6 mutation in BK channels reveals beta1 subunit effects on intrinsic and voltage-dependent gating. *The Journal of General Physiology, 128*(6), 731–744.

Wang, Y. W., Ding, J. P., Xia, X. M., & Lingle, C. J. (2002). Consequences of the stoichiometry of Slo1 alpha and auxiliary beta subunits on functional properties of large-conductance Ca^{2+}-activated K^+ channels. *The Journal of Neuroscience, 22*(5), 1550–1561.

Wang, L., & Sigworth, F. J. (2009). Structure of the BK potassium channel in a lipid membrane from electron cryomicroscopy. *Nature, 461*(7261), 292–295.

Wei, A., Solaro, C., Lingle, C., & Salkoff, L. (1994). Calcium sensitivity of BK-type KCa channels determined by a separable domain. *Neuron, 13*(3), 671–681.

Wilkens, C. M., & Aldrich, R. W. (2006). State-independent block of BK channels by an intracellular quaternary ammonium. *The Journal of General Physiology, 128*(3), 347–364.

Wu, R. S., Chudasama, N., Zakharov, S. I., Doshi, D., Motoike, H., Liu, G., et al. (2009). Location of the beta 4 transmembrane helices in the BK potassium channel. *The Journal of Neuroscience, 29*(26), 8321–8328.

Wu, D., Delaloye, K., Zaydman, M. A., Nekouzadeh, A., Rudy, Y., & Cui, J. (2010). State-dependent electrostatic interactions of S4 arginines with E1 in S2 during Kv7.1 activation. *The Journal of General Physiology, 135*(6), 595–606.

Wu, R. S., Liu, G., Zakharov, S. I., Chudasama, N., Motoike, H., Karlin, A., et al. (2013). Positions of beta2 and beta3 subunits in the large-conductance calcium- and voltage-activated BK potassium channel. *The Journal of General Physiology, 141*(1), 105–117.

Wu, R. S., & Marx, S. O. (2010). The BK potassium channel in the vascular smooth muscle and kidney: Alpha- and beta-subunits. *Kidney International, 78*(10), 963–974.

Wu, Y., Xiong, Y., Wang, S., Yi, H., Li, H., Pan, N., et al. (2009). Intersubunit coupling in the pore of BK channels. *The Journal of Biological Chemistry, 284*(35), 23353–23363.

Wu, Y., Yang, Y., Ye, S., & Jiang, Y. (2010). Structure of the gating ring from the human large-conductance Ca(2+)-gated K(+) channel. *Nature, 466*(7304), 393–397.

Xia, X. M., Fakler, B., Rivard, A., Wayman, G., Johnson-Pais, T., Keen, J. E., et al. (1998). Mechanism of calcium gating in small-conductance calcium-activated potassium channels. *Nature, 395*(6701), 503–507.

Xia, X. M., Zeng, X., & Lingle, C. J. (2002). Multiple regulatory sites in large-conductance calcium-activated potassium channels. *Nature, 418*(6900), 880–884.

Xu, W., Liu, Y., Wang, S., McDonald, T., Van Eyk, J. E., Sidor, A., et al. (2002). Cytoprotective role of Ca^{2+}-activated K^+ channels in the cardiac inner mitochondrial membrane. *Science, 298*(5595), 1029–1033.

Yan, J., & Aldrich, R. W. (2010). LRRC26 auxiliary protein allows BK channel activation at resting voltage without calcium. *Nature, 466*(7305), 513–516.

Yang, H., Hu, L., Shi, J., & Cui, J. (2006). Tuning magnesium sensitivity of BK channels by mutations. *Biophysical Journal, 91*(8), 2892–2900.

Yang, H., Hu, L., Shi, J., Delaloye, K., Horrigan, F. T., & Cui, J. (2007). Mg2+ mediates interaction between the voltage sensor and cytosolic domain to activate BK channels.

Proceedings of the National Academy of Sciences of the United States of America, 104(46), 18270–18275.

Yang, J., Krishnamoorthy, G., Saxena, A., Zhang, G., Shi, J., Yang, H., et al. (2010). An epilepsy/dyskinesia-associated mutation enhances BK channel activation by potentiating Ca2+ sensing. *Neuron, 66*(6), 871–883.

Yang, H., Shi, J., Zhang, G., Yang, J., Delaloye, K., & Cui, J. (2008). Activation of Slo1 BK channels by Mg2+ coordinated between the voltage sensor and RCK1 domains. *Nature Structural & Molecular Biology, 15*(11), 1152–1159.

Yang, H., Zhang, G., & Cui, J. (2015). BK channels: Multiple sensors, one activation gate. *Frontiers in Physiology, 6*, 29.

Yarov-Yarovoy, V., Baker, D., & Catterall, W. A. (2006). Voltage sensor conformations in the open and closed states in ROSETTA structural models of K(+) channels. *Proceedings of the National Academy of Sciences of the United States of America, 103*(19), 7292–7297.

Ye, S., Li, Y., Chen, L., & Jiang, Y. (2006). Crystal structures of a ligand-free MthK gating ring: Insights into the ligand gating mechanism of K^+ channels. *Cell, 126*(6), 1161–1173.

Yellen, G. (1984). Ionic permeation and blockade in Ca^{2+}-activated K^+ channels of bovine chromaffin cells. *The Journal of General Physiology, 84*(2), 157–186.

Yu, F. H., & Catterall, W. A. (2004). The VGL-chanome: A protein superfamily specialized for electrical signaling and ionic homeostasis. *Science's STKE: Signal Transduction Knowledge Environment, 2004*(253), re15.

Yuan, P., Leonetti, M. D., Hsiung, Y., & MacKinnon, R. (2012). Open structure of the Ca2+ gating ring in the high-conductance Ca^{2+}-activated K^+ channel. *Nature, 481*(7379), 94–97.

Yuan, P., Leonetti, M. D., Pico, A. R., Hsiung, Y., & MacKinnon, R. (2010). Structure of the human BK channel Ca^{2+}-activation apparatus at 3.0 A resolution. *Science, 329*(5988), 182–186.

Yue, D. T. (2004). The dawn of high-resolution structure for the queen of ion channels. *Neuron, 42*(3), 357–359.

Yusifov, T., Javaherian, A. D., Pantazis, A., Gandhi, C. S., & Olcese, R. (2010). The RCK1 domain of the human BKCa channel transduces Ca2+ binding into structural rearrangements. *The Journal of General Physiology, 136*(2), 189–202.

Yusifov, T., Savalli, N., Gandhi, C. S., Ottolia, M., & Olcese, R. (2008). The RCK2 domain of the human BKCa channel is a calcium sensor. *Proceedings of the National Academy of Sciences of the United States of America, 105*(1), 376–381.

Zagotta, W. N., Hoshi, T., & Aldrich, R. W. (1994a). Shaker potassium channel gating. III: Evaluation of kinetic models for activation. *The Journal of General Physiology, 103*(2), 321–362.

Zagotta, W. N., Hoshi, T., Dittman, J., & Aldrich, R. W. (1994b). Shaker potassium channel gating. II: Transitions in the activation pathway. *The Journal of General Physiology, 103*(2), 279–319.

Zakharov, S. I., Morrow, J. P., Liu, G., Yang, L., & Marx, S. O. (2005). Activation of the BK (SLO1) potassium channel by mallotoxin. *The Journal of Biological Chemistry, 280*(35), 30882–30887.

Zemen, B. G., Lai, M. H., Whitt, J. P., Khan, Z., Zhao, G., & Meredith, A. L. (2015). Generation of Kcnma1fl-tdTomato, a conditional deletion of the BK channel alpha subunit in mouse. *Physiological Reports, 3*(11).

Zeng, X. H., Xia, X. M., & Lingle, C. J. (2005). Divalent cation sensitivity of BK channel activation supports the existence of three distinct binding sites. *The Journal of General Physiology, 125*(3), 273–286.

Zhang, G., Huang, S. Y., Yang, J., Shi, J., Yang, X., Moller, A., et al. (2010). Ion sensing in the RCK1 domain of BK channels. *Proceedings of the National Academy of Sciences of the United States of America, 107*(43), 18700–18705.

Zhang, M., Liu, J., & Tseng, G. N. (2004). Gating charges in the activation and inactivation processes of the HERG channel. *The Journal of General Physiology*, *124*(6), 703–718.
Zhang, J., & Yan, J. (2014). Regulation of BK channels by auxiliary gamma subunits. *Frontiers in Physiology*, *5*, 401.
Zhou, M., Morais-Cabral, J. H., Mann, S., & MacKinnon, R. (2001). Potassium channel receptor site for the inactivation gate and quaternary amine inhibitors. *Nature*, *411*(6838), 657–661.
Zhou, Y., Tang, Q. Y., Xia, X. M., & Lingle, C. J. (2010). Glycine311, a determinant of paxilline block in BK channels: A novel bend in the BK S6 helix. *The Journal of General Physiology*, *135*(5), 481–494.
Zhou, Y., Xia, X. M., & Lingle, C. J. (2011). Cysteine scanning and modification reveal major differences between BK channels and Kv channels in the inner pore region. *Proceedings of the National Academy of Sciences of the United States of America*, *108*(29), 12161–12166.
Zhou, Y., Xia, X. M., & Lingle, C. J. (2015). Cadmium-cysteine coordination in the BK inner pore region and its structural and functional implications. *Proceedings of the National Academy of Sciences of the United States of America*, *112*(16), 5237–5242.
Zhou, Y., Zeng, X. H., & Lingle, C. J. (2012). Barium ions selectively activate BK channels via the Ca^{2+}-bowl site. *Proceedings of the National Academy of Sciences of the United States of America*, *109*(28), 11413–11418.
Zhu, W., Varga, Z., & Silva, J. R. (2015). Molecular motions that shape the cardiac action potential: Insights from voltage clamp fluorometry. *Progress in Biophysics and Molecular Biology*, *120*(1–3), 3–17.

CHAPTER TWO

Modulation of BK Channel Function by Auxiliary Beta and Gamma Subunits

Q. Li, J. Yan[1]
The University of Texas MD Anderson Cancer Center, Houston, TX, United States
[1]Corresponding author: e-mail address: jyan1@mdanderson.org

Contents

Abstract

The large-conductance, Ca^{2+}- and voltage-activated K^+ (BK) channel is ubiquitously expressed in mammalian tissues and displays diverse biophysical or pharmacological characteristics. This diversity is in part conferred by channel modulation with different regulatory auxiliary subunits. To date, two distinct classes of BK channel auxiliary subunits have been identified: β subunits and γ subunits. Modulation of BK channels by the four auxiliary β (β1–β4) subunits has been well established and intensively investigated over the past two decades. The auxiliary γ subunits, however, were identified only very recently, which adds a new dimension to BK channel regulation and improves our understanding of the physiological functions of BK channels in various tissues and cell types. This chapter will review the current understanding of BK channel modulation by auxiliary β and γ subunits, especially the latest findings.

International Review of Neurobiology, Volume 128
ISSN 0074-7742
http://dx.doi.org/10.1016/bs.irn.2016.03.015

1. INTRODUCTION

The large-conductance, calcium- and voltage-activated potassium (BK, also known as MaxiK, Slo1, or $K_{Ca}1.1$) channel is a unique member of the potassium channel family. The BK channel has exceptionally large single-channel conductance (200–300 pS) that is 10–20 times larger than most other K^+ channels. BK channel activity is dually regulated by two independent physiological signals, membrane voltage, and intracellular free Ca^{2+} $(Ca^{2+}{}_i)$, and therefore has a powerful integrative role in regulating cellular excitability and calcium signaling in electrically excitable cells (Ghatta, Nimmagadda, Xu, & O'Rourke, 2006; Salkoff, Butler, Ferreira, Santi, & Wei, 2006). BK channels are ubiquitously expressed in most types of mammalian tissues and cells and are critically involved in various physiological processes. In central nervous system neurons, BK channels mediate the repolarization and fast afterhyperpolarization of action potentials (Shao, Halvorsrud, Borg-Graham, & Storm, 1999; Womack & Khodakhah, 2002), shape dendritic Ca^{2+} spikes (Golding, Jung, Mickus, & Spruston, 1999), and regulate neurotransmitter release at presynaptic terminals (Hu et al., 2001; Raffaelli, Saviane, Mohajerani, Pedarzani, & Cherubini, 2004; Samengo, Curro, Barrese, Taglialatela, & Martire, 2014; Xu & Slaughter, 2005; see Chapter "BK Channels in Neurons" by Barth and Contet). Neuronal BK channels are involved in motor coordination (Sausbier et al., 2004), learning and memory (Matthews & Disterhoft, 2009; Springer, Burkett, & Schrader, 2014; Typlt et al., 2013; Ye, Jalini, Mylvaganam, & Carlen, 2010), the brain's intrinsic rhythmicity of the circadian clock (Farajnia, Meijer, & Michel, 2015; Meredith et al., 2006; Montgomery, Whitt, Wright, Lai, & Meredith, 2013; Pitts, Ohta, & McMahon, 2006) and respiration (Onimaru, Ballanyi, & Homma, 2003; Zavala-Tecuapetla, Aguileta, Lopez-Guerrero, Gonzalez-Marin, & Pena, 2008; Zhao, Hulsmann, Winter, Dutschmann, & Richter, 2006), frequency tuning of the cochlear hair cell (Fettiplace & Fuchs, 1999), pain modulation (Cao, Chen, Li, & Pan, 2012; Chen, Cai, & Pan, 2009; Waxman & Zamponi, 2014; Zhang, Mok, Lee, Charbonnet, & Gold, 2012), and neuroprotection in pathological conditions (Mancini et al., 2014; Runden-Pran, Haug, Storm, & Ottersen, 2002; Shen, Kishimoto, Linden, & Sapirstein, 2007; Zhang, Xie, et al., 2009). Defects or dysregulation in human neuronal BK channels can cause epilepsy and paroxysmal dyskinesia (Brenner et al., 2005; Du et al., 2005) and are implicated in mental retardation (Deng et al., 2013; Higgins, Hao, Kosofsky, & Rajadhyaksha, 2008), autism (Laumonnier et al., 2006), and schizophrenia (Zhang, Li, Zhou, & Xing, 2006). BK channels also control contractile tone of almost all types of smooth muscle cells, playing a central role in the regulation of vascular blood pressure (Brenner, Perez, et al., 2000), urinary bladder function (Meredith, Thorneloe, Werner, Nelson, & Aldrich, 2004), and erectile function (Werner, Zvara, Meredith, Aldrich, & Nelson, 2005).

BK channels are homotetramers of the pore-forming, Ca^{2+}-, and voltage-sensing α subunits (BKα). The BKα (~130 kDa) contains seven-transmembrane (TM) segments (S0–S6), a short extracellular N-terminus, and a large cytosolic C-terminus composed of two RCK (regulating conductance of K^+) domains (Fig. 1A). Similar to other voltage-gated K^+ channels, the S1–S4 TM segments form the voltage-sensor domain (VSD), and the S5 and S6 TM segments form the pore-gate domain (PGD). Ca^{2+} and Mg^{2+} sensitivity is conferred by the two RCK domains in the cytosolic C-terminus. The RCK2 domain contains a well-defined Ca^{2+}-binding site called the Ca^{2+} bowl, formed by a string of negatively charged residues (Schreiber & Salkoff, 1997; Xia, Zeng, & Lingle, 2002; Yuan, Leonetti, Hsiung, & MacKinnon, 2012; Yusifov, Savalli, Gandhi, Ottolia, & Olcese, 2008). The RCK1 domain is believed to be involved in the formation of another Ca^{2+}-binding site (Xia et al., 2002; Zhang et al., 2010) and a Mg^{2+}-binding site which also involves residues located in the S0–S1 linker (D99) and S2–S3 loop (N172) (Shi et al., 2002; Xia et al., 2002; Yang et al., 2008). The electric (voltage) and chemical (Ca^{2+} or Mg^{2+} binding) energies are converted to mechanical forces through their sensory domains to induce

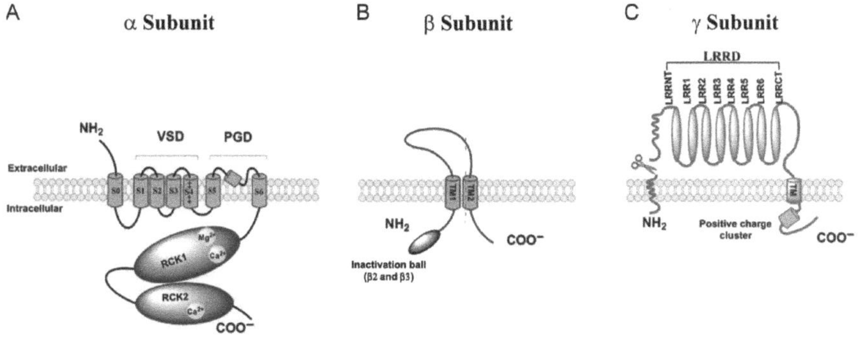

Fig. 1 Schematic structure and membrane topology of BK channel α (A), β (B), and γ (C) subunits and their models of assembly. *VSD*, voltage-sensor domain; *PGD*, pore-gate domain; *RCK*, regulator of K^+ conductance; *NH_2*, aminoterminus; *COO^-*, carboxyterminus; *LRRD*, leucine-rich repeat domain; *S* or *TM*, transmembrane segment.

conformational change in the PGD, which toggles between the "closed state" and "open state" to control K^+ flux (Horrigan & Aldrich, 1999; Horrigan, Cui, & Aldrich, 1999).

Unlike many other mammalian potassium channels, all BKα subunits are encoded by a single gene (*Slo1*, *KCNMA1*). Native BK channels exhibit a wide range of biophysical, pharmacological, and functional properties that differ among various cell types (Ding, Li, & Lingle, 1998; Gessner et al., 2005; Jones, Gray-Keller, Art, & Fettiplace, 1999; Xia, Ding, & Lingle, 1999), at different stages of development (Carvalho-de-Souza, Varanda, Tostes, & Chignalia, 2013), and in different physiological or pathological conditions (Hu et al., 2011; Manzanares et al., 2015; Tao et al., 2015). BK channels with slightly different biophysical properties can be produced by alternative splicing of *KCNMA1* precursor mRNA at several different sites (Glauser, Johnson, Aldrich, & Goodman, 2011; Ramanathan, Michael, Jiang, Hiel, & Fuchs, 1999; Saito, Nelson, Salkoff, & Lingle, 1997; Schreiber, Yuan, & Salkoff, 1999; Yu, Upadhyaya, & Atkinson, 2006). BK channel function is differently and potently modulated by auxiliary subunits. To date, four types of beta (β) auxiliary subunits (β1–β4, encoded by the *KCNMB1–4* genes) (Behrens et al., 2000; Brenner, Jegla, Wickenden, Liu, & Aldrich, 2000; Knaus, Folander, et al., 1994; Meera, Wallner, & Toro, 2000; Orio, Rojas, Ferreira, & Latorre, 2002; Wallner, Meera, & Toro, 1999; Xia et al., 1999; Xia, Ding, & Lingle, 2003; Xia, Ding, Zeng, Duan, & Lingle, 2000; Zeng, Xia, & Lingle, 2008) and four types of gamma (γ) auxiliary subunits (γ1–γ4, encoded by the *LRRC26*, *LRRC52*, *LRRC55*, and *LRRC38* genes) (Yan & Aldrich, 2010, 2012) have been identified (Table 1). These auxiliary subunits affect nearly all aspects of the BK channel's biophysical and pharmacological properties, including the apparent voltage dependence (Cox & Aldrich, 2000; Wang & Brenner, 2006; Yan & Aldrich, 2010, 2012) and Ca^{2+} sensitivity (Brenner, Jegla, et al., 2000; McManus et al., 1995; Nimigean & Magleby, 1999; Xia et al., 1999), gating kinetics, inactivation (Brenner, Jegla, et al., 2000; Xia et al., 1999, 2000), ion current rectification (Xia et al., 2000; Zeng, Xia, & Lingle, 2003), and sensitivity to extracellular modulators (Almassy & Begenisich, 2012; Valverde et al., 1999). Regulation by auxiliary subunits is therefore a key mechanism of BK channel functional diversity that adapts the channel for diverse needs in different mammalian tissues and cell types (Table 1). The modulation of BK channels by β subunits has been extensively investigated and comprehensively reviewed elsewhere (Sun, Zaydman, & Cui, 2012; Torres, Granados, & Latorre, 2014; Torres,

Table 1 Characteristics of BK Channel Auxiliary Subunits

		Effect on BK Channels Function						
Subunit	Gene Symbol (Human)	Ca^{2+} Sensitivity	Voltage Sensitivity	Inactivation	Deactivation	References	Tissue Expression	Physiologic Relevance
BKβ1	KCNMB1	Increase	Decrease	No inactivation	Slow	Wallner et al. (1995), Tseng-Crank et al. (1996), Cox and Aldrich (2000), and Orio and Latorre (2005)	Smooth muscle, kidney, urinary bladder, and brain (Knaus, Garcia-Calvo, Kaczorowski, & Garcia, 1994; Liu et al., 2010; Tseng-Crank et al., 1996)	Blood pressure, vascular smooth muscle function, bladder function, alcohol dependence, and tolerance (Amberg, Bonev, Rossow, Nelson, & Santana, 2003; Bukiya, Liu, & Dopico, 2009; Kreifeldt, Cates-Gatto, Roberts, & Contet, 2015; Kreifeldt, Le, Treistman, Koob, & Contet, 2013; Werner, Knorn, Meredith, Aldrich, & Nelson, 2007; Werner, Meredith, Aldrich, & Nelson, 2008)
BKβ2	KCNMB2	Increase	No significant decrease	Fast and complete	Slow	Orio and Latorre (2005), Lee, Shi, and Cui (2010), Wallner et al. (1999), Xia et al. (1999, 2000), and Bentrop, Beyermann, Wissmann, and Fakler (2001)	Pancreas, kidney, spleen, ovary, and brain (Brenner, Jegla, et al., 2000; Uebele et al., 2000; Wallner et al., 1999)	The adrenal medullary chromaffin cells (CCs) function (Martinez-Espinosa, Yang, Gonzalez-Perez, Xia, & Lingle, 2014)

Continued

Table 1 Characteristics of BK Channel Auxiliary Subunits—cont'd

Subunit	Gene Symbol (Human)		Effect on BK Channels Function: Ca^{2+} Sensitivity	Voltage Sensitivity	Inactivation	Deactivation	References	Tissue Expression	Physiologic Relevance
BKβ3	KCNMB3	β3a	No effect	Increase	Incomplete, comparing to hβ3a, mβ3a: more rapid, more complete	No effect	Uebele et al. (2000), Brenner, Jegla, et al. (2000), Orio and Latorre (2005), Zeng, Benzinger, Xia, and Lingle (2007), and Zeng et al. (2008)	Spleen, placenta, pancreas, heart, and kidney (Brenner, Jegla, et al., 2000; Uebele et al., 2000; Xia et al., 2000)	Generalized epilepsy (Lorenz, Heils, Kasper, & Sander, 2007)
		β3b	No effect	Decrease, while mβ3b: gating shift more to negative potentials at a given Ca^{2+} concentration	Inactivation at a given Ca^{2+} concentration, hβ3b: rapid inactivation; mβ3b: no inactivation	Slow	Uebele et al. (2000), Xia et al. (2000), Zeng et al. (2008), and Lingle, Zeng, Ding, and Xia (2001)	Spleen, pancreas, kidney, heart, testes, brain, placenta, lung, and liver (Brenner, Jegla, et al., 2000; Uebele et al., 2000; Xia et al., 2000)	
		β3c	No effect	No effect	Incomplete	No effect	Uebele et al. (2000) and Brenner, Jegla, et al. (2000)	Spleen, pancreas, liver, kidney, prostate, placenta, ovary, brain, and lung (Uebele et al., 2000; Xia et al., 2000)	
		β3d	Not known	Not known	Not known	Not known	Uebele et al. (2000) and Zeng et al. (2008)	Spleen, pancreas, kidney, testes, Lung, and brain (Brenner, Jegla, et al., 2000; Uebele et al., 2000)	

BKβ4	KCNMB4	Inhibit channel at low Ca^{2+}; activate channel at high Ca^{2+}	Decrease	No inactivation	Slow	Brenner, Jegla, et al. (2000), Meera et al. (2000), Jin, Weiger, and Levitan (2002), and Wang, Rothberg, and Brenner (2006)	Brain, neuronal tissue, kidney, and bladder smooth muscle (Behrens et al., 2000; Brenner, Jegla, et al., 2000; Chen & Petkov, 2009)	Epileptic phenotype, alcohol dependence, and tolerance (Brenner et al., 2005; Kreifeldt et al., 2015, 2013; Martin et al., 2008)
BKγ1	LRRC26	No significant effect	Largest increase	No inactivation	Slow	Yan and Aldrich (2010, 2012)	Prostate, salivary glands, trachea, thyroid gland, thymus, cerebellum, brain (whole), aorta, mucosa, and fetal brain (Yan & Aldrich, 2012)	Airway hydration, vasodilation, and cancer (Evanson, Bannister, Leo, & Jaggar, 2014; Liu et al., 2012; Manzanares et al., 2014)
BKγ2	LRRC52	No significant decrease	Increase	No inactivation	No significant effect	Yan and Aldrich (2012)	Testis, skeletal muscle, placenta, and sperm cells (Yan & Aldrich, 2012)	Fertility deficit (Yang, Zeng, Zhou, Xia, & Lingle, 2011; Zeng, Yang, Xia, Liu, & Lingle, 2015)
BKγ3	LRRC55	Decrease	Increase	No inactivation	Slow	Yan and Aldrich (2012)	Fetal brain, brain (whole), mitral cell layers of olfactory bulb, liver, and spleen (Dolan et al., 2007; Yan & Aldrich, 2012)	Not known
BKγ4	LRRC38	No effect	Increase	No inactivation	No effect	Yan and Aldrich (2012)	Adrenal gland, skeletal muscle, thymus, cerebellum, brain (whole), testis, and spleen (Yan & Aldrich, 2012)	Not known

Morera, Carvacho, & Latorre, 2007). This chapter will review the literature of BK channel modulation by auxiliary β and γ subunits, including a comparison of these two different types of auxiliary subunits, specifically their modulatory effects and underlying mechanisms.

2. DISCOVERY

Four different β subunits (β1–β4) have been cloned and identified in mammals. The first β subunit was identified as a binding partner of BKα in the BK channel complexes purified from bovine tracheal smooth muscle by extensive conventional chromatography together with sucrose gradient centrifugation or by immunoprecipitation (Garcia-Calvo et al., 1994; Knaus, Eberhart, Kaczorowski, & Garcia, 1994; Knaus, Folander, et al., 1994; Knaus, Garcia-Calvo, et al., 1994). In these early experiments, charybdotoxin (ChTX), which is a peptide blocker of BK channels, was radiolabeled as a tool for BK channel complex or protein detection, and ChTX was found to be attached to a β subunit upon cross-linking. This smooth muscle-specific auxiliary protein was later named the β1 subunit. With advancements in molecular biology, the other three family members were discovered thereafter by sequence similarity and molecular cloning (Behrens et al., 2000; Brenner, Jegla, et al., 2000; Meera et al., 2000; Uebele et al., 2000; Wallner et al., 1999; Xia et al., 1999).

The first representative of the γ subunit family was identified relatively more recently. An unusual type of K^+ current was initially noticed in lymph node carcinoma of prostate (LNCaP) cells, which showed a Kv-like low half-activation voltage ($V_{1/2}$) of ~30 mV in the absence of intracellular calcium ($[Ca^{2+}]_i$) but had many characteristics of BK channels (Gessner et al., 2005). These BK-like features included large single-channel conductance, activation by Ca^{2+} and Mg^{2+}, and sensitivity to specific BK channel activator and blockers (Gessner et al., 2005). The range of voltages needed for channel activation for this endogenous BK-like channel in LNCaP cells was shifted to the hyperpolarization direction by more than 120 mV compared with human BKα channels expressed in HEK-293 cells. Therefore, it was concluded that the LNCaP cell contains a special BK channel or a BK-like K^+ channel which was designated as BK_L (Gessner et al., 2005). Later, it was demonstrated by another research group that the LNCaP cells did express BKα at the protein level, as detected by the BKα antibody, and the BKα existed in a normal zero-splicing form (NCBI accession NP_002238), according to reverse transcriptase *PCR* and sequencing of mRNA

(Yan & Aldrich, 2010). A proteomic approach was then used to immunopurify the channel complex, and mass spectrometry was used to identify potential novel interacting partners that may drastically modify the BK channel's gating property (Yan & Aldrich, 2010). A 35-kDa leucine-rich repeat-containing protein, LRRC26, was specifically identified in the BKα pull-down components. Knockdown of this protein in LNCaP cells resulted in a complete loss of the BK channel's property of being activated at low voltage in the absence of calcium. Meanwhile, overexpression of LRRC26 in another prostate cancer cell line, PC3, which lacks endogenous LRRC26 expression, converted the endogenous typical BKα channels into the low-voltage-activated LNCaP-type BK channels. In addition, it was shown in a heterologous expression system (HEK-293 cells) that LRRC26 was specifically associated with BKα, as detected by reciprocal coimmunoprecipitation, and shifted the conductance–voltage (*G–V*) relationship of BK channels to the hyperpolarization direction by ~140 mV, as was seen in the LNCaP cells. LRRC26 is structurally and functionally distinct from the four β subunits and thus was considered a new type of BK channel auxiliary subunit. Later, three other structurally related leucine-rich repeat-containing proteins, LRRC52, LRRC55, and LRRC38, were also reported as able to modulate BK channels when coexpressed heterologously with BKα in HEK-293 cells (Yan & Aldrich, 2012). These proteins have also been shown to produce marked shifts in the ranges of voltages needed for channel activation in the hyperpolarizing direction, although the shifts are smaller than those produced by LRRC26.

3. STRUCTURAL CHARACTERISTICS

BK channel β subunits are a family of small double-pass membrane proteins (20–30 kDa) (Behrens et al., 2000; Brenner, Jegla, et al., 2000; Uebele et al., 2000). Four types of BKβ genes (*KCNMB1-4*) have been cloned in mammals (Marty, 1981; Orio et al., 2002; Xia et al., 2003, 2000; Zeng et al., 2008). Among the four β subunits, β1 shares 53% similarity with β2 in amino acid sequence similarity, and 37% with β3, whereas less than 20% with β4 (Behrens et al., 2000; Wallner et al., 1999; Xia et al., 2000) (Fig. 2A). The four β subunits display similar topology containing short N- and C-termini both on the intracellular side, two TM helices, and a large extracellular loop (116–128 amino acid residues) connecting the two TM segments (Orio et al., 2002) (Figs. 1B and 2A). The loop

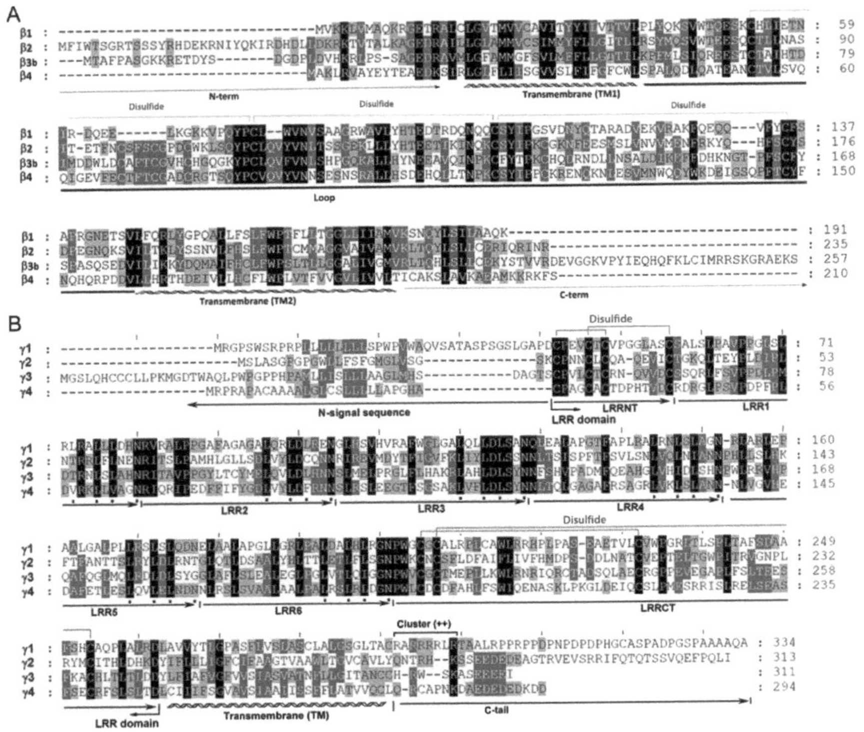

Fig. 2 The protein sequence alignment of human BK channel β (A) and γ (B) subunits. The cysteine pairs for potential disulfide formation are based on previous publications (Brenner, Jegla, et al., 2000; Yan & Aldrich, 2012).

has three or four putative glycosylation sites and multiple pairs of conserved disulfide-forming cysteine residues (Fig. 2A).

The four γ subunits have similar molecular weights of about 35 kDa. They are type I single-span membrane proteins containing a classic N-terminal cleavable signal peptide for extracellular localization of the N-terminal LRR domain in the mature proteins (Fig. 1C). The signal peptide region was found to be absent in the mature protein. Mutations in this region caused the signal peptide to be retained in the expressed protein and led to a loss of modulatory function in the γ1 subunit, suggesting that proper maturation guided by the signal peptide region is critical for the function of γ subunits (Yan & Aldrich, 2012). The mature proteins of the four γ subunits all contain a single-transmembrane domain, an N-terminal extracellular LRRD (~240 amino acids), and a short C-terminal tail (Fig. 1C). The four γ subunits share an overall sequence similarity of 35–40%, which is comparable with that of the four β subunits (Fig. 2B).

The LRR domain in many proteins is known to provide a structural framework for protein–protein interactions, typically through the concave β-sheet side (Kobe & Kajava, 2001). The LRR domains of γ subunits all contain six LRR units and two cysteine-rich regions, a small one called LRRNT, capped on the N-terminal side, and a large one called LRRCT, capped on the C-terminal side. As in many other LRR-containing proteins, each LRR unit in the γ subunits consists of 24 residues and has a classic consensus sequence of LxxLxLxxN (where x can be any amino acid). Based on structural modeling (Yan & Aldrich, 2012), the LRR domain is a banana-shaped structure with a curved parallel β-sheet lining the inner circumference and small helices or turns flanking the convex circumference, formed by six LRR units stacked together in the middle (Fig. 3A). Each LRR unit forms a β-strand lining the concave face and a short α-helix connected by loops flanking the outer circumference. The hydrophobic core of the LRR domain is tightly packed by the parallel inward-pointing leucine residues, shielded by the LRRCT and LRRNT caps on the N- and C-terminal ends. Both LRRNT and LRRCT contain two pairs of fully conserved cysteine residues that in total potentially form four disulfide linkages in the favorable oxidizing extracellular environment. Consistent with their predicted extracellular location, the LRR domains of the γ subunits all contain single or multiple consensus N-glycosylation sites: Asn-Xaa-Ser/Thr, where Xaa is not a proline. For the γ1 subunit, N147Q mutation and enzymatic removal of the N-linked glycan by PNGase F resulted in the

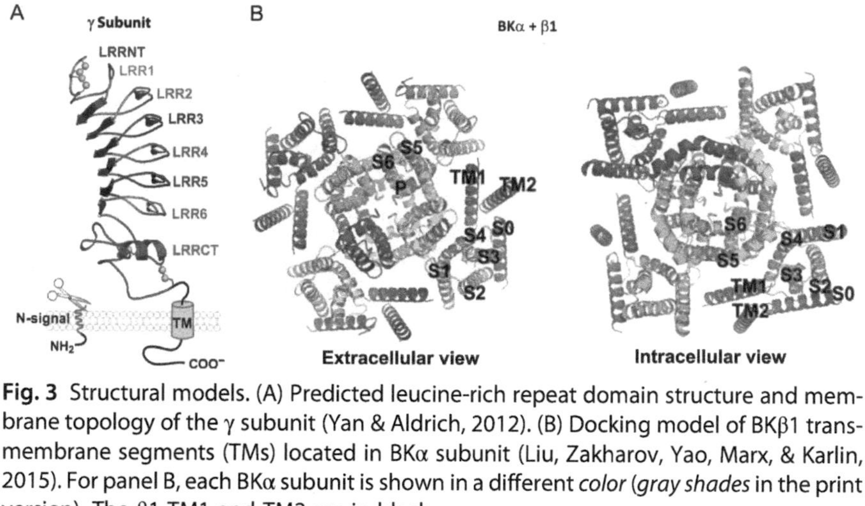

Fig. 3 Structural models. (A) Predicted leucine-rich repeat domain structure and membrane topology of the γ subunit (Yan & Aldrich, 2012). (B) Docking model of BKβ1 transmembrane segments (TMs) located in BKα subunit (Liu, Zakharov, Yao, Marx, & Karlin, 2015). For panel B, each BKα subunit is shown in a different *color* (*gray shades* in the print version). The β1 TM1 and TM2 are in black.

disappearance of an upper glycosylated-mass band on SDS-PAGE (Yan & Aldrich, 2012).

The protein sequences in the LRR domains of γ subunits are closely related but become divergent in the transmembrane and intracellular C-terminal tail regions (Fig. 2B). The single-transmembrane segments of the γ subunits are well predicted from their hydrophobicity and the presence of charged residues on both sides, particularly multiple positively charged residues on the intracellular side following the general "positive-inside rule" for membrane insertion and orientation of membrane proteins. For the C-terminal tail regions, in addition to the cluster of positively charged residues adjacent to the transmembrane domain, it is interesting to note that the rest of the amino acid sequence is enriched in proline residues (11 out of 36 residues) for γ1 and enriched in acidic residues for γ2, γ3, and γ4.

In the absence of direct structural information, a group led by Marx and Karlin have used a biochemical method to engineer disulfide linkage to approximately localize the various structural parts of the β subunits relative to the different BKα TM helices within the BKα/β complex. It was found that the first TM helix (TM1) of β1 was localized near the S1 and S2 helix of α subunits, and the second TM helix (TM2) near the S0 helix. This places the extracellular ends of β1 TM helices within the crevice formed by VSDs of two adjacent α subunits (Liu et al., 2010, 2008). Later, similar interacting patterns between β2, β3, or β4 subunits with BKα subunits were also reported (Wu et al., 2009, 2013). Recently, the cytoplasmic ends of β1 TM1 and TM2 were found to be adjacent and located between the S2–S3 loop of one α subunit and S1 of a neighboring α subunit, but not adjacent to S0, suggesting that the interaction between BKα and the β1 subunit is more complex than previously thought (Fig. 3B) (Liu et al., 2015). As discussed later, it is believed that the extracellular loop of the β subunit is in proximity to the extracellular mouth of the BKα pore gate because of its effect on the ion flow through the channel's extracellular mouth (Fig. 1B) (Gruslova, Semenov, & Wang, 2012; Zeng et al., 2003). Currently, very little is known about the exact protein regions and residues in BKα that directly interact with β subunits to mediate channel modulation. There is no report yet on the structural location of the γ subunit in the BKα/γ complex. The structural information obtained from the biochemical studies on the location of the β subunits in the BKα/β channel complex might provide an initial biochemical basis from which to decode the molecular mechanisms of BK channel modulation by auxiliary subunits.

The BKα subunit is universally expressed across different species of animals including invertebrates *Drosophila melanogaster* and *Caenorhabditis elegans*. The BK channel β subunits are present in both mammal and nonmammal animals such as chicken and turtle. The γ subunits (LRRC26 and its three paralogs) belong to a previously grouped "Elron" subfamily of the "extracellular LRR"-containing proteins that are mainly present in mammals (Dolan et al., 2007), suggesting that in evolution, the acquisition of BK channel proteins occurred in the order of BKα, β, and γ subunits. Although the β and γ subunits belong to distinct protein families with no amino acid sequence similarity, they all share some common structural features. First, they are all single or double membrane-spanning proteins with a large extracellular domain and a relatively much smaller intracellular region. Our previous study showed that the TM domain in γ1, as in the β subunits, is essential for the domain's modulatory function and physical association with BKα (Yan & Aldrich, 2010). Second, they all contain multiple pairs of disulfide-bond-forming cysteine residues on their extracellular sides. Third, as plasma membrane proteins with an extracellular soluble domain, they are all subject to modification by N-glycosylation. Both glycosylation and extracellular disulfide linkages in the β subunits were reported to directly affect the BK channels' voltage dependence (Hagen & Sanders, 2006) or ion flow through the channel's extracellular mouth (Zeng et al., 2003), respectively.

4. MODULATION OF THE BK CHANNEL'S BIOPHYSICAL PROPERTIES BY β SUBUNITS

The four β subunits display different and complex effects on apparent calcium and voltage sensitivities, macroscopic current kinetics, and pharmacological sensitivities, which involve multiple distinct mechanisms (Fig. 4). The β1 and β2 subunits overall induce slowing of the macroscopic kinetics and an increase in apparent calcium and voltage sensitivity (Behrens et al., 2000; Brenner, Jegla, et al., 2000; Contreras, Neely, Alvarez, Gonzalez, & Latorre, 2012; Orio & Latorre, 2005; Savalli, Kondratiev, de Quintana, Toro, & Olcese, 2007). The β2 and some splice variants of β3 subunits also cause rapid inactivation through their intracellular N-termini (Uebele et al., 2000; Xia et al., 1999, 2000). In contrast with the classic shaker Kv inactivation peptide, the intrinsically disordered N-terminal peptide of β3 subunit inactivated BK channels in two steps, which involves a stereospecific binding interaction that precedes blockade

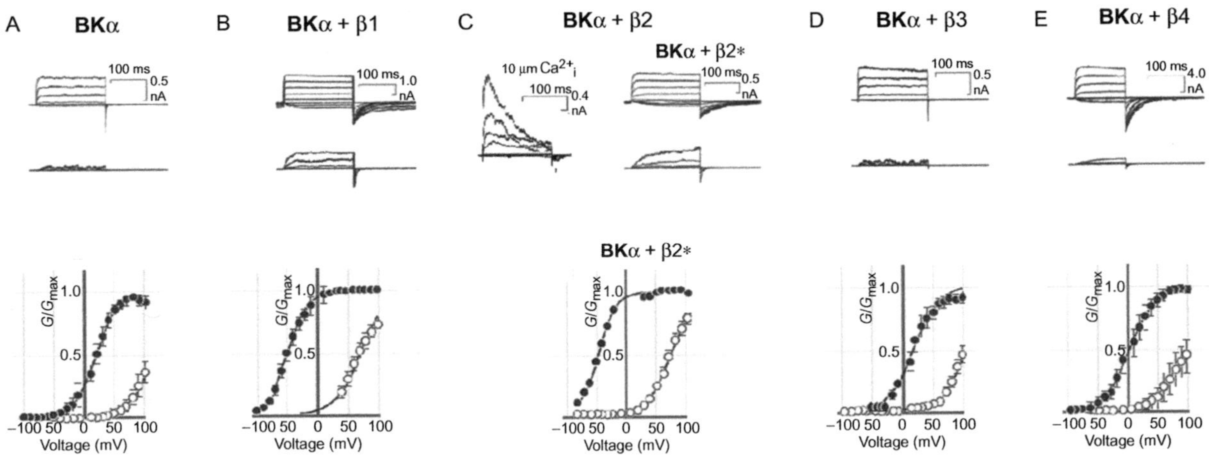

Fig. 4 The voltage- and Ca^{2+}-dependent activities of human BKα alone (A) or together with β1 (B), β2 (C), β3 (D), or β4 (E). The current traces were recorded in an inside-out patch from HEK-293 cells at 10 μM (*upper*) or 1 μM $[Ca^{2+}]_i$ (*lower*). The normalized conductance–voltage relations (*G–V*) for the activation of BK channels were plotted at 10 μM $[Ca^{2+}]_i$ (•) and 1 μM $[Ca^{2+}]_i$ (○). *Inactivation was removed by exposing the patch to 1 mg/mL trypsin for 60 s. *Modified from Pongs, O., & Schwarz, J. R. (2010). Ancillary subunits associated with voltage-dependent K^+ channels.* Physiological Reviews, 90, *755–796.*

(Gonzalez-Perez, Zeng, Henzler-Wildman, & Lingle, 2012; Lingle et al., 2001; Xia et al., 2000). The β3 subunits generate rectifying outward currents regulated by their extracellular loops (Xia et al., 2000; Zeng et al., 2003). The brain-specific β4 subunit, in addition to greatly slowing activation and deactivation kinetics, reduces apparent calcium sensitivity in low $Ca^{2+}{}_i$ conditions but increases apparent sensitivity in high $[Ca^{2+}]_i$ conditions (Behrens et al., 2000; Brenner, Jegla, et al., 2000). The β1, β2, and β4 subunits were also found to modulate BK channels by altering BKα expression and trafficking. It was reported that when coexpressed in HEK-293 cells, β1 could reduce the steady-state BKα surface expression and alter the diffused intracellular expression of BKα to a pattern of punctate cytoplasmic localization that overlapped with the β1 expression (Toro et al., 2006). A similar effect on BK channel expression was also observed with the β2 subunit (Lv et al., 2008; Zarei et al., 2007). Expression of β4 subunit was also found to reduce surface expression of BK channels in mouse CA3 neurons which was mediated by a C-terminal ER retention sequence (Shruti et al., 2012). In contrast, the N-terminal domain of β1 was reported to stimulate trafficking of VEDEC (a BKα C-terminal isoform) channels to the plasma membrane (Kim, Zou, Ridgway, & Dryer, 2007). A recent study showed that the palmitoylated β4 subunit regulated surface expression of BK channels through masking a trafficking motif (REVEDEC) at the C-terminus of the BKα subunit (Chen et al., 2013).

With mathematical modeling and simulation, the mechanisms of action of different β subunits were investigated by multiple laboratories within the framework of the well-established BK channel allosteric HCA (Horrigan et al., 1999) or HA (Horrigan & Aldrich, 2002) models of voltage- and Ca^{2+}-dependent gating. According to the HA model (Fig. 5) (Horrigan & Aldrich, 2002), the activation or open probability (P_o) of BK channels is independently affected by voltage and Ca^{2+}, which can be calculated or described by eight gating parameters: L_0 and Z_L were referred to as equilibrium constant and associated gating charge for the channel pore's closed ↔ open transition; J_0 and Z_J were referred to as equilibrium constant and associated gating charge for the voltage sensors' resting ↔ activated transition; D was considered the allosteric coupling factors between the pore gate and the voltage sensors; C was designated as the allosteric coupling factor between the pore gate and the calcium sensors, and K_d as the elementary Ca^{2+} dissociation constant when the channel is closed and voltage sensors are not activated; E was referred to as the allosteric coupling factor between the voltage and the calcium sensors, which was considered to be very weak in

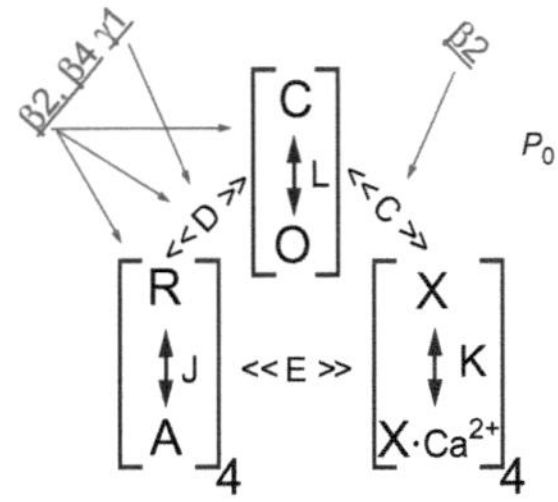

$$P_0 = \frac{L(1 + KC + JD + JKCDE)^4}{L(1 + KC + JD + JKCD)^4 + (1 + J + K + JKE)^4} \quad (1)$$

where $L = L_{0\exp}\left(-\frac{z_L V}{kT}\right)$;

$J = J_{0\exp}\left(-\frac{z_J V}{kT}\right)$;

$K = \frac{[Ca^{2+}]}{K_D}$

Fig. 5 The allosteric HA model of BK channels and the possible effects of the β and γ subunits. The three processes of channel opening (C-O), voltage-sensor activation (R-A), and Ca^{2+}-binding transition (X-XCa^{2+}) are linked, respectively, by the allosteric coupling factors C, D, and E. The possible gating processes or parameters affected by the β and γ subunits are indicated by *gray arrows*, which are based on previous reports (Contreras et al., 2012; Ma, Lou, & Horrigan, 2006; Orio & Latorre, 2005; Wang & Brenner, 2006; Yan & Aldrich, 2012; Yang et al., 2008). *The HA model was taken from Horrigan, F. T., & Aldrich, R. W. (2002). Coupling between voltage sensor activation, Ca^{2+} binding and channel opening in large conductance (BK) potassium channels.* The Journal of General Physiology, 120, *267–305.*

the case of the BK channel (Horrigan & Aldrich, 2002). The β1 subunit was found to affect multiple gating processes including the voltage-sensor activation (J_0), the intrinsic opening of pore gate (L_0), and the coupling between voltage sensors and pore gate (D) but with only minor effect on the Ca^{2+} sensor and its coupling the pore gate (C) (Bao & Cox, 2005; Cox & Aldrich, 2000; Orio & Latorre, 2005; Wang & Brenner, 2006). The apparent increase in sensitivity to Ca^{2+} is largely caused by modulation on other gating parameters (Ma et al., 2006; Orio & Latorre, 2005; Wang & Brenner, 2006; Yang et al., 2008). The β4 subunit appears to be essentially similar to the β1 subunit in reducing the intrinsic pore gate opening and enhancing voltage-sensor activation, which ultimately results in a shift of the G–V relationship to the depolarizing direction at low Ca^{2+} but to the hyperpolarizing direction at high Ca^{2+} ($\geq$~10 μM) (Wang et al., 2006). By measuring gating currents, the β1, β2, and β4 but not β3 subunits were found to stabilize the BK voltage sensor in the active conformation (Contreras et al., 2012).

Over the past two decades, progress has been made in structurally and functionally identifying the potential protein regions and residues involved in BK channel modulation by β subunits. Studies of chimeric and mutant β subunits indicated that the cytosolic N- and C-termini of the BKβ1 subunit were crucial to altering the channel's intrinsic opening and voltage-sensor activation (Orio et al., 2006; Wang et al., 2006). It was recently found that substituting two lysine residues (ie, K3 and K4) in the N-terminus of β1 can virtually abolish the effects of β1 on voltage-sensor activation (Castillo et al., 2015). The N-termini of the β2, β3a, and β3c subunits contain an inactivating domain which blocks the entrance to the intracellular pore and inactivates the BK channels (Li et al., 2007; Wallner et al., 1999; Xia et al., 1999, 2003; Zeng et al., 2003; Zhang, Zeng, Xia, & Lingle, 2009; Zhang, Zhou, Ding, Xia, & Lingle, 2006). Different from the simple one-step occlusion mechanism for Kv channel blockade by N-terminal peptides which may mainly involve hydrophobic interactions with the channel pore, the BK channel inactivation by auxiliary β subunits involves two distinguishable kinetic steps (Lingle et al., 2001), a stereospecific binding interaction that precedes blockade (Gonzalez-Perez et al., 2012). Although the sites of stereospecific interactions between the β subunit N-terminus and BK channel pore have not been identified, it was proposed that the relatively larger inner pore of BK channels than Kv channels might require some stereospecific binding to achieve the affinity necessary for BK channel inhibition (Gonzalez-Perez et al., 2012). The extracellular loop and the disulfide bridges within the loop were found to be important in β3 subunit-induced current rectification (Xia et al., 2000; Zeng et al., 2003). The lysine-rich ring (K137, K141, K147, and K150) on the extracellular loop of β2 also was involved in causing the outward rectification (Chen et al., 2008). These results suggest close proximity between the extracellular loop of the β subunits and the extracellular mouth of the channel pore. Several residues (Y74, S104, Y105, and I106) on the extracellular loop of β1, which are also conserved in other β subunits, were also found to be important in the β1 subunit's modulatory effect on voltage-dependence activation of BK channel (Gruslova et al., 2012). It remains unclear whether the β subunit's extracellular loop can directly interact and modulate the BK channel or indirectly affect the channel properties by affecting the overall structure of the β subunit. Compared with the other regions, the function of the two TM segments in the β subunits has been relatively less studied. Their likely interposed location between VSDs of adjacent BKα subunits suggests a role in modulating voltage-sensor activation. The two TM segments in β1 subunit were functionally replaceable by those from the β2 subunit (Orio et al., 2006) but not from the β4 subunit even replaced individually (Kuntamallappanavar, Toro, & Dopico, 2014). The latter showed that both TM segments are required to maintain the characteristic modulatory function of the β1 subunit (Kuntamallappanavar et al., 2014).

On the BKα subunit, by using chimeric BKα channels from different species, it was found that the unique S0 TM segment, as well as the extracellular N-terminus of BKα, was important for BK channel modulation by β1 and β2 subunits (Lee et al., 2010; Morrow et al., 2006; Wallner, Meera, & Toro, 1996). The β1 and β2 subunits seem to be different in regulating the various functional domains of the BKα, although both affect the apparent voltage and Ca^{2+} sensitivity. Mutations on BKα that affected voltage-sensor activation altered the apparent changes in Ca^{2+} sensitivity induced by the β1 subunit (Yang et al., 2008). In contrast, β2 mainly affected Ca^{2+} sensitivity, presumably by altering the allosteric coupling between the Ca^{2+} sensors and the pore gate. Mutations in the BKα voltage sensor were found to have little effect on β2's modulatory effect on Ca^{2+} sensitivity (Yang et al., 2008). Instead, mutations on the N-terminus of RCK1 and on the linker region between S6 TM segment and RCK1 had a large impact on β2's but not β1's modulation on the BK channel's Ca^{2+} sensitivity (Lee et al., 2010). However, experiments with voltage-clamp fluorometry correlated the β2 subunit with changes in voltage sensors (Savalli et al., 2007). The β1 and β2 subunits were also reported to alter the interactions between bound Mg^{2+} and R213 in the voltage sensor and to disrupt the engineered disulfide-bond formation between the voltage sensor and C-terminal intracellular domain (Sun et al., 2013).

5. MODULATION OF THE BK CHANNEL'S BIOPHYSICAL PROPERTIES BY γ SUBUNITS

Unlike the complex effects and mechanisms of different β subunits on many aspects of BK channel gating, the actions of the γ subunits appear to be remarkable in mechanistic simplicity and modulatory magnitudes (Yan & Aldrich, 2010; Zhang & Yan, 2014). The four γ subunits have distinct capabilities in shifting the voltage dependence of BK channel activation toward hyperpolarizing voltages by approximately 140 mV (γ1), 100 mV (γ2), 50 mV (γ3), and 20 mV (γ4) in the absence of Ca^{2+} (Fig. 6) (Yan & Aldrich, 2010, 2012). The gating shift produced by the γ1 subunit is equivalent to the effect of ~10 μM $[Ca^{2+}]_i$ on BK channels formed by BKα alone. However, Ca^{2+} and Mg^{2+} sensitivities were shown to be largely unaffected by the γ1 subunit (Yan & Aldrich, 2010). In the presence of Ca^{2+}, the γ1–3 subunits all caused slight changes in the slope of the relationship between $Q_{app}V_{1/2}$ (Q_{app}, apparent gating charge obtained from a Boltzmann fit of the G–V curve for voltage dependence of channel activation) and log $[Ca^{2+}]_i$ with some Ca^{2+}-dependent reduction in their $V_{1/2}$ shifting capabilities, suggesting some small but noticeable effects on the apparent calcium

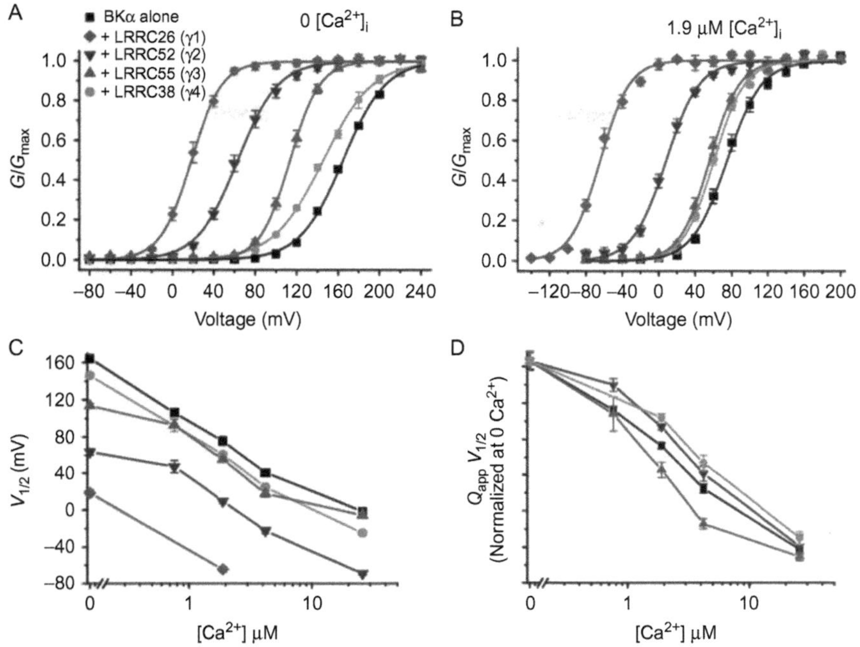

Fig. 6 Modulatory effects of the γ subunits on BK channels heterologously expressed in HEK-293 cells. (A and B) Voltage dependence of the BK channel activation for the BKα alone or together with γ1, γ2, γ3, or γ4 in the virtual absence of $[Ca^{2+}]_i$ (A) or in the presence of 1.9 μM $[Ca^{2+}]_i$ (B). (C) Plot of $V_{1/2}$ vs $[Ca^{2+}]_i$ (log scale after break). (D) Plot of $Q_{app}V_{1/2}$ vs $[Ca^{2+}]_i$ normalized at 0 $[Ca^{2+}]_i$. *This figure is taken from Yan, J., & Aldrich, R. W. (2012). BK potassium channel modulation by leucine-rich repeat-containing proteins.* Proceedings of the National Academy of Sciences of the United States of America, 109, *7917–7922.*

sensitivity of BK channel activation (Yan & Aldrich, 2010). It is unclear whether the observed Ca^{2+} effects on the function of γ subunits were mediated by the Ca^{2+} effects on the BKα or on the γ subunits.

The mechanistic actions of the γ1 subunit were investigated and analyzed within the framework of the HA model in the absence of Ca^{2+}. By measuring the kinetics and open probabilities of the channels at very negative voltages to achieve good estimates of the Z_L and L_0 parameters, it was found that the pore's gating parameters L_0 and Z_L are largely unaffected by the γ1 subunit (Yan & Aldrich, 2010). The lack of effect on L_0 also suggests that the γ1 subunit is not a Ca^{2+}-like ligand that directly activates BK channels independently of voltage-sensor activation. By simulation with changes in other gating parameters, the γ1 subunit's modulatory effect was found to be best explained by an approximate 20-fold increase in the allosteric coupling D

factor (Yan & Aldrich, 2010) (Fig. 5). According to this study, the γ1 subunit may mainly affect the coupling between voltage sensors and the pore. The γ subunits may serve as good tools to study BK channel gating mechanisms because little is currently known about the molecular basis underlying allosteric coupling between voltage sensors and the pore in this voltage- and ligand-gated channel. This earlier analysis, which was performed in the absence of Ca^{2+} and assumed that only one gating parameter was affected in simulation, was not comprehensive enough to allow complete assessment of some other gating parameters, such as some slight changes in the voltage-sensor parameters, J_0 and Z_J. Additionally, a later study observed some effects of the intracellular Ca^{2+} on the modulatory function of the γ subunits (Yan & Aldrich, 2012).

To understand the molecular mechanisms of BK channel regulation by auxiliary γ subunits, it is critical to identify key structural elements underlying their channel modulatory functions. By swapping structural elements among γ subunits and by mutations, we recently found that the differences in the various γ subunit-induced shifts of the BK channel $V_{1/2}$ are primarily determined by their single TM segments during the approximate −100-mV shift in $V_{1/2}$, in which the γ1 and γ2 TMs produced low $V_{1/2}$ BK channels, while the γ3 and γ4 TM domains all resulted in high $V_{1/2}$ channels (Fig. 7) (Li, Fan, Kwak, & Yan, 2015). We also found that their intracellular C-tails, particularly the juxta-membrane positively charged cluster regions that contain multiple positively charged amino acids, further adjust the modulatory functions of the four γ subunits by conferring to the BK channels an additional approximate −40 to −50 mV shift in $V_{1/2}$ from the γ1 and γ3 C-tails (Li et al., 2015) (Fig. 7). In a more recent study, we investigated in detail the structure and function in BK channel modulation by the γ1 subunit's peptide region (~40 amino acids) encompassing the single TM segment and the adjacent poly-Arg cluster (Li, Guan, Yen, Zhang, & Yan, 2016). We demonstrated that this peptide region, independent of the N-terminal LRR domain and the rest of the C-terminal tail, was sufficient to fully modulate BK channels. We found that Phe273 and its neighboring residues in the middle of the TM segment play a key role in BK channel association and modulation and that a minimum of three Arg residues in the charged cluster are required for the γ1 subunit's modulatory function. Allosteric coupling between the TM segment and the intracellular positively charged cluster was also observed. We concluded that the TM segment is a key molecular determinant for channel association and modulation, and the intracellular positively charged cluster is involved mainly in channel association likely

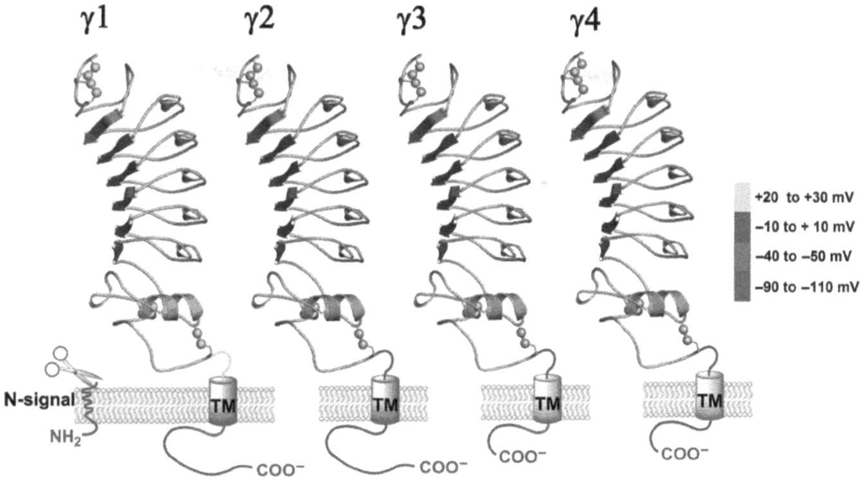

Fig. 7 Schematic of the structure, membrane topology, and generalized relative functional contributions of individual structural elements in the BK channel γ subunits. The different *colors* (*gray shades* in the print version) indicate the structural elements' relative contributions (in millivolts, mV) to the γ subunit-induced shifts of voltage dependence of BK channel activation. *This figure is taken from Li, Q., Fan, F., Kwak, H. R., & Yan, J. (2015). Molecular basis for differential modulation of BK channel voltage-dependent gating by auxiliary gamma subunits.* The Journal of General Physiology, 145, *543–554.*

through its TM anchoring effect (Li et al., 2016). These new findings provide insights into the structure–function relationship of the γ subunits for understanding their potent modulatory effects on BK channels.

6. MODULATION OF THE BK CHANNEL'S PHARMACOLOGICAL PROPERTIES BY β AND γ SUBUNITS

As a result of decades of exploration, numerous endogenous and synthetic BK channel modulators have been identified and developed, some of which show BK auxiliary subunit-dependent characteristics. Generally, the β subunits confer alteration in sensitivity of the BK channel to peptide toxins, in which the extracellular loop of β subunits plays a critical role (Chen et al., 2008; Hanner et al., 1998; Meera et al., 2000; Xia et al., 1999). The four residues (Leu^{90}, Tyr^{91}, Thr^{93}, and Glu^{94}) of BKβ1 were found to be crucial for the enhanced affinity of ChTX, and the lysine-rich ring (K137, K141, K147, and K150) of subunit β2 could reduce the ChTX sensitivity to BK channels. In addition, the β2 and β3 subunit were reported to reduce the degrees of blockade by ChTX (Xia et al., 1999, 2000; Zeng et al., 2008), together with a change in the magnitude and kinetics of the blocking reaction (Ding et al., 1998). However, the β4-complexed BK channel is extremely resistant to both iberiotoxin (IbTX) and ChTX (Bergeron & Bingham, 2012; Meera et al., 2000; Schneider, Rogowski, Krueger, & Blaustein, 1989). This kind of resistance is related with the so-called helmet structure constituted by BKα and three residues in extracellular loop (K120, R121, and K125) of the β4 subunit, which can block the entry of ChTX by both electrostatic interaction and the limited space of the pore area (Gan et al., 2008). Moreover, compared with BKα alone, the BKα/β1 channel, but not the BKα/β4 channel, is more sensitive to the blockade by slotoxin (αKTx1.11). But the binding between slotoxin and the BKα/β4 channel seemed to be irreversible in an unknown mechanism (Garcia-Valdes, Zamudio, Toro, & Possani, 2001). In contrast, martentoxin (Shi et al., 2008; Tao, Shi, Liu, & Ji, 2012) and conopeptide Vt3.1 (Li et al., 2014) were identified as the more selective blockers for the BKα/β4 channel, but not BKα/β1 channel, when compared with BKα channel alone. The β subunits were also reported to regulate BK channel modulation by ethanol. In heterologous expression system (HEK-293 cells), the β1 subunit caused blockade of ethanol-induced potentiation of the BK channels (Feinberg-Zadek & Treistman, 2007; Martin et al., 2004). In the supraoptic nucleus neurons, the β4-complexed axonal terminal BK channels were highly potentiated by ethanol, but the β1-complexed somatic and dendritic channels were insensitive to ethanol (Dopico, Widmer, Wang, Lemos, & Treistman, 1999; Wynne, Puig, Martin, & Treistman, 2009). Similarly, the difference in sensitivity of BK channels to ethanol in the dendrites and cell bodies of nucleus accumbens medium spiny neurons can be explained by the differential expression of the β1 subunit in these two compartments (Martin et al., 2004). In contrast to the nervous system, ethanol caused inhibition of BK currents in myocytes that led to cerebrovascular constriction, in which the β1 subunit played a major role (Bukiya, Liu, et al., 2009). This group recently identified K361 on BKα as a putative binding site for ethanol (Bukiya, Kuntamallappanavar, et al., 2014). Further studies will be needed to clarify the different effects of different β subunits on the BK channel's responses to ethanol. A detailed review of BK channel modulation by ethanol can be found in Chapter "Modulation of BK Channels by Ethanol" by Dopico.

The earliest reported activator, dehydrosoyasaponin-I (DHS-I, the potent compounds extracted from *Desmodium adscendens*) activates BK channels only when coexpressed with the β1 subunit in *Xenopus oocytes*

(McManus et al., 1993, 1995), and such β1-dependent activation was also observed in smooth muscle membranes (Bukiya, Patil, Li, Miller, & Dopico, 2012). The xenoestrogen, tamoxifen, and the endogenous steroid hormone 17β-estradiol were also found to activate the BK channel in a β1-dependent manner (De Wet et al., 2006; Dick, Rossow, Smirnov, Horowitz, & Sanders, 2001; Duncan, 2005; Valverde et al., 1999). Likewise, the β4-complexed BK channel was more sensitive to corticosterone for channel function potentiation by steroids, whereas the β2-complexed channel was more sensitive to dehydroepiandrosterone (King et al., 2006). The endogenous lithocholate was reported to selectively recognize the steroid-sensing site (T169, L172, and L173) in TM2 of the β1 subunit to activate BK channels (Bukiya, McMillan, Parrill, & Dopico, 2008; Bukiya, Singh, Parrill, & Dopico, 2011; Bukiya, Vaithianathan, Toro, & Dopico, 2009). T169 in TM2 of the β1 subunit was also involved in the activation of BKα/β1 channels by nonsteroidal lithocholic acid (Bukiya et al., 2013). Current research has shown that lipid agents can act through β subunits, although the underlying mechanisms are unclear. Omega-3 docosahexaenoic acid could increase BK channel activity in a β1 and β4 subunit-dependent manner, while not in a β2 or γ1 subunit-dependent manner (Hoshi et al., 2013). In addition, phosphatidylinositol 4,5-biphosphate (PIP2) was reported to inhibit macroscopic currents of BKα and BKα/γ1 channels but increase macroscopic currents of BKα/β1 and BKα/β4 channels without altering those of BKα/β2 channels (Tian et al., 2015). Endogenous leukotrienes (LTB4) can potently activate BK channels when coexpressed with the β1 subunit in *X. oocytes*, which makes LTB4 a good template for the future development of β1-specific BK channel activators (Bukiya, McMillan, et al., 2014). A recent study reported that bis-(1,3-dibutylbarbituricacid)trimethine oxonol [DiBAC4(3)] and *N*-arylbenzamide can activate the BK channel in β1 subunit dependence, but DiBAC4(3) partially blocked the BKα/β2 channel's currents (Kirby, Martelli, Calderone, McKay, & Lawson, 2013; Morimoto et al., 2007). Additionally, HBD2 (human β-defensin 2) could activate BK channels via interactions with Leu41 and Gln43 of the β1 extracellular loop (Liu et al., 2013). A recent report showed that the newly identified BK channel opener GoSlo-SR-5-130, but not its analogue GoSlo-SR-5-6, required the presence of β1 or β4 subunit to achieve its the maximal modulatory effects on BK channels (Large et al., 2015).

Little has been done to investigate the interaction between the newly identified auxiliary γ subunits and the BK channel's modulators.

Nevertheless, the BK channel in LNCaP cells was found to be still sensitive to NS1619, ChTX, IbTX, paxilline, and penitrem A (Gessner et al., 2005), suggesting that most known pharmacological properties of the BK channels are likely retained in the presence of the endogenous γ1 subunit. Interestingly, the γ1 subunit can inhibit the effect of some BK activator (Almassy & Begenisich, 2012). In native salivary gland parotid acinar cells, the endogenous γ1 subunit blocked the activating effect of mallotoxin but not NS1619 (Almassy & Begenisich, 2012). A similar blocking effect of the γ1 subunit on mallotoxin action was also observed in HEK-293 cells when the γ1 subunit was heterologously coexpressed with BKα. It was proposed that mallotoxin may displace the γ1 subunit instead of lacking accessibility to the binding site (Almassy & Begenisich, 2012). Further biophysical studies and biochemical binding assays will be needed to clarify the detailed mechanisms. NS1619 was recently shown to bind to the S6/RCK linker region (Gessner et al., 2012), but little is known about the mallotoxin binding site. Identification of the mallotoxin binding site may complement our understanding of the actions of γ subunits on BK channels.

7. STOICHIOMETRY

When the β1 subunit was first identified from bovine tracheal and aortic smooth muscle with immunoprecipitation, the isolated BK channel complex was found to contain an octameric assembly of BKα and BKβ1 subunits in 1:1 stoichiometry (Knaus, Eberhart, et al., 1994; Knaus, Garcia-Calvo, et al., 1994). In a classic model of tetrameric ion channel complex, auxiliary subunits bind to the pore-forming α subunit with a fourfold symmetry so that the regulatory effect is incremental upon variation in the relative molecular ratio of the auxiliary subunit to core subunit. Consistently, the β subunits regulated the voltage dependence of BK channel activation in the titration-dependent manner in that the *G–V* curves shifted in a parallel manner as a function of the injected ratio of β to α subunit mRNA in *X. oocytes* (Wang, Ding, Xia, & Lingle, 2002). Variation in stoichiometry of BKα and β subunits likely contributed to the heterogeneity in BK channel activity in rat chromaffin cells (Ding et al., 1998; Xia et al., 1999) and turtle auditory hair cells (Jones et al., 1999). As the newly identified BK channel auxiliary subunits, no biochemical or stoichiometric study has been reported on the assembly of BKα and γ subunits.

A recent study indicated that the regulatory mechanism of the γ1 subunit may be fundamentally different from that of the β subunit (Gonzalez-Perez,

Xia, & Lingle, 2014). In contrast to β subunits (Gonzalez-Perez et al., 2014), the γ1 subunit caused the voltage dependence of channel activation to be either fully shifted or unchanged, independent of the molar ratio of the injected BKα:γ1 RNA to *X. oocytes*, although the ratio of these two populations of channels varied (Gonzalez-Perez et al., 2014). It is unknown whether one γ1 subunit per channel complex is sufficient to fully modulate BK channels. Alternatively, the γ1 subunit may preferably exist in a tetrameric form when forming a complex with BKα. Additional studies will be required to determine the detailed mechanisms, particularly the stoichiometry and the interaction sites between the BKα and γ subunits in the tetrameric channel complex.

8. PHYSIOLOGICAL AND PATHOLOGICAL RELEVANCE OR ROLES

The BK channel β and γ auxiliary subunits have been reported to be involved in tissue-specific functions in health or disease (Table 1). An early study of epigenetic epidemiology showed that the individuals carrying E65K polymorphism in β1 subunit experienced low prevalence of diastolic hypertension and cardiovascular disease compared to those with normal β1 subunit (Fernandez-Fernandez et al., 2004). The mRNA level of β1 subunit was markedly decreased in the vascular smooth muscle of patients with acquired hypertension (Amberg et al., 2003). Studies with knockout mice showed that the presence of the β1 subunit was critical to normal BK channel function in vascular smooth muscle cells (Werner et al., 2007, 2008). One recent report showed that β1 deficiency exacerbated vascular fibrosis and remodeling (Xu et al., 2015). The knockout mice of β4 subunit showed signs of an epileptic phenotype (Brenner et al., 2005), and the β2 knockout mice had an increased tendency toward spontaneous burst firing in the adrenal medullary chromaffin cells (Martinez-Espinosa et al., 2014). A β3 (delA750) mutant, which led to truncation of the C-terminus, was reported to be associated with a form of generalized epilepsy (Lorenz et al., 2007). The results obtained from the β subunit KO mice so far have used method of global ablation of the β subunit in the whole animal body which could complicate the results and interpretations because of functional communications between different tissues and organs. Studies with tissue- or cell-specific KO mice will be needed to further clarify the physiological functions of different β subunits in different cells or tissues.

The tissue-specific distribution patterns of the four γ subunits at the mRNA level had been investigated with use of TaqMan quantitative PCR in various human tissues (Yan & Aldrich, 2012). The γ1 subunit was highly expressed in the salivary glands, prostate, and trachea, whereas γ2 (LRRC52) was found predominantly in the testes, and γ3 (LRRC55) was found primarily in the nervous system. The γ4 (LRRC38) subunit was observed mainly in skeletal muscle, adrenal glands, and the thymus. These results suggest that like β subunits, γ subunits have different tissue-specific distributions to fit the diverse functional requirements of various tissues and cell types (Yan & Aldrich, 2012). The γ1 subunit's endogenous functional regulation of BK channels has been confirmed in prostate and salivary gland cells (Almassy & Begenisich, 2012; Yan & Aldrich, 2010). The physiological roles of the γ1 subunit in prostate and salivary glands remain to be determined. Conceivably, constitutive activation of BK channels might be required for K^+ flow-mediated fluid secretion in these nonexcitable tissues. A very recent study suggested that the γ1 subunit in airway epithelial cells may participate in BK channel-mediated airway hydration for effective mucociliary clearance (Manzanares et al., 2014). The K^+ flow through the apically expressed BK channels in airway epithelial cells provides an electrochemical driving gradient for Cl^- secretion and thus plays a role in airway hydration. It was found that both the mRNA level of the γ1 subunit and the sensitivity of BK channels to mallotoxin were decreased after IFN-γ treatment, suggesting that the γ1 subunit might be involved in IFN-γ-mediated reduction in BK channel activity and the resulting mucociliary dysfunction (Manzanares et al., 2014). The γ1 subunit under a different name (CAPC) was reported to be able to suppress tumor growth and metastasis, which may likely involve ion channel-independent function (Liu et al., 2012). The enhanced K^+ channel activity generally promoted cancer cell proliferation (Pardo & Stuhmer, 2014). It will be necessary to determine whether the association of the γ1 subunit with BK channels will affect tumor growth because of change in BK channel activity.

Because of the drastic activating effect caused by the γ1 subunit, expression of this protein even at low levels might exert significant effect on BK channel currents. For example, a low level of mRNA expression of the γ1 subunit has been detected in aorta cells (Yan & Aldrich, 2012). A very recent study reported that knockdown of γ1 subunit expression in rat cerebral artery myocytes led to reductions in the apparent voltage/Ca^{2+} sensitivity, in current frequency and amplitude of the BK channels, and in the extents of BK channel-specific inhibitor-induced vasoconstriction and activator-induced vasodilation (Evanson et al., 2014). This study suggested that the γ1 subunit may play broad physiological roles that are not limited to nonexcitable cells. In excitable cells, the voltage and Ca^{2+} sensitivities of the BK channels are more finely tuned to be properly responsive to different levels of voltage and Ca^{2+} in different cell types; therefore, even a low expression of this potent BK channel modulator might exert a significant physiological effect. It is worth noting that the γ1 subunit is also expressed in fetal brain tissue (Yan & Aldrich, 2012) and that the γ1 subunit might participate in maintaining proper neuronal excitability in the fetal nervous system during early development. The recent finding of independent association and function of the β (β2) and γ (γ1) subunits in the same functional BK channel complex (Gonzalez-Perez, Xia, & Lingle, 2015) suggests that the BK channel function can be extremely diversified across different tissues or cells types when multiple auxiliary subunits (β and γ) are coexpressed.

9. PERSPECTIVES

The four β subunits have been found to confer the BK channel a high diversity in the channel's biophysical properties. Over the past two decades, BK channel modulation by β subunits has been extensively investigated. Studies in both humans and knockout mice have led to an appreciation of the physiological and pathological significance of the BK channel β subunits. With research efforts using heterologous expression systems in vitro, we have overall achieved a good understanding of the mechanisms underlying the β subunits' modulatory functions conferring the BK channel inactivation and alterations in ion permeation and toxin sensitivity, which involve more specific regions of interactions between the β and BKα subunits. Significant progress has been made in understanding the mechanisms involved in the β subunits' effect on BK channel voltage and Ca^{2+} dependence of channel activation, including the identification of altered gating processes and the important regions and residues involved on the β subunits. However, it remains poorly understood how the β subunit interacts with the BK channel, particularly the interacting amino acid residues on the BKα, to modulate the voltage and/or Ca^{2+} sensitivity. This is in parallel with our very limited understanding of the detailed molecular mechanisms underlying BK channel activation by voltage and Ca^{2+}, which is partly due to the lack of atomic structure, particularly the structure that includes the channel's TM domain.

Our understanding of BK channel modulation by γ subunits is still in its very early stage. In particular, very little is known about the physiologic functions and the structural basis underlying the regulatory mechanisms of γ subunits. The few published studies examining the modulatory mechanisms and physiological functions of BK channel γ subunits have mainly focused on the γ1 subunit. The regulation of BK channels by γ2–4 subunits has so far been demonstrated only in the heterologous expression system. It will be important to determine whether γ2–4 subunits also play any functional or physiological role in BK channel modulation in vivo. The mouse γ2 subunit has been found to function as an accessory subunit of the sperm-specific mouse Slo3 channels (Yang et al., 2011), and its ablation in mice was recently found to cause a severe fertility impairment (Zeng et al., 2015). It will be intriguing to determine whether in any stage of germ cells the γ2 subunit also modulates BK channels and whether Slo3 and BK channels can form functional heterotetrameric channels. When the BK and Slo3 channels were coexpressed in a heterologous expression system (*X. oocytes*), only two distinct BK-like and Slo3-like channels were observed, arguing against the presence of functional heterotetrameric channels (Yang, Zeng, Xia, & Lingle, 2009). BK channels and Slo3 belong to the Slo channel family, which also includes two more distantly related Na^+-activated channels, Slo2.1 (slick) and Slo2.2 (slack). It remains an open question whether the γ subunits may broadly function as auxiliary proteins of the Slo channel family.

Effective BK channel openers have been sought or explored to treat a variety of diseases such as stroke, epilepsy, psychoses, bladder overactivity, erectile dysfunction, asthma, arterial hypertension, ischemic heart disease, and gastric hypermotility (Nardi & Olesen, 2008; Chapter "Developing the Molecular Pharmacology of BK Channels for Therapeutic Benefit" by Kaczorowski and Garcia). Although the widely used BK channel opener NS1619 can give an approximate −40-mV shift in $V_{1/2}$ at a high concentration (30 μM) (Gessner et al., 2012), its specificity was recently questioned because of its direct inhibiting effect on the sarcoplasmic/endoplasmic reticulum Ca^{2+}-ATPase (SERCA) (Wrzosek, 2014) in addition to a previous report on its stimulatory effect on Ca^{2+} release from caffeine/ryanodine-sensitive intracellular store (Yamamura, Ohi, Muraki, Watanabe, & Imaizumi, 2001). Currently, no BK channel-targeted drug has been approved for clinical use, in spite of extensive academic and pharmaceutical efforts over the past two decades. Drug development specifically targeting auxiliary β or γ subunits of BK channels is likely to be more effective in modulating BK channel function, while minimizing the global adverse effects originated from the ubiquitously expressed BKα subunit. Deciphering the biochemical mechanisms underlying BK channel activation by auxiliary subunits will be useful for the development of new BK channel-targeted drugs.

ACKNOWLEDGMENT

This work is supported by NIH grant NS078152 (J.Y.).

REFERENCES

Almassy, J., & Begenisich, T. (2012). The LRRC26 protein selectively alters the efficacy of BK channel activators. *Molecular Pharmacology*, *81*, 21–30.

Amberg, G. C., Bonev, A. D., Rossow, C. F., Nelson, M. T., & Santana, L. F. (2003). Modulation of the molecular composition of large conductance, Ca^{2+} activated K^{+} channels in vascular smooth muscle during hypertension. *The Journal of Clinical Investigation*, *112*, 717–724.

Bao, L., & Cox, D. H. (2005). Gating and ionic currents reveal how the BKCa channel's Ca^{2+} sensitivity is enhanced by its beta1 subunit. *The Journal of General Physiology*, *126*, 393–412.

Behrens, R., Nolting, A., Reimann, F., Schwarz, M., Waldschutz, R., & Pongs, O. (2000). hKCNMB3 and hKCNMB4, cloning and characterization of two members of the large-conductance calcium-activated potassium channel beta subunit family. *FEBS Letters*, *474*, 99–106.

Bentrop, D., Beyermann, M., Wissmann, R., & Fakler, B. (2001). NMR structure of the "ball-and-chain" domain of KCNMB2, the beta 2-subunit of large conductance Ca^{2+}- and voltage-activated potassium channels. *The Journal of Biological Chemistry*, *276*, 42116–42121.

Bergeron, Z. L., & Bingham, J. P. (2012). Scorpion toxins specific for potassium (K^{+}) channels: A historical overview of peptide bioengineering. *Toxins (Basel)*, *4*, 1082–1119.

Brenner, R., Chen, Q. H., Vilaythong, A., Toney, G. M., Noebels, J. L., & Aldrich, R. W. (2005). BK channel beta4 subunit reduces dentate gyrus excitability and protects against temporal lobe seizures. *Nature Neuroscience*, *8*, 1752–1759.

Brenner, R., Jegla, T. J., Wickenden, A., Liu, Y., & Aldrich, R. W. (2000). Cloning and functional characterization of novel large conductance calcium-activated potassium channel beta subunits, hKCNMB3 and hKCNMB4. *The Journal of Biological Chemistry*, *275*, 6453–6461.

Brenner, R., Perez, G. J., Bonev, A. D., Eckman, D. M., Kosek, J. C., Wiler, S. W., et al. (2000). Vasoregulation by the beta1 subunit of the calcium-activated potassium channel. *Nature*, *407*, 870–876.

Bukiya, A. N., Kuntamallappanavar, G., Edwards, J., Singh, A. K., Shivakumar, B., & Dopico, A. M. (2014). An alcohol-sensing site in the calcium- and voltage-gated, large conductance potassium (BK) channel. *Proceedings of the National Academy of Sciences of the United States of America*, *111*, 9313–9318.

Bukiya, A. N., Liu, J., & Dopico, A. M. (2009). The BK channel accessory beta1 subunit determines alcohol-induced cerebrovascular constriction. *FEBS Letters*, *583*, 2779–2784.

Bukiya, A. N., McMillan, J. E., Fedinec, A. L., Patil, S. A., Miller, D. D., Leffler, C. W., et al. (2013). Cerebrovascular dilation via selective targeting of the cholane steroid-recognition site in the BK channel beta1-subunit by a novel nonsteroidal agent. *Molecular Pharmacology*, *83*, 1030–1044.

Bukiya, A. N., McMillan, J., Liu, J., Shivakumar, B., Parrill, A. L., & Dopico, A. M. (2014). Activation of calcium- and voltage-gated potassium channels of large conductance by leukotriene B4. *The Journal of Biological Chemistry*, *289*, 35314–35325.

Bukiya, A. N., McMillan, J., Parrill, A. L., & Dopico, A. M. (2008). Structural determinants of monohydroxylated bile acids to activate beta 1 subunit-containing BK channels. *Journal of Lipid Research*, *49*, 2441–2451.

Bukiya, A. N., Patil, S. A., Li, W., Miller, D. D., & Dopico, A. M. (2012). Calcium- and voltage-gated potassium (BK) channel activators in the 5beta-cholanic acid-3alpha-ol analogue series with modifications in the lateral chain. *ChemMedChem*, *7*, 1784–1792.

Bukiya, A. N., Singh, A. K., Parrill, A. L., & Dopico, A. M. (2011). The steroid interaction site in transmembrane domain 2 of the large conductance, voltage- and calcium-gated potassium (BK) channel accessory beta1 subunit. *Proceedings of the National Academy of Sciences of the United States of America*, *108*, 20207–20212.

Bukiya, A. N., Vaithianathan, T., Toro, L., & Dopico, A. M. (2009). Channel beta2-4 subunits fail to substitute for beta1 in sensitizing BK channels to lithocholate. *Biochemical and Biophysical Research Communications*, *390*, 995–1000.

Cao, X. H., Chen, S. R., Li, L., & Pan, H. L. (2012). Nerve injury increases brain-derived neurotrophic factor levels to suppress BK channel activity in primary sensory neurons. *Journal of Neurochemistry*, *121*, 944–953.

Carvalho-de-Souza, J. L., Varanda, W. A., Tostes, R. C., & Chignalia, A. Z. (2013). BK channels in cardiovascular diseases and aging. *Aging and Disease*, *4*, 38–49.

Castillo, K., Contreras, G. F., Pupo, A., Torres, Y. P., Neely, A., Gonzalez, C., et al. (2015). Molecular mechanism underlying beta1 regulation in voltage- and calcium-activated potassium (BK) channels. *Proceedings of the National Academy of Sciences of the United States of America*, *112*, 4809–4814.

Chen, L., Bi, D., Tian, L., McClafferty, H., Steeb, F., Ruth, P., et al. (2013). Palmitoylation of the beta4-subunit regulates surface expression of large conductance calcium-activated potassium channel splice variants. *The Journal of Biological Chemistry*, *288*, 13136–13144.

Chen, S. R., Cai, Y. Q., & Pan, H. L. (2009). Plasticity and emerging role of BKCa channels in nociceptive control in neuropathic pain. *Journal of Neurochemistry*, *110*, 352–362.

Chen, M., Gan, G., Wu, Y., Wang, L., Wu, Y., & Ding, J. (2008). Lysine-rich extracellular rings formed by hbeta2 subunits confer the outward rectification of BK channels. *PLoS One*, *3*, e2114.

Chen, M., & Petkov, G. V. (2009). Identification of large conductance calcium activated potassium channel accessory beta4 subunit in rat and mouse bladder smooth muscle. *The Journal of Urology*, *182*, 374–381.

Contreras, G. F., Neely, A., Alvarez, O., Gonzalez, C., & Latorre, R. (2012). Modulation of BK channel voltage gating by different auxiliary beta subunits. *Proceedings of the National Academy of Sciences of the United States of America*, *109*, 18991–18996.

Cox, D. H., & Aldrich, R. W. (2000). Role of the beta1 subunit in large-conductance Ca^{2+}-activated K^{+} channel gating energetics. Mechanisms of enhanced Ca^{2+} sensitivity. *The Journal of General Physiology*, *116*, 411–432.

De Wet, H., Allen, M., Holmes, C., Stobbart, M., Lippiat, J. D., & Callaghan, R. (2006). Modulation of the BK channel by estrogens: Examination at single channel level. *Molecular Membrane Biology*, *23*, 420–429.

Deng, P. Y., Rotman, Z., Blundon, J. A., Cho, Y., Cui, J., Cavalli, V., et al. (2013). FMRP regulates neurotransmitter release and synaptic information transmission by modulating action potential duration via BK channels. *Neuron*, *77*, 696–711.

Dick, G. M., Rossow, C. F., Smirnov, S., Horowitz, B., & Sanders, K. M. (2001). Tamoxifen activates smooth muscle BK channels through the regulatory beta 1 subunit. *The Journal of Biological Chemistry*, *276*, 34594–34599.

Ding, J. P., Li, Z. W., & Lingle, C. J. (1998). Inactivating BK channels in rat chromaffin cells may arise from heteromultimeric assembly of distinct inactivation-competent and non-inactivating subunits. *Biophysical Journal*, *74*, 268–289.

Dolan, J., Walshe, K., Alsbury, S., Hokamp, K., O'Keeffe, S., Okafuji, T., et al. (2007). The extracellular leucine-rich repeat superfamily; a comparative survey and analysis of evolutionary relationships and expression patterns. *BMC Genomics*, *8*, 320.

Dopico, A. M., Widmer, H., Wang, G., Lemos, J. R., & Treistman, S. N. (1999). Rat supraoptic magnocellular neurones show distinct large conductance, Ca^{2+}-activated K^{+} channel subtypes in cell bodies versus nerve endings. *The Journal of Physiology*, *519*(Pt. 1), 101–114.

Du, W., Bautista, J. F., Yang, H., Diez-Sampedro, A., You, S. A., Wang, L., et al. (2005). Calcium-sensitive potassium channelopathy in human epilepsy and paroxysmal movement disorder. *Nature Genetics*, *37*, 733–738.

Duncan, R. K. (2005). Tamoxifen alters gating of the BK alpha subunit and mediates enhanced interactions with the avian beta subunit. *Biochemical Pharmacology*, *70*, 47–58.

Evanson, K. W., Bannister, J. P., Leo, M. D., & Jaggar, J. H. (2014). LRRC26 is a functional BK channel auxiliary gamma subunit in arterial smooth muscle cells. *Circulation Research*, *115*, 423–431.

Farajnia, S., Meijer, J. H., & Michel, S. (2015). Age-related changes in large-conductance calcium-activated potassium channels in mammalian circadian clock neurons. *Neurobiology of Aging*, *36*, 2176–2183.

Feinberg-Zadek, P. L., & Treistman, S. N. (2007). Beta-subunits are important modulators of the acute response to alcohol in human BK channels. *Alcoholism, Clinical and Experimental Research*, *31*, 737–744.

Fernandez-Fernandez, J. M., Tomas, M., Vazquez, E., Orio, P., Latorre, R., Senti, M., et al. (2004). Gain-of-function mutation in the KCNMB1 potassium channel subunit is associated with low prevalence of diastolic hypertension. *The Journal of Clinical Investigation*, *113*, 1032–1039.

Fettiplace, R., & Fuchs, P. A. (1999). Mechanisms of hair cell tuning. *Annual Review of Physiology*, *61*, 809–834.

Gan, G., Yi, H., Chen, M., Sun, L., Li, W., Wu, Y., et al. (2008). Structural basis for toxin resistance of beta4-associated calcium-activated potassium (BK) channels. *The Journal of Biological Chemistry*, *283*, 24177–24184.

Garcia-Calvo, M., Knaus, H. G., McManus, O. B., Giangiacomo, K. M., Kaczorowski, G. J., & Garcia, M. L. (1994). Purification and reconstitution of the high-conductance, calcium-activated potassium channel from tracheal smooth muscle. *The Journal of Biological Chemistry*, *269*, 676–682.

Garcia-Valdes, J., Zamudio, F. Z., Toro, L., & Possani, L. D. (2001). Slotoxin, alphaKTx1.11, a new scorpion peptide blocker of MaxiK channels that differentiates between alpha and alpha + beta (beta1 or beta4) complexes. *FEBS Letters*, *505*, 369–373.

Gessner, G., Cui, Y. M., Otani, Y., Ohwada, T., Soom, M., Hoshi, T., et al. (2012). Molecular mechanism of pharmacological activation of BK channels. *Proceedings of the National Academy of Sciences of the United States of America*, *109*, 3552–3557.

Gessner, G., Schonherr, K., Soom, M., Hansel, A., Asim, M., Baniahmad, A., et al. (2005). BKCa channels activating at resting potential without calcium in LNCaP prostate cancer cells. *The Journal of Membrane Biology*, *208*, 229–240.

Ghatta, S., Nimmagadda, D., Xu, X., & O'Rourke, S. T. (2006). Large-conductance, calcium-activated potassium channels: Structural and functional implications. *Pharmacology & Therapeutics*, *110*, 103–116.

Glauser, D. A., Johnson, B. E., Aldrich, R. W., & Goodman, M. B. (2011). Intragenic alternative splicing coordination is essential for Caenorhabditis elegans slo-1 gene function. *Proceedings of the National Academy of Sciences of the United States of America*, *108*, 20790–20795.

Golding, N. L., Jung, H. Y., Mickus, T., & Spruston, N. (1999). Dendritic calcium spike initiation and repolarization are controlled by distinct potassium channel subtypes in CA1 pyramidal neurons. *The Journal of Neuroscience*, *19*, 8789–8798.

Gonzalez-Perez, V., Xia, X. M., & Lingle, C. J. (2014). Functional regulation of BK potassium channels by gamma1 auxiliary subunits. *Proceedings of the National Academy of Sciences of the United States of America*, *111*, 4868–4873.

Gonzalez-Perez, V., Xia, X. M., & Lingle, C. J. (2015). Two classes of regulatory subunits coassemble in the same BK channel and independently regulate gating. *Nature Communications*, *6*, 8341.

Gonzalez-Perez, V., Zeng, X. H., Henzler-Wildman, K., & Lingle, C. J. (2012). Stereospecific binding of a disordered peptide segment mediates BK channel inactivation. *Nature*, *485*, 133–136.

Gruslova, A., Semenov, I., & Wang, B. (2012). An extracellular domain of the accessory beta1 subunit is required for modulating BK channel voltage sensor and gate. *The Journal of General Physiology*, *139*, 57–67.

Hagen, B. M., & Sanders, K. M. (2006). Deglycosylation of the beta1-subunit of the BK channel changes its biophysical properties. *American Journal of Physiology Cell Physiology*, *291*, C750–C756.

Hanner, M., Vianna-Jorge, R., Kamassah, A., Schmalhofer, W. A., Knaus, H. G., Kaczorowski, G. J., et al. (1998). The beta subunit of the high conductance calcium-activated potassium channel. Identification of residues involved in charybdotoxin binding. *The Journal of Biological Chemistry*, *273*, 16289–16296.

Higgins, J. J., Hao, J., Kosofsky, B. E., & Rajadhyaksha, A. M. (2008). Dysregulation of large-conductance Ca^{2+}-activated K^{+} channel expression in nonsyndromal mental retardation due to a cereblon p.R419X mutation. *Neurogenetics*, *9*, 219–223.

Horrigan, F. T., & Aldrich, R. W. (1999). Allosteric voltage gating of potassium channels II. Mslo channel gating charge movement in the absence of Ca^{2+}. *The Journal of General Physiology*, *114*, 305–336.

Horrigan, F. T., & Aldrich, R. W. (2002). Coupling between voltage sensor activation, Ca^{2+} binding and channel opening in large conductance (BK) potassium channels. *The Journal of General Physiology*, *120*, 267–305.

Horrigan, F. T., Cui, J., & Aldrich, R. W. (1999). Allosteric voltage gating of potassium channels I. Mslo ionic currents in the absence of Ca^{2+}. *The Journal of General Physiology*, *114*, 277–304.

Hoshi, T., Wissuwa, B., Tian, Y., Tajima, N., Xu, R., Bauer, M., et al. (2013). Omega-3 fatty acids lower blood pressure by directly activating large-conductance Ca(2)(+)-dependent K^{+} channels. *Proceedings of the National Academy of Sciences of the United States of America*, *110*, 4816–4821.

Hu, H., Shao, L. R., Chavoshy, S., Gu, N., Trieb, M., Behrens, R., et al. (2001). Presynaptic Ca^{2+}-activated K^{+} channels in glutamatergic hippocampal terminals and their role in spike repolarization and regulation of transmitter release. *The Journal of Neuroscience*, *21*, 9585–9597.

Hu, X. Q., Xiao, D., Zhu, R., Huang, X., Yang, S., Wilson, S., et al. (2011). Pregnancy upregulates large-conductance Ca^{2+}-activated K^{+} channel activity and attenuates myogenic tone in uterine arteries. *Hypertension*, *58*, 1132–1139.

Jin, P., Weiger, T. M., & Levitan, I. B. (2002). Reciprocal modulation between the alpha and beta 4 subunits of hSlo calcium-dependent potassium channels. *Journal of Biological Chemistry*, *277*, 43724–43729.

Jones, E. M., Gray-Keller, M., Art, J. J., & Fettiplace, R. (1999). The functional role of alternative splicing of Ca^{2+}-activated K^{+} channels in auditory hair cells. *Annals of the New York Academy of Sciences*, *868*, 379–385.

Kim, E. Y., Zou, S., Ridgway, L. D., & Dryer, S. E. (2007). Beta1-subunits increase surface expression of a large-conductance Ca^{2+}-activated K^{+} channel isoform. *Journal of Neurophysiology*, *97*, 3508–3516.

King, J. T., Lovell, P. V., Rishniw, M., Kotlikoff, M. I., Zeeman, M. L., & McCobb, D. P. (2006). Beta2 and beta4 subunits of BK channels confer differential sensitivity to acute modulation by steroid hormones. *Journal of Neurophysiology*, *95*, 2878–2888.

Kirby, R. W., Martelli, A., Calderone, V., McKay, N. G., & Lawson, K. (2013). Large conductance Ca^{2+}-activated K^{+} channel (BKCa) activating properties of a series of novel N-arylbenzamides: Channel subunit dependent effects. *Bioorganic & Medicinal Chemistry*, *21*, 4186–4191.

Knaus, H. G., Eberhart, A., Kaczorowski, G. J., & Garcia, M. L. (1994). Covalent attachment of charybdotoxin to the beta-subunit of the high conductance Ca^{2+}-activated K^{+} channel. Identification of the site of incorporation and implications for channel topology. *The Journal of Biological Chemistry*, *269*, 23336–23341.

Knaus, H. G., Folander, K., Garcia-Calvo, M., Garcia, M. L., Kaczorowski, G. J., Smith, M., et al. (1994). Primary sequence and immunological characterization of beta-subunit of high conductance Ca^{2+}-activated K^{+} channel from smooth muscle. *The Journal of Biological Chemistry*, *269*, 17274–17278.

Knaus, H. G., Garcia-Calvo, M., Kaczorowski, G. J., & Garcia, M. L. (1994). Subunit composition of the high conductance calcium-activated potassium channel from smooth muscle, a representative of the mSlo and slowpoke family of potassium channels. *The Journal of Biological Chemistry*, *269*, 3921–3924.

Kobe, B., & Kajava, A. V. (2001). The leucine-rich repeat as a protein recognition motif. *Current Opinion in Structural Biology*, *11*, 725–732.

Kreifeldt, M., Cates-Gatto, C., Roberts, A. J., & Contet, C. (2015). BK channel beta1 subunit contributes to behavioral adaptations elicited by chronic intermittent ethanol exposure. *Alcoholism, Clinical and Experimental Research*, *39*, 2394–2402.

Kreifeldt, M., Le, D., Treistman, S. N., Koob, G. F., & Contet, C. (2013). BK channel beta1 and beta4 auxiliary subunits exert opposite influences on escalated ethanol drinking in dependent mice. *Frontiers in Integrative Neuroscience*, 7, 105.

Kuntamallappanavar, G., Toro, L., & Dopico, A. M. (2014). Both transmembrane domains of BK beta1 subunits are essential to confer the normal phenotype of beta1-containing BK channels. *PLoS One*, *9*, e109306.

Large, R. J., Kshatri, A., Webb, T. I., Roy, S., Akande, A., Bradley, E., et al. (2015). Effects of the novel BK (KCa 1.1) channel opener GoSlo-SR-5-130 are dependent on the presence of BKbeta subunits. *British Journal of Pharmacology*, *172*, 2544–2556.

Laumonnier, F., Roger, S., Guerin, P., Molinari, F., M'Rad, R., Cahard, D., et al. (2006). Association of a functional deficit of the BKCa channel, a synaptic regulator of neuronal excitability, with autism and mental retardation. *The American Journal of Psychiatry*, *163*, 1622–1629.

Lee, U. S., Shi, J., & Cui, J. (2010). Modulation of BK channel gating by the ss2 subunit involves both membrane-spanning and cytoplasmic domains of Slo1. *The Journal of Neuroscience*, *30*, 16170–16179.

Li, M., Chang, S., Yang, L., Shi, J., McFarland, K., Yang, X., et al. (2014). Conopeptide Vt3.1 preferentially inhibits BK potassium channels containing beta4 subunits via electrostatic interactions. *The Journal of Biological Chemistry*, *289*, 4735–4742.

Li, Q., Fan, F., Kwak, H. R., & Yan, J. (2015). Molecular basis for differential modulation of BK channel voltage-dependent gating by auxiliary gamma subunits. *The Journal of General Physiology*, *145*, 543–554.

Li, Q., Guan, X., Yen, K., Zhang, J., & Yan, J. (2016). The single transmembrane segment determines the modulatory function of the BK channel auxiliary γ subunit. *The Journal of General Physiology*, *147*, 337–351.

Li, H., Yao, J., Tong, X., Guo, Z., Wu, Y., Sun, L., et al. (2007). Interaction sites between the Slo1 pore and the NH2 terminus of the beta2 subunit, probed with a three-residue sensor. *The Journal of Biological Chemistry*, *282*, 17720–17728.

Lingle, C. J., Zeng, X. H., Ding, J. P., & Xia, X. M. (2001). Inactivation of BK channels mediated by the NH(2) terminus of the beta3b auxiliary subunit involves a two-step mechanism: Possible separation of binding and blockade. *The Journal of General Physiology*, *117*, 583–606.

Liu, G., Niu, X., Wu, R. S., Chudasama, N., Yao, Y., Jin, X., et al. (2010). Location of modulatory beta subunits in BK potassium channels. *The Journal of General Physiology*, *135*, 449–459.

Liu, X. F., Xiang, L., Zhang, Y., Becker, K. G., Bera, T. K., & Pastan, I. (2012). CAPC negatively regulates NF-kappaB activation and suppresses tumor growth and metastasis. *Oncogene*, *31*, 1673–1682.

Liu, G., Zakharov, S. I., Yang, L., Wu, R. S., Deng, S. X., Landry, D. W., et al. (2008). Locations of the beta1 transmembrane helices in the BK potassium channel. *Proceedings of the National Academy of Sciences of the United States of America*, *105*, 10727–10732.

Liu, G., Zakharov, S. I., Yao, Y., Marx, S. O., & Karlin, A. (2015). Positions of the cytoplasmic end of BK alpha S0 helix relative to S1-S6 and of beta1 TM1 and TM2 relative to S0-S6. *The Journal of General Physiology*, *145*, 185–199.

Liu, R., Zhang, Z., Liu, H., Hou, P., Lang, J., Wang, S., et al. (2013). Human beta-defensin 2 is a novel opener of Ca^{2+}-activated potassium channels and induces vasodilation and hypotension in monkeys. *Hypertension*, *62*, 415–425.

Lorenz, S., Heils, A., Kasper, J. M., & Sander, T. (2007). Allelic association of a truncation mutation of the KCNMB3 gene with idiopathic generalized epilepsy. *American Journal of Medical Genetics. Part B, Neuropsychiatric Genetics*, *144B*, 10–13.

Lv, C., Chen, M., Gan, G., Wang, L., Xu, T., & Ding, J. (2008). Four-turn alpha-helical segment prevents surface expression of the auxiliary hbeta2 subunit of BK-type channel. *The Journal of Biological Chemistry*, *283*, 2709–2715.

Ma, Z., Lou, X. J., & Horrigan, F. T. (2006). Role of charged residues in the S1-S4 voltage sensor of BK channels. *The Journal of General Physiology*, *127*, 309–328.

Mancini, M., Soldovieri, M. V., Gessner, G., Wissuwa, B., Barrese, V., Boscia, F., et al. (2014). Critical role of large-conductance calcium- and voltage-activated potassium channels in leptin-induced neuroprotection of N-methyl-D-aspartate-exposed cortical neurons. *Pharmacological Research*, *87*, 80–86.

Manzanares, D., Krick, S., Baumlin, N., Dennis, J. S., Tyrrell, J., Tarran, R., et al. (2015). Airway surface dehydration by transforming growth factor beta (TGF-beta) in cystic fibrosis is due to decreased function of a voltage-dependent potassium channel and can be rescued by the drug pirfenidone. *The Journal of Biological Chemistry*, *290*, 25710–25716.

Manzanares, D., Srinivasan, M., Salathe, S. T., Ivonnet, P., Baumlin, N., Dennis, J. S., et al. (2014). IFN-gamma-mediated reduction of large-conductance, Ca^{2+}-activated, voltage-dependent K^+ (BK) channel activity in airway epithelial cells leads to mucociliary dysfunction. *American Journal of Physiology. Lung Cellular and Molecular Physiology*, *306*, L453–L462.

Martin, G. E., Hendrickson, L. M., Penta, K. L., Friesen, R. M., Pietrzykowski, A. Z., Tapper, A. R., et al. (2008). Identification of a BK channel auxiliary protein controlling molecular and behavioral tolerance to alcohol. *Proceedings of the National Academy of Sciences of the United States of America*, *105*, 17543–17548.

Martin, G., Puig, S., Pietrzykowski, A., Zadek, P., Emery, P., & Treistman, S. (2004). Somatic localization of a specific large-conductance calcium-activated potassium channel subtype controls compartmentalized ethanol sensitivity in the nucleus accumbens. *The Journal of Neuroscience*, *24*, 6563–6572.

Martinez-Espinosa, P. L., Yang, C., Gonzalez-Perez, V., Xia, X. M., & Lingle, C. J. (2014). Knockout of the BK beta2 subunit abolishes inactivation of BK currents in mouse adrenal chromaffin cells and results in slow-wave burst activity. *The Journal of General Physiology, 144*, 275–295.

Marty, A. (1981). Ca-dependent K channels with large unitary conductance in chromaffin cell membranes. *Nature, 291*, 497–500.

Matthews, E. A., & Disterhoft, J. F. (2009). Blocking the BK channel impedes acquisition of trace eyeblink conditioning. *Learning & Memory, 16*, 106–109.

McManus, O. B., Harris, G. H., Giangiacomo, K. M., Feigenbaum, P., Reuben, J. P., Addy, M. E., et al. (1993). An activator of calcium-dependent potassium channels isolated from a medicinal herb. *Biochemistry, 32*, 6128–6133.

McManus, O. B., Helms, L. M., Pallanck, L., Ganetzky, B., Swanson, R., & Leonard, R. J. (1995). Functional role of the beta subunit of high conductance calcium-activated potassium channels. *Neuron, 14*, 645–650.

Meera, P., Wallner, M., & Toro, L. (2000). A neuronal beta subunit (KCNMB4) makes the large conductance, voltage- and Ca^{2+}-activated K^{+} channel resistant to charybdotoxin and iberiotoxin. *Proceedings of the National Academy of Sciences of the United States of America, 97*, 5562–5567.

Meredith, A. L., Thorneloe, K. S., Werner, M. E., Nelson, M. T., & Aldrich, R. W. (2004). Overactive bladder and incontinence in the absence of the BK large conductance Ca^{2+}-activated K^{+} channel. *The Journal of Biological Chemistry, 279*, 36746–36752.

Meredith, A. L., Wiler, S. W., Miller, B. H., Takahashi, J. S., Fodor, A. A., Ruby, N. F., et al. (2006). BK calcium-activated potassium channels regulate circadian behavioral rhythms and pacemaker output. *Nature Neuroscience, 9*, 1041–1049.

Montgomery, J. R., Whitt, J. P., Wright, B. N., Lai, M. H., & Meredith, A. L. (2013). Mis-expression of the BK K^{+} channel disrupts suprachiasmatic nucleus circuit rhythmicity and alters clock-controlled behavior. *American Journal of Physiology Cell Physiology, 304*, C299–C311.

Morimoto, T., Sakamoto, K., Sade, H., Ohya, S., Muraki, K., & Imaizumi, Y. (2007). Voltage-sensitive oxonol dyes are novel large-conductance Ca^{2+}-activated K^{+} channel activators selective for beta1 and beta4 but not for beta2 subunits. *Molecular Pharmacology, 71*, 1075–1088.

Morrow, J. P., Zakharov, S. I., Liu, G., Yang, L., Sok, A. J., & Marx, S. O. (2006). Defining the BK channel domains required for beta1-subunit modulation. *Proceedings of the National Academy of Sciences of the United States of America, 103*, 5096–5101.

Nardi, A., & Olesen, S. P. (2008). BK channel modulators: A comprehensive overview. *Current Medicinal Chemistry, 15*, 1126–1146.

Nimigean, C. M., & Magleby, K. L. (1999). The beta subunit increases the Ca^{2+} sensitivity of large conductance Ca^{2+}-activated potassium channels by retaining the gating in the bursting states. *The Journal of General Physiology, 113*, 425–440.

Onimaru, H., Ballanyi, K., & Homma, I. (2003). Contribution of Ca^{2+}-dependent conductances to membrane potential fluctuations of medullary respiratory neurons of newborn rats in vitro. *The Journal of Physiology, 552*, 727–741.

Orio, P., & Latorre, R. (2005). Differential effects of beta 1 and beta 2 subunits on BK channel activity. *The Journal of General Physiology, 125*, 395–411.

Orio, P., Rojas, P., Ferreira, G., & Latorre, R. (2002). New disguises for an old channel: MaxiK channel beta-subunits. *News in Physiological Sciences, 17*, 156–161.

Orio, P., Torres, Y., Rojas, P., Carvacho, I., Garcia, M. L., Toro, L., et al. (2006). Structural determinants for functional coupling between the beta and alpha subunits in the Ca^{2+}-activated K^{+} (BK) channel. *The Journal of General Physiology, 127*, 191–204.

Pardo, L. A., & Stuhmer, W. (2014). The roles of K^{+} channels in cancer. *Nature Reviews. Cancer, 14*, 39–48.

Pitts, G. R., Ohta, H., & McMahon, D. G. (2006). Daily rhythmicity of large-conductance Ca^{2+}-activated K^{+} currents in suprachiasmatic nucleus neurons. *Brain Research*, *1071*, 54–62.

Raffaelli, G., Saviane, C., Mohajerani, M. H., Pedarzani, P., & Cherubini, E. (2004). BK potassium channels control transmitter release at CA3-CA3 synapses in the rat hippocampus. *The Journal of Physiology*, *557*, 147–157.

Ramanathan, K., Michael, T. H., Jiang, G. J., Hiel, H., & Fuchs, P. A. (1999). A molecular mechanism for electrical tuning of cochlear hair cells. *Science*, *283*, 215–217.

Runden-Pran, E., Haug, F. M., Storm, J. F., & Ottersen, O. P. (2002). BK channel activity determines the extent of cell degeneration after oxygen and glucose deprivation: A study in organotypical hippocampal slice cultures. *Neuroscience*, *112*, 277–288.

Saito, M., Nelson, C., Salkoff, L., & Lingle, C. J. (1997). A cysteine-rich domain defined by a novel exon in a slo variant in rat adrenal chromaffin cells and PC12 cells. *The Journal of Biological Chemistry*, *272*, 11710–11717.

Salkoff, L., Butler, A., Ferreira, G., Santi, C., & Wei, A. (2006). High-conductance potassium channels of the SLO family. *Nature Reviews. Neuroscience*, *7*, 921–931.

Samengo, I., Curro, D., Barrese, V., Taglialatela, M., & Martire, M. (2014). Large conductance calcium-activated potassium channels: Their expression and modulation of glutamate release from nerve terminals isolated from rat trigeminal caudal nucleus and cerebral cortex. *Neurochemical Research*, *39*, 901–910.

Sausbier, M., Hu, H., Arntz, C., Feil, S., Kamm, S., Adelsberger, H., et al. (2004). Cerebellar ataxia and Purkinje cell dysfunction caused by Ca^{2+}-activated K^{+} channel deficiency. *Proceedings of the National Academy of Sciences of the United States of America*, *101*, 9474–9478.

Savalli, N., Kondratiev, A., de Quintana, S. B., Toro, L., & Olcese, R. (2007). Modes of operation of the BKCa channel beta2 subunit. *The Journal of General Physiology*, *130*, 117–131.

Schneider, M. J., Rogowski, R. S., Krueger, B. K., & Blaustein, M. P. (1989). Charybdotoxin blocks both Ca-activated K channels and Ca-independent voltage-gated K channels in rat brain synaptosomes. *FEBS Letters*, *250*, 433–436.

Schreiber, M., & Salkoff, L. (1997). A novel calcium-sensing domain in the BK channel. *Biophysical Journal*, *73*, 1355–1363.

Schreiber, M., Yuan, A., & Salkoff, L. (1999). Transplantable sites confer calcium sensitivity to BK channels. *Nature Neuroscience*, *2*, 416–421.

Shao, L. R., Halvorsrud, R., Borg-Graham, L., & Storm, J. F. (1999). The role of BK-type Ca^{2+}-dependent K^{+} channels in spike broadening during repetitive firing in rat hippocampal pyramidal cells. *The Journal of Physiology*, *521*(Pt. 1), 135–146.

Shen, Y., Kishimoto, K., Linden, D. J., & Sapirstein, A. (2007). Cytosolic phospholipase A(2) alpha mediates electrophysiologic responses of hippocampal pyramidal neurons to neurotoxic NMDA treatment. *Proceedings of the National Academy of Sciences of the United States of America*, *104*, 6078–6083.

Shi, J., He, H. Q., Zhao, R., Duan, Y. H., Chen, J., Chen, Y., et al. (2008). Inhibition of martentoxin on neuronal BK channel subtype (alpha + beta4): Implications for a novel interaction model. *Biophysical Journal*, *94*, 3706–3713.

Shi, J., Krishnamoorthy, G., Yang, Y., Hu, L., Chaturvedi, N., Harilal, D., et al. (2002). Mechanism of magnesium activation of calcium-activated potassium channels. *Nature*, *418*, 876–880.

Shruti, S., Urban-Ciecko, J., Fitzpatrick, J. A., Brenner, R., Bruchez, M. P., & Barth, A. L. (2012). The brain-specific beta4 subunit downregulates BK channel cell surface expression. *PLoS One*, *7*, e33429.

Springer, S. J., Burkett, B. J., & Schrader, L. A. (2014). Modulation of BK channels contributes to activity-dependent increase of excitability through MTORC1 activity in CA1 pyramidal cells of mouse hippocampus. *Frontiers in Cellular Neuroscience*, *8*, 451.

Sun, X., Shi, J., Delaloye, K., Yang, X., Yang, H., Zhang, G., et al. (2013). The interface between membrane-spanning and cytosolic domains in Ca^{2+}-dependent K^+ channels is involved in beta subunit modulation of gating. *The Journal of Neuroscience, 33*, 11253–11261.

Sun, X., Zaydman, M. A., & Cui, J. (2012). Regulation of voltage-activated K^+ channel gating by transmembrane beta subunits. *Frontiers in Pharmacology, 3*, 63.

Tao, X., Lin, M. T., Thorington, G. U., Wilson, S. M., Longo, L. D., & Hessinger, D. A. (2015). Long-term hypoxia increases calcium affinity of BK channels in ovine fetal and adult cerebral artery smooth muscle. *American Journal of Physiology. Heart and Circulatory Physiology, 308*, H707–H722.

Tao, J., Shi, J., Liu, Z. R., & Ji, Y. H. (2012). Martentoxin: A unique ligand of BK channels. *Sheng Li Xue Bao, 64*, 355–364.

Tian, Y., Ullrich, F., Xu, R., Heinemann, S. H., Hou, S., & Hoshi, T. (2015). Two distinct effects of PIP2 underlie auxiliary subunit-dependent modulation of Slo1 BK channels. *The Journal of General Physiology, 145*, 331–343.

Toro, B., Cox, N., Wilson, R. J., Garrido-Sanabria, E., Stefani, E., Toro, L., et al. (2006). KCNMB1 regulates surface expression of a voltage and Ca^{2+}-activated K^+ channel via endocytic trafficking signals. *Neuroscience, 142*, 661–669.

Torres, Y. P., Granados, S. T., & Latorre, R. (2014). Pharmacological consequences of the coexpression of BK channel alpha and auxiliary beta subunits. *Frontiers in Physiology, 5*, 383.

Torres, Y. P., Morera, F. J., Carvacho, I., & Latorre, R. (2007). A marriage of convenience: Beta-subunits and voltage-dependent K^+ channels. *The Journal of Biological Chemistry, 282*, 24485–24489.

Tseng-Crank, J., Godinot, N., Johansen, T. E., Ahring, P. K., Strobaek, D., Mertz, R., et al. (1996). Cloning, expression, and distribution of a Ca^{2+}-activated K^+ channel beta-subunit from human brain. *Proceedings of the National Academy of Sciences of the United States of America, 93*, 9200–9205.

Typlt, M., Mirkowski, M., Azzopardi, E., Ruettiger, L., Ruth, P., & Schmid, S. (2013). Mice with deficient BK channel function show impaired prepulse inhibition and spatial learning, but normal working and spatial reference memory. *PLoS One, 8*, e81270.

Uebele, V. N., Lagrutta, A., Wade, T., Figueroa, D. J., Liu, Y., McKenna, E., et al. (2000). Cloning and functional expression of two families of beta-subunits of the large conductance calcium-activated K^+ channel. *The Journal of Biological Chemistry, 275*, 23211–23218.

Valverde, M. A., Rojas, P., Amigo, J., Cosmelli, D., Orio, P., Bahamonde, M. I., et al. (1999). Acute activation of Maxi-K channels (hSlo) by estradiol binding to the beta subunit. *Science, 285*, 1929–1931.

Wallner, M., Meera, P., Ottolia, M., Kaczorowski, G. J., Latorre, R., Garcia, M. L., et al. (1995). Characterization of and modulation by a beta-subunit of a human maxi KCa channel cloned from myometrium. *Receptors & Channels, 3*, 185–199.

Wallner, M., Meera, P., & Toro, L. (1996). Determinant for beta-subunit regulation in high-conductance voltage-activated and Ca^{2+}-sensitive K^+ channels: An additional transmembrane region at the N terminus. *Proceedings of the National Academy of Sciences of the United States of America, 93*, 14922–14927.

Wallner, M., Meera, P., & Toro, L. (1999). Molecular basis of fast inactivation in voltage and Ca^{2+}-activated K^+ channels: A transmembrane beta-subunit homolog. *Proceedings of the National Academy of Sciences of the United States of America, 96*, 4137–4142.

Wang, B., & Brenner, R. (2006). An S6 mutation in BK channels reveals beta1 subunit effects on intrinsic and voltage-dependent gating. *The Journal of General Physiology, 128*, 731–744.

Wang, Y. W., Ding, J. P., Xia, X. M., & Lingle, C. J. (2002). Consequences of the stoichiometry of Slo1 alpha and auxiliary beta subunits on functional properties of large-conductance Ca^{2+}-activated K^+ channels. *The Journal of Neuroscience, 22*, 1550–1561.

Wang, B., Rothberg, B. S., & Brenner, R. (2006). Mechanism of beta4 subunit modulation of BK channels. *The Journal of General Physiology*, *127*, 449–465.

Waxman, S. G., & Zamponi, G. W. (2014). Regulating excitability of peripheral afferents: Emerging ion channel targets. *Nature Neuroscience*, *17*, 153–163.

Werner, M. E., Knorn, A. M., Meredith, A. L., Aldrich, R. W., & Nelson, M. T. (2007). Frequency encoding of cholinergic- and purinergic-mediated signaling to mouse urinary bladder smooth muscle: Modulation by BK channels. *American Journal of Physiology. Regulatory, Integrative and Comparative Physiology*, *292*, R616–R624.

Werner, M. E., Meredith, A. L., Aldrich, R. W., & Nelson, M. T. (2008). Hypercontractility and impaired sildenafil relaxations in the BKCa channel deletion model of erectile dysfunction. *American Journal of Physiology Regulatory, Integrative and Comparative Physiology*, *295*, R181–R188.

Werner, M. E., Zvara, P., Meredith, A. L., Aldrich, R. W., & Nelson, M. T. (2005). Erectile dysfunction in mice lacking the large-conductance calcium-activated potassium (BK) channel. *The Journal of Physiology*, *567*, 545–556.

Womack, M. D., & Khodakhah, K. (2002). Characterization of large conductance Ca^{2+}-activated K^+ channels in cerebellar Purkinje neurons. *The European Journal of Neuroscience*, *16*, 1214–1222.

Wrzosek, A. (2014). The potassium channel opener NS1619 modulates calcium homeostasis in muscle cells by inhibiting SERCA. *Cell Calcium*, *56*, 14–24.

Wu, R. S., Chudasama, N., Zakharov, S. I., Doshi, D., Motoike, H., Liu, G., et al. (2009). Location of the beta 4 transmembrane helices in the BK potassium channel. *The Journal of Neuroscience*, *29*, 8321–8328.

Wu, R. S., Liu, G., Zakharov, S. I., Chudasama, N., Motoike, H., Karlin, A., et al. (2013). Positions of beta2 and beta3 subunits in the large-conductance calcium- and voltage-activated BK potassium channel. *The Journal of General Physiology*, *141*, 105–117.

Wynne, P. M., Puig, S. I., Martin, G. E., & Treistman, S. N. (2009). Compartmentalized beta subunit distribution determines characteristics and ethanol sensitivity of somatic, dendritic, and terminal large-conductance calcium-activated potassium channels in the rat central nervous system. *The Journal of Pharmacology and Experimental Therapeutics*, *329*, 978–986.

Xia, X. M., Ding, J. P., & Lingle, C. J. (1999). Molecular basis for the inactivation of Ca^{2+}- and voltage-dependent BK channels in adrenal chromaffin cells and rat insulinoma tumor cells. *The Journal of Neuroscience*, *19*, 5255–5264.

Xia, X. M., Ding, J. P., & Lingle, C. J. (2003). Inactivation of BK channels by the NH2 terminus of the beta2 auxiliary subunit: An essential role of a terminal peptide segment of three hydrophobic residues. *The Journal of General Physiology*, *121*, 125–148.

Xia, X. M., Ding, J. P., Zeng, X. H., Duan, K. L., & Lingle, C. J. (2000). Rectification and rapid activation at low Ca^{2+} of Ca^{2+}-activated, voltage-dependent BK currents: Consequences of rapid inactivation by a novel beta subunit. *The Journal of Neuroscience*, *20*, 4890–4903.

Xia, X. M., Zeng, X., & Lingle, C. J. (2002). Multiple regulatory sites in large-conductance calcium-activated potassium channels. *Nature*, *418*, 880–884.

Xu, H., Garver, H., Fernandes, R., Phelps, J. T., Harkema, J. J., Galligan, J. J., et al. (2015). BK channel beta1-subunit deficiency exacerbates vascular fibrosis and remodelling but does not promote hypertension in high-fat fed obesity in mice. *Journal of Hypertension*, *33*, 1611–1623.

Xu, J. W., & Slaughter, M. M. (2005). Large-conductance calcium-activated potassium channels facilitate transmitter release in salamander rod synapse. *The Journal of Neuroscience*, *25*, 7660–7668.

Yamamura, H., Ohi, Y., Muraki, K., Watanabe, M., & Imaizumi, Y. (2001). BK channel activation by NS-1619 is partially mediated by intracellular Ca^{2+} release in smooth muscle cells of porcine coronary artery. *British Journal of Pharmacology*, *132*, 828–834.

Yan, J., & Aldrich, R. W. (2010). LRRC26 auxiliary protein allows BK channel activation at resting voltage without calcium. *Nature, 466*, 513–516.
Yan, J., & Aldrich, R. W. (2012). BK potassium channel modulation by leucine-rich repeat-containing proteins. *Proceedings of the National Academy of Sciences of the United States of America, 109*, 7917–7922.
Yang, C. T., Zeng, X. H., Xia, X. M., & Lingle, C. J. (2009). Interactions between beta subunits of the KCNMB family and Slo3: Beta4 selectively modulates Slo3 expression and function. *PLoS One, 4*, e6135.
Yang, C., Zeng, X. H., Zhou, Y., Xia, X. M., & Lingle, C. J. (2011). LRRC52 (leucine-rich-repeat-containing protein 52), a testis-specific auxiliary subunit of the alkalization-activated Slo3 channel. *Proceedings of the National Academy of Sciences of the United States of America, 108*, 19419–19424.
Yang, H., Zhang, G., Shi, J., Lee, U. S., Delaloye, K., & Cui, J. (2008). Subunit-specific effect of the voltage sensor domain on Ca^{2+} sensitivity of BK channels. *Biophysical Journal, 94*, 4678–4687.
Ye, H., Jalini, S., Mylvaganam, S., & Carlen, P. (2010). Activation of large-conductance Ca^{2+}-activated K^{+} channels depresses basal synaptic transmission in the hippocampal CA1 area in APP (swe/ind) TgCRND8 mice. *Neurobiology of Aging, 31*, 591–604.
Yu, J. Y., Upadhyaya, A. B., & Atkinson, N. S. (2006). Tissue-specific alternative splicing of BK channel transcripts in Drosophila. *Genes, Brain, and Behavior, 5*, 329–339.
Yuan, P., Leonetti, M. D., Hsiung, Y., & MacKinnon, R. (2012). Open structure of the Ca^{2+} gating ring in the high-conductance Ca^{2+}-activated K^{+} channel. *Nature, 481*, 94–97.
Yusifov, T., Savalli, N., Gandhi, C. S., Ottolia, M., & Olcese, R. (2008). The RCK2 domain of the human BKCa channel is a calcium sensor. *Proceedings of the National Academy of Sciences of the United States of America, 105*, 376–381.
Zarei, M. M., Song, M., Wilson, R. J., Cox, N., Colom, L. V., Knaus, H. G., et al. (2007). Endocytic trafficking signals in KCNMB2 regulate surface expression of a large conductance voltage and Ca^{2+}-activated K^{+} channel. *Neuroscience, 147*, 80–89.
Zavala-Tecuapetla, C., Aguileta, M. A., Lopez-Guerrero, J. J., Gonzalez-Marin, M. C., & Pena, F. (2008). Calcium-activated potassium currents differentially modulate respiratory rhythm generation. *The European Journal of Neuroscience, 27*, 2871–2884.
Zeng, X. H., Benzinger, G. R., Xia, X. M., & Lingle, C. J. (2007). BK channels with beta3a subunits generate use-dependent slow afterhyperpolarizing currents by an inactivation-coupled mechanism. *The Journal of Neuroscience, 27*, 4707–4715.
Zeng, X. H., Xia, X. M., & Lingle, C. J. (2003). Redox-sensitive extracellular gates formed by auxiliary beta subunits of calcium-activated potassium channels. *Nature Structural Biology, 10*, 448–454.
Zeng, X., Xia, X. M., & Lingle, C. J. (2008). Species-specific differences among KCNMB3 BK beta3 auxiliary subunits: Some beta3 N-terminal variants may be primate-specific subunits. *The Journal of General Physiology, 132*, 115–129.
Zeng, X. H., Yang, C., Xia, X. M., Liu, M., & Lingle, C. J. (2015). SLO3 auxiliary subunit LRRC52 controls gating of sperm KSPER currents and is critical for normal fertility. *Proceedings of the National Academy of Sciences of the United States of America, 112*, 2599–2604.
Zhang, G., Huang, S. Y., Yang, J., Shi, J., Yang, X., Moller, A., et al. (2010). Ion sensing in the RCK1 domain of BK channels. *Proceedings of the National Academy of Sciences of the United States of America, 107*, 18700–18705.
Zhang, L., Li, X., Zhou, R., & Xing, G. (2006). Possible role of potassium channel, big K in etiology of schizophrenia. *Medical Hypotheses, 67*, 41–43.
Zhang, X. L., Mok, L. P., Lee, K. Y., Charbonnet, M., & Gold, M. S. (2012). Inflammation-induced changes in BK(Ca) currents in cutaneous dorsal root ganglion neurons from the adult rat. *Molecular Pain, 8*, 37.

Zhang, H., Xie, M., Schools, G. P., Feustel, P. F., Wang, W., Lei, T., et al. (2009). Tamoxifen mediated estrogen receptor activation protects against early impairment of hippocampal neuron excitability in an oxygen/glucose deprivation brain slice ischemia model. *Brain Research*, *1247*, 196–211.

Zhang, J., & Yan, J. (2014). Regulation of BK channels by auxiliary gamma subunits. *Frontiers in Physiology*, *5*, 401.

Zhang, Z., Zeng, X. H., Xia, X. M., & Lingle, C. J. (2009). N-terminal inactivation domains of beta subunits are protected from trypsin digestion by binding within the antechamber of BK channels. *The Journal of General Physiology*, *133*, 263–282.

Zhang, Z., Zhou, Y., Ding, J. P., Xia, X. M., & Lingle, C. J. (2006). A limited access compartment between the pore domain and cytosolic domain of the BK channel. *The Journal of Neuroscience*, *26*, 11833–11843.

Zhao, M. G., Hulsmann, S., Winter, S. M., Dutschmann, M., & Richter, D. W. (2006). Calcium-regulated potassium currents secure respiratory rhythm generation after loss of glycinergic inhibition. *The European Journal of Neuroscience*, *24*, 145–154.

CHAPTER THREE

Posttranscriptional and Posttranslational Regulation of BK Channels

M.J. Shipston[1], L. Tian
Centre for Integrative Physiology, College of Medicine & Veterinary Medicine, University of Edinburgh, Edinburgh, United Kingdom
[1]Corresponding author: e-mail address: mike.shipston@ed.ac.uk

Contents

Abstract

Large conductance calcium- and voltage-activated potassium (BK) channels are ubiquitously expressed and play an important role in the regulation of an eclectic array of physiological processes. Their diverse functional role requires channels with a wide variety of properties even though the pore-forming α-subunit is encoded by a single gene, *KCNMA1*. To achieve this, BK channels exploit some of the most fundamental posttranscriptional and posttranslational mechanisms that allow proteomic diversity to be generated from a single gene. These include mechanisms that diversify mRNA variants and abundance such as alternative pre-mRNA splicing, editing, and control by miRNA. The BK channel is also subject to a diverse array of posttranslational modifications including protein phosphorylation, lipidation, glycosylation, and ubiquitination to control the number, properties, and regulation of BK channels in specific cell types. Importantly, "cross talk" between these posttranscriptional and posttranslational modifications

International Review of Neurobiology, Volume 128
ISSN 0074-7742
http://dx.doi.org/10.1016/bs.irn.2016.02.012

typically converge on disordered domains of the BK channel α-subunit. This allows both wide physiological diversity to be generated and a diversity of mechanisms to allow conditional regulation of BK channels and is emerging as an important determinant of BK channel function in health and disease.

1. INTRODUCTION

The pore-forming α-subunit of large conductance calcium- and voltage-activated potassium (BK) channels is encoded by a single gene in all species (*KCNMA*1 in humans, *Kcnma1* in rodents, *slo* in *Drosophila*) (Contreras et al., 2013; Salkoff, Butler, Ferreira, Santi, & Wei, 2006). BK channels are ubiquitously expressed and control a wide diversity of physiological processes: from control of blood pressure to neuronal excitability and endocrine hormone secretion. Thus, BK channels are implicated in a number of major diseases of the nervous, vascular, and endocrine systems including epilepsy, autism, hypertension, obesity, and stress-related disorders.

While *KCNMA1* represents the only known calcium- and voltage-activated potassium channel gene in any genome, the functional properties of native BK channels vary considerably between tissues and even in distinct subcellular compartments of individual cells. Indeed, endogenous BK channels show considerable diversity in calcium and voltage sensitivity as well as rates of activation, inactivation, and deactivation. Furthermore, BK channel activity can be controlled by an eclectic array of cellular signaling pathways, and the same pathway may have diametrically opposite effects on BK channel function even in the same cell. The key to understanding how such physiological diversity is generated from a single gene represents a fundamental challenge in modern physiology and is largely determined by fundamental mechanisms that control proteomic diversity. Moreover, defining the rules by which BK channels are controlled under normal physiological conditions, and how this may go wrong, is key to our understanding of the role and potential therapeutic utility of these channels in a wide diversity of human disorders.

The assembly of the BK channel pore-forming α-subunit with two classes of transmembrane accessory subunits, the β- and γ-subunits, respectively (see chapter "Modulation of BK Channel Function by Auxiliary Beta and Gamma Subunits" by Yan; Contreras et al., 2013; Orio, Rojas, Ferreira, & Latorre, 2002; Zhang & Yan, 2014), is important determinants of BK channel diversity. However, over the last decade, a diverse array of fundamental post-transcriptional and posttranslational mechanisms, which directly regulate the pore-forming subunit, has also been identified as critical determinants of BK channel functional diversity (Fig. 1). Major posttranscriptional mechanisms that control BK channel mRNA diversity include alternative precursor messenger RNA (pre-mRNA) splicing, RNA editing, and subsequent control of mRNA abundance and translation by microRNAs (miRNAs). BK channels are potently regulated by an array of posttranslational mechanisms, which result in the covalent attachment of a diverse range of moieties. These include addition of phosphate groups (phosphorylation), lipids (lipidation), carbohydrates (glycosylation), and small proteins (ubiquitination). A large body of data not only supports control by these pre- and posttranslational mechanisms, but, more importantly, it also reveals the complex interplay between these mechanisms to specify and diversify BK channel properties and function. For

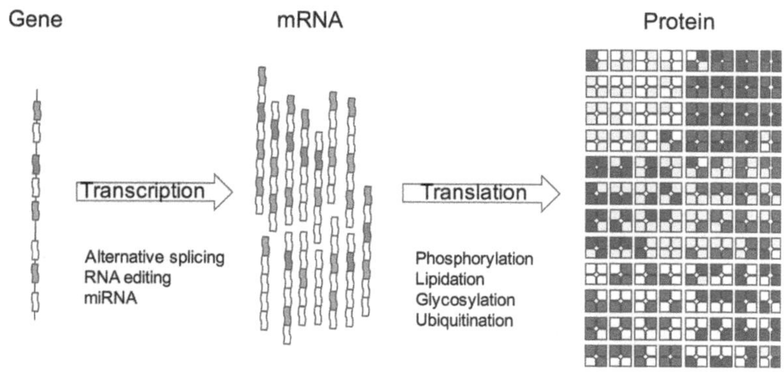

Fig. 1 Posttranscriptional and posttranslational mechanisms generate proteomic and physiological diversity of BK channels. BK channel pore-forming α-subunits are encoded by a single gene (*KCNMA1* in humans). The BK channel gene is composed of multiple constitutively expressed (*open*) exons and a variety of alternatively spliced (*colored*) exons that are interspersed with variable intronic regions in the DNA. Multiple posttranscriptional mechanisms, including alternative pre-mRNA splicing, RNA editing, and miRNA, regulate the expression and diversity of alternatively spliced mRNAs that may be translated into a number of different splice variant polypeptides generating diversity. Posttranslation modifications, such as phosphorylation, lipidation, glycosylation, and ubiquitination, can differentially alter the function of BK channels, which assemble as tetramers of α-subunits, leading to considerable expansion of proteomic and physiological diversity. In addition to these fundamental mechanisms, BK channel α-subunits can also assemble with regulatory β- and γ-subunits as well as an eclectic array of other interacting proteins to further expand diversity.

example, many alternatively spliced inserts introduce sites for posttranslational modification (PTM) such as phosphorylation and lipidation (Fig. 2). Indeed, a common theme is the convergence of posttranscriptional and posttranslational mechanisms on domains of the BK channel, including the S0–S1 intracellular loop and linker between RCK1 and RCK2, that display intrinsic disorder (Peng, Sakai, Kurgan, Sokolowski, & Uversky, 2014). Due to their flexibility, disordered domains are typically not captured in crystal structures and are lacking in the available structures of the BK channel C-terminus (Wu, Yang, Ye, & Jiang, 2010; Yuan, Leonetti, Pico, Hsiung, & MacKinnon, 2010). Intrinsically disordered domains allow dynamic fine-tuning of protein behavior while allowing core properties of proteins, often encoded by more structured domains, which are captured in crystal structures, to be retained. In this regard, as the core properties of BK channel calcium and voltage sensing are

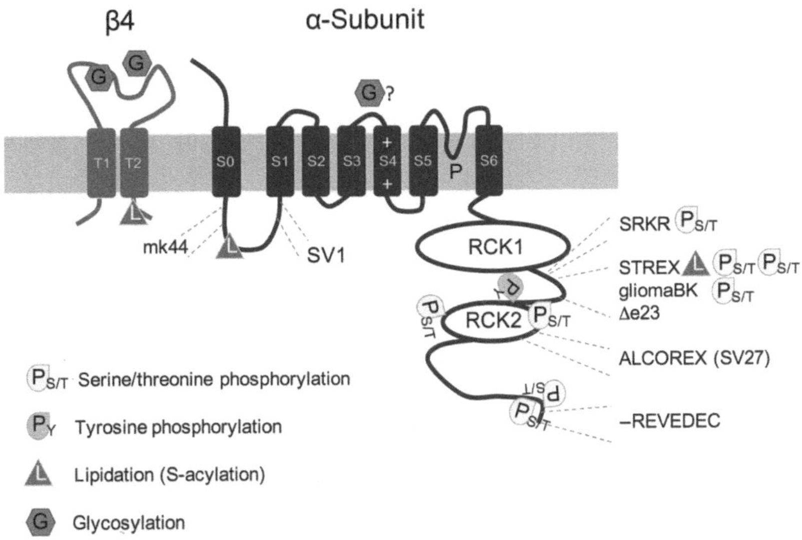

Fig. 2 Selected alternative splice variants and posttranslational modifications of BK channel subunits. Schematic of BK channel α-subunits and the β4-subunit illustrating selected experimentally validated splice variants and posttranslational mechanisms highlighted in this chapter. The pore-forming α-subunit is subject to extensive alternative mRNA splicing in particular in the intracellular C-terminus. Several splice variants introduce sites of posttranslational modification in addition to those identified in constitutive domains of the channel. Selected sites of regulation by members of the AGC family of protein kinases (serine/threonine phosphorylation) or src family tyrosine phosphorylation are indicated, lipidation (S-acylation) and glycosylation are shown for both the α-subunit and β4-subunit. Sites of ubiquitination are poorly defined and are not illustrated.

complex, yet highly conserved mechanisms, the convergence of many of these mechanisms in intrinsically disordered domains likely allows introduction/removal of regulatory mechanisms without disrupting the core properties of channel function.

1.1 Spatiotemporal Control of BK Channel Number and Activity

In any consideration of the functional role of posttranscriptional and post-translational regulation of BK channels, it is always important to bear in mind the spatiotemporal control of channel number and activity (Fig. 3). The physiological function of a BK channel in any system depends crucially on both the number and the activity of individual channels at the target membrane. Important in this regard is that channel number can be controlled by a variety of mechanisms from gene expression through channel protein expression and assembly to trafficking, recycling, and final degradation of the channel protein. Once at the target membrane, the activity and gating characteristics of the channel are imperative for function. Importantly, posttranscriptional and posttranslational mechanisms can control multiple facets of the BK channel life cycle. In many cases, there is an overlap between the time domains over which different mechanisms that control

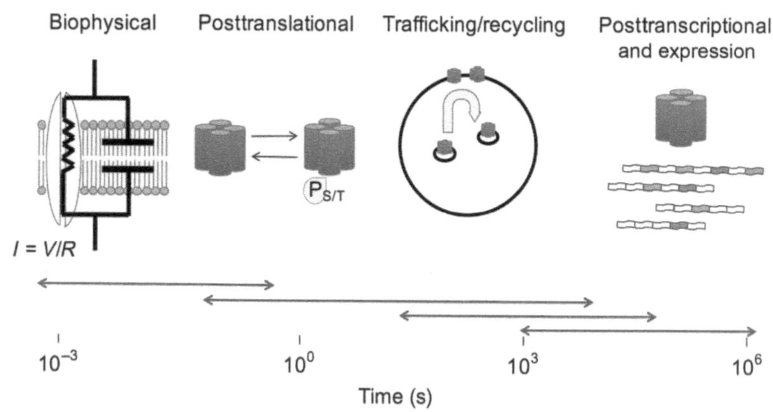

Fig. 3 Schematic illustrating the overlapping time domains of BK channel regulation. Biophysical regulation of ion channels operates on the submillisecond to second timescale (eg, rates of activation and deactivation in response to a voltage step). Posttranslational regulation can span a broad time domain from several milliseconds to control of activity and function over hours. Trafficking and recycling of ion channels at target membranes can operate not only on fast (second) timescales but also in much longer time domains. While transcription and posttranscriptional mechanisms per se can be rapid, the functional impact on channel expression, properties, and regulation is generally much slower, on the order of tens of minutes.

channel number or activity operate (Fig. 3). For example, trafficking is often considered a relatively slow process (minutes to hours); however, recycling of BK channel subunits at the plasma membrane can occur on the timescale of seconds, dependent upon the cellular system (Leo et al., 2014). Conversely, PTM is generally assumed to control channel function on a shorter timescale. However, PTMs can control channel function from the millisecond to hours timescale. In addition to the temporal control of function, the spatial localization of BK channels is also important for function. For example, whether BK channels are localized close to, or distant from an extracellular or intracellular source of calcium, or localized in different cellular microenvironments will determine their functional role. Thus, mechanisms that determine either the temporal or spatial control of BK channels may have a profound effect on channel function. It should be noted that in most cases we know relatively little about the spatiotemporal dynamics of BK channel control.

2. POSTTRANSCRIPTIONAL REGULATION

Most eukaryotic genes have multiple short exons, interspersed with introns of highly variable length. After being transcribed, exons must be spliced together to produce functional mRNA that is subsequently translated into protein. In mammalian species, the BK channel α-subunit is encoded by >30 exons with the gene spanning several megabases of DNA (Fodor & Aldrich, 2009). The three major posttranscriptional mechanisms that control mRNA abundance and diversity: alternative pre-mRNA splicing, RNA editing, and control by miRNAs are all exploited to diversify BK channel properties. Of course, additional pretranscriptional mechanisms including alternative promoter usage may also be used to further diversify mRNA transcripts.

2.1 Alternative Pre-mRNA Splicing

Eukaryotic genes are transcribed into a pre-mRNA that is processed by the spliceosomal machinery to excise intervening RNA sequences (introns) from the exons, which are spliced together to form the mature mRNA (Fig. 1). Alternative splicing can be dynamically regulated and allows multiple mRNAs, and thus multiple polypeptide isoforms, to be generated from the same gene. Alternative pre-mRNA splicing is now widely recognized as the "norm" for most eukaryotic genes and represents one of the most fundamental posttranscriptional mechanisms to generate physiological diversity.

In mammals, the large majority of pre-mRNAs are subject to alternative splicing events and splicing is particularly prevalent in ion channels and other signaling proteins.

2.1.1 Control of Alternative Splicing of the Pore-Forming α-Subunit

Since the cloning of the *Drosophila* and mouse genes encoding BK channels in the early 1990s, alternative splicing has been recognized as an important mechanism to generate functional diversity of the pore-forming α-subunit from worms to man (Atkinson, Robertson, & Ganetzky, 1991; Butler, Tsunoda, McCobb, Wei, & Salkoff, 1993). More than 1500 distinct splice variants may be predicted to be generated from *KCNMA*1, although additional mechanisms, as yet not fully understood, likely limit the number of exon combinations in transcripts (Glauser, Johnson, Aldrich, & Goodman, 2011). Constitutive exons are largely conserved between vertebrates and invertebrates and typically encode core features such as the pore, voltage sensor, and calcium-binding domains. Alternatively, spliced exons are poorly conserved between vertebrates and invertebrates; however, *sites* of alternative splicing are largely conserved across phyla. These sites of splicing are typically located at functionally established domain boundaries with multiple sites of splicing distributed across the entire coding sequence of the intracellular C-terminus (Fodor & Aldrich, 2009). This suggests that alternative splicing is important for controlling communication between functional domains of BK channels (Johnson et al., 2011). Indeed, alternative splicing has been reported to modify multiple aspects of BK channel properties and function including: gating, functional coupling to accessory subunits, trafficking to target membranes, regulation by PTMs, and differential control by other signaling molecules. Importantly, alternative splicing is highly dynamically controlled both in space and time.

One of the best-studied examples of alternative splicing controlling such diverse properties and functions of BK channels is the alternatively spliced *st*ress-*re*gulated *ex*on (STREX) that is highly conserved in vertebrates. STREX is one of a number of exons that can be included at a single site of splicing in the disordered flexible linker between the two intracellular Regulator of Conductance of K^+ (RCK) domains in the intracellular C-terminus (Figs. 2 and 4). Inclusion of the STREX exon is dynamically regulated during development and its expression is highly plastic in adult cells through hormonal control, as well as intrinsic cell excitability and calcium signaling (Xie & Black, 2001; Xie & McCobb, 1998). Moreover, studies in bovine chromaffin cells clearly demonstrate how the complex

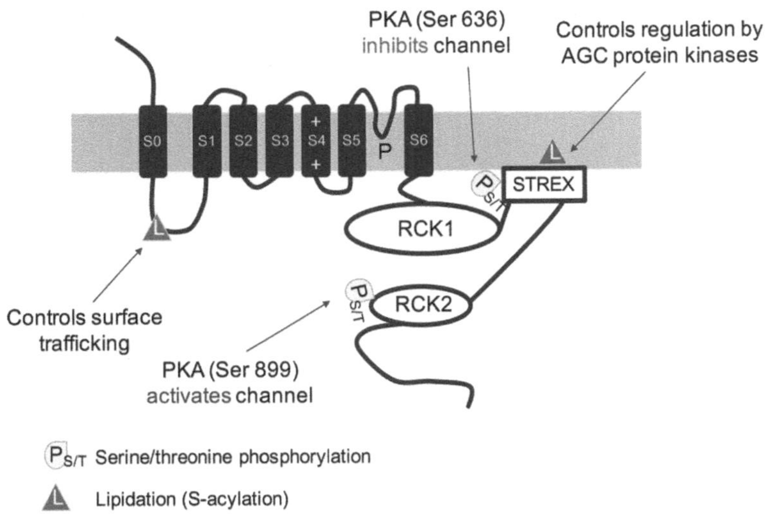

Fig. 4 A posttranslational modification (PTM) can control distinct BK channel properties, or have diametrically opposite effects, by regulating distinct domains of the α-subunit. Schematic of the pore-forming α-subunit of BK channel illustrating the STREX spice variant. The functional effect of phosphorylation or lipidation (S-acylation) on channel properties and function is dependent upon the domain in the α-subunit that is regulated. For example, S-acylation of the S0–S1 loop controls channel trafficking and surface expression, whereas inclusion of the alternatively spliced STREX insert introduces an additional site of S-acylation that determines STREX channel regulation by AGC family protein kinases. The same PTM can also regulate the channel in diametrically opposite ways as exemplified by protein kinase A (PKA)-dependent phosphorylation. PKA phosphorylation of the conserved C-terminal Ser 899 site promotes BK channel activation. However, inclusion of STREX introduces an additional PKA site (Ser 636) that acts in a dominant fashion resulting in inhibition of STREX channels by PKA. This exemplifies the conditional nature of BK channel regulation by the cross talk of both posttranscriptional and posttranslational mechanisms.

interplay between multiple hormone systems can dynamically control STREX splicing and cellular excitability (Lai & McCobb, 2002). In this system, exposure to adrenal glucocorticoids suppresses STREX inclusion, whereas adrenal androgens, including dehydroepiandrosterone, androstenedione, and testosterone, promote STREX inclusion.

For most splicing events in BK channels the cellular mechanisms controlling splicing are very poorly understood. However, at least for STREX, we are beginning to gain some insights into the repression of STREX inclusion upon activation of calcium-/calmodulin-dependent protein kinase IV (CaMKIV) as a result of cellular depolarization and influx of calcium (Xie & Black, 2001). CaMKIV phosphorylates a single serine, Ser-513, in the fourth RNA recognition binding motif in heterogeneous ribonucleoprotein L (hnRNP L1), which enhances interaction of hnRNP L with the CamKIV-responsive RNA element in the STREX exon. This inhibits binding of the large subunit of the U2 auxillary factor U2AF65, thereby inhibiting spliceosomal assembly (Liu et al., 2012). hnRNP L is upregulated in many brain cancers, which may explain the observation that STREX expression is repressed in these cancers whereas other alternatively spliced exons, such as the gliomaBK exon that can be included at the same site of splicing, are upregulated (Liu, Chang, Reinhart, Sontheimer, & Chang, 2002).

2.1.2 Alternative Splicing Controls the Biophysical Properties of BK Channels and Their Regulation by PTMs, Accessory Subunits, and Other Signaling Molecules

Inclusion of the STREX exon results in dramatic changes in BK channel phenotype including slowing of deactivation and a significant leftward shift in the voltage for half-maximal activation (Chen et al., 2005; Xie & McCobb, 1998). The mechanistic basis for these effects is poorly understood, but likely reflects the effect of STREX inclusion on the conformation of the disordered linker between RCK1 and RCK2.

In common with alternatively spliced inserts in many proteins, the STREX insert introduces additional sites of PTM, including for phosphorylation and S-acylation (Figs. 2 and 4). Inclusion of the STREX insert switches channel regulation by AGC family protein kinases (the cAMP-dependent (PKA), cGMP-dependent (PKG), and diacylglycerol/calcium-activated (PKC) protein kinases, respectively). For example, STREX inclusion results in channels that are inhibited, rather than activated, by cAMP-dependent protein kinase A (PKA) through introduction of an additional PKA consensus site (Tian et al., 2004, 2001). STREX inclusion also conditionally determines inhibition by protein kinase C (PKC), most likely by controlling phosphorylation of a PKC site downstream of the site of STREX inclusion (Zhou et al., 2012). Indeed, a common feature of alternative splicing in the BK channels is to act as a switch to determine control by reversible protein phosphorylation, as exemplified in other C-terminal splice variants such as those containing the gliomaBK or SRKR inserts, and the bovine BK isoforms A and B (Liu et al., 2002; Shelley, Whitt, Montgomery, & Meredith, 2013; Zhou et al., 2001). This likely reflects the high prevalence of splicing sites within disordered domains of the channels (Peng et al., 2014) that are also targets for PTMs.

Alternative splicing also controls channel regulation by a wide diversity of other signaling molecules. Inclusion of the STREX insert confers sensitivity to inhibition by acute hypoxia, through a mechanism that is independent of redox regulation and that requires highly conserved cysteine residues in a Cys-Ser-Cys motif located downstream of the *S*-acylated cysteines (McCartney et al., 2005). Inclusion of the STREX insert also increases the sensitivity of BK channels to inhibition by oxidation. In rat BK channels, expression of the STREX α-subunit variant in HEK293 cells results in channels that are potently inhibited by the oxidizing agent thimerosal. In contrast, α-subunits lacking the STREX insert are only weakly sensitive to oxidation (Erxleben et al., 2002). This differential sensitivity to oxidation may be a consequence of the high density of cysteine amino acids in the STREX. However, as STREX is included immediately upstream of the cluster of amino acids –CXXCH– important for heme regulation and behaves as a thiol/disulphide switch (Yi, Morgan, & Ragsdale, 2010), inclusion of STREX may affect REDOX sensitivity of this motif. STREX inclusion also confers activation by cerebrosides that are active components of the Chinese herbal medicine, Baifuzi, and may provide protection against ischemic stroke (Chi et al., 2010; Zhou et al., 2014). Conversely, STREX inclusion reduces the acute stimulatory effect of alcohol on BK channel activity, with variants lacking STREX being robustly activated by alcohol (Pietrzykowski et al., 2008). Other C-terminal variants also confer sensitivity to other signaling pathways. For example, unoprostone, a synthetic docosanoid that is used in glaucoma therapy to regulate intraocular pressure, is a potent activator of human BK channels that contain a C-terminal 27 amino acid insert known as *alco*hol-*r*egulated *ex*on (ALCOREX, see below) located immediately upstream of the calcium bowl (Yu et al., 2015). BK channels that lack this insert are largely insensitive to unoprostone, in general, accordance with the finding that unoprostone activates BK channels only in the presence of calcium without affecting voltage sensitivity. Inclusion of this same insert also results in channels being inhibited by nonhydrolyzable ATP analogues (Clark, Hall, & Shipston, 1999) and generates channels with faster activation kinetics (Ha et al., 2000). Splicing can also modify the functional coupling with regulatory subunits. Inclusion of STREX modifies distinct aspects of the functional coupling with regulatory β4-subunits dependent upon local calcium and phosphorylation status (Petrik & Brenner, 2007). Different N-terminal splice variants also modify functional coupling with β1-subunits (Erxleben et al., 2002; Lorca et al., 2014).

2.1.3 Splicing and Control of BK Channel Trafficking

Alternative splicing plays an important role in controlling cell surface expression, as well as endomembrane targeting of BK channels; however, the mechanisms controlling trafficking are diverse. The simplest mechanisms include splicing events resulting in channels that are either prematurely truncated and/or subjected to endoproteolytic cleavage. Two examples from the N- and C-terminal domains of BK channels, respectively, exemplify this process. Inclusion of the N-terminal mk44 alternatively spliced insert in the S0–S1 intracellular loop results in endoproteolytic cleavage of the channel just after S0 (Korovkina, Fergus, Holdiman, & England, 2001). The N-terminal proteolytic fragment is trafficked to the plasma membrane; however, the much larger C-terminal region of the channel from S1 is retained in the ER. In myometrial smooth muscle, an elevation of intracellular calcium results in trafficking of the C-terminal region to the plasma membrane where it can now interact with the N-terminal domain to reconstitute functional BK channel activity.

A splice variant insert (Δe23) located at the same site of splicing as STREX between the RCK domains results in a frameshift and introduction of a premature stop codon (Chen et al., 2005). This results in channels lacking the C-terminus following RCK1 and acts as a dominant negative of cell surface expression and may be important for preventing BK channel surface expression in cells expressing high levels of the truncated variant such as cardiac myocytes (Chen et al., 2005). Other variants at this site can disrupt acidic trafficking motifs within the RCK1–RCK2 linker and prevent BK channel exit from the endoplasmic reticulum (ER; Chen et al., 2010).

Other mechanisms include incorporation of splice inserts that introduce additional trafficking signals. For example, inclusion of the SV1 variant in the N-terminal S0–S1 loop results in inclusion of an ER retention signal (–CVLF–) that prevents efficient export from the ER (Zarei et al., 2001, 2004). A large number of additional C-terminal splice variants, including variants that modify the extreme C-terminus, introduce or modify a variety of trafficking motifs that control cell surface expression (Kim, Ridgway, Zou, Chiu, & Dryer, 2007; Ma et al., 2007). Importantly, these C-terminal variants, including the variant ending in –REVEDEC also determine whether regulatory β4-subunits can enhance or suppress surface expression of BK channels (Chen et al., 2013). In addition to controlling surface expression, recent data have also revealed a role for alternative splicing in controlling targeting of BK channel variants to other subcellular compartments in a cell-specific fashion. In adult cardiac myocytes, the

C-terminal –REVEDEC splice variant is one of three C-terminal splice variants expressed. Overexpression of the –REVEDEC variant, but not the other two variants, results in specific targeting of BK channels to mitochondria in rat cardiomyocytes (Singh et al., 2013). An intriguing variation of splicing controlling subcellular localization of BK channels is the identification of an intron-retaining splice variant in rat hippocampal neurones (Bell et al., 2008). In this system, retention of intron 16 (i16 variant) was found in ~10% of total BK channel α-subunit mRNA transcripts. Intron 16 precedes the exons encoding the cytoplasmic linker between RCK1 and RCK2. The cytoplasmic i16 variant mRNA located in puncta in the cell body and was expressed in a gradient across the somatodendritic compartment, with highest expression in the first 50 μm of the proximal segment of the dendrite. Intriguingly, specific knockdown of i16 containing mRNA transcripts in hippocampal neurones reduced the amount of BK channel α-subunit protein colocalizing with the actin cytoskeleton in dendritic spines. This effect on dendritic BK channel localization was associated with a dramatic increase in action potential reducing the maximum number of action potentials evoked by a depolarizing stimulus. As accommodation was enhanced as depolarization was increased, this suggests that the i16 intron-retaining transcripts controls the ability of hippocampal neurones to encode different stimulus levels (Bell et al., 2008).

2.1.4 Splicing and the Control of Regulatory β-Subunits

Although the effects of alternative pre-mRNA splicing on the properties of the pore-forming α-subunit have been well explored, the role of alternative splicing of regulatory β- or γ-subunits is relatively poorly understood. However, alternative splicing can have dramatic effects on regulatory subunits, as exemplified for the functional diversity of β2- and β3-subunits. In neurones and many endocrine cells, fast-inactivating BK channels result from coassembly of the α-subunit with regulatory β2-subunits. Two splice variants of β2 have been identified in human and rodent pancreas, with the β2a variant conferring fast inactivation whereas β2b lacks inactivation (Ohya, Fujimori, Kimura, Yamamura, & Imaizumi, 2010). A similar situation is seen with β3-subunits, which have at least four distinct splice variants that modify the short intracellular N-terminus (Xia, Ding, & Lingle, 1999). Splicing includes or excludes an N-terminal domain that confers rapid inactivation on BK channels that assemble with β3 (Hu et al., 2003). Thus β3-containing BK channels can display nonactivating or rapidly inactivating currents dependent upon the β3 splice variant expressed.

2.2 miRNA Control of Abundance and Diversity

The posttranscriptional control of mRNA abundance is tightly regulated by small (19–25 nt) noncoding miRNAs belonging to a larger class of endogenous repressors of gene expression. miRNAs bind to miRNA recognition elements located in the 3′ untranslated region (UTR) of target mRNAs. In cultured hypothalamic and striatal neurons, miRNA-9 (miR-9) is upregulated by exposure to alcohol resulting in selective suppression of mRNAs containing a specific 3′ UTR. This results in a reduction in the complexity of mRNA splice variants expressed, including loss of variants containing a C-terminal splice insert (ALCOREX) (Pietrzykowski et al., 2008). The resultant remodeling of mRNA splice variants is thought to contribute to the development of tolerance to alcohol by favoring expression of BK channel STREX splice variants. STREX channels are relatively insensitive to alcohol, whereas channels not including the STREX insert are robustly activated by alcohol in this system. Thus, miRNA-mediated posttranscriptional regulation of BK channel splice variant expression may provide a mechanism for drug adaptation and cellular plasticity in response to a variety of environmental challenges (Pietrzykowski et al., 2008; Tatro et al., 2013). Other miRNAs are likely to target BK channels. Indeed, in ovarian cancer cells, upregulation of miR-31 is correlated with a decrease in *Kcnma*1 expression (Samuel et al., 2015).

2.3 RNA Editing

RNA editing of pre-mRNAs is prevalent in ligand- and voltage-activated ion channels involved in the control of neuronal excitability. RNA editing results in the hydrolytic deamination of adenine to inosine residues (A–I editing) in pre-mRNAs carried out by adenosine deaminases acting on RNA (ADARs). As inosines prefer to base pair with cytosine, the inosine is "read" by the cellular translational machinery as guanosine. This allows recoding of the resultant open reading frame to generate proteins with altered function/properties. Multiple A–I editing sites have been reported in *Drosophila* BK channels, modifying both untranslated and coding regions of the mRNA. One site in the coding region modifies the amino acid sequence of the extracellular vestibule of the channel pore (Graveley et al., 2011; Hoopengardner, Bhalla, Staber, & Reenan, 2003). However, RNA editing in mammals is significantly reduced compared to invertebrates and thus the extent to which A–I editing is important for functional variation in mammalian BK channels is poorly described and understood (Graveley et al., 2011; Hoopengardner et al., 2003).

3. POSTTRANSLATIONAL MODIFICATION

More than 200 different PTMs of proteins have been defined and the pore-forming α-subunit of BK channels is subject to an eclectic array of different PTMs that control channel properties and function. Here, we outline four major modes of PTM in mammalian BK channels as exemplars of the diversity of regulatory mechanisms and their impact on channel function and behavior: protein phosphorylation, lipidation, glycosylation, and ubiquitination. These examples illustrate the functional effects of adding diverse moieties to the BK channel from small highly charged phosphate groups (phosphorylation) through lipids (lipidation) and carbohydrates (glycosylation) to small proteins (ubiquitination; Figs. 1 and 2). Furthermore, these exemplars begin to illustrate the complex "cross talk" between these diverse PTMs in the control of BK channels with some PTMs being conditionally dependent upon either other PTMs or mechanisms of posttranscriptional control.

The discussion below focuses on the control of BK channels by direct PTM regulation of the pore-forming and regulatory subunits in mammalian systems. However, as BK channels can assemble with an eclectic array of other proteins (see chapters "Modulation of BK Channel Function by Auxiliary Beta and Gamma Subunits" by Yan and "Protein Network Interacting with BK Channels" by Kim; Berkefeld, Fakler, & Schulte, 2010; Lu et al., 2005; Singh et al., 2016) the convergence of signaling pathways on these proteins is likely an important, yet poorly explored, potential mode of regulation.

3.1 Reversible Protein Phosphorylation

Reversible protein phosphorylation represents one of the most ubiquitous PTMs to control protein function. Indeed, there are more than 500 genes encoding protein kinases (Manning, Whyte, Martinez, Hunter, & Sudarsanam, 2002), and more than 200 genes encoding protein phosphatases (Sacco et al., 2012), in the human genome, supporting the widespread use of this mechanism in controlling diverse aspects of physiology. Discussion of the full extent to which different protein kinases and protein phosphatase signaling pathways can control BK channel activity or trafficking is beyond the scope of this chapter. However, some key exemplars illustrate the diversity and complexity of BK channel regulation by reversible phosphorylation. Indeed, not only may the same signaling pathways have diametrically opposite effects on BK channel trafficking and/or activity, but phosphorylation also interacts widely with both posttranscriptional and other posttranslational mechanisms to diversify BK channel properties (Figs. 1, 2, and 4).

3.1.1 Regulation by Reversible Serine–Threonine Protein Phosphorylation

3.1.1.1 Control of Channel Activity via Regulation of Pore-Forming α-Subunits

The classical serine/threonine AGC family of protein kinases and their cognate protein phosphatases have long been recognized as important modulators of BK channel function. The AGC family includes PKA, the calcium-, and diacylglycerol-activated PKC family, and cGMP-activated protein kinase G (PKG). Cognate phosphatases included protein phosphatase 1 (PP1) and protein phosphatase 2A (PP2A).

Indeed, some of the very first evidence for ion channel regulation by phosphorylation came from the pioneering work on BK channels in *Helix aspersa* (Ewald, Williams, & Levitan, 1985) with evidence for a key role of dephosphorylation from later studies of BK channels in endocrine cells and reconstituted BK channels from rat brain incorporated into bilayers (Reinhart & Levitan, 1995; White et al., 1993; White, Schonbrunn, & Armstrong, 1991).

In a large number of different systems, BK channels are activated by PKA-dependent phosphorylation; however, there are clear exceptions as in some neurones and many endocrine cells native BK channels are inhibited by PKA. Using largely site-directed mutagenesis approaches coupled with radioactive phosphate incorporation assays, one explanation for the molecular basis of this divergent regulation has been revealed to involve distinct PKA sites of phosphorylation being included as a result of alterative pre-mRNA splicing. In mammalian systems, BK channels are phosphorylated by PKA at a conserved C-terminal serine (Ser 899 in mouse BK channel) that is located within a short unstructured sequence near the calcium bowl of RCK2 (Nara, Dhulipala, Wang, & Kotlikoff, 1998; Tian et al., 2004, 2001; Fig. 4). Site-directed mutagenesis of this site prevents PKA-dependent phosphorylation of channels and loss of activation of BK channels by PKA. Intriguingly, as a tetrameric protein complex, PKA-dependent activation of BK channels follows an "all or nothing" rule in which the PKA site in each subunit of the tetramer has to be phosphorylated for channel activation (Tian et al., 2004; Fig. 5). This mechanism has some interesting consequences as it may explain why in many systems, such as Purkinje neurones, PKA-mediated phosphorylation only transiently activates the channel

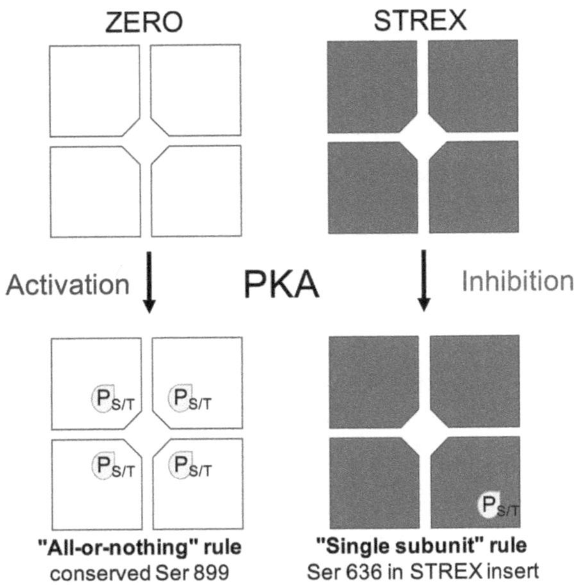

Fig. 5 BK channel as a phosphorylation abacus: increasing diversity by exploiting tetrameric assembly of α-subunits. Schematic of the tetrameric assembly of ZERO and STREX splice variant α-subunits to generate functional channels. Protein kinase A phosphorylation of a single conserved PKA site (Ser 899) in the C-terminus of the ZERO variant stimulates channel activation. Channel activation is only achieved when each subunit of the tetramer is phosphorylated. Thus, PKA activation follows an "all or nothing" rule. In contrast, PKA phosphorylation of Ser 636 located in the alternatively spliced STREX exon in the C-terminus results in inhibition of STREX channel activity. Inhibition only requires phosphorylation of a single subunit and thus follows a "single subunit" rule. Thus, BK channels can perform computation by acting as phosphorylation abacus. This has important functional consequences for regulation as it would be predicted that PKA activation is more sensitive to phosphatase regulation and that PKA activation is, in general, more transient than the more sustained PKA inhibition (ie, because of the "all or nothing" rule of PKA activation, a phosphatase would only have to dephosphorylate one subunit in the tetramer to prevent activation).

(Widmer, Rowe, & Shipston, 2003). In this case, the model would predict that a cognate phosphatase would reverse channel activation by only having to remove phosphate from one subunit. That is, the balance of phosphatase to PKA regulation favors phosphatase control.

Conversely, inhibition of BK channels by PKA is largely restricted to BK channel variants that include the STREX splice variant insert (Chen et al., 2005; Tian et al., 2001). Inclusion of STREX in murine BK channels introduces an additional serine residue (Ser 636) for PKA phosphorylation at the start of the *S*-acylated STREX insert (Figs. 2 and 4). Thus the *functional* effect of PKA phosphorylation of the α-subunit is dependent upon the channel domain being phosphorylated. This may sound intuitive but of course has major implications for our ability to *interpret* global proteomic assays of PTM of the BK channel, as for any other protein.

Importantly, analysis of channel phosphorylation stoichiometry reveals that only one STREX insert in the tetramer needs to be phosphorylated for channel inhibition (Tian et al., 2004). This "single subunit" rule (Fig. 5) may also explain why PKA inhibition is typically sustained—in this case, a phosphatase would have to ensure all four subunits are dephosphorylated. From this simple set of "all or nothing" (activation) and "single subunit" (inhibition) rules, one might predict that heterotetramers of channels with STREX subunits would also be inhibited. However, it turns out that in heterotetramers assembled from α-subunits containing the STREX insert together with α-subunits lacking this insert (ZERO variant) PKA phosphorylation of STREX results in inhibition only when the conserved C-terminal PKA site is dephosphorylated in all subunits. This makes physiological sense due to the "noise" of alternative splicing—if this rule did not apply, most BK channels would be inhibited in cells expressing even a low abundance of STREX splice variants. This likely provides a mechanism to allow discrete activation or inhibitor phenotypes to be conferred, as can be seen in the same cell. Thus, alternative splicing acts as a molecular switch to specify regulation by the same signaling pathways.

Such conditional regulation of phosphorylation is a recurrent and important theme in BK channel regulation. Regulation of BK channels by other serine/threonine kinases such as PKC and AMP-activated protein kinase (AMPK) is also dependent upon the interaction between alternative splicing and phosphorylation. In most, but not all systems, activation of PKC results in BK channel inhibition. For most variants of BK channels, PKC-dependent inhibition requires phosphorylation of two conserved PKC phosphorylation sites in the C-terminus (Zhou et al., 2010). Importantly, this regulation reveals another type of conditional regulation. In this case, phosphorylation of the extreme C-terminal PKC site (Ser 1151) is dependent upon the prior phosphorylation of a conserved serine at Ser 695. An additional level of control is also conferred by inclusion of the STREX insert upstream of Ser 695. In this case, inclusion of an *S*-acylated STREX insert prevents PKC inhibition likely by precluding phosphorylation of Ser 695. However, if STREX is deacylated, Ser 695 is accessible and thus PKC can now inhibit STREX variant channels (Zhou et al., 2012).

Another example is provided by studies analyzing the regulation of BK channels by AMPK (Ross et al., 2011). In HEK293 cells, AMPK inhibits ZERO but not STREX splice variant channels. ZERO and STREX channels are phosphorylated by AMPK at a conserved, yet unidentified, phosphorylation site. In addition, STREX channels are phosphorylated by AMPK at an additional site (Ser 657) within the alternatively spliced STREX insert. This suggests that phosphorylation of the Ser 657 site prevents the functional inhibitory effect at the conserved site. Indeed, mutation of Ser 657 to alanine not only reduced the stoichiometry of STREX channel phosphorylation back to that of ZERO alone but also restored the inhibitory effect of AMPK phosphorylation on STREX channel activity. Thus, Ser 657 phosphorylation by AMPK appears to prevent the functional effect of AMPK phosphorylation on sites outside of the STREX insert. These conditional modes of phosphorylation regulation, dependent upon the presence or absence of specific alternatively spliced inserts, have emerged as a very common theme in BK channels. Indeed, several additional C-terminal BK channel variants determine how phosphorylation impacts channel activity and function (Shelley et al., 2013; Zhou et al., 2001).

Phosphorylation may also conditionally control BK regulation by other, distinct, signaling pathways. For example, in pituitary GH4C1 cells, PKA- and PKG-dependent phosphorylation of endogenous BK channels differentially controls their sensitivity to the gasotransmitter, H_2S (Sitdikova, Fuchs, Kainz, Weiger, & Hermann, 2014). In GH4C1 cells, prior phosphorylation of endogenous channels with PKG results in a "priming" effect that promotes subsequent BK channel activation in response to H_2S. Conversely, prior phosphorylation by PKA blunts the ability of H_2S to activate the same BK channels. These effects of phosphorylation are likely to be dependent on the molecular composition of BK channels in cells as PKA did not prevent the stimulatory action of H_2S on endogenous BK channels from the related GH3 cell line (Sitdikova et al., 2014). However, how phosphorylation controls sensitivity to H_2S, perhaps by controlling conformations of the channel that are sensitive to H_2S regulation, remain to be defined.

BK channels are also regulated by a number of other serine/threonine kinases with sites of phosphorylation largely conserved in mammalian species. In most systems investigated, activation of PKG results in stimulation of BK channel activity. While this regulation has been long recognized (Alioua et al., 1998), specific sites of PKG phosphorylation on the BK channel have only been recently revealed unequivocally. Phosphoproteomic analysis of BK channels expressed in rat A7r5 myocytes revealed multiple sites of PKG phosphorylation including sites previously implicated in the cytosolic C-terminus including Ser 691, Ser 873, and Ser 1112 (Kyle, Hurst, Swayze, Sheng, & Braun, 2013). Site-directed mutagenesis supports a model in which all three sites appear to be required for functional activation by PKG. Intriguingly, Ser 873 in rat channels is analogous to the C-terminal site phosphorylated by PKA (see Ser 899 earlier) and is also required for PKA-mediated activation of channel activity in other species (Nara et al., 1998; Tian et al., 2004). In contrast, Ser 1112 is located toward the extreme C-terminal domain immediately upstream of a site of alternative splicing and previously reported to be required for channel activation by PKG (Zhou et al., 2001). While other serine/threonine kinases, such as calcium-/calmodulin-dependent kinase II (CaMKII), have also been reported to control BK channel function, at least some of these effects are species specific and depend on the evolutionary conservation of Ser/Thr residues. For example, in bovine BK channels, phosphorylation of Thr 107 in the intracellular S0–S1 loop by CaMKII results in channel activation and modifies the response of the channel to regulation by alcohol (Liu, Asuncion-Chin, Liu, & Dopico, 2005). However, Thr 107 is not conserved in mouse or human BK channels and thus represents a species-specific mode of control. Moreover, other kinase signaling pathways such as ERK and p38 MAPK have been reported to control BK channel activity but whether this is through direct phosphorylation of the BK channel per se remains unclear (Li et al., 2006).

The above examples of serine/threonine phosphorylation generally imply that control is through discrete sites and confers an essentially "*digital*" mode of regulation by protein kinases. However, phosphoproteomic mapping of rat brain BK channels reveals a highly complex and extensive phosphorylation state of native BK channels with more than 30 Ser/Thr residues basally phosphorylated in rat brain (Yan et al., 2008). Although the kinases/phosphatases that control this multitude of novel sites are poorly characterized, site-directed mutagenesis of some of these key residues also reveals an *analogue* mode of BK channel regulation. In this case, increasing the number of sites that can be phosphorylated results in a *graded* shift in voltage for half-maximal activation of the channels. Thus, by acting as a phosphorylation "abacus," the BK channel may be tuned both digitally and in an analogue fashion by distinct phosphorylation cascades.

3.1.1.2 Control of Regulatory Subunits by Serine/Threonine Phosphorylation

In contrast to the extensive regulation of α-subunits by phosphorylation, the role of phosphorylation in controlling regulatory subunit function is poorly understood. Several studies have suggested that the presence of the β1- or β4-subunit can modify BK channel regulation by phosphorylation (Dworetzky et al., 1996; Petrik & Brenner, 2007; Velàzquez-Marrero, Seale, Treistman, & Martin, 2014). For example, coexpression of the β4-subunit with the ZERO channel variant enhances PKA-mediated activation of the channels in isolated inside-out patches from HEK293 cells (Petrik & Brenner, 2007; Velàzquez-Marrero et al., 2014) and reported to increase STREX variant sensitivity to dephosphorylation by alkaline phosphatase (Petrik & Brenner, 2007). More intriguingly, while site-directed mutation of the proposed PKA consensus serine (S899 in ZERO) to alanine abolishes PKA activation of the ZERO variant expressed alone, PKA is still able to activate channels where the S899A ZERO and β4-subunit are coexpressed (Velàzquez-Marrero et al., 2014). Whether this results from β4-subunits per se being a target for PKA phosphorylation in this system, or assembly with β4 allows other sites on the ZERO α-subunit to be made available for regulation by PKA, remain to be determined. Moreover, in these same studies the presence of β4 switched the effect of inhibiting CaMKII activity in patches on channel activity. With ZERO alone, pharmacological inhibition of CaMKII resulted in channel activation. In contrast, inhibition of CaMKII inhibited ZERO channels coassembled with β4 (Velàzquez-Marrero et al., 2014). Taken together with the PKA data, this suggests that the presence of β4 can dramatically affect how BK channels are regulated by phosphorylation.

However, a direct role for phosphorylation of regulatory subunits per se is limited. Studies with the human β4-subunit suggest that multiple phosphorylation sites on β4 determine functional coupling with α-subunits (Jin, Weiger, Wu, & Levitan, 2002). For example, site-directed mutagenesis of three residues (Thr 11 and Ser 17 in the N-terminus and Ser 210 in the C-terminus) to alanine prevents the shift in BK channel activation kinetics and voltage dependence that results from inhibition of an okadaic-sensitive protein phosphatase in HEK293 cells. This alanine mutation reduces, but does not completely eliminate β4 phosphorylation in HEK293 cells but has no effect on the ability of the β4-subunit to coimmunoprecipitate with the α-subunit. Conversely, phosphomimetic mutation of these residues (to negatively charged Asp or Glu residues) suppresses the effects of the β4-subunit on channel activation, deactivation, and voltage dependence.

These data suggest that phosphorylation of these residues controls functional, but not physical, coupling to α-subunits. Whether these sites are important for the changes in PKA, or CaMKII phosphorylation upon assembly of β4-subunits with α-subunits, discussed earlier, remain to be determined.

3.1.1.3 Control of Channel Trafficking

Relatively few studies have examined the role of phosphorylation in dynamically controlling BK channel trafficking. Several recent studies support a role for rapid control of trafficking by serine/threonine phosphorylation; however, the precise molecular mechanisms are poorly defined and may not involve phosphorylation of the pore forming or regulatory subunits per se. For example, in vascular smooth muscle cells (VSMCs), activation of PKA or PKG results in rapid (<1 min) anterograde trafficking of rab11a-positive recycling endosomes to the plasma membrane (Leo et al., 2014). These endosomes provide an intracellular store of β1-subunits and thus provide a mechanism to increase the ratio of β1-subunits to α-subunits at the plasma membrane. Thus, PKA- and PKG-mediated enhancement of BK current in VSMC may, in part, involve control of retrograde trafficking and changes in assembly of α- and β1-subunits at the plasma membrane. This would be predicted to promote vasodilation.

AMPK has also been reported to control surface expression of BK channels (Föller et al., 2012). Overexpression of constitutively active AMPK ($AMPK^{R70Q}$), but not an inactive AMPK mutant, increases the surface expression of BK channels expressed in *Xenopus oocytes*. This leads to an enhancement of BK current without effects on channel gating properties per se. Although the effect is to increase surface expression, whether this requires AMPK-dependent phosphorylation of the BK channel α-subunit, or some other component involved in channel trafficking, is not known. In mice with a genetic deletion of AMPK, BK currents and channel expression are suppressed in outer hair cells of the cochlea. This has led to the suggestion that AMPK upregulation of BK channels in the cochlea may help protect the inner ear from damage following acoustic overstimulation (Föller et al., 2012).

3.1.1.4 Protein Serine/Threonine Dephosphorylation

While the vast majority of data have examined protein kinase-mediated regulation of BK channels, phosphorylation is of course rapidly and fully reversibly mediated via a family of serine/threonine phosphatases.

Importantly, dephosphorylation is not a simple "passive" control of BK channels that have been previously phosphorylated; rather, it can be dynamically regulated itself. Indeed, some of the first direct evidence for BK channel regulation by dephosphorylation (by PP2A) revealed dephosphorylation as a target for neuropeptide regulation of BK channels in pituitary GH4C1 cells (White et al., 1993, 1991). Somatostatin, via $G_{i/o}$ protein-coupled receptors, promotes BK channel activation in GH4C1 cells through stimulation of PP2A. PP2A dephosphorylates the BK channel resulting in robust activation of the channel, most likely through dephosphorylation of sites involved in PKA-dependent inhibition of BK channels in GH4C1 cells (White et al., 1991). In this system, atrial natriuretic factor, acting through its cell surface guanylate cyclase receptor, stimulates a rapid elevation of cGMP and activation of PKG, which results in a PKG phosphorylation-dependent stimulation of PP2A and subsequent activation of BK channels (White et al., 1993). Intriguingly, dephosphorylation of BK channels in pituitary corticotroph cells is also a key mechanism for glucocorticoid-mediated regulation of BK channels via a genomic pathway (Shipston, Kelly, & Antoni, 1996; Tian, Knaus, & Shipston, 1998). In this model, glucocorticoids activate PP2A-dependent dephosphorylation to antagonize PKA-dependent inhibition of BK channel activity, a mechanism that can be recapitulated in HEK293 cells. Thus, "active" regulation of protein dephosphorylation by diverse cellular signaling pathways represents an important pathway for control of BK channels in endocrine cells.

Dephosphorylation is a key regulator of BK channel function in many systems including in the nervous system (Reinhart, Chung, Martin, Brautigan, & Levitan, 1991). Again, the effects of dephosphorylation on BK channel activity can be complex—likely dependent upon the site(s) of phosphorylation dephosphorylated on the BK channel and the specific phosphatase involved. A startling exemplar of such interplay is the control of a population of BK channels in murine cerebellar Purkinje neurones (Widmer et al., 2003). In this system, PKA-dependent phosphorylation stimulates the activity of a population of BK channels with low calcium/voltage sensitivity. However, this PKA-mediated activation is conditional upon dephosphorylation of a PKC site regulated by PP1. PKC itself inhibits these BK channels, whereas the PKA site is dephosphorylated by PP2A. Thus, in Purkinje neurons, dephosphorylation can either activate or inhibit BK channel activity depending upon the phosphorylation site being controlled: inhibition of PP2A would promote BK channel activation whereas inhibition of PP1 would promote BK channel inhibition (Widmer et al., 2003).

3.1.2 Reversible Tyrosine Phosphorylation

BK channel α-subunits have been reported to be basally phosphorylated on tyrosine residues when expressed in HEK293 cells (Ling, Woronuk, Sy, Lev, & Braun, 2000) and regulated upon overexpression of a variety of tyrosine kinases, including members of the src family (c-src and Hck) and the calcium-dependent tyrosine kinase Pyk2 (Ling, Sheng, & Braun, 2004; Ling et al., 2000). However, other tyrosine kinases, such as the receptor tyrosine kinase erbB2, regulate BK channel activity indirectly through changes in cellular calcium handling (Olsen, Weaver, Ritch, & Sontheimer, 2005).

Even for tyrosine kinases that have been shown to increase phosphorylation of the α-subunit, the functional effect of tyrosine phosphorylation may either increase or decrease channel activity. This, in part, likely reflects the fact that different kinases may regulate distinct sites and that the sites of tyrosine phosphorylation are poorly characterized. For example, approaches employing site-directed mutagenesis in conjunction with kinase overexpression revealed that a C-terminal tyrosine (Tyr 766) is required for c-src-mediated phosphorylation and activation of BK channels, but not for regulation by PyK2 (Ling et al., 2004, 2000). In contrast, in other systems, c-src overexpression inhibited BK channel activity through a mechanism that involved increased α-subunit phosphorylation, based on phosphotyrosine antibody analysis, but the tyrosine residue(s) involved were not identified (Alioua et al., 2002).

In most studies, the *functional* effects of exogenous tyrosine kinases are typically only observed at micromolar levels of intracellular free calcium, conditions under which the cytoskeleton may be disrupted. This may be important as, in studies analyzing the role of endogenous tyrosine kinases and phosphatases intimately associated with the channel complex, an additional mode of tyrosine kinase/phosphatase regulation of BK channels that is *independent* of α-subunit phosphorylation per se but dependent upon regulation of the cytoskeleton has been reported (Tian, McClafferty, Chen, & Shipston, 2008). These studies revealed that one of the major tyrosine-phosphorylated adapter proteins in cells, cortactin, is a molecular bridge between BK channel α-subunits and the actin cytoskeleton. At low (submicromolar) calcium concentrations, tyrosine phosphorylation of cortactin regulates the coupling of BK channels with the cytoskeleton, a process dynamically controlled by endogenous src family tyrosine kinases and phosphatases, as revealed using pharmacological manipulation of these pathways. The control of BK channel activity through tyrosine kinase regulation of cortactin is independent of tyrosine phosphorylation of the α-subunit.

Taken together, these studies may suggest a dual mechanism for tyrosine kinase regulation of BK channels. At low calcium concentration, tyrosine kinases may control BK channel activity through control of cytoskeletal dynamics, via cortactin. When calcium concentration is elevated into the micromolar range, that would be predicted to result in cytoskeletal reorganization, *functional* regulation of BK channels may now be switched and exerted through regulation of the α-subunit directly.

3.2 Lipidation

Lipidation is the general term for the covalent attachment of a lipid to a protein. Although many forms of lipidation exist, the only truly reversible post-translational protein lipidation (S-acylation) has recently been revealed to control multiple aspects of BK channel function. In addition, an atypical internal myristoylation mechanism has been reported to control BK channel trafficking.

3.2.1 Reversible S-Acylation

Protein S-acylation, the covalent addition of a fatty acid, most typically the C16 palmitate, to a cysteine residue via a thioester bond is the only truly dynamic and fully reversible posttranslational lipid modification of proteins. S-acylation is catalyzed by a family of zinc-dependent acyltransferases (zDHHCs, 24 genes in humans) and reversed by a growing family of acylthioesterases (Chamberlain & Shipston, 2015).

In a proteome-wide analysis of S-acylated proteins in synaptosomes, the BK channel α-subunit was identified as an endogenous *S*-acylated protein in rat brain (Kang et al., 2008). Subsequent site-directed mutagenic and biochemical analysis has revealed that the BK channel α-subunit may be *S*-acylated at two distinct locations: the conserved intracellular S0–S1 linker and within the alternatively spliced STREX insert in the linker between RCK1 and RCK2 (Jeffries et al., 2010; Tian, Jeffries, et al., 2008; Figs. 2 and 4).

S-acylation of the intracellular disordered S0–S1 linker is catalyzed primarily by zDHHC23 at a cluster of cysteine residues (Cys 53 and Cys 56, in mouse; Tian, McClafferty, Knaus, Ruth, & Shipston, 2012). Knockdown of zDHHC23 or site-directed mutagenesis of Cys53:Cys56 to alanine reduces cell surface expression of the α-subunit and enhances retention in the ER. The site is deacylated by acylthioesterase LYPLA1, but not LYPLA2, resulting in enhanced retention in the *trans*-Golgi network and a reduction in channel colocalization in recycling endosomes (Tian et al., 2012). S-acylation of the S0–S1 loop has no effect on intrinsic calcium/voltage sensitivity; however, deacylated channels also display enhanced lateral diffusion in the plasma membrane (Kim, Lee, Lee, Won, & Park, 2014). Whether the reduced trafficking and mobility is a function of the ability of the *S*-acylated S0–S1 loop to act as an additional membrane anchor, or allows BK channels to assemble with other trafficking/cytoskeletal proteins, remains to be determined.

In contrast to the role of S0–S1 loop S-acylation in controlling BK channel trafficking, S-acylation of the alternatively spliced STREX insert controls channel calcium-voltage sensitivity and regulation by the classical AGC family of protein kinases (Jeffries, Tian, McClafferty, & Shipston, 2012; Tian, Jeffries, et al., 2008; Tian, McClafferty, Jeffries, & Shipston, 2010). This again exemplifies, as for PKA regulation of BK channels earlier, that the functional effect of a PTM is of course dependent upon the domain within the BK channel that is regulated (Fig. 4).

siRNA knockdown approaches reveal that S-acylation of the murine variant of STREX is mediated by multiple zDHHC members, including the plasma membrane-localized zDHHC5 and the Golgi-localized zDHHC17. STREX is *S*-acylated at a cysteine doublet Cys645:Cys646 that is immediately downstream of the PKA consensus phosphorylation serine residue (Ser 636) that mediates PKA-dependent inhibition of STREX channels (Figs. 2 and 4). S-acylation of Cys645:Cys646 allows the normally cytosolic C-terminus of the BK channel RCK1–RCK2 linker to interact with the plasma membrane (Tian, Jeffries, et al., 2008). BK channels deacylated at this site (by site-directed mutagenesis or knockdown of the cognate zDHHCs), results in channels with a significant rightward shift in STREX variant voltage for half-maximal activation, close to values obtained with the ZERO variant that does not include the STREX insert. Moreover, in deacylated STREX variants, PKA-dependent phosphorylation no longer inhibits BK channels, a mechanism that appears to involve mutual control of STREX domain interaction with the plasma membrane (Tian, Jeffries, et al., 2008). S-acylation promotes association of the STREX domain with the plasma membrane, whereas PKA phosphorylation, by introducing negative charge into the otherwise disordered polybasic linker introduced by STREX inclusion, promotes STREX dissociation from the plasma membrane. S-acylation does not affect STREX channel regulation by redox, which acts on the STREX domain, or by PKG phosphorylation, which controls sites outside of the STREX insert. However, the S-acylation status also determines STREX channel sensitivity to PKC phosphorylation (Zhou et al., 2012). STREX channels *S*-acylated at Cys645:Cys646 are resistant to regulation by PKC. In contrast, deacylation of Cys645:Cys646 results in channels that are now inhibited by PKC phosphorylation. This mechanism is proposed to result from a conserved conditional PKC phosphorylation site just downstream of the STREX insert being accessible to PKC phosphorylation only when the STREX insert is deacylated and not associated with the plasma membrane. These complex interactions between S-acylation, alternative splicing, and phosphorylation demonstrate the potential for dramatic differences in channel phenotype between different cell types and physiological conditions.

In addition to S-acylation of the pore-forming α-subunit, regulatory and other proteins of the BK channel may be *S*-acylated. In the mammalian brain, the regulatory β4-subunit is *S*-acylated at a cysteine residue juxtaposed to the second transmembrane domain (Chen et al., 2013). Again, the major effect of β4-subunit S-acylation is to control surface trafficking of the α-subunit rather than its functional coupling and gating properties. Mutation of the β4-subunit *S*-acylated cysteine prevents β4-subunit enhancement of selected BK channel α-subunit splice variants by retaining the channels in the ER. A number of signaling proteins reported to assemble with BK channels in other systems, such as the β2 adrenergic receptor and the A-kinase anchoring protein AKAP79/150, have been shown, or predicted, to be *S*-acylated (Shipston, 2014). However, the functional consequence of S-acylation of other proteins in the BK channel macromolecular complex is currently unknown.

3.2.2 Atypical Internal Protein Myristoylation

Classical protein myristoylation is irreversible and occurs most commonly cotranslationally by the covalent addition of myristic acid, via an amide bond, at N-terminal glycine amino acids. Myristoylation has also been reported, in a few cases, to occur posttranslationally at an internal glycine residue exposed after peptide cleavage, for example, by caspases. However, in a few rare examples, including BK channels, myristate has been reported to be attached posttranslationally to internal serine/threonine or tyrosine residues via a hydroxyester linkage (Alioua, Li, Wu, Stefani, & Toro, 2011). In human BK channels, multiple serine and threonine residues in both the intracellular S0–S1 linker and S4–S5 linkers contribute to >90%

of myristate incorporation based on site-directed mutagenesis analysis. The functional consequence of this internal myristoylation is to reduce cell surface expression of the channel, most likely through enhanced internalization via a clathrin-dependent pathway. How this internal myristoylation is controlled is not known; however, as it is largely irreversible, it may provide a final "endocytic" signal to reduce cell surface expression.

3.3 N-Linked Glycosylation

N-linked glycosylation of the extracellular domains of many proteins, including ion channels, typically occurs in two stages. Firstly, covalent attachment of an oligosaccharide to the amide nitrogen of asparagine (N) amino acids in the nascent polypeptide chain of proteins occurs in the ER (core glycosylation). These core-glycosylated proteins may be further processed with addition of new sugars occurring in the Golgi complex leading to mature, heavily glycosylated proteins. Glycosylation can control a number of diverse ion channels with effects on ion channel folding, stability, trafficking, and function.

Although mammalian BK channel α-subunits have a single consensus N-glycosylation site (Asn 200 in human) in the short extracellular loop between S3 and S4, direct evidence for N-glycosylation is largely lacking in most systems studied. N-glycosylation has been detected in the extracellular N-terminal domain preceding S0 in Drosophila channels but this is not conserved in mammalian BK channels (Meera, Wallner, Song, & Toro, 1997). Site-directed mutation of Asn 200 to alanine in human BK channels has no detectable effect on surface delivery or apical localization in polarized MDCK cells (Bravo-Zehnder et al., 2000). Although this point mutant had a significant leftshifted voltage for half-maximal activation over a range of calcium concentrations (from 200 nM to 30 μM), whether this is a result of deglycosylation per se or an effect of the mutation of Asn to Ala is not known. The latter is a possibility as Asn 200 is very close to the third transmembrane segment and studies in native BK channels have revealed little effect of enzymatic deglycosylation on BK channel α-subunits, for example, in murine colonic smooth muscle cells lacking the β1-subunit (Hagen & Sanders, 2006).

In contrast, β-subunits of BK channels, including β1 and β4, are N-glycosylated in the large extracellular loop (Jin, Weiger, & Levitan, 2002; Knaus, Garcia-Calvo, Kaczorowski, & Garcia, 1994; Wallner,

Meera, & Toro, 1996). The human β4-subunit is glycosylated at two residues Asn 53 and Asn 90 in the extracellular loop. Enzymatic deglycosylation assays suggest that β4 can exist in both an immature, core-glycosylated form as well as a mature, highly glycosylated form with the latter being enhanced upon coexpression with the α-subunit (Jin, Weiger, & Levitan, 2002). Site-directed mutagenesis of both asparagine residues to alanine has no effect on the physical assembly with α-subunits or surface trafficking of the β4-subunit. However, while coexpression of α-subunits with β4-subunits largely prevents channel blockade by iberiotoxin (IbTx), deglycosylated β4-subunits result in channels that have an enhanced sensitivity to inhibition by IbTx, similar to the inhibitory effect of IbTx on α-subunits alone. The extracellular loop of β4 is thought to prevent access of IbTx to the extracellular mouth of the pore, suggesting that glycosylation may compete with IbTx binding or promote a conformation of the extracellular loop that shields the pore from the toxin.

β1-subunits are also heavily glycosylated in vascular and colonic smooth muscle cells (Hagen & Sanders, 2006; Knaus et al., 1994). Enzymatic deglycosylation of BK channels using PNGase F in inside-out patches from colonic smooth muscle cells resulted in an increase in channel open probability and a leftward shift in the voltage dependence of activation (Hagen & Sanders, 2006). In contrast, BK channels in smooth muscle cells from β1 knockout mice were unaffected by enzymatic deglycosylation. Taken together, these data suggest that glycosylation is important for the functional coupling of β1-subunits with α-subunits, in contrast to that observed with β4.

3.4 Ubiquitination

Ubiquitination of proteins typically acts as a cellular signal to identify proteins for degradation by the 26*S* proteosomal complex and has been implicated in the control of diverse cellular processes. Ubiquitin, a small (8.5 kDa) protein, is controlled by a diverse family of more than 600 E3 ubiquitin ligases in humans (Li et al., 2008). In the rodent brain, BK channel α-subunits assemble as a complex with the CRL4A E3 ubiquitin ligase via its receptor substrate cereblon (Liu et al., 2014). Subsequent polyubiquitination of the BK channel at multiple sites in the C-terminus results in BK channel retention in the ER. Inhibition of BK channel ubiquitination through pharmacological inhibition the CRL4A ligase activity, using the small-molecule thalidomide that binds to cereblon, or genetic mutation of cereblon results in enhanced surface expression of BK channels. In mice, this gain of BK channel function leads to enhanced hippocampal neuronal excitability and induction of epileptic seizures.

The β1-subunit is also reported to undergo ubiquitination in VSMCs in response to elevations in glucose in vitro or from arteries of streptozocin-induced diabetic mice (Yi et al., 2014). β1-subunit ubiquitination is mediated by the muscle-specific RING finger 1 (MurF1) ubiquitin E3 ligase and leads to enhanced β1-subunit proteolysis. This would be predicted to depolarize VSMCs and thus enhance vasoconstriction.

4. CONCLUSIONS AND PERSPECTIVES

BK channels are subject to modulation by a wide range of the major posttranscriptional and posttranslational mechanisms that allow considerable expansion of proteomic and functional diversity in BK channels. Importantly, these mechanisms show considerable cross talk and provide multiple modes of "conditional" regulation of BK channels (Figs. 2 and 4). This is exemplified by the number of alternatively spliced inserts of the α-subunit that introduce, or modify, sites for PTM in particular by phosphorylation and lipidation (Fig. 2) and the fact that the same PTM can have different, and sometimes opposite, functional effects dependent upon the domain of the channel being controlled (Fig. 4). Many of these sites are in intrinsically disordered domains of the channel that couple functional domains of the channel, allowing "fine-tuning" of the core and unique properties of BK channels regulated by calcium and voltage.

Although we are now beginning to probe some of the fundamental rules of cross talk and how these pathways may control diversity of BK channel properties, we are only on the start of understanding their importance and impact in physiological systems in health and disease. To this end, we have a large "catalogue" of interacting mechanisms but there are many challenges for the field if we are to exploit these therapeutically.

First, we know very little about the spatiotemporal dynamics of these modes of regulation and how they impact the two important modes of controlling BK channel function: controlling the number of BK channels at target membranes, and their activity and regulation once at the membrane. Importantly, the time domains for biophysical, trafficking, and/or expression and regulation by these pathways overlap. Thus, understanding whether effects are dependent upon trafficking or regulation of function in situ are critical for devising strategies to manipulate BK channel physiological function.

Second, in relatively few endogenous systems do we understand how these mechanisms integrate to provide cell-specific modes of regulation and how this may be modified during development or in response to physiological, or pathological, challenge. For example, we lack many of the key tools, such as site-specific antibodies, to probe whether a specific site on the channel is posttranslationally modified or how a protein isoform is expressed or spatially organized in specific cell types.

Third, the physiological impact of many of these mechanisms at a systems level is very poorly understood. For example, genetically modified animals in which specific sites of PTM or splice variant expression has been manipulated in specific cell types are largely not available.

The diverse physiological mechanisms that BK channels control, coupled with the array of disorders with which BK channel dysfunction has been associated, make targeting the fundamental processes that BK channels regulate an attractive therapeutic target. By deciphering the fundamental rules by which posttranscriptional and posttranslational mechanisms control BK channel diversity and interrogating the impact of these mechanisms to control systems level function promises to open a new vista in BK channel physiology in health and disease.

ACKNOWLEDGMENTS

This chapter is dedicated to the memory and work of Dr Lijun Tian who passed away during the writing of this chapter. Work in the authors' laboratory is generously supported by the Medical Research Council, British Heart Foundation and Diabetes UK.

REFERENCES

Alioua, A., Li, M., Wu, Y., Stefani, E., & Toro, L. (2011). Unconventional myristoylation of large-conductance Ca^{2+}-activated K^+ channel (Slo1) via serine/threonine residues regulates channel surface expression. *Proceedings of the National Academy of Sciences of the United States of America*, *108*, 10744–10749.

Alioua, A., Mahajan, A., Nishimaru, K., Zarei, M. M., Stefani, E., & Toro, L. (2002). Coupling of c-Src to large conductance voltage- and Ca^{2+}-activated K^+ channels as a new mechanism of agonist-induced vasoconstriction. *Proceedings of the National Academy of Sciences of the United States of America*, *99*, 14560–14565.

Alioua, A., Tanaka, Y., Wallner, M., Hofmann, F., Ruth, P., Meera, P., et al. (1998). The large conductance, voltage-dependent, and calcium-sensitive K^+ channel, Hslo, is a target of cGMP-dependent protein kinase phosphorylation *in vivo*. *Journal of Biological Chemistry*, *273*, 32950–32956.

Atkinson, N. S., Robertson, G. A., & Ganetzky, B. (1991). A component of calcium-activated potassium channels encoded by the Drosophila slo locus. *Science, 253*, 551–555.

Bell, T. J., Miyashiro, K. Y., Sul, J.-Y., McCullough, R., Buckley, P. T., Jochems, J., et al. (2008). Cytoplasmic BK(Ca) channel intron-containing mRNAs contribute to the intrinsic excitability of hippocampal neurons. *Proceedings of the National Academy of Sciences of the United States of America, 105*, 1901–1906.

Berkefeld, H., Fakler, B., & Schulte, U. (2010). Ca^{2+}-activated K^{+} channels: From protein complexes to function. *Physiological Reviews, 90*, 1437–1459.

Bravo-Zehnder, M., Orio, P., Norambuena, A., Wallner, M., Meera, P., Toro, L., et al. (2000). Apical sorting of a voltage- and Ca^{2+}-activated K^{+} channel alpha-subunit in Madin-Darby canine kidney cells is independent of N-glycosylation. *Proceedings of the National Academy of Sciences of the United States of America, 97*, 13114–13119.

Butler, A., Tsunoda, S., McCobb, D. P., Wei, A., & Salkoff, L. (1993). mSlo, a complex mouse gene encoding "maxi" calcium-activated potassium channels. *Science, 261*, 221–224.

Chamberlain, L. H., & Shipston, M. J. (2015). The physiology of protein S-acylation. *Physiological Reviews, 95*, 341–376.

Chen, L., Bi, D., Tian, L., McClafferty, H., Steeb, F., Ruth, P., et al. (2013). Palmitoylation of the β4-subunit regulates surface expression of large conductance calcium-activated potassium channel splice variants. *Journal of Biological Chemistry, 288*, 13136–13144.

Chen, L., Jeffries, O., Rowe, I. C. M., Liang, Z., Knaus, H.-G., Ruth, P., et al. (2010). Membrane trafficking of large conductance calcium-activated potassium channels is regulated by alternative splicing of a transplantable, acidic trafficking motif in the RCK1-RCK2 linker. *Journal of Biological Chemistry, 285*, 23265–23275.

Chen, L., Tian, L., MacDonald, S. H.-F., McClafferty, H., Hammond, M. S. L., Huibant, J.-M., et al. (2005). Functionally diverse complement of large conductance calcium- and voltage-activated potassium channel (BK) alpha-subunits generated from a single site of splicing. *Journal of Biological Chemistry, 280*, 33599–33609.

Chi, S., Cai, W., Liu, P., Zhang, Z., Chen, X., Gao, L., et al. (2010). Baifuzi reduces transient ischemic brain damage through an interaction with the STREX domain of BKCa channels. *Cell Death & Disease, 1*, e13.

Clark, A. G., Hall, S. K., & Shipston, M. J. (1999). ATP inhibition of a mouse brain large-conductance K^{+} (mslo) channel variant by a mechanism independent of protein phosphorylation. *Journal of Physiology, 516*, 45–53.

Contreras, G. F., Castillo, K., Enrique, N., Carrasquel-Ursulaez, W., Castillo, J. P., Milesi, V., et al. (2013). A BK (Slo1) channel journey from molecule to physiology. *Channels, 7*, 442–458.

Dworetzky, S. I., Boissard, C. G., Lum-Ragan, J. T., Mckay, M. C., Post-Munson, D. J., Trojnacki, J. T., et al. (1996). Phenotypic alteration of a human BK (hSlo) channel by hSlobeta subunit coexpression: Changes in blocker sensitivity, activation/relaxation and inactivation kinetics, and protein kinase A modulation. *Journal of Neuroscience, 16*, 4543–4550.

Erxleben, C., Everhart, A. L., Romeo, C., Florance, H., Bauer, M. B., Alcorta, D. A., et al. (2002). Interacting effects of N-terminal variation and strex exon splicing on slo potassium channel regulation by calcium, phosphorylation, and oxidation. *Journal of Biological Chemistry, 277*, 27045–27052.

Ewald, D. A., Williams, A., & Levitan, I. B. (1985). Modulation of single Ca^{2+}-dependent K^{+}-channel activity by protein phosphorylation. *Nature, 315*, 503–506.

Fodor, A. A., & Aldrich, R. W. (2009). Convergent evolution of alternative splices at domain boundaries of the BK channel. *Annual Review of Physiology, 71*, 19–36.

Föller, M., Jaumann, M., Dettling, J., Saxena, A., Pakladok, T., Munoz, C., et al. (2012). AMP-activated protein kinase in BK-channel regulation and protection against hearing loss following acoustic overstimulation. *The FASEB Journal, 26*, 4243–4253.

Glauser, D. A., Johnson, B. E., Aldrich, R. W., & Goodman, M. B. (2011). Intragenic alternative splicing coordination is essential for Caenorhabditis elegans slo-1 gene function. *Proceedings of the National Academy of Sciences of the United States of America, 108*, 20790–20795.

Graveley, B. R., Brooks, A. N., Carlson, J. W., Duff, M. O., Landolin, J. M., Yang, L., et al. (2011). The developmental transcriptome of Drosophila melanogaster. *Nature, 471*, 473–479.

Ha, T. S., Jeong, S. Y., Cho, S. W., Jeon, H. K., Roh, G. S., Choi, W. S., et al. (2000). Functional characteristics of two BK_{Ca} channel variants differentially expressed in rat brain tissues. *European Journal of Biochemistry, 267*, 910–918.

Hagen, B. M., & Sanders, K. M. (2006). Deglycosylation of the beta1-subunit of the BK channel changes its biophysical properties. *American Journal of Physiology: Cell Physiology, 291*, C750–C756.

Hoopengardner, B., Bhalla, T., Staber, C., & Reenan, R. (2003). Nervous system targets of RNA editing identified by comparative genomics. *Science, 301*, 832–836.

Hu, S., Labuda, M. Z., Pandolfo, M., Goss, G. G., McDermid, H. E., & Ali, D. W. (2003). Variants of the KCNMB3 regulatory subunit of maxi BK channels affect channel inactivation. *Physiological Genomics, 15*, 191–198.

Jeffries, O., Geiger, N., Rowe, I. C. M., Tian, L., McClafferty, H., Chen, L., et al. (2010). Palmitoylation of the S0-S1 linker regulates cell surface expression of voltage- and calcium-activated potassium (BK) channels. *Journal of Biological Chemistry, 285*, 33307–33314.

Jeffries, O., Tian, L., McClafferty, H., & Shipston, M. J. (2012). An electrostatic switch controls palmitoylation of the large conductance voltage- and calcium-activated potassium (BK) channel. *Journal of Biological Chemistry, 287*, 1468–1477.

Jin, P., Weiger, T. M., & Levitan, I. B. (2002a). Reciprocal modulation between the alpha and beta 4 subunits of hSlo calcium-dependent potassium channels. *Journal of Biological Chemistry, 277*, 43724–43729.

Jin, P., Weiger, T. M., Wu, Y., & Levitan, I. B. (2002b). Phosphorylation-dependent functional coupling of hSlo calcium-dependent potassium channel and its hβ4 subunit. *Journal of Biological Chemistry, 277*, 10014–10020.

Johnson, B. E., Glauser, D. A., Dan-Glauser, E. S., Halling, D. B., Aldrich, R. W., & Goodman, M. B. (2011). Alternatively spliced domains interact to regulate BK potassium channel gating. *Proceedings of the National Academy of Sciences of the United States of America, 108*, 20784–20789.

Kang, R., Wan, J., Arstikaitis, P., Takahashi, H., Huang, K., Bailey, A. O., et al. (2008). Neural palmitoyl-proteomics reveals dynamic synaptic palmitoylation. *Nature, 456*, 904–909.

Kim, S., Lee, B.-C., Lee, A.-R., Won, S., & Park, C.-S. (2014). Effects of palmitoylation on the diffusional movement of BKCa channels in live cells. *FEBS Letters, 588*(5), 713–719.

Kim, E. Y., Ridgway, L. D., Zou, S., Chiu, Y.-H., & Dryer, S. E. (2007). Alternatively spliced C-terminal domains regulate the surface expression of large conductance calcium-activated potassium channels. *Neuroscience, 146*, 1652–1661.

Knaus, H.-G., Garcia-Calvo, M., Kaczorowski, G. J., & Garcia, M. L. (1994). Subunit composition of the high conductance calcium-activated potassium channel from smooth muscle, a representative of the mSlo and slowpoke family of potassium channels. *Journal of Biological Chemistry, 269*, 3921–3924.

Korovkina, V. P., Fergus, D. J., Holdiman, A. J., & England, S. K. (2001). Characterization of a novel 132-bp exon of the human maxi-K channel. *American Journal of Physiology: Cell Physiology, 281*, C361–C367.

Kyle, B. D., Hurst, S., Swayze, R. D., Sheng, J., & Braun, A. P. (2013). Specific phosphorylation sites underlie the stimulation of a large conductance, $Ca^{(2+)}$-activated

$K^{(+)}$ channel by cGMP-dependent protein kinase. *The FASEB Journal*, *27*, 2027–2038.

Lai, G.-J., & McCobb, D. P. (2002). Opposing actions of adrenal androgens and glucocorticoids on alternative splicing of Slo potassium channels in bovine chromaffin cells. *Proceedings of the National Academy of Sciences of the United States of America*, *99*, 7722–7727.

Leo, M. D., Bannister, J. P., Narayanan, D., Nair, A., Grubbs, J. E., Gabrick, K. S., et al. (2014). Dynamic regulation of β1 subunit trafficking controls vascular contractility. *Proceedings of the National Academy of Sciences of the United States of America*, *111*, 2361–2366.

Li, W., Bengtson, M. H., Ulbrich, A., Matsuda, A., Reddy, V. A., Orth, A., et al. (2008). Genome-wide and functional annotation of human E3 ubiquitin ligases identifies MULAN, a mitochondrial E3 that regulates the organelle's dynamics and signaling. *PLoS One*, *3*, e1487.

Li, D., Wang, Z., Sun, P., Jin, Y., Lin, D.-H., Hebert, S. C., et al. (2006). Inhibition of MAPK stimulates the Ca^{2+}-dependent big-conductance K channels in cortical collecting duct. *Proceedings of the National Academy of Sciences of the United States of America*, *103*, 19569–19574.

Ling, S., Sheng, J.-Z., & Braun, A. P. (2004). The calcium-dependent activity of large-conductance, calcium-activated K+ channels is enhanced by Pyk2- and Hck-induced tyrosine phosphorylation. *American Journal of Physiology: Cell Physiology*, *287*, C698–C706.

Ling, S., Woronuk, G., Sy, L., Lev, S., & Braun, A. P. (2000). Enhanced activity of a large conductance, calcium-sensitive K^+ channel in the presence of Src tyrosine kinase. *Journal of Biological Chemistry*, *275*, 30683–30689.

Liu, J., Asuncion-Chin, M., Liu, P., & Dopico, A. M. (2005). CaM kinase II phosphorylation of slo Thr107 regulates activity and ethanol responses of BK channels. *Nature Neuroscience*, *9*, 41–49.

Liu, X., Chang, Y., Reinhart, P. H., Sontheimer, H., & Chang, Y. (2002). Cloning and characterization of glioma BK, a novel BK channel isoform highly expressed in human glioma cells. *Journal of Neuroscience*, *22*, 1840–1849.

Liu, G., Razanau, A., Hai, Y., Yu, J., Sohail, M., Lobo, V. G., et al. (2012). A conserved serine of heterogeneous nuclear ribonucleoprotein L (hnRNP L) mediates depolarization-regulated alternative splicing of potassium channels. *Journal of Biological Chemistry*, *287*, 22709–22716.

Liu, J., Ye, J., Zou, X., Xu, Z., Feng, Y., Zou, X., et al. (2014). CRL4A(CRBN) E3 ubiquitin ligase restricts BK channel activity and prevents epileptogenesis. *Nature Communications*, *5*, 3924.

Lorca, R. A., Stamnes, S. J., Pillai, M. K., Hsiao, J. J., Wright, M. E., & England, S. K. (2014). N-terminal isoforms of the large-conductance Ca^{2+}-activated K^+ channel are differentially modulated by the auxiliary β1-subunit. *Journal of Biological Chemistry*, *289*, 10095–10103.

Lu, R., Alioua, A., Kumar, Y., Eghbali, M., Stefani, E., & Toro, L. (2005). MaxiK channel partners: Physiological impact. *The Journal of Physiology*, *570*, 65–72.

Ma, D., Nakata, T., Zhang, G., Hoshi, T., Li, M., & Shikano, S. (2007). Differential trafficking of carboxyl isoforms of Ca^{2+}-gated (Slo1) potassium channels. *FEBS Letters*, *581*, 1000–1008.

Manning, G., Whyte, D. B., Martinez, R., Hunter, T., & Sudarsanam, S. (2002). The protein kinase complement of the human genome. *Science*, *298*, 1912–1934.

McCartney, C. E., McClafferty, H., Huibant, J.-M., Rowan, E. G., Shipston, M. J., & Rowe, I. C. M. (2005). A cysteine-rich motif confers hypoxia sensitivity to mammalian large conductance voltage- and Ca-activated K (BK) channel alpha-subunits. *Proceedings of the National Academy of Sciences of the United States of America*, *102*, 17870–17876.

Meera, P., Wallner, M., Song, M., & Toro, L. (1997). Large conductance voltage- and calcium-dependent K^+ channel, a distinct member of voltage-dependent ion channels with seven N-terminal transmembrane segments (S0-S6), an extracellular N terminus, and an intracellular (S9-S10) C terminus. *Proceedings of the National Academy of Sciences of the United States of America*, *94*, 14066–14071.

Nara, M., Dhulipala, P. D., Wang, Y. X., & Kotlikoff, M. I. (1998). Reconstitution of beta-adrenergic modulation of large conductance, calcium-activated potassium (maxi-K) channels in Xenopus oocytes. Identification of the cAMP-dependent protein kinase phosphorylation site. *Journal of Biological Chemistry*, *273*, 14920–14924.

Ohya, S., Fujimori, T., Kimura, T., Yamamura, H., & Imaizumi, Y. (2010). Novel spliced variants of large-conductance $Ca^{(2+)}$-activated $K^{(+)}$-channel β(2)-subunit in human and rodent pancreas. *Journal of Pharmacological Sciences*, *114*, 198–205.

Olsen, M., Weaver, A., Ritch, P., & Sontheimer, H. (2005). Modulation of glioma BK channels via erbB2. *Journal of Neuroscience Research*, *81*, 179–189.

Orio, P., Rojas, P., Ferreira, G., & Latorre, R. (2002). New disguises for an old channel: MaxiK channel beta-subunits. *News in Physiological Sciences*, *17*, 156–161.

Peng, Z., Sakai, Y., Kurgan, L., Sokolowski, B., & Uversky, V. (2014). Intrinsic disorder in the BK channel and its interactome. *PLoS One*, *9*, e94331.

Petrik, D., & Brenner, R. (2007). Regulation of STREX exon large conductance, calcium-activated potassium channels by the β4 accessory subunit. *Neuroscience*, *149*, 789–803.

Pietrzykowski, A. Z., Friesen, R. M., Martin, G. E., Puig, S. I., Nowak, C. L., Wynne, P. M., et al. (2008). Posttranscriptional regulation of BK channel splice variant stability by miR-9 underlies neuroadaptation to alcohol. *Neuron*, *59*(2), 274–287.

Reinhart, P. H., Chung, S., Martin, B. L., Brautigan, D. L., & Levitan, I. B. (1991). Modulation of calcium-activated potassium channels from rat brain by protein kinase A and phosphatase 2A. *Journal of Neuroscience*, *11*, 1627–1635.

Reinhart, P. H., & Levitan, I. B. (1995). Kinase and phosphatase activities intimately associated with a reconstituted calcium-dependent potassium channel. *Journal of Neuroscience*, *15*, 4572–4579.

Ross, F. A., Rafferty, J. N., Dallas, M. L., Ogunbayo, O., Ikematsu, N., McClafferty, H., et al. (2011). Selective expression in carotid body type I cells of a single splice variant of the large conductance calcium- and voltage-activated potassium channel confers regulation by AMP-activated protein kinase. *Journal of Biological Chemistry*, *23*, 4543–4551.

Sacco, F., Gherardini, P. F., Paoluzi, S., Saez-Rodriguez, J., Helmer-Citterich, M., Ragnini-Wilson, A., et al. (2012). Mapping the human phosphatome on growth pathways. *Molecular Systems Biology*, *8*, 603.

Salkoff, L., Butler, A., Ferreira, G., Santi, C., & Wei, A. (2006). High-conductance potassium channels of the SLO family. *Nature Reviews. Neuroscience*, 7, 921–931.

Samuel, P., Pink, R. C., Caley, D. P., Currie, J. M. S., Brooks, S. A., & Carter, D. R. F. (2015). Over-expression of miR-31 or loss of KCNMA1 leads to increased cisplatin resistance in ovarian cancer cells. *Tumour Biology*. http://dx.doi.org/10.1007/s13277-015-4081-z.

Shelley, C., Whitt, J. P., Montgomery, J. R., & Meredith, A. L. (2013). Phosphorylation of a constitutive serine inhibits BK channel variants containing the alternate exon "SRKR" *Journal of General Physiology*, *142*, 585–598.

Shipston, M. J. (2014). S-acylation dependent post-translational cross-talk regulates large conductance calcium- and voltage-activated potassium (BK) channels. *Frontiers in Physiology*, *5*, 281.

Shipston, M. J., Kelly, J. S., & Antoni, F. A. (1996). Glucocorticoids block protein kinase A inhibition of calcium-activated potassium channels. *Journal of Biological Chemistry*, *271*, 9197–9200.

Singh, H., Li, M., Hall, L., Chen, S., Sukur, S., Lu, R., et al. (2016). MaxiK channel interactome reveals its interaction with GABA transporter 3 and heat shock protein 60 in the mammalian brain. *Neuroscience*, *317*, 76–107.

Singh, H., Lu, R., Bopassa, J. C., Meredith, A. L., Stefani, E., & Toro, L. (2013). MitoBK (Ca) is encoded by the Kcnma1 gene, and a splicing sequence defines its mitochondrial location. *Proceedings of the National Academy of Sciences of the United States of America*, *110*, 10836–10841.

Sitdikova, G. F., Fuchs, R., Kainz, V., Weiger, T. M., & Hermann, A. (2014). Phosphorylation of BK channels modulates the sensitivity to hydrogen sulfide (H2S). *Frontiers in Physiology*, *5*, 431.

Tatro, E. T., Hefler, S., Shumaker-Armstrong, S., Soontornniyomkij, B., Yang, M., Yermanos, A., et al. (2013). Modulation of BK channel by microRNA-9 in neurons after exposure to HIV and methamphetamine. *Journal of Neuroimmune Pharmacology*, *8*, 1210–1223.

Tian, L., Coghill, L. S., McClafferty, H., MacDonald, S. H.-F., Antoni, F. A., Ruth, P., et al. (2004). Distinct stoichiometry of BK_{Ca} channel tetramer phosphorylation specifies channel activation and inhibition by cAMP-dependent protein kinase. *Proceedings of the National Academy of Sciences of the United States of America*, *101*, 11897–11902.

Tian, L., Duncan, R. R., Hammond, M. S., Coghill, L. S., Wen, H., Rusinova, R., et al. (2001). Alternative splicing switches potassium channel sensitivity to protein phosphorylation. *Journal of Biological Chemistry*, *276*, 7717–7720.

Tian, L., Jeffries, O., McClafferty, H., Molyvdas, A., Rowe, I. C. M., Saleem, F., et al. (2008). Palmitoylation gates phosphorylation-dependent regulation of BK potassium channels. *Proceedings of the National Academy of Sciences of the United States of America*, *105*, 21006–21011.

Tian, L., Knaus, H.-G., & Shipston, M. J. (1998). Glucocorticoid regulation of calcium-activated potassium channels mediated by serine/threonine protein phosphatase. *Journal of Biological Chemistry*, *273*, 13531–13536.

Tian, L., McClafferty, H., Chen, L., & Shipston, M. J. (2008). Reversible tyrosine protein phosphorylation regulates large conductance voltage- and calcium-activated potassium channels via cortactin. *Journal of Biological Chemistry*, *283*, 3067–3076.

Tian, L., McClafferty, H., Jeffries, O., & Shipston, M. J. (2010). Multiple palmitoyltransferases are required for palmitoylation-dependent regulation of large conductance calcium- and voltage-activated potassium channels. *Journal of Biological Chemistry*, *285*, 23954–23962.

Tian, L., McClafferty, H., Knaus, H.-G., Ruth, P., & Shipston, M. J. (2012). Distinct acyl protein transferases and thioesterases control surface expression of calcium-activated potassium channels. *Journal of Biological Chemistry*, *287*, 14718–14725.

Velàzquez-Marrero, C., Seale, G. E., Treistman, S. N., & Martin, G. E. (2014). Large conductance voltage- and Ca^{2+}-gated potassium (BK) channel β4 subunit influences sensitivity and tolerance to alcohol by altering its response to kinases. *Journal of Biological Chemistry*, *289*, 29261–29272.

Wallner, M., Meera, P., & Toro, L. (1996). Determinant for beta-subunit regulation in high-conductance voltage-activated and $Ca^{(2+)}$-sensitive K^+ channels: An additional transmembrane region at the N terminus. *Proceedings of the National Academy of Sciences of the United States of America*, *93*, 14922–14927.

White, R. E., Lee, A. B., Shcherbatko, A. D., Lincoln, T. M., Schonbrunn, A., & Armstrong, D. L. (1993). Potassium channel stimulation by natriuretic peptides through cGMP-dependent dephosphorylation. *Nature*, *361*(6409), 263–266.

White, R. E., Schonbrunn, A., & Armstrong, D. L. (1991). Somatostatin stimulates $Ca^{(2+)}$-activated K+ channels through protein dephosphorylation. *Nature*, *351*, 570–573.

Widmer, H. A., Rowe, I. C. M., & Shipston, M. J. (2003). Conditional protein phosphorylation regulates BK channel activity in rat cerebellar Purkinje neurons. *Journal of Physiology, 552*, 379–391.

Wu, Y., Yang, Y., Ye, S., & Jiang, Y. (2010). Structure of the gating ring from the human large-conductance $Ca^{(2+)}$-gated $K^{(+)}$ channel. *Nature, 466*, 393–397.

Xia, X.-M., Ding, J.-P., & Lingle, C. J. (1999). Molecular basis for the inactivation of Ca^{2+}- and voltage-dependent BK channels in adrenal chromaffin cells and rat insulinoma tumor cells. *Journal of Neuroscience, 19*, 5255–5264.

Xie, J., & Black, D. L. (2001). A CaMK IV responsive RNA element mediates depolarization-induced alternative splicing of ion channels. *Nature, 410*, 936–939.

Xie, J., & McCobb, D. P. (1998). Control of alternative splicing of potassium channels by stress hormones. *Science, 280*, 443–446.

Yan, J., Olsen, J. V., Park, K.-S., Li, W., Bildl, W., Schulte, U., et al. (2008). Profiling the phospho-status of the BK_{Ca} channel alpha subunit in rat brain reveals unexpected patterns and complexity. *Molecular & Cellular Proteomics*, 7, 2188–2198.

Yi, L., Morgan, J. T., & Ragsdale, S. W. (2010). Identification of a thiol/disulfide redox switch in the human BK channel that controls its affinity for heme and CO. *Journal of Biological Chemistry, 285*, 20117–20127.

Yi, F., Wang, H., Chai, Q., Wang, X., Shen, W.-K., Willis, M. S., et al. (2014). Regulation of large conductance Ca^{2+}-activated K^{+} (BK) channel β1 subunit expression by muscle RING finger protein 1 in diabetic vessels. *Journal of Biological Chemistry, 289*, 10853–10864.

Yu, L., Eaton, A. F., Yue, Q., Bao, H.-F., Ma, H.-P., Cuppoletti, J., et al. (2015). Unoprostone activation of BK (KCa1.1) channel splice variants. *Biochimica et Biophysica Acta, 1848*, 2859–2867.

Yuan, P., Leonetti, M. D., Pico, A. R., Hsiung, Y., & MacKinnon, R. (2010). Structure of the human BK channel Ca^{2+}-activation apparatus at 3.0 A resolution. *Science, 329*, 182–186.

Zarei, M. M., Eghbali, M., Alioua, A., Song, M., Knaus, H.-G., Stefani, E., et al. (2004). An endoplasmic reticulum trafficking signal prevents surface expression of a voltage- and Ca^{2+}-activated K^{+} channel splice variant. *Proceedings of the National Academy of Sciences of the United States of America, 101*, 10072–10077.

Zarei, M. M., Zhu, N., Alioua, A., Eghbali, M., Stefani, E., & Toro, L. (2001). A novel maxiK splice variant exhibits dominant-negative properties for surface expression. *Journal of Biological Chemistry, 276*, 16232–16239.

Zhang, J., & Yan, J. (2014). Regulation of BK channels by auxiliary γ subunits. *Frontiers in Physiology, 5*, 401.

Zhou, X. B., Arntz, C., Kamm, S., Motejlek, K., Sausbier, U., Wang, G. X., et al. (2001). A molecular switch for specific stimulation of the BK_{Ca} channel by cGMP and cAMP kinase. *Journal of Biological Chemistry, 276*, 43239–43245.

Zhou, X., Wulfsen, I., Korth, M., McClafferty, H., Lukowski, R., Shipston, M. J., et al. (2012). Palmitoylation and membrane association of the stress axis regulated insert (STREX) controls BK channel regulation by protein kinase C. *Journal of Biological Chemistry, 287*, 32161–32171.

Zhou, X.-B., Wulfsen, I., Utku, E., Sausbier, U., Sausbier, M., Wieland, T., et al. (2010). Dual role of protein kinase C on BK channel regulation. *Proceedings of the National Academy of Sciences of the United States of America, 107*, 8005–8010.

Zhou, L., Zhang, Y.-J., Gao, L.-J., Ye, Y., Qi, J.-H., & Qi, Z. (2014). Structure-activity relationship of Baifuzi-cerebrosides on BKCa channel activation. *European Journal of Medicinal Chemistry, 75*, 301–307.

CHAPTER FOUR

Protein Network Interacting with BK Channels

H. Kim[1], K.H. Oh
Chicago Medical School, Rosalind Franklin University of Medicine and Science, North Chicago, IL, United States
[1]Corresponding author: e-mail address: hongkyun.kim@rosalindfranklin.edu

Contents

International Review of Neurobiology, Volume 128
ISSN 0074-7742
http://dx.doi.org/10.1016/bs.irn.2016.03.003

Abstract

The large conductance, voltage- and calcium-dependent potassium BK channel, also termed BK_{Ca}, Maxi-K, $K_{Ca}1.1$, or SLO1, regulates a wide array of key biological processes that are mediated by calcium ions. This broad regulatory function of BK channels requires the interactions of BK channels with diverse sets of proteins that have different molecular or cellular functions. Proteins that interact with BK channels range from the auxiliary subunits of BK channels to cytoskeletal proteins responsible for BK channel trafficking and localization. Furthermore, depending on cellular contexts, BK channels interact with proteins modifying channel properties, and with proteins that allow cell-specific responses. In this chapter, we review some of the relevant studies investigating the protein interaction network of BK channels with a focus on the biological roles of the interaction.

1. INTRODUCTION

The large conductance, voltage- and calcium-dependent potassium BK channel, is cooperatively activated by membrane depolarization and an increase in intracellular calcium ions. BK channel activation leads to hyperpolarization and thereby negatively regulates calcium channels. This unique physiological property is well suited for fine-tuning a wide array of calcium-controlled physiological processes in excitable cells, including muscle excitation, neurotransmitter release, and hormone release (see chapters "BK Channels in Neurons" by Barth and Contet, "BK Channels in the Pituitary and Hypothalamus" by Duncan and Shipston, and "BK Channels in the Vascular System" by Krishnamoorthy-Natarajan and Koide). BK channels are expressed not only in the plasma membrane but also in intracellular membranes, such as nuclear, mitochondrial, and lysosomal membranes, further underscoring the important roles that BK channels play in many cellular functions (see these chapters). In this review, we will limit our discussions to the BK channels localized to the plasma membrane. BK channels consist of a pore-forming α subunit and auxiliary β and γ subunits. The α subunit is sufficient for channel function. The β and γ subunits are often, but not always, associated with the α subunit and modulate the channel properties (Lee & Cui, 2009; Yan & Aldrich, 2010). The α subunit is encoded by a single gene, named Slo1, Kcnma1, or KCNMA1 depending on species. The β and γ subunits are each encoded by four different genes (β1–β4, γ1–γ4). Both β and γ subunits are expressed in a tissue- or cell-type-specific manner. The roles of the β and γ subunits will be discussed in depth in this issue (see chapter "Modulation of BK Channel Function by Auxiliary Beta and Gamma Subunits" by Yan and Li). The BK channel α subunit has typical characteristics of voltage-gated potassium channels, which include a voltage-sensor and a pore-forming loop domain. Unlike other voltage-gated potassium channels, however, the BKα subunit possesses an extramembrane-spanning sequence and a long cytoplasmic C-terminal region. The extramembrane-spanning sequence is known to interact with the β subunit (Morrow et al., 2006). The C-terminal cytoplasmic region, which represents two-thirds of the entire length of the channel, possesses two RCK (regulator of conductance for K^+) domains. Calcium (or magnesium) ions bind to these RCK domains and mediate a conformational change to facilitate channel opening. In addition to its contribution to calcium-dependent gating, the cytoplasmic C-terminal region of the BK channel α subunit undergoes extensive pre-mRNA splicing and posttranslational modifications (see chapter "Posttranscriptional and Posttranslational Regulation of BK Channels" by Shipston and Tian). Because of its size and variability in sequence, the cytoplasmic C-terminal region serves as binding sites for most of the interacting partners of BK channels (Fig. 1). Specific regions that are important for binding other proteins are summarized in Table 1.

Identification of interacting partners of a protein sheds light on its functions and regulatory mechanisms. Two main approaches that have mainly been employed to identify the interacting partners are a binary approach and a cocomplex approach. The binary approach identifies a direct physical

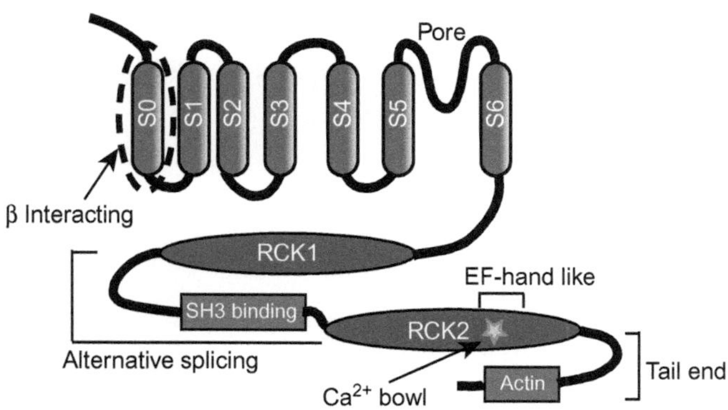

Fig. 1 The main binding sites of the BK channel α subunit for its interacting partners.

Table 1 BK Channel-Interacting Proteins

BK Channel-Interacting Protein	Interacting Motif	The Nature of Interaction	References
(a) BK channel subunits			
β1, β2, β3, and β4 subunits	S0 transmembrane	Direct	Giangiacomo et al. (1995) and Knaus et al. (1994)
γ Subunits		Direct	Yan and Aldrich (2010) and Yan and Aldrich (2012)
(b) Voltage-gated calcium channels			
L-type Cav1.2	?	Direct	Berkefeld, Sailer, Bildl, et al. (2006) and Grunnet and Kaufmann (2004)
P/Q-type Cav2.1	?	Direct	Berkefeld et al. (2006)
N-type Cav2.2	?	Direct	Berkefeld et al. (2006) and Loane, Lima, and Marrion (2007)
Cavβ1	Noncanonical SH3 binding motif	Direct	Zou, Jha, Kim, and Dryer (2008a)
(c) Cytoskeletal proteins			
Actin	Tail end	Direct	Zou, Jha, Kim, and Dryer (2008b)
Filamin A	Tail end of BKα or β subunit	Direct	Kim, Ridgway, and Dryer (2007)
Membrane associated guanylated kinase with interverted orientation 1 (MAGI-1)	Tail end	Direct	Ridgway, Kim, and Dryer (2009)
Cortactin	Noncanonical SH3 binding motif	Direct	Tian et al. (2006)
Microtubule-associated protein 1A (MAP-1A)	RCK2 and tail end	Direct	Park et al. (2004)
Dynamin-1	?	?	Gorini et al. (2010)

Table 1 BK Channel-Interacting Proteins—cont'd

BK Channel-Interacting Protein	Interacting Motif	The Nature of Interaction	References
β-Catenin	Calcium bowl	Direct	Bian et al. (2011) and Lesage, Hibino, and Hudspeth (2004)
Ankyrin Repeat Family A (ANKRA)	Tail end	Direct	Lim and Park (2005)
α-Catulin	RCK1 and RCK2 domains	Direct	Oh et al. (2015)
Dystrobrevin	Via α-catulin	Indirect	Oh et al. (2015)
Synaptopodin	Tail end	Direct	Kim, Suh, Chiu, and Dryer (2010)
(d) Signaling			
RACK1	Tail end	Direct	Isacson, Lu, Karas, and Cox (2007) and Surguchev, Bai, Joshi, and Navaratnam (2012)
Caveolin-1, -2, and -3	Tail end consensus sequence	Direct	Brainard, Miller, Martens, and England (2005), Suzuki, Yamamura, Ohya, and Imaizumi (2013b), and Wang et al. (2005)
Cereblon	RCK1 or RCK2 domain	Direct	Jo, Lee, Song, Jung, and Park (2005)
Syntaxin-1A	S0–S1 loop and the C-terminus	Direct	Cibulsky, Fei, and Levitan (2005) and Ling, Sheng, Braun, and Braun (2003)
GABA transporter 3	?	Direct	Singh, Li, Hall, et al. (2016)
Heat shock protein 60	?	Direct	Singh et al. (2016)
Apolipoprotein A1	?	Direct	Sokolowski et al. (2009)

"?" denotes not known.

interaction between two proteins and the cocomplex approach identifies components of a protein complex that is formed through both direct and indirect interaction. These two approaches are complementary and both are required to build a comprehensive interaction network. The most frequently used method for the binary approach has been a yeast two-hybrid screen using cDNA library constructed from various sources, such as different species, tissues, and organelles. The most common cocomplex approach employs affinity purification and subsequent identification of the components of the cocomplex through mass spectrometry. A large-scale proteomics approach utilizing computational methodologies can lead to complex interactomes, which shed light on regulatory mechanisms and biological pathways that a protein of interest is involved in. Another approach worthwhile to mention is genetic screening for modifiers. This screening approach identifies gene loci that suppress or enhance the phenotype of specific mutants that have a defect on the gene of interest. This approach can yield proteins that directly or indirectly interact with a protein of interest. While the genetic approach is powerful, it is limited to a small number of model organisms due to the costs and technical difficulty.

Currently, hundreds of proteins are reported to interact with BK channels in various systems in vitro and/or in vivo. These studies have provided insights into the mechanisms that regulate the BK channel functions and into the biological pathways mediated by BK channels. In the first part, we will focus on the binary interactions, and in the second part, we will review interactomes revealed by cocomplex methods. For the first part, we loosely categorized the interacting partners based on their cellular and molecular steps where they regulate ultimate BK channel function, although some interactions may impact on multiple steps: (1) increases in cytoplasmic calcium ions through calcium entry/release, (2) trafficking/targeting of BK channels to the plasma membrane, (3) localization of BK channels to specific plasma membrane compartments, (4) removal or degradation of BK channels, and (5) modulation of BK channel properties. In this chapter, we review many, but not all, proteins that function in the above steps, and discuss their contribution to physiological, cellular, and molecular functions of BK channels. In addition, we review the possible contribution of the BK channel interactome, which was assembled based on mass spectrometry and bioinformatics data, to our understanding of BK channel function.

2. CALCIUM CHANNELS THAT FUNCTIONALLY COUPLE WITH BK CHANNELS

A key characteristic of BK channels is their ability to respond to calcium ions. Thus, it is not surprising that BK channels are functionally coupled with calcium channels that mediate the increase of intracellular calcium ions. The main function of BK channels in this context is to prevent calcium channels from further activation and to limit calcium influx. Consistent with such functional coupling between BK channels and calcium channels, there are many studies that support the physical interaction between them. Given the importance of calcium channels in BK channel function, we will first discuss the interaction between BK channels and calcium channels.

2.1 Voltage-Gated Calcium Channels

The affinity for calcium ions of the RCK domains of BK channels is relatively low; the effective calcium ion concentration to activate BK channels is approximately ~10 μM. Such high concentrations are observed only near the entry and release sites of calcium ions due to the enormous cellular calcium buffering capacity. For this reason, BK channels have been thought to be intimately associated with calcium channels by direct physical interaction or indirectly by their close proximity in their localization.

Calcium influx in excitable cells occurs primarily through voltage-gated calcium (Cav) channels that are activated in response to membrane depolarization. Cav channels are grouped into Cav1 (L-type), Cav2 (N-, P/Q-, and R-type), and Cav3 (T-type) based on their pharmacological and biophysical properties. Cav channels consist of the pore-forming α subunit and auxiliary β, γ, and $\alpha_2\delta$ subunits. The auxiliary subunits modulate the biophysical properties, trafficking, and localization of Cav channels.

The first evidence for physical interaction between Cav and BK channels came from immunoprecipitation in rat brain membrane preparations (Grunnet & Kaufmann, 2004). The α subunit of BK channels coimmunoprecipitated with L-type calcium channels, and vice versa. This experiment was further refined and improved by Berkefeld et al. (2006). Affinity purification with BKα-specific antibodies on solubilized plasma membrane-enriched protein fractions prepared from rat brain extracts led to the identification of high-molecular-weight complexes of BK channels, ranging from 0.65 to 3 MDa. This size exceeded the size of the tetrameric BK channel complex consisting of four α (~120 kDa) and β (~30 kDa) subunits. To identify additional components from the high-molecular-weight BK channel complexes, total protein fractions eluted from affinity purification was subjected to nanoflow liquid chromatography tandem mass spectrometry (nano-LC MS/MS). MS spectra identified two BKβ subunits, BKβ2 and BKβ4, which are highly expressed in the CNS. In addition, this approach identified Cav1.2, Cav2.1, and Cav2.2, which encode the pore-forming subunits of the L-, P/Q-, and N-type Cav, respectively, and three Cavβ subunits, Cavβ1b, Cavβ2, and Cavβ3, indicating that these Cav and BK channels are coassembled together. When Cav2.1 and BK channels were heterologously expressed in *Xenopus* oocytes, they also exhibited the functional coupling in an electrophysiological analysis. The current output in giant inside-out patches excised from oocytes expressing BK and Cav2.1 channels in response to step depolarization exhibited a biphasic pattern; an initial inward Ca^{2+} current was followed by an outward K^{+} current, indicating that calcium influx fed to BK channels. Importantly, the functional properties of the BK-Cav coupling matched the criteria of calcium nanodomains, with an estimated distance of less than 10 nm. However, Cav2.3, a R-type Cav channel, and the Cavβ4 were not detected in the eluates resulting from affinity purification of BK channels. Coexpression of BK and Cav2.3 channels in heterologous cells did not result in calcium-mediated BK channel activation in response to step depolarization. These results indicate that BK channels associate with only specific subsets of Cav channels.

The interaction between BK and Cav channels appears to be mediated via the direct interaction between the α subunits of BK and Cav channels. Coexpression of the α subunits of BK and Cav channels in the absence of the respective auxiliary subunits in heterologous cells was sufficient for reciprocal coimmunoprecipitation of these channels. However, it has not been defined which regions of these two channels are responsible for direct interaction. It will be very interesting to identify specific regions of Cav channels, particularly given the specificity of Cav channels that can interact with BK channels. For instance, R-type Cav2.3 was not coeluted in two independent BK channel immunoprecipitation experiments by Berkefeld et al. (2006) and Loane et al. (2007). Furthermore, Loane et al. showed that the functional, and potentially physical, association of BK channels with N-type Cav2.2 is stronger than with L-type Cav1.2 and P/Q-type Cav2.1 (Loane et al., 2007).

While the above studies showed that the α subunits of BK channels and Cav directly interact with each other, there is also evidence that the C-terminus of the α subunit of BK channels interacts with the Cavβ1 (Zou, Jha, Kim, & Dryer, 2008a). Cavβ1 was identified as an interacting protein for BK channels in a yeast two-hybrid screen from a chick ciliary ganglion (CG) cDNA library. In vitro characterization showed that the calcium bowl and SH3 domain-binding motifs of SLO1 directly interacts with Cavβ1 in the absence of the Cav channel α subunit. Cavβ1 reduced calcium sensitivity of BK channels without affecting surface expression of BK channels in HEK293T cells. Overexpression of Cavβ1 inhibited calcium-activated potassium current while increasing the L-type calcium current in cultured chick CG neurons. This suggests that Cavβ may further increase calcium current by inhibiting BK channels, the negative regulator of calcium channels, while directly potentiating the calcium channels.

It appears reasonable to conclude that BK channels and a subset of Cav channels are physically and functionally coupled. This raises new questions: how do they become physically coupled? Do they coassemble before being transported to the plasma membrane? Is their localization dependent on each other? In our recent *Caenorhabditis elegans* study (Oh et al., 2015), we showed that BK channel localization was slightly compromised in select cholinergic motor neurons by a null mutation in *unc-2*, the gene encoding the P/Q-type Cav channel, while UNC-2 localization at presynaptic terminals was not altered in the absence of BK channels. Therefore, the localization of Cav channels at the plasma membrane is independent of BK channels, but the localization of BK channels is influenced by Cav channels. There is a lot to learn about how BK channels are trafficked from where they are synthesized to the final destination at the calcium nanodomains of the plasma membranes.

2.2 Other Calcium Channels

It is well established that BK channels are functionally linked to other calcium channels responsible for calcium release from the ER (eg, ryanodine receptors and IP_3 receptors). However, biochemical evidence of physical interactions between BK channels and those intracellular calcium channels is limited. In smooth muscle cells, ryanodine receptors cause local, transient calcium release events from the ER. These spontaneous calcium release events lead to the activation of nearby BK channels, which induces membrane hyperpolarization (Nelson et al., 1995; Perez, Bonev, Patlak, & Nelson, 1999). This potassium current is called the spontaneous transient outward current (STOC) and is critical for determining vascular muscle tone (Bolton & Imaizumi, 1996). Accordingly, blocking STOC depolarizes smooth muscle cells and induces constriction (Knot, Standen, & Nelson, 1998). Despite physiological evidence, the physical interaction of ryanodine receptors and BK channels has not been clearly defined. While immunostaining experiments demonstrated that these two proteins are colocalized (Lifshitz et al., 2011), how they are brought together is currently unknown.

Vascular smooth muscle cells respond to vasoactive ligands, which activate G protein-coupled receptors that are associated with G_q/phospholipase C. This signal transduction pathway produces inositol 1,4,5-trisphosphate (IP_3). IP_3 binds to IP_3 receptor in sarcoplasmic reticulum (SR) membrane, which induces calcium releases from the SR and facilitates cell contraction. IP_3 increases calcium sensitivity of BK channels through the interaction between IP_3 receptors and BK channels (Zhao et al., 2010). However, the conclusive evidence that BK channels directly interact with IP_3 receptors is lacking, although FRET and immunoprecipitation experiments demonstrated that IP_3 receptors and BK channels are found in close proximity (Weaver, Olsen, McFerrin, & Sontheimer, 2007; Zhao et al., 2010). It was previously shown that TRPC3 (canonical transient receptor potential type 3), which is a nonselective cation channel found in the caveolae of the plasma membrane, is in close spatial proximity to IP_3 receptors, which is present in the SR (Adebiyi, Narayanan, & Jaggar, 2011). This spatial proximity facilitates functional coupling of TRPC3 with IP_3 receptors. Another independent study demonstrated that TRPC3 are colocalized with BK channels, which are also found in caveolae, in differentiated podocytes (Kim, Alvarez-Baron, & Dryer, 2009). Together, these results provide the possibility that BK channels, TRPC3, and IP_3 receptors may form a subcellular domain that regulates calcium signaling.

3. ACTIN CYTOSKELETON

Numerous studies have demonstrated a close association of BK channel expression and its activity with actin cytoskeleton. BK channel activity is often modulated by extracellular signals, such as growth factors and mechanical stretch, and one of the mechanisms that transduce the extracellular signals to the BK channel activity appears to be through the actin cytoskeleton. In this section, we review some of the studies that link BK channels to the actin cytoskeleton.

3.1 Actin

Several independent studies have demonstrated that actin directly and indirectly interacts with BK channels (Tian et al., 2006; Zou, Jha, Kim, & Dryer, 2008b). For instance, actin directly binds to the distal cytoplasmic tail of the α subunit (Zou, Jha, Kim, & Dryer, 2008b). This binding appears to promote the normal surface expression of BK channels in HEK293T cells and chick CG. α and γ Actin were also shown to associate with BK channels in human myometrial smooth muscle cells (hMSMCs) and immunoelectron microscopy showed that both proteins localize to the cell surface caveolae in human myometrium. Won et al. reported that cytoskeleton limits lateral movement of BK channels when they were expressed in COS-7 cells by transfection of cDNA for rat BK channel α subunit and this effect was dependent on the C-terminal actin-binding domain of BKα (Won, Lee, & Park, 2011). Thus, actin cytoskeleton appears to be required for normal level of surface expression and localization at the plasma membrane.

Not only actin cytoskeleton affects BK channel expression and localization on the cell surface, but also it participates in regulating BK channel activity. Disruption of actin cytoskeleton by cytochalasin D and latrunculin A led to an increase in the open probability of the BK channel in cultured hMSMCs. Similarly, O'Malley et al. showed that destabilization of actin cytoskeleton activates BK channels in cultured hippocampal neurons of 1–3-day-old rat in response to leptin and insulin (O'Malley & Harvey, 2007; O'Malley, Irving, & Harvey, 2005). However, contrary to this study, Huang et al. reported that destabilization of actin cytoskeleton resulted in shortening of open half-life and reduction of opening frequency of BK channel in hippocampal CA1 pyramidal neurons isolated from male adult rat (Huang, Rao, Sun, & Gong, 2002). The discrepancy may come from the source of the neurons; while Huang et al. used isolated hippocampal CA1 pyramidal neurons from adult rats without further culturing them, O'Malley et al. used whole hippocampi of 1–3-day-old rat pubs to isolate neurons and the neurons were in culture for 4–14 days before the experiments were performed. Thus, there are many variables between these two studies, for example, the age, types of neurons, and cultured or not for a prolonged time.

It was shown that ethanol inhibits BK channel activity at a lower cytoplasmic Mg^{2+} concentration, while stimulating it at a higher concentration. This Mg^{2+}-dependent ethanol effect on BK channel activity requires intact actin cytoskeleton as destabilization of actin cytoskeleton abolishes the Mg^{2+} dependence of ethanol effect on BK channel activity in cultured embryonic hippocampal neurons and HEK cells (Marrero, Treistman, & Lemos, 2015). It is not clear, though, whether this effect is due to the Mg^{2+}'s effect on BK channel or on actin filament.

Stabilization of actin cytoskeleton inhibits stretch-induced activation of BK channels in rabbit coronary artery smooth muscle cells, indicating that dynamic actin cytoskeleton is required for the transduction of stretch signal to BK channel activity (Piao, Ho, & Earm, 2003). Together, BK channel expression, localization, and channel activity are intimately associated with actin cytoskeleton.

3.2 Synaptopodin

Synaptopodin was reported to interact and colocalize with BKα in mouse glomeruli (Kim et al., 2010). Coexpression of synaptopodin with BKα increases the steady-state expression level of BKα at the plasma membrane and macroscopic currents in HEK293T cells. Conversely, knockdown of synaptopodin decreases the surface expression and the channel current in a podocyte cell line. Synaptopodin associates with actin-binding proteins including Rho. Inhibition of Rho activity, which disrupts actin cytoskeleton, or treatment of cytochasin D resulted in a reduction of channel current density (surface expression) in podocytes.

3.3 Filamin A

In a yeast two-hybrid screen, filamin A, an actin-binding protein, was identified as a protein that interacts with BK channels independently of actin binding (Kim, Ridgway, & Dryer, 2007a). Filamin A forms perpendicular actin cross-links and contributes to changes in cell shape, stiffness, and motility. It was known that the cell surface expression of a number of receptors and ion channels is reduced in one of the melanoma cell lines that do not express filamin A. BK channel surface expression was also significantly reduced in this cell line, compared to another cell line that expresses filamin A, without any measurable difference in the total BK channel expression levels. Transient transfection of filamin A into this cell line restored BK channel expression in the plasma membrane. Hence, filamin A is necessary for the trafficking of BK channels from an intracellular compartment to the surface. However, contrary to this study, filamin A was reported to enhance dynamin-mediated endocytosis of the hyperpolarization activated cyclic nucleotide-gated channel 1 (HCN1) in adult hippocampal neurons, thereby negatively regulate the HCN1 channel surface expression

(Noam et al., 2014). It needs to be further investigated whether the seemingly opposite effects of filamin A are cell-type specific or channel specific. Thus, actin cytoskeletal networks may play important roles in regulating BK channel density at the plasma membrane by complex, yet-to-be elucidated mechanisms.

4. TARGETING/TRAFFICKING OF BK CHANNELS TO THE PLASMA MEMBRANE

The control of channel density at the plasma membrane is one of the major mechanisms to regulate BK channel function. Identification of BK channel-interacting proteins indicates that actin cytoskeletal networks, microtubule networks, and the machinery controlling trafficking from the ER to the plasma membrane play roles in BK channel surface expression.

4.1 Membrane-Associated Guanylate Kinase with Inverted Orientation Protein 1

Contrary to actin and filamin A that promote the BK channel surface expression, MAGI-1 (membrane-associated guanylate kinase with inverted orientation protein 1) inhibits BK channel expression on the cell surface. In a yeast two-hybrid screen, MAGI-1, a member of the membrane-associated guanylate kinase homologue family (MAGUK), was identified as a protein that interacts with the C-terminal proximal end of the BK channel α subunit (Ridgway et al., 2009). MAGI-1 possesses six PDZ domains and two WW domains and a guanylate kinase-like domain and interacts with actin cytoskeleton through an actin regulator, AFD/afardin. Similarly to other members of MAGUK family proteins, it is hypothesized that MAGI-1 forms a multiprotein complex near the plasma membrane that includes cell adhesion proteins, cytoskeletal proteins, and ion channels (Hruska-Hageman, Benson, Leonard, Price, & Welsh, 2004; Tanemoto, Toyohara, Abe, & Ito, 2008). However, contrary to other channels, coexpression of MAGI-1 reduced the surface expression of BK channels in podocytes (Ridgway et al., 2009). While the mechanism by which MAGI-1 reduces the levels of BK channel surface expression is not clear, this highlights that cytoskeletal proteins can regulate the surface levels of BK channels.

4.2 α-Tubulin and Microtubule-Associated Protein 1A

The light chain of microtubule-associated protein 1A (MAP1A) was identified as a BK channel binding protein in a yeast two-hybrid screen

(Park et al., 2004). MAP1A is mainly localized to the dendrite area and critical for activity-dependent branching and dendritic arbor stabilization (Szebenyi, Bollati, Bisbal, et al., 2005). While MAP1A does not appear to be essential for microtubule assembly, it may play a critical role in the maintenance of microtubule stability. Its absence in mice results in a decrease in the microtubule network particularly in the somatodendritic compartment of cerebellar Purkinje neurons, thus causing Purkinje cell degeneration and ataxia in adult mice (Liu, Lee, & Ackerman, 2015). Because mice-lacking BK channels show cerebellar ataxia and Purkinje cell dysfunction (Sausbier, Hu, Arntz, et al., 2004), MAP1A may play a role in BK channel localization or trafficking in Purkinje cells, although it has not been examined or reported whether BK channel localization or expression is altered in mice-lacking MAP1A. The light chain of MAP1A associates both actin cytoskeleton and microtubule network, but it is not known whether MAP1A links BK channels to actin cytoskeleton or microtubule network. MAP1A was shown to play a role in maintaining synaptic surface expression of Cav2.2 in hippocampal neurons in an actin cytoskeleton-dependent manner. Given the functional and physical coupling of BK channels with Cav2.2, MAP1A may tether both BK channels and Cav2.2 to the actin cytoskeleton.

BK channel was also shown to colocalize with microtubule networks in primary cultured astrocytes and in the cerebral cortex of newborn mice, and to bind to α-tubulin in an in vitro pull-down assay with brain extracts (Ou et al., 2009). How BK channels are tethered to the microtubule network is not known. Although MAP1A could be a candidate, Ou et al. failed to detect the association between MAP1A with BK channels in their study. This discrepancy also may lie at the tissue specificity or methodological issues. Ou et al. proposed that the microtubule network serves as a BK channel intracellular reservoir, where BK channel release to the plasma membrane is controlled by intracellular calcium. It is an intriguing hypothesis that needs further investigation. Microtubule dynamics may also modulate physical properties of BK channel, as Piao et al. reported that BK channel opening is increased by destabilization of microtubules and is decreased by stabilization of microtubules in rabbit coronary artery smooth muscle cells (Piao et al., 2003).

4.3 Trafficking from the Endoplasmic Reticulum: Cereblon

Cereblon was identified in a yeast two-hybrid screen as a protein that binds to either of the RCK domains of the BK channel (Jo et al., 2005). Cereblon functions as a substrate receptor protein for the CRL4A E3 ubiquitin ligase complex, which consists of DDB1, Cul4A, and Roc1 (Angers et al., 2006; Ito et al., 2010). Intriguingly, thalidomide, a small compound that used to be prescribed as a sedative and hypnotic but causes a malformation in limbs during development, inhibits, or alters the ubiquitin ligase activity of cereblon-CRL4A by direct binding to cereblon. BK channels are ubiquitinated in the endoplasmic reticulum by the cereblon-CRL4A E3 ubiquitin ligase (Liu et al., 2014). Ubiquitinated BK channels are not targeted for proteasome-mediated degradation, they are rather retained in the endoplasmic reticulum and could be released to the cell surface as functional channels upon deubiquitination. Blocking BK channel ubiquitination, disrupting the interaction between BK channels and cereblon, or treating with thalidomide can increase the levels of the cell surface BK channels. Consistent with the negative regulatory role of ubiquitination in BK channel surface expression, inactivation of cereblon-CRL4A in mice results in hypersensitivity to seizure induction. Similarly, gain-of-function mutations of BK channels in humans are known to cause generalized epilepsy (Du, Bautista, Yang, et al., 2005). Thus, cereblon-CRL4A E3 ubiquitin ligase plays a negative role in regulating the levels of BK channels on the cell surface. As ubiquitination of BK channels by cereblon-CRL4A sequesters BK channels in the ER, a deubiquitinase that counters the cereblon-CRL4A activity and thus promotes BK channel trafficking to the plasma membrane must be present in the ER. This deubiquitinase, if identified, will provide another regulating point for BK channel function.

5. PROTEINS THAT AFFECT THE LOCALIZATION OF BK CHANNELS TO SPECIFIC PLASMA MEMBRANE COMPARTMENTS

BK channels are not localized uniformly throughout the plasma membrane of neurons. An electron microscopy (EM) study in neurons showed that BK channels form two distinct pools: one pool, in which BK channels scattered throughout the neuronal plasma membrane, and the other pool that consists of clustered BK channels (Indriati et al., 2013; Kaufmann, Ferraguti, Fukazawa, et al., 2009; Kaufmann, Kasugai, Ferraguti, & Storm, 2010). An immuno-EM study in hippocampal and cerebellar Purkinje neurons showed that while Cav2.1 and BK channels both exhibited clustered and scattered localization in the neuronal plasma membrane, clustered, but not scattered, Cav2.1 channels were well colocalized with BK channels (Indriati et al., 2013). These results suggest that the two, clustered and scattered, pools of BK channels may differ in their coupling to specific calcium channels, and thus engage in different functions. Several proteins that have been reported to interact with BK channels are implicated in BK channel localization at specific plasma membrane compartments.

5.1 β-Catenin

A yeast two-hybrid screen using the C-terminus of BK channel α subunit as a bait led to the identification of β-catenin (Lesage et al., 2004). The detailed biochemical analysis showed that the calcium bowl located within the RCK2 domain interacts with β-catenin. In HEK293 human embryonic kidney cell line, β-catenin was colocalized with BK channels, and its overexpression increased the surface expression of BK channels (Bian et al., 2011). β-Catenin is a component of the cadherin cell adhesion complex, which mediates cell–cell adhesion and Wnt signaling. Cell adhesion complex regulates neuronal differentiation and migration, neurite outgrowth, synapse formation, and synaptic plasticity (Missler, Sudhof, & Biederer, 2012). Adhesion junctions are formed by the interaction between E-cadherins on two neighboring cells. β-Catenin links the adhesion junctions to the actin cytoskeleton through its interaction with α-catenin, which recruits a number of cytoskeletal proteins, including F-actin (Ratheesh & Yap, 2012). β-Catenin interacts with a synaptic protein, Lin7/Velis/MALS, whose interaction partner Lin2/CASK binds Cav channels (Maximov, Sudhof, & Bezprozvanny, 1999). Thus, the cell adhesion complex may play a role in bringing together BK channels and Cav channels for neural development and function.

5.2 Caveolins

The plasma membrane is a dynamic and heterogeneous lipid bilayer with various microdomains. One such microdomain is lipid rafts. Lipid rafts can be biochemically isolated based on their detergent-resistant property, which is attributed to their high content in cholesterol and sphingolipids. Caveolae are a special type of lipid rafts that has an omega-shaped invagination of the plasma membrane (Parton & del Pozo, 2013). Caveolae are involved in signal transduction, mechanosensing, and clathrin-independent endocytosis.

Caveolins are the main integral proteins of caveolae and serve as scaffolding proteins for caveolae. Immunoprecipitation experiments showed that whereas caveolin-1 and -2 associate with BK channels in smooth muscle cells (Brainard et al., 2005; Wang et al., 2005), caveolin-3 interacts with BK channels in skeletal muscle cells (Suzuki, Yamamura, Ohya, & Imaizumi, 2013b). Caveolins binds to other proteins through the interaction with a loosely defined consensus sequence that includes two or three aromatic amino acids. The tail end of the BK channel α subunit possesses this consensus binding sequence for caveolins (Alioua et al., 2008; Lu et al., 2010; Suzuki, Yamamura, Ohya, & Imaizumi, 2013b). Despite several independent observations that caveolins interact with BK channels, the exact role of caveolins in BK channel function is controversial. Earlier studies showed that treatment of uterus smooth muscle cells with a cholesterol-depleting agent, cyclodextrin, or caveolin-1 knockdown by small interference RNA in aortic endothelial cells led to a disruption of caveolae, and increased BK channel function (Brainard et al., 2005; Wang et al., 2005). However, a more recent study showed that the absence of caveolin-1 in mesenteric artery smooth muscle cells resulted in reduced levels of the BK channel/L-type Cav channel complex, and a reduced activation of BK channels in response to calcium entry via L-type calcium channels (Suzuki, Yamamura, Ohya, & Imaizumi, 2013a). It was postulated that caveolae promotes the formation of the BK channel/L-type Cav channel complex by serving as a platform that fosters the accumulation of both BK channels and Cav1.2. Further investigations are required to resolve whether the discrepancy in the two experiments results from cell-type-specific responses or experimental conditions.

5.3 Syntaxin-1A

Syntaxin-1A is localized at the plasma membrane of presynaptic neuronal terminals, and together with other synaptic proteins, mediates the fusion of synaptic vesicles to the plasma membrane. Coimmunoprecipitation of BK channels in hippocampal neurons identified syntaxin-1A as a BK channel-interacting protein (Ling et al., 2003). In vitro interaction experiments demonstrated that only syntaxin-1A, but not other syntaxins, interacts with BK channels. The interaction between syntaxin-1A and BK channels was further confirmed by two independent studies (Cibulsky et al., 2005; Gorini et al., 2010). Two opposite physiological roles of the interaction have been proposed. First, the interaction between BK channels and syntaxin-1A appears to increase the sensitivity of BK channels to calcium ions, which could prevent excessive release of neurotransmitters at presynaptic terminals

(Ling et al., 2003). Second, the interaction of syntaxin-1A with BK channels leads to the inhibition of BK channel activity by increasing the voltage necessary for channel opening and by slowing the rate of activation at a constant calcium concentration (Cibulsky et al., 2005). Given that syntaxin-1A was also reported to interact with N-type calcium channels, how three proteins interact, or compete, with one another at presynaptic terminals may have an impact on neurotransmitter release.

5.4 α-Catulin and the Dystrophin Complex

To identify binding partners of BK channels, yeast two-hybrid screens and immunoprecipitation/mass spectrometry have been widely used. As an alternative to these approaches, we developed a *C. elegans* forward genetic screen that can identify proteins regulating BK channel trafficking or localization (Abraham, Oh, Sancar, Richmond, & Kim, 2010). This genetic screen specifically took advantage of a locomotor phenotype of a *slo-1* gain-of-function mutant, *ky399* (Troemel, Sagasti, & Bargmann, 1999). This mutant has a conversion of glutamate to lysine at the amino acid position 350 (E350K) between the last transmembrane domain and RCK1 (Davies et al., 2003). A ring of eight negatively charged glutamate residues in this region controls the conductance of BK channels, because it is located at the entrance to the intracellular vestibule (Brelidze, Niu, & Magleby, 2003). It was shown that the absence of negative charge in this region increases the conductance. Indeed, *ky399* mutation causes a delayed inactivation of BK channels and results in sluggish, uncoordinated movements (Davies et al., 2003). Mutations in genes that are normally required for BK channel function, trafficking, or localization are expected to suppress the overactive BK channel phenotypes by reducing BK channel functions. The screen successfully yielded α-catulin, which has significant homology to α-catenin and vinculin (Abraham et al., 2010). Because these two proteins interact with several cytoskeletal proteins, including F-actin and paxilin, α-catulin is likely to interact with other cytoskeletal proteins. Intriguingly, our yeast two-hybrid screen with α-catulin as a bait identified dystrobrevin as an interacting protein (Oh, Abraham, van Hengel, et al., 2012). Dystrobrevin is a known component of the dystrophin complex, a macromolecular protein complex present in the plasma membrane. Mutations in genes encoding components of the dystrophin complex are implicated in different forms of muscular dystrophy. Consistent with the involvement of dystrobrevin, we found that loss of dystrophin, α-catulin, or dystrobrevin in *C. elegans* abolishes the clustering of BK channels in muscle cells without significantly impacting the total BK channel protein expression (Abraham et al., 2010; Kim, Pierce-Shimomura, et al., 2009; Sancar et al., 2011). In neurons, however, the dystrophin does not play a role in BK channel localization. Yet, α-catulin and dystrobrevin are critical for BK channel localization at presynaptic terminals. Loss-of-function mutations in the BK channel gene in *C. elegans* prolong evoked presynaptic response, when electrophysiological recordings at the neuromuscular junctions are carried out. Similarly, defect in BK channel localization at presynaptic terminals due to α-catulin mutation causes an increase in evoked presynaptic currents (Abraham et al., 2010). These results indicate that BK channel localization at presynaptic terminals is critical for normal channel function. Further characterization of BK channel localization in cholinergic neurons indicates that BK channels are organized as a macromolecular complex and localized by α-catulin, whose localization depends in part on dystrobrevin (Oh et al., 2015).

6. PROTEINS THAT MEDIATE THE REMOVAL OR DEGRADATION OF BK CHANNELS: DYNAMIN-1

The overall levels of BK channels at the plasma membrane can have an impact on cell excitability. In addition to the efficiency of BK channel trafficking to the plasma membrane, changes in BK channel removal from the plasma membrane will alter the levels of BK channels at the plasma membrane. However, the mechanisms by which BK channels are removed from the plasma membrane have not been well studied. One potential mechanism is dynamin-1-mediated endocytosis.

Dynamin-1 was identified from tandem mass spectrometry (MS/MS) analysis of proteins that were coimmunoprecipitated with BK channels in rodent brain extracts (Gorini et al., 2010). Endocytosis is a process that internalizes a portion of the plasma membrane along with membrane proteins embedded in it. Dynamin is localized at the sites of endocytosis and is involved in the formation and scission of invaginated nascent vesicles from the plasma membrane (Ferguson & De Camilli, 2012). Endocytosis provides a mechanism that can control the abundance of membrane proteins, such as G protein-coupled receptors and ion channels. Indeed, it has been demonstrated that a potassium channel, Kir2.3, is internalized from the plasma membrane via vesicle-mediated endocytosis in a dynamin-dependent manner (Mason, Jacobs, & Welling, 2008). It was also reported that a fraction of

BK channels could be found in the endosome when expressed in HEK293T cells (Toro et al., 2006). Thus, the density of BK channels could be regulated by endocytosis through the interaction with dynamin, although the direct interaction sites between BK channels and dynamin have not been demonstrated. Additional experiments that determine whether dynamin-1 indeed regulates the level of BK channels at presynaptic terminals are required to verify the functional significance of this interaction.

7. PROTEINS THAT ALTER CHANNEL FUNCTION

Considering that BK channels are an important regulator of a wide range of calcium-dependent processes, it is perhaps not surprising that BK channels themselves are under the control of many proteins that alter channel function, such as protein kinases or ubiquitin. In this section, proteins that were reported to alter BK channel function directly without affecting channel trafficking or surface expression are described.

7.1 Cortical Actin Binding Protein

A noncanonical SH3 domain-binding motif is positioned between the RCK1 and RCK2 domains of the BK channel α subunit (Fig. 1). To identify proteins that bind to this motif, SH3 domain arrays derived from human proteins were screened using an overlay assay (Tian et al., 2006). From this screen, cortactin (cortical actin-binding protein) was identified. Cortactin is an SH3 adaptor protein that promotes actin cytoskeleton rearrangement or polymerization. Given its role as an adaptor, cortactin was postulated to connect the actin cytoskeleton to BK channels at the plasma membrane. Disruption of the SH3-mediated interaction increased the open probability of BK channels in inside-out and cell-attached patch electrophysiology experiments. However, how cortactin alters the BK channel open probability has yet to be determined,

7.2 Ankyrin Repeat Family A

Ankyrin repeat family A (ANKRA) was identified as a BK channel-interacting protein in a yeast two-hybrid screen using a rat brain library (Lim & Park, 2005). Ankyrin repeats are commonly found in proteins that mediate the attachment of transmembrane proteins to cortical cytoskeleton networks. By extension, ANKRA may act as an adapter to distinct membrane proteins and recruit specific modulating proteins near the target membrane proteins, including ion channels and membrane receptors (Lim & Park, 2005). However, while ANKRA alters the gating kinetics of BK channel in a calcium-dependent manner, it does not alter the levels of BK channels on the surface in heterologously expressed cells. Interestingly, in addition to BK channels, both ANKRA and MAGI-1 can interact with LRP2/megalin (low-density lipoprotein receptor-related protein 2), a 600-kDa transmembrane protein with a large extracellular domain (Patrie, Drescher, Goyal, Wiggins, & Margolis, 2001; Rader, Orlando, Lou, & Farquhar, 2000). While all three proteins are prominently expressed in epithelial cells, such as kidney cells, they are also expressed in other cells, including neurons. Recent phenotype analyses of LRP2 KO mice have shown that in addition to neurodevelopmental deficits, the LRP2 KO mice exhibited abnormalities in epithelial tissues, including kidney and lung, due to defects in endocytosis of ligands (Spoelgen et al., 2005; Willnow et al., 1996). It needs to be further investigated whether trafficking or endocytosis of BK channels is altered in these KO mice.

7.3 Protein Kinases and Phosphatases

Modification of BK channel activity by phosphorylation-dependent signaling pathways has been well documented. Since more details are described in "Posttranscriptional and Posttranslational Regulation of BK Channels" by Shipston and Tian, we will briefly overview the interaction of protein kinases with BK channels. The main protein kinases include serine/threonine kinases (cAMP- and cGMP-dependent kinases (PKA and PKG) and protein kinase C (PKC) and calmodulin-dependent kinase II (CaMKII)), and tyrosine kinases (Src, FAK, and Syk). Some of these kinases were reported to directly interact with BK channels. In case of protein kinase A, a noncanonical, leucine-zipper region of the BK channel, which is located within the RCK1 domain, serves as an anchor for a PKA-associated complex (Tian, Coghill, MacDonald, Armstrong, & Shipston, 2003). This anchor site appears to be important for the modification of BK channel activity by PKA; blocking the interaction by alanine substitution in the BK channel α subunit or injection of a leucine-zipper peptide into HEK293 cells ectopically expressing BK channels eliminated the regulation of BK channels by PKA. Conversely, protein phosphatases A1 and A2 that reverse the action of PKA were shown to have an opposite effect of PKA on BK channel (Tian, Knaus, & Shipston, 1998; Zhou, Ruth, Schlossmann, Hofmann, & Korth, 1996). In addition to serine/threonine kinases, the tyrosine kinases focal adhesion kinase (FAK) and Syk were reported to interact with BK channels in bone cells (Rezzonico, Cayatte, Bourget-Ponzio, et al., 2003; Rezzonico et al., 2002). An increased activity of BK channels under hypotonic conditions was correlated with the increased association of FAK with BK channels in these bone cells. However, the contribution of tyrosine phosphorylation to BK channel activity was not demonstrated in this study.

7.4 Receptor Activated C Kinase 1

Receptor activated C kinase 1 (RACK1) was identified in a yeast two-hybrid screen of a human aorta library using the tail end of the BK channel α subunit as bait (Isacson et al., 2007). The interaction between RACK1 and BK channels was also observed in hair cells (Surguchev et al., 2012). RACK1 is a scaffolding protein that possesses the tryptophan-aspartate repeat (WD) domain, which is also found in the β subunit of G protein (Gβ). The name RACK1 was originated from its association with an active form of PKC. It was proposed that the interaction of BK channels with RACK1 mediates the observed increase in the BK channel expression on the cell surface in response to the PKC activation (Surguchev et al., 2012). However, RACK1 interacts with numerous proteins independently of PKC directly or indirectly to integrate distinct signaling inputs for various cellular functions. The proteins that are known to interact with RACK1 include protein kinases (MAP kinase, PKC, and protein tyrosine kinase Src and Fyn), ribosomal proteins, and proteins that control cell death/survival and cell proliferation (Gandin, Senft, Topisirovic, & Ronai, 2013). Intriguingly, the protein tyrosine kinases Src and Fyn are also known to interact with BK channels. Thus, it is possible that the BK channel surface expression is mediated through cross-talks between different kinases and RACK1.

7.5 Apolipoprotein A1

Apolipoprotein A1 (ApoA1) was identified as a BK channel-interacting protein in a yeast two-hybrid screen using a chick cochlea cDNA library (Sokolowski et al., 2009) and also in interactome study using mouse cochlea (Kathiresan, Harvey, Orchard, Sakai, & Sokolowski, 2009). The interaction was verified by reciprocal coimmunoprecipitation and colocalization in both studies. In electrophysiological study with transfected HECK293 cells showed that ApoA1 slows down the activation and deactivation kinetics of BK channels in a calcium-independent manner.

Mutation in transmembrane serine protease 3 (Tmprss3) results in deafness in both human and mouse (Guipponi, Vuagniaux, Wattenhofer, et al., 2002; Scott, Kudoh, Wattenhofer, et al., 2001). Protein networks in wild-type and Tmprss3 mutant mouse cochlea were compared using differential proteomics and bioinformatics. BKα was emerged as the nod that connects the most proteins that are differentially expressed in Tmprss3 mutant cochlea (Molina et al., 2013). BKα was absent in the neck of the inner hair cells in the Tmprss3 mutant cochlea. ApoA1 was also significantly down regulated in Tmprss3 mutant cochlea. Given the direct interaction between ApoA1 and BKα, it is possible that ApoA1 may also contribute to the stable expression of BKα in cochlea.

8. EXAMPLES OF BK CHANNEL INTERACTOMES

Biochemical isolation of cocomplex and subsequent mass spectrometry identified proteins that directly or indirectly associate with BK channels. These studies provide the foundation for building global interactomes, which shed light on the biological pathways and regulatory networks of BK channel function. In this section, we will review some of the studies that took cocomplex approaches.

8.1 Hemoxygenase 2

Inhibition of BK channel activity by hypoxia plays important roles in normal physiology, and possibly in pathophysiology, of many tissues. With a cocomplex approach using immunoprecipitation and mass spectrometry, Williams et al. showed that hemoxygenase 2 (HO-2) associates with BK channel complex in transfected HECK293 cells (Williams et al., 2004). Using O2 and NADPH as substrates, HO-2 produces iron, biliverdin, and CO. Under normoxic condition, HO-2 activates BK channels through its catalytic products CO and biliverdin. Hypoxic condition leads to lower CO production, which depresses BK channel activity, leading to the hyperexcitation. Thus, HO-2 acts as a sensor for oxygen and transduces the signal to BK channel through CO production. Supporting the role of CO in BK channel activation, it was shown that CO activation of BK channels requires an aspartic acid and two histidine residues in the RCK1 domain (Hou, Xu, Heinemann, & Hoshi, 2008).

8.2 Myelin Binding Protein

Kim et al. used a cocomplex approach to identify proteins that bind to the tail region of the BK channel (Kim, Jo, Song, Park, & Park, 2007). They fused the tail region (C-terminal region including RCK2 domain) of rat BK channel to GST, and used the resulting GST-BK-tail protein to isolate coprecipitated proteins in rat brain lysates. The coprecipitated proteins were in turn identified by MudPIT (multidimensional protein identification technology), which identifies components of protein mixtures in solution. In this study, they identified 53 proteins and among them myelin binding protein (MBP) was the major protein and further investigated. They showed that MBP increased the calcium sensitivity, but not the voltage sensing, of BK channel. Based on the interaction between calmodulin (CaM) and MBP, they proposed that MBP recruits CaM to the vicinity of BK channels, thereby influencing the calcium sensitivity of BK channels.

8.3 Isoform-Specific-Interacting Proteins

There are over 20 splice variants of mammalian BK channels. Ma et al. examined 3 C-terminal splice variants, BKα-DEC, BKα-ERL, and BKα-VYR (Ma et al., 2007). When transfected into COS-7 cells, BKα-DEC was mostly found inside of the cells. They concluded that BKα-DEC is less efficient in trafficking from the ER to the plasma membrane, because BKα-DEC colocalizes with an ER protein, calreticulin. However, Singh et al. found that BKα-DEC localizes to mitochondria in cardiomyocytes (Singh et al., 2013). The variant used by Ma et al. includes 11 additional amino acids. It is not known if the extra 11 amino acids or the cell type used in the experiment contributed to the difference. In an effort to uncover the molecular mechanism underlying this trafficking deficiency, Ma et al. employed proteomic approach to identify proteins that associate with the unique C-terminal sequences of BKα-DEC. They used a GST fusion protein to pull-down proteins associating with the C-terminus of BKα-DEC from mouse brain lysates. This construct also included a small fragment that is common in all three isoforms. They identified actin, tubulin, and protein phosphatase 2A (PP2A). In this study, PP2A was further confirmed as the BKα-DEC-specific binding partner. It needs to be further investigated if the dephosphorylation by PP2A is directly responsible for the inefficient trafficking of the BKα-DEC in the transfected cells. Nonetheless, this study suggests that the diverse functional and biophysical properties might be achieved by recruiting isoform-specific protein complexes.

8.4 BK Channel Interactomes

BK channel interactomes can be built based on known physical interactions that involve BK channels. Proteins that directly interact with BK channels can be used as a framework for delineating proteins that make second-order interactions with BK channels (Kathiresan et al., 2009). These second-order interactions can be obtained from other available interactome datasets that do not involve BK channels. The resulting BK channel interactome can be further expanded to include the third- and fourth-order interactions. We overview two examples that successfully used this strategy to build BK channel interactome.

8.4.1 Interactome in Mouse Whole Brain

Singh et al. performed immunoprecipitation, one-dimensional gel electrophoresis, and LC/MS/LS to identify proteins that coprecipitate with BK channels (Singh et al., 2016). Immunoprecipitation experiments were performed in wild-type and control BKα knockout mouse brain lysates using control or anti-Maxi-K rabbit polyclonal IgG antibodies. In wild-type mouse brain lysates they identified 255 proteins specific to the Maxi-K antibodies, but not to the control antibodies. In addition, they identified 911 proteins unique to wild-type mouse brain lysate, but not to BKα knockout brain lysate, in immunoprecipitation with the Maxi-K antibodies. The majority of the identified proteins localize in the cytoplasm, plasma membrane, nucleus, and mitochondria, while only a small fraction of them are distributed in the Golgi apparatus, endoplasmic reticulum, and extracellular proteins. They further validated the interactions of BK channels with HSP60 and GABA transporter 3 (GAT3) using coimmunoprecipitation and colocalization analyses in transfected HEK293T cells and in rat primary hippocampal neurons. This study also demonstrated the presence of BK channels in the mitochondria of brain by showing colocalization of BK channels with mitochondrial markers. Although it was not demonstrated experimentally, the BK channel antibody used in this study appears to precipitate isoforms present in multiple organelles, as they identified binding proteins from all of the cellular compartments. This study provided a list of potential binding partners that can be further exploited in building interactomes using bioinformatics tools.

8.4.2 Interactome of Synaptic Proteins

A BK channel interactome was also built in the context of synaptic proteins. Gorini et al. carried out coimmunoprecipitation experiments in a mouse cortical membrane preparation using antibodies recognizing BK channels

(Gorini et al., 2010). Coprecipitated proteins were identified by either western blot analysis or LC–MS/MS. In addition to syntaxin-1A, this study identified several proteins found in presynaptic terminals, including dynamin-1, PI-3 kinase, γ-tubulin complex components, and Na^+/K^+-transporting ATPase. With additional coimmunoprecipitation experiments using antibodies recognizing other presynaptic proteins, they assembled a protein–protein interaction network of BK channels, whose nodes include calcium channels, cytoskeletal proteins, and components of the synaptic vesicle fusion/recycle machinery, such as syntaxin-1A, SNAP-25 (synaptosome-associated protein of 25 kDa), VAMP-2 (vesicle-associated membrane protein-2), and dynamin-1.

8.4.3 Interactome in Cochlea

To identify proteins that interact with BK channels in cochlea, the coaffinity purification method was used in a low-stringency condition that could pull down additional proteins that are directly or indirectly associated with BK channels. These proteins were separated by 2D gel, and individual protein spots on the gel were identified using mass spectrometry. With this approach, Kathiresan et al. identified 174 putative binding partners from mouse cochlea (Kathiresan et al., 2009) and Sokolowski et al. identified 110 putative partners in chick cochlea (Sokolowski, Orchard, Harvey, Sridhar, & Sakai, 2011). By combining their experimental data with other interactome data, additional second-order interactors with BK channels were identified and BK channel interactomes were assembled.

In mouse cochlea, Kathiresan et al. used 174 primary interacting proteins, to identify 199 secondary proteins, which mediate 234 protein–protein interactions, and then built a network from these primary and secondary interactions (Kathiresan et al., 2009). The network consists of 199 nodes and 234 edges. Eighty-seven percent of the proteins in this network are linked together and form one large global network, and the rest of the proteins form 12 small networks of five or less nodes. The large network consisted of 12 major hubs with 10 central proteins, CaM, α-tubulin, ATP synthase, β-subunit, calretiulin, chromobox homolog 1, γ-actin, NMDA receptor, protein kinase ε, protein SET, and ubiquitin. The majority of the primary interacting proteins are involved in metabolism and trafficking/scaffolding.

In chick cochlea, Sokolowski et al. identified 110 putative BK-associated proteins (BKAPs) and verified the interaction of 16 proteins by reciprocal coimmunoprecipitation with chick cochlea lysates (Sokolowski et al., 2011). By incorporating informatics tools, they built the interactome with 193 independent proteins. The interactome included eight major hubs, containing a central protein connected to six or more partners, some of which are linked to the larger global network. Three of the hubs have connections to apoptosis, other three contribute to neuronal function, and the remaining two are involved in metabolic processes or calcium binding. This interactome revealed a possible link of BK channels to cell survival and apoptosis. In CHO cells transfected with a BKα variant, BK-DEC, they showed that knockdown of select BKAPs that are associated with life/death signals alters the expression of BK-DEC. Studies on the BK channel interactomes in both mouse and chick cochlea provided new insights into previously unrecognized biological processes that BK channels might participate in, such as metabolisms and cell survival.

A large number of binding partners for BK channels can be explained in part by the presence of intrinsic disorder regions (IDRs) in BKα. IDRs have no defined structure and are able to adopt different conformation upon interacting with different partners (Oldfield & Dunker, 2014). Peng et al. examined IDRs in 27 BKα variants from mouse cochlea and BKα-associated proteins (Peng, Sakai, Kurgan, Sokolowski, & Uversky, 2014). The C-terminus, the linker region between the RCK1 and RCK2 domains, the linker between first two transmembrane domains, and a small patch within the RCK2 domain are predicted to have IDRs. Coincidently, these IDRs are interaction sites for numerous known BKα binding partners and show sequence diversity due to alternative splicing. This computational analysis thus suggested that the IDRs of BKα provide protein–protein interaction interfaces necessary for a variety of protein binding.

While BK channel interactome has been useful to unravel the novel relationship between BK channels and other unsuspected biological processes, caution is warranted in interpreting the resulting data. It is possible that biologically significant interactions are mixed with false positives. Furthermore, some interactions of BK channels with their partners may form a highly stable complex, whereas others may be transient with high dissociation constants. Given that all of the interactions are not equal, it would be helpful to introduce weighted coefficients to BK channel interactome by ranking the interactions based on the strength and confidence (Kiemer, Costa, Ueffing, & Cesareni, 2007).

9. DISCUSSION

Extensive studies on BK channel interactome have identified hundreds of proteins thus far. Are all of these reported interactions biologically significant in vivo? It will not be definitively known until further in vivo experimental data are available for many of the reported interactions. However, at least two possibilities, cell-type- and isoform-specific interactomes and possible experimental artifacts, may explain this seemingly vast number of BK channel-interacting proteins.

A number of proteins are capable of binding to the same domain of the BK channel at least in vitro in isolation. However, due to the physical constraint, not all of the reported BK channel-interacting proteins are likely to be engaged at a given time. A cell-type-specific proteome is likely to limit the BK channel interactome. For example, the interactions of BK channels with synaptic proteins are likely to be present in neurons but not in muscles. It is also likely that BK channels interact with distinct sets of proteins in different cellular compartments. For example, at presynaptic terminals, BK channels form a complex with CaV channels, and in the ER, BK channels interact with the trafficking machinery. Considering the intrinsic disorder of BKα, cellular environments in different compartments will play very important roles in binding partner selection.

It is also possible that some of the reported interactions with BK channel may not be significant in vivo. A majority of the BK channel-interacting proteins were identified in vitro by either biochemical purification, such as coimmunoprecipitation or yeast two-hybrid screens. Disruption of cell structure during biochemical purification has a potential to produce artificial interactions that may not occur in vivo due to the compartmentalization of the interacting proteins. Yeast two-hybrid screens are also intrinsically prone to produce false-positive interactions. Thus, interactions identified on the basis of proteomics or yeast two-hybrid screens must be verified by independent experimental evidence that involves functional assays in vivo. Changes of the BK channel properties can be measured after disrupting the interaction between BK channels and its partner in native cells. Heterologous expression system may not provide relevant cellular milieu due to different proteomes compared to native cells. If model organisms that lack BK channel-interacting partners (eg, *C. elegans*, *Drosophila*, and mice) are available, the contribution of specific interactions to BK channel function can be evaluated at the physiological and behavioral levels. In the case that BK channel-interacting proteins interact with multiple proteins, the physiological and behavioral outcomes of the null mutant animals may or may not correlate with BK channel function. Thus, in vitro and in vivo experimental evidence must be considered to evaluate the significance of the interaction between BK channels and its partner proteins.

In summary, studies on the protein networks interacting with BK channels shed a light on the regulatory mechanisms that allow BK channels to integrate multiple cellular signaling pathways. Given that BK channels have been linked to many distinct diseased conditions, including epilepsy, alcohol dependence, autism spectrum disorders, and hypertension, assembling elaborate, complete, and tissue-specific BK channel interactomes will help us to expand our knowledge on how altered BK channel functions contribute to the diseased conditions. This knowledge can be used for the design of pharmaceutical interventions that block specific regulatory mechanisms by targeting the interaction of BK channels and their interacting proteins.

ACKNOWLEDGMENTS

The authors apologize to colleagues whose work could not be cited due to space limitation. Work in the laboratory of the authors was supported by a DePaul-RFUMS pilot grant, and a grant from the National Institute of Health.

REFERENCES

Abraham, L. S., Oh, H. J., Sancar, F., Richmond, J. E., & Kim, H. (2010). An alpha-catulin homologue controls neuromuscular function through localization of the dystrophin complex and BK channels in *Caenorhabditis elegans*. *PLoS Genetics*, *6*, e1001077.

Adebiyi, A., Narayanan, D., & Jaggar, J. H. (2011). Caveolin-1 assembles type 1 inositol 1,4,5-trisphosphate receptors and canonical transient receptor potential 3 channels into a functional signaling complex in arterial smooth muscle cells. *Journal of Biological Chemistry*, *286*, 4341–4348.

Alioua, A., Lu, R., Kumar, Y., Eghbali, M., Kundu, P., Toro, L., et al. (2008). Slo1 caveolin-binding motif, a mechanism of caveolin-1-Slo1 interaction regulating Slo1 surface expression. *Journal of Biological Chemistry*, *283*, 4808–4817.

Angers, S., Li, T., Yi, X., MacCoss, M. J., Moon, R. T., & Zheng, N. (2006). Molecular architecture and assembly of the DDB1-CUL4A ubiquitin ligase machinery. *Nature*, *443*, 590–593.

Berkefeld, H., Sailer, C. A., Bildl, W., et al. (2006). BKCa-Cav channel complexes mediate rapid and localized Ca^{2+}-activated K^{+} signaling. *Science*, *314*, 615–620.

Bian, S., Bai, J. P., Chapin, H., Le Moellic, C., Dong, H., Caplan, M., et al. (2011). Interactions between beta-catenin and the HSlo potassium channel regulates HSlo surface expression. *PloS One*, *6*, e28264.

Bolton, T. B., & Imaizumi, Y. (1996). Spontaneous transient outward currents in smooth muscle cells. *Cell Calcium*, *20*, 141–152.

Brainard, A. M., Miller, A. J., Martens, J. R., & England, S. K. (2005). Maxi-K channels localize to caveolae in human myometrium: A role for an actin-channel-caveolin

complex in the regulation of myometrial smooth muscle K^+ current. *American Journal of Physiology. Cell Physiology, 289*, C49–C57.
Brelidze, T. I., Niu, X., & Magleby, K. L. (2003). A ring of eight conserved negatively charged amino acids doubles the conductance of BK channels and prevents inward rectification. *Proceedings of National Academy of Science of the United States of America, 100*, 9017–9022.
Cibulsky, S. M., Fei, H., & Levitan, I. B. (2005). Syntaxin-1A binds to and modulates the Slo calcium-activated potassium channel via an interaction that excludes syntaxin binding to calcium channels. *Journal of Neurophysiology, 93*, 1393–1405.
Davies, A. G., Pierce-Shimomura, J. T., Kim, H., VanHoven, M. K., Thiele, T. R., Bonci, A., et al. (2003). A central role of the BK potassium channel in behavioral responses to ethanol in *C. elegans*. *Cell, 115*, 655–666.
Du, W., Bautista, J. F., Yang, H., et al. (2005). Calcium-sensitive potassium channelopathy in human epilepsy and paroxysmal movement disorder. *Nature Genetics, 37*, 733–738.
Ferguson, S. M., & De Camilli, P. (2012). Dynamin, a membrane-remodelling GTPase. *Nature Reviews Molecular Cell Biology, 13*, 75–88.
Gandin, V., Senft, D., Topisirovic, I., & Ronai, Z. A. (2013). RACK1 function in cell motility and protein synthesis. *Genes & Cancer, 4*, 369–377.
Giangiacomo, K. M., Garcia-Calvo, M., Knaus, H. G., Mullmann, T. J., Garcia, M. L., & McManus, O. (1995). Functional reconstitution of the large-conductance, calcium-activated potassium channel purified from bovine aortic smooth muscle. *Biochemistry, 34*, 15849–15862.
Gorini, G., Ponomarva, O., Shores, K. S., Person, M. D., Harris, R. A., & Mayfield, R. D. (2010). Dynamin-1 co-associates with native mouse brain BKCa channels: Proteomics analysis of synaptic protein complexes. *FEBS Letters, 584*, 845–851.
Grunnet, M., & Kaufmann, W. A. (2004). Coassembly of big conductance Ca^{2+}-activated K^+ channels and L-type voltage-gated Ca^{2+} channels in rat brain. *Journal of Biological Chemistry, 279*, 36445–36453.
Guipponi, M., Vuagniaux, G., Wattenhofer, M., et al. (2002). The transmembrane serine protease (TMPRSS3) mutated in deafness DFNB8/10 activates the epithelial sodium channel (ENaC) in vitro. *Human Molecular Genetics, 11*, 2829–2836.
Hou, S., Xu, R., Heinemann, S. H., & Hoshi, T. (2008). The RCK1 high-affinity Ca^{2+} sensor confers carbon monoxide sensitivity to Slo1 BK channels. *Proceedings of National Academy of Science of the United States of America, 105*, 4039–4043.
Hruska-Hageman, A. M., Benson, C. J., Leonard, A. S., Price, M. P., & Welsh, M. J. (2004). PSD-95 and Lin-7b interact with acid-sensing ion channel-3 and have opposite effects on H^+-gated current. *Journal of Biological Chemistry, 279*, 46962–46968.
Huang, H., Rao, Y., Sun, P., & Gong, L. W. (2002). Involvement of actin cytoskeleton in modulation of Ca(2+)-activated K(+) channels from rat hippocampal CA1 pyramidal neurons. *Neuroscience Letters, 332*, 141–145.
Indriati, D. W., Kamasawa, N., Matsui, K., Meredith, A. L., Watanabe, M., & Shigemoto, R. (2013). Quantitative localization of Cav2.1 (P/Q-type) voltage-dependent calcium channels in Purkinje cells: Somatodendritic gradient and distinct somatic coclustering with calcium-activated potassium channels. *The Journal of Neuroscience, 33*, 3668–3678.
Isacson, C. K., Lu, Q., Karas, R. H., & Cox, D. H. (2007). RACK1 is a BKCa channel binding protein. *American Journal of Physiology. Cell Physiology, 292*, C1459–C1466.
Ito, T., Ando, H., Suzuki, T., Ogura, T., Hotta, K., Imamura, Y., et al. (2010). Identification of a primary target of thalidomide teratogenicity. *Science, 327*, 1345–1350.
Jo, S., Lee, K. H., Song, S., Jung, Y. K., & Park, C. S. (2005). Identification and functional characterization of cereblon as a binding protein for large-conductance calcium-activated potassium channel in rat brain. *Journal of Neurochemistry, 94*, 1212–1224.

Kathiresan, T., Harvey, M., Orchard, S., Sakai, Y., & Sokolowski, B. (2009). A protein interaction network for the large conductance Ca(2+)-activated K(+) channel in the mouse cochlea. *Molecular & Cellular Proteomics, 8*, 1972–1987.
Kaufmann, W. A., Ferraguti, F., Fukazawa, Y., et al. (2009). Large-conductance calcium-activated potassium channels in purkinje cell plasma membranes are clustered at sites of hypolemmal microdomains. *The Journal of Comparative Neurology, 515*, 215–230.
Kaufmann, W. A., Kasugai, Y., Ferraguti, F., & Storm, J. F. (2010). Two distinct pools of large-conductance calcium-activated potassium channels in the somatic plasma membrane of central principal neurons. *Neuroscience, 169*, 974–986.
Kiemer, L., Costa, S., Ueffing, M., & Cesareni, G. (2007). WI-PHI: A weighted yeast interactome enriched for direct physical interactions. *Proteomics, 7*, 932–943.
Kim, E. Y., Alvarez-Baron, C. P., & Dryer, S. E. (2009). Canonical transient receptor potential channel (TRPC)3 and TRPC6 associate with large-conductance Ca^{2+}-activated K^{+} (BKCa) channels: Role in BKCa trafficking to the surface of cultured podocytes. *Molecular Pharmacology, 75*, 466–477.
Kim, H., Jo, S., Song, H. J., Park, Z. Y., & Park, C. S. (2007). Myelin basic protein as a binding partner and calmodulin adaptor for the BKCa channel. *Proteomics, 7*, 2591–2602.
Kim, H., Pierce-Shimomura, J. T., Oh, H. J., Johnson, B. E., Goodman, M. B., & McIntire, S. L. (2009). The dystrophin complex controls bk channel localization and muscle activity in *Caenorhabditis elegans. PLoS Genetics, 5*, e1000780.
Kim, E. Y., Ridgway, L. D., & Dryer, S. E. (2007). Interactions with filamin A stimulate surface expression of large-conductance Ca^{2+}-activated K^{+} channels in the absence of direct actin binding. *Molecular Pharmacology, 72*, 622–630.
Kim, E. Y., Suh, J. M., Chiu, Y. H., & Dryer, S. E. (2010). Regulation of podocyte BK(Ca) channels by synaptopodin, Rho, and actin microfilaments. *American Journal of Physiology. Renal Physiology, 299*, F594–F604.
Knaus, H. G., Folander, K., Garcia-Calvo, M., Garcia, M. L., Kaczorowski, G. J., Smith, M., et al. (1994). Primary sequence and immunological characterization of beta-subunit of high conductance Ca(2+)-activated K^{+} channel from smooth muscle. *Journal of Biological Chemistry, 269*, 17274–17278.
Knot, H. J., Standen, N. B., & Nelson, M. T. (1998). Ryanodine receptors regulate arterial diameter and wall [Ca^{2+}] in cerebral arteries of rat via Ca^{2+}-dependent K^{+} channels. *The Journal of Physiology, 508*(Pt 1), 211–221.
Lee, U. S., & Cui, J. (2009). {beta} Subunit-specific modulations of BK channel function by a mutation associated with epilepsy and dyskinesia. *The Journal of Physiology, 587*, 1481–1498.
Lesage, F., Hibino, H., & Hudspeth, A. J. (2004). Association of beta-catenin with the alpha-subunit of neuronal large-conductance Ca^{2+}-activated K^{+} channels. *Proceedings of National Academy of Science of the United States of America, 101*, 671–675.
Lifshitz, L. M., Carmichael, J. D., Lai, F. A., Sorrentino, V., Bellve, K., Fogarty, K. E., et al. (2011). Spatial organization of RYRs and BK channels underlying the activation of STOCs by Ca(2^{+}) sparks in airway myocytes. *Journal of General Physiology, 138*, 195–209.
Lim, H. H., & Park, C. S. (2005). Identification and functional characterization of ankyrin-repeat family protein ANKRA as a protein interacting with BKCa channel. *Molecular Biology of the Cell, 16*, 1013–1025.
Ling, S., Sheng, J. Z., Braun, J. E., & Braun, A. P. (2003). Syntaxin 1A co-associates with native rat brain and cloned large conductance, calcium-activated potassium channels in situ. *The Journal of Physiology, 553*, 65–81.
Liu, Y., Lee, J. W., & Ackerman, S. L. (2015). Mutations in the microtubule-associated protein 1A (Map1a) gene cause Purkinje cell degeneration. *The Journal of Neuroscience, 35*, 4587–4598.

Liu, J., Ye, J., Zou, X., Xu, Z., Feng, Y., Zou, X., et al. (2014). CRL4A(CRBN) E3 ubiquitin ligase restricts BK channel activity and prevents epileptogenesis. *Nature Communications*, *5*, 3924.

Loane, D. J., Lima, P. A., & Marrion, N. V. (2007). Co-assembly of N-type Ca^{2+} and BK channels underlies functional coupling in rat brain. *Journal of Cell Science*, *120*, 985–995.

Lu, T., Zhang, D. M., Wang, X. L., He, T., Wang, R. X., Chai, Q., et al. (2010). Regulation of coronary arterial BK channels by caveolae-mediated angiotensin II signaling in diabetes mellitus. *Circulation Research*, *106*, 1164–1173.

Ma, D., Nakata, T., Zhang, G., Hoshi, T., Li, M., & Shikano, S. (2007). Differential trafficking of carboxyl isoforms of Ca^{2+}-gated (Slo1) potassium channels. *FEBS Letters*, *581*, 1000–1008.

Marrero, H. G., Treistman, S. N., & Lemos, J. R. (2015). Ethanol effect on BK channels is modulated by magnesium. *Alcoholism, Clinical and Experimental Research*, *39*, 1671–1679.

Mason, A. K., Jacobs, B. E., & Welling, P. A. (2008). AP-2-dependent internalization of potassium channel Kir2.3 is driven by a novel di-hydrophobic signal. *Journal of Biological Chemistry*, *283*, 5973–5984.

Maximov, A., Sudhof, T. C., & Bezprozvanny, I. (1999). Association of neuronal calcium channels with modular adaptor proteins. *Journal of Biological Chemistry*, *274*, 24453–24456.

Missler, M., Sudhof, T. C., & Biederer, T. (2012). Synaptic cell adhesion. *Cold Spring Harbor Perspectives in Biology*, *4*, a005694.

Molina, L., Fasquelle, L., Nouvian, R., Salvetat, N., Scott, H. S., Guipponi, M., et al. (2013). Tmprss3 loss of function impairs cochlear inner hair cell Kcnma1 channel membrane expression. *Human Molecular Genetics*, *22*, 1289–1299.

Morrow, J. P., Zakharov, S. I., Liu, G., Yang, L., Sok, A. J., & Marx, S. O. (2006). Defining the BK channel domains required for beta1-subunit modulation. *Proceedings of National Academy of Science of the United States of America*, *103*, 5096–5101.

Nelson, M. T., Cheng, H., Rubart, M., Santana, L. F., Bonev, A. D., Knot, H. J., et al. (1995). Relaxation of arterial smooth muscle by calcium sparks. *Science*, *270*, 633–637.

Noam, Y., Ehrengruber, M. U., Koh, A., Feyen, P., Manders, E. M., Abbott, G. W., et al. (2014). Filamin A promotes dynamin-dependent internalization of hyperpolarization-activated cyclic nucleotide-gated type 1 (HCN1) channels and restricts Ih in hippocampal neurons. *Journal of Biological Chemistry*, *289*, 5889–5903.

Oh, K. H., Abraham, L. S., Gegg, C., Silvestri, C., Huang, Y. C., Alkema, M. J., et al. (2015). Presynaptic BK channel localization is dependent on the hierarchical organization of alpha-catulin and dystrobrevin and fine-tuned by CaV2 calcium channels. *BMC Neuroscience*, *16*, 26.

Oh, H. J., Abraham, L. S., van Hengel, J., et al. (2012). Interaction of alpha-catulin with dystrobrevin contributes to integrity of dystrophin complex in muscle. *Journal of Biological Chemistry*, *287*, 21717–21728.

Oldfield, C. J., & Dunker, A. K. (2014). Intrinsically disordered proteins and intrinsically disordered protein regions. *Annual Review of Biochemistry*, *83*, 553–584.

O'Malley, D., & Harvey, J. (2007). MAPK-dependent actin cytoskeletal reorganization underlies BK channel activation by insulin. *The European Journal of Neuroscience*, *25*, 673–682.

O'Malley, D., Irving, A. J., & Harvey, J. (2005). Leptin-induced dynamic alterations in the actin cytoskeleton mediate the activation and synaptic clustering of BK channels. *FASEB Journal*, *19*, 1917–1919.

Ou, J. W., Kumar, Y., Alioua, A., Sailer, C., Stefani, E., & Toro, L. (2009). Ca^{2+}- and thromboxane-dependent distribution of MaxiK channels in cultured astrocytes: From microtubules to the plasma membrane. *Glia*, *57*, 1280–1295.

Park, S. M., Liu, G., Kubal, A., Fury, M., Cao, L., & Marx, S. O. (2004). Direct interaction between BKCa potassium channel and microtubule-associated protein 1A. *FEBS Letters*, *570*, 143–148.

Parton, R. G., & del Pozo, M. A. (2013). Caveolae as plasma membrane sensors, protectors and organizers. *Nature Reviews. Molecular Cell Biology*, *14*, 98–112.

Patrie, K. M., Drescher, A. J., Goyal, M., Wiggins, R. C., & Margolis, B. (2001). The membrane-associated guanylate kinase protein MAGI-1 binds megalin and is present in glomerular podocytes. *Journal of the American Society of Nephrology : JASN*, *12*, 667–677.

Peng, Z., Sakai, Y., Kurgan, L., Sokolowski, B., & Uversky, V. (2014). Intrinsic disorder in the BK channel and its interactome. *PloS One*, *9*, e94331.

Perez, G. J., Bonev, A. D., Patlak, J. B., & Nelson, M. T. (1999). Functional coupling of ryanodine receptors to KCa channels in smooth muscle cells from rat cerebral arteries. *Journal of General Physiology*, *113*, 229–238.

Piao, L., Ho, W. K., & Earm, Y. E. (2003). Actin filaments regulate the stretch sensitivity of large-conductance, Ca^{2+}-activated K^+ channels in coronary artery smooth muscle cells. *Pflügers Archiv/European Journal of Physiology*, *446*, 523–528.

Rader, K., Orlando, R. A., Lou, X., & Farquhar, M. G. (2000). Characterization of ANKRA, a novel ankyrin repeat protein that interacts with the cytoplasmic domain of megalin. *Journal of the American Society of Nephrology : JASN*, *11*, 2167–2178.

Ratheesh, A., & Yap, A. S. (2012). A bigger picture: Classical cadherins and the dynamic actin cytoskeleton. *Nature Reviews. Molecular Cell Biology*, *13*, 673–679.

Rezzonico, R., Cayatte, C., Bourget-Ponzio, I., et al. (2003). Focal adhesion kinase pp125FAK interacts with the large conductance calcium-activated hSlo potassium channel in human osteoblasts: Potential role in mechanotransduction. *Journal of Bone and Mineral Research*, *18*, 1863–1871.

Rezzonico, R., Schmid-Alliana, A., Romey, G., Bourget-Ponzio, I., Breuil, V., Breittmayer, V., et al. (2002). Prostaglandin E2 induces interaction between hSlo potassium channel and Syk tyrosine kinase in osteosarcoma cells. *Journal of Bone and Mineral Research*, *17*, 869–878.

Ridgway, L. D., Kim, E. Y., & Dryer, S. E. (2009). MAGI-1 interacts with Slo1 channel proteins and suppresses Slo1 expression on the cell surface. *American Journal of Physiology. Cell Physiology*, *297*, C55–C65.

Sancar, F., Touroutine, D., Gao, S., Oh, H. J., Gendrel, M., Bessereau, J. L., et al. (2011). The dystrophin-associated protein complex maintains muscle excitability by regulating Ca^{2+}-dependent K^+ (BK) channel localization. *Journal of Biological Chemistry*, *286*, 33501–33510.

Sausbier, M., Hu, H., Arntz, C., et al. (2004). Cerebellar ataxia and Purkinje cell dysfunction caused by Ca^{2+}-activated K^+ channel deficiency. *Proceedings of National Academy of Science of the United States of America*, *101*, 9474–9478.

Scott, H. S., Kudoh, J., Wattenhofer, M., et al. (2001). Insertion of beta-satellite repeats identifies a transmembrane protease causing both congenital and childhood onset autosomal recessive deafness. *Nature Genetics*, *27*, 59–63.

Singh, H., Li, M., Hall, L., et al. (2016). MaxiK channel interactome reveals its interaction with GABA transporter 3 and heat shock protein 60 in the mammalian brain. *Neuroscience*, *317*, 76–107.

Singh, H., Lu, R., Bopassa, J. C., Meredith, A. L., Stefani, E., & Toro, L. (2013). MitoBK (Ca) is encoded by the Kcnma1 gene, and a splicing sequence defines its mitochondrial location. *Proceedings of National Academy of Science of the United States of America*, *110*, 10836–10841.

Sokolowski, B., Duncan, R. K., Chen, S., Karolat, J., Kathiresan, T., & Harvey, M. (2009). The large-conductance Ca(2+)-activated K(+) channel interacts with the apolipoprotein ApoA1. *Biochemical and Biophysical Research Communications*, *387*, 671–675.

Sokolowski, B., Orchard, S., Harvey, M., Sridhar, S., & Sakai, Y. (2011). Conserved BK channel-protein interactions reveal signals relevant to cell death and survival. *PloS One*, *6*, e28532.

Spoelgen, R., Hammes, A., Anzenberger, U., Zechner, D., Andersen, O. M., Jerchow, B., et al. (2005). LRP2/megalin is required for patterning of the ventral telencephalon. *Development*, *132*, 405–414.

Surguchev, A., Bai, J. P., Joshi, P., & Navaratnam, D. (2012). Hair cell BK channels interact with RACK1, and PKC increases its expression on the cell surface by indirect phosphorylation. *American Journal of Physiology. Cell Physiology*, *303*, C143–C150.

Suzuki, Y., Yamamura, H., Ohya, S., & Imaizumi, Y. (2013a). Caveolin-1 facilitates the direct coupling between large conductance Ca^{2+}-activated K^+ (BKCa) and Cav1.2 Ca^{2+} channels and their clustering to regulate membrane excitability in vascular myocytes. *Journal of Biological Chemistry*, *288*, 36750–36761.

Suzuki, Y., Yamamura, H., Ohya, S., & Imaizumi, Y. (2013b). Direct molecular interaction of caveolin-3 with KCa1.1 channel in living HEK293 cell expression system. *Biochemical and Biophysical Research Communications*, *430*, 1169–1174.

Szebenyi, G., Bollati, F., Bisbal, M., et al. (2005). Activity-driven dendritic remodeling requires microtubule-associated protein 1A. *Current Biology*, *15*, 1820–1826.

Tanemoto, M., Toyohara, T., Abe, T., & Ito, S. (2008). MAGI-1a functions as a scaffolding protein for the distal renal tubular basolateral K^+ channels. *Journal of Biological Chemistry*, *283*, 12241–12247.

Tian, L., Chen, L., McClafferty, H., Sailer, C. A., Ruth, P., Knaus, H. G., et al. (2006). A noncanonical SH3 domain binding motif links BK channels to the actin cytoskeleton via the SH3 adapter cortactin. *FASEB Journal*, *20*, 2588–2590.

Tian, L., Coghill, L. S., MacDonald, S. H., Armstrong, D. L., & Shipston, M. J. (2003). Leucine zipper domain targets cAMP-dependent protein kinase to mammalian BK channels. *Journal of Biological Chemistry*, *278*, 8669–8677.

Tian, L., Knaus, H. G., & Shipston, M. J. (1998). Glucocorticoid regulation of calcium-activated potassium channels mediated by serine/threonine protein phosphatase. *Journal of Biological Chemistry*, *273*, 13531–13536.

Toro, B., Cox, N., Wilson, R. J., Garrido-Sanabria, E., Stefani, E., Toro, L., et al. (2006). KCNMB1 regulates surface expression of a voltage and Ca^{2+}-activated K^+ channel via endocytic trafficking signals. *Neuroscience*, *142*, 661–669.

Troemel, E. R., Sagasti, A., & Bargmann, C. I. (1999). Lateral signaling mediated by axon contact and calcium entry regulates asymmetric odorant receptor expression in C. elegans. *Cell*, *99*, 387–398.

Wang, X. L., Ye, D., Peterson, T. E., Cao, S., Shah, V. H., Katusic, Z. S., et al. (2005). Caveolae targeting and regulation of large conductance Ca(2+)-activated K^+ channels in vascular endothelial cells. *Journal of Biological Chemistry*, *280*, 11656–11664.

Weaver, A. K., Olsen, M. L., McFerrin, M. B., & Sontheimer, H. (2007). BK channels are linked to inositol 1,4,5-triphosphate receptors via lipid rafts: A novel mechanism for coupling [Ca(2+)](i) to ion channel activation. *Journal of Biological Chemistry*, *282*, 31558–31568.

Williams, S. E., Wootton, P., Mason, H. S., Bould, J., Iles, D. E., Riccardi, D., et al. (2004). Hemoxygenase-2 is an oxygen sensor for a calcium-sensitive potassium channel. *Science*, *306*, 2093–2097.

Willnow, T. E., Hilpert, J., Armstrong, S. A., Rohlmann, A., Hammer, R. E., Burns, D. K., et al. (1996). Defective forebrain development in mice lacking gp330/megalin. *Proceedings of National Academy of Science of the United States of America*, *93*, 8460–8464.

Won, S., Lee, B. C., & Park, C. S. (2011). Functional effects of cytoskeletal components on the lateral movement of individual BKCa channels expressed in live COS-7 cell membrane. *FEBS Letters*, *585*, 2323–2330.

Yan, J., & Aldrich, R. W. (2010). LRRC26 auxiliary protein allows BK channel activation at resting voltage without calcium. *Nature, 466*, 513–516.

Yan, J., & Aldrich, R. W. (2012). BK potassium channel modulation by leucine-rich repeat-containing proteins. *Proceedings of National Academy of Science of the United States of America, 109*, 7917–7922.

Zhao, G., Neeb, Z. P., Leo, M. D., Pachuau, J., Adebiyi, A., Ouyang, K., et al. (2010). Type 1 IP3 receptors activate BKCa channels via local molecular coupling in arterial smooth muscle cells. *Journal of General Physiology, 136*, 283–291.

Zhou, X. B., Ruth, P., Schlossmann, J., Hofmann, F., & Korth, M. (1996). Protein phosphatase 2A is essential for the activation of Ca^{2+}-activated K^+ currents by cGMP-dependent protein kinase in tracheal smooth muscle and Chinese hamster ovary cells. *Journal of Biological Chemistry, 271*, 19760–19767.

Zou, S., Jha, S., Kim, E. Y., & Dryer, S. E. (2008a). The beta 1 subunit of L-type voltage-gated Ca^{2+} channels independently binds to and inhibits the gating of large-conductance Ca^{2+}-activated K^+ channels. *Molecular Pharmacology, 73*, 369–378.

Zou, S., Jha, S., Kim, E. Y., & Dryer, S. E. (2008b). A novel actin-binding domain on Slo1 calcium-activated potassium channels is necessary for their expression in the plasma membrane. *Molecular Pharmacology, 73*, 359–368.

CHAPTER FIVE

Functional Role of Mitochondrial and Nuclear BK Channels

B. Li, T.-M. Gao[1]
State Key Laboratory of Organ Failure Research, Key Laboratory of Psychiatric Disorders of Guangdong Province, School of Basic Medical Sciences, Southern Medical University, Guangzhou, China
[1]Corresponding author: e-mail address: tgao@smu.edu.cn

Contents

Abstract

BK channels are important for the regulation of many cell functions. The significance of plasma membrane BK channels in the control of action potentials, resting membrane potential, and neurotransmitter release is well established; however, the composition and functions of mitochondrial and nuclear BK (nBK) channels are largely unknown. In this chapter, we summarize the recent findings on the subcellular localization, biophysical, and pharmacological properties of mitochondrial and nBK channels and discuss their molecular identity and physiological functions.

Large-conductance Ca^{2+}-activated potassium channels (BK channels) are ubiquitously expressed in tissues, including peripheral neurons, brain, and muscle tissues (Brenner et al., 2005; Faber & Sah, 2003a; Girouard et al., 2010; Heppner, Bonev, & Nelson, 1997; Kang, Huguenard, & Prince, 1996; McCobb et al., 1995; Nelson et al., 1995; Ramanathan, Michael, Jiang, Hiel, & Fuchs, 1999; Sausbier et al., 2004; Scholz, Gruss, & Vogel, 1998; Smith, Nelson, & Du Lac, 2002; Tseng-Crank et al., 1994), and in

International Review of Neurobiology, Volume 128
ISSN 0074-7742
http://dx.doi.org/10.1016/bs.irn.2016.03.018

nonexcitable cells, such as glia, bone, kidney, and endothelial cells (Filosa et al., 2006; Henney et al., 2009; Li, Sun, Simard, Wang, & Chai, 2009; Morita, Hanaoka, Morales, Montrose-Rafizadeh, & Guggino, 1997; Papassotiriou et al., 2000; Ransom & Sontheimer, 2001). Owing to their large conductance for potassium ion efflux, many studies have focused on the electrophysiological properties and functions of BK channels in the plasma membrane. Many studies have demonstrated the importance of BK channels in the regulation of action potentials, resting membrane potential, and neurotransmitter release (Dufer et al., 2011; Filosa et al., 2006; Kent & Meredith, 2008; Lai et al., 2014; Meredith, Thorneloe, Werner, Nelson, & Aldrich, 2004; Meredith et al., 2006; Montgomery, Whitt, Wright, Lai, & Meredith, 2013; Pyott et al., 2007; Ruttiger et al., 2004; Sausbier et al., 2005, 2004, 2007; Typlt et al., 2013; Werner, Zvara, Meredith, Aldrich, & Nelson, 2005), as well as their indirect effects, such as on cell volume, apoptosis, and cell proliferation (see review Faber & Sah, 2003b).

Increasing evidence has indicated that BK channels are also localized in intracellular organelles (Singh, Stefani, & Toro, 2012), such as mitochondria (Douglas et al., 2006; Gu, Pamenter, Siemen, Sun, & Haddad, 2014; Ohya, Kuwata, Sakamoto, Muraki, & Imaizumi, 2005; Siemen, Loupatatzis, Borecky, Gulbins, & Lang, 1999; Singh et al., 2016, 2013; Xu et al., 2002; see review Balderas, Zhang, Stefani, & Toro, 2015), nuclei (Gobeil et al., 2002; Hu et al., 2012; Li et al., 2014; Maruyama, Shimada, & Taniguchi, 1995), and lysosomes (Cao et al., 2015). In this review, we discuss the subcellular localization, electrophysiological properties, and physiological functions of mitochondrial and nuclear BK channels (nBK channel), particularly the findings in two studies based on BK channel pore-forming α subunit knockout mice (Li et al., 2014; Singh et al., 2013). Additionally, we compare these properties with those of BK channels in plasma membranes.

1. MITOCHONDRIAL BK CHANNELS

Mitochondria are considered the "power houses" of cells because they supply necessary biological energy via the enzymatic oxidation of substrates in the Krebs cycle. Structurally, mitochondria are double-membrane-bound organelles, consisting of an outer membrane (OMM) and an inner membrane (IMM), which are composed of phospholipids and proteins.

As transmembrane proteins, ion channels are found in both OMMs and IMMs (see excellent review Szabo & Zoratti, 2014). For example, voltage-dependent anion channels (VDAC) in the OMM regulate ion and protein movement across the OMM and participate in cell apoptosis. Various potassium channels (K_{ATP} channels, Kv7 channels, BK channels, etc.), calcium-permeable channels (MCU), and chloride channels (CLIC4, IMAC, CLIC5, Ca^{2+}-activated Cl channels) have been found in the IMM (Szabo & Zoratti, 2014), where protein complexes of the electron transport system and the ATP synthase complex also reside. These ion channels mediate the ion flux between the intermembrane space and the matrix, which contains the enzymes that are responsible for metabolic processes, such as the citric acid cycle and fatty acid beta-oxidation.

BK channel is one of the many types of potassium channels located in the IMM (O'Rourke, 2007; Szewczyk, Jarmuszkiewicz, & Kunz, 2009; Szewczyk et al., 2006). Ca^{2+}- and voltage-dependent potassium channels with a conductance of 295 pS were originally described in mitoplasts (ie, mitochondria devoid of OMM) of the human glioma cell line LN229 using the patch-clamp technique by Siemen et al. (1999). Since then, mitochondrial BK channels (mitoBK channels) have been detected in excitable and nonexcitable cell types of multiple species, including guinea pig and rat ventricular myocytes, rat astrocytes, and human endothelial cells (Table 1) (Cheng et al., 2008, 2011; Gu et al., 2007; Ohya et al., 2005; Siemen et al., 1999; Xu et al., 2002). BK channels have also been detected using channel reconstitution in mitochondrial lipid bilayers using patch-clamp techniques (Bednarczyk, Koziel, et al., 2013; Fahanik-Babaei, Eliassi, Jafari, et al., 2011; Fahanik-Babaei, Eliassi, & Saghiri, 2011; Skalska et al., 2009). Most of these channels have a conductance ranging from 200 to 370 pS. However, one study reported a channel conductance of 564 pS in brain mitochondrial bilayers (Fahanik-Babaei, Eliassi, & Saghiri, 2011) (Table 1).

1.1 Biophysical and Pharmacological Properties of mitoBK Channels

The three major biophysical properties of BK channels are (1) large-conductance (>200 pS), (2) voltage-dependence, and (3) Ca^{2+}-dependence. All mitoBK channels summarized in Table 1 exhibit these features. The mitoBK channel conductance range of 211–564 pS is comparable to that of BK channels in the plasma membrane.

Table 1 mitoBK and nBK Channels Properties

	Cell Type	Methods	Conductance[a] (pS)	Pharmacology	References
mitoBK channels	Human glioma cell line (LN229)	Mitoplast (patch clamp)	295	ChTX	Siemen et al. (1999)
	Guinea pig ventricular myocytes	Mitoplast (patch clamp)	307	ChTX	Xu et al. (2002)
	Rat ventricular myocytes	Mitoplast (patch clamp)	270	Paxilline	Ohya et al. (2005)
	Human glioma cell line (LN229 and LN405)	Mitoplast (patch clamp)	276	ChTX	Gu et al. (2007)
	Rat astrocytes	Mitoplast (patch clamp)	295	IbTX	Cheng et al. (2008) and Cheng, Gulbins, and Siemen (2011)
	Rat brain	Mitochondria lipid bilayers (patch clamp)	265	ChTX	Skalska et al. (2009)
	Rat brain	Mitochondria lipid bilayers (patch clamp)	211	IbTX, ChTX	Fahanik-Babaei, Eliassi, Jafari, et al. (2011)
	Rat brain	Mitochondria lipid bilayers (patch clamp)	565	IbTX, ChTX	Fahanik-Babaei, Eliassi, and Saghiri (2011)
	Human endothelial cells line (EA. hy926)	Mitoplast (patch clamp)	270	IbTX, paxilline, NS1619, NS11021	Bednarczyk, Koziel, Jarmuszkiewicz, and Szewczyk (2013)

	Mouse ventriculocytes, brain	Intact cells, isolated mitochondria (immunostaining, Western blot, mass spectrometry)		NS1619	Singh et al. (2013)
nBK channels	Rat pancreatic acinar cells	Isolated nucleus (patch clamp)	200		Maruyama et al. (1995)
	Piglet brain endothelial cells	Isolated nucleus (immunostaining)		IbTX, NS1619	Gobeil et al. (2002)
	Human fibroblast-like synoviocytes	Intact cell (immunostaining)			Hu et al. (2012)
	Mouse cultured hippocampal neurons	Intact cell, isolated nucleus (patch clamp, immunostaining, Western blot)	217	Paxilline	Li et al. (2014)

[a]Some conductance were modified from Singh et al. (2012) as some of the values were calculated from published figures.

Similar to BK channels in the plasma membrane, the open probability of mitoBK channels is regulated by Ca^{2+} and voltage, though to different degrees. For example, the open probability of BK channels in the plasma membrane is very low for a low Ca^{2+} concentration (0–1 μM Ca^{2+}), even when the membrane is depolarized. However, the open probability for mitoBK channels is maintained at ~0.8 for +20 mV and 0 Ca^{2+} (Fahanik-Babaei, Eliassi, & Saghiri, 2011). The open probability reaches ~0.9 for 100 μM Ca^{2+} when the membrane potential ranges from −40 to +40 mV (Fahanik-Babaei, Eliassi, & Saghiri, 2011). The dependence of the open probability of mitoBK channels on Ca^{2+} and voltage also varies among studies depending on species and experimental conditions. For instance, in guinea pig ventricular myocytes, the open probability is ~0.9 for a voltage range of −60 to +60 and 0.5 μM Ca^{2+} (Xu et al., 2002), but in rat ventricular myocytes, the open probability is only 0.0087 for similar conditions (Ohya et al., 2005).

Pharmacological properties are another important feature of mitoBK channels. All specific blockers/inhibitors, such as iberiotoxin (IbTX), charybdotoxin (ChTX), and paxilline, and agonists of plasma membrane BK (pmBK) channels are applicable to mitoBK channels. The best-studied class of mitoBK channel openers is benzimidazolones, including NS1619, NS004, NS1604, NS11021, and NS1643 (Nardi & Olesen, 2008), of which NS1619 and NS11021 are the most commonly used agonists. The specificity of these drugs to BK channels is always a concern, especially when the drugs are used at high concentrations (O'Rourke, 2007; Szewczyk et al., 2006). For example, paxilline can inhibit calcium-dependent ATPases when used at 5–50 μM (Bilmen, Wootton, & Michelangeli, 2002). In rat cerebral microsomes, paxilline inhibits 1,4,5-triphosphate (InsP3)-dependent calcium release (Longland, Dyer, & Michelangeli, 2000). In isolated mitochondria of rat hearts, NS1619 was shown to inhibit complex I function (Kicinska & Szewczyk, 2004).

In some conditions, mitoBK channels show different sensitivities to blockers. For example, in mitochondrial lipid bilayers, some mitoBK channels are insensitive to ChTX, but sensitive to IbTX and the nonspecific potassium channel blocker 4-AP (Fahanik-Babaei, Eliassi, Jafari, et al., 2011). These observations indicate the need for caution when characterizing mitoBK functions based on pharmacological properties.

Peptide toxins, such as IbTX and ChTX, can be used not only to regulate channel activity but also to identify mitoBK channel orientation in IMMs. In on-mitoplast patch recordings, Siemen et al. showed that ChTX could completely block mitoBK channel activity (Siemen et al., 1999), suggesting that the N-terminus of the mitoBK channel (the extracellular side of the pmBK channel) is oriented toward the intermembrane space because ChTX binds to the extracellular pores of the pmBK channel. This hypothesis is further supported by findings that ChTX or IbTX applied to the patch pipette completely abolishes mitoBK channel activity (Cheng et al., 2008, 2011; Gu et al., 2007; Skalska et al., 2009; Xu et al., 2002).

Variation in the conductance, open probability, and pharmacological properties of mitoBK channels could be due to several factors. Different cell types host different mitoBK channel isoforms. Although BK channel pore-forming α subunits are encoded by a single *Kcnma1* gene, the gene exhibits extensive alternative splicing, which yields different isoforms. Some exons are constitutively expressed in all cell types, whereas others are expressed in a subset of cell types. Different exon combinations are related to differences in channel structure and function (Fodor & Aldrich, 2009; Glauser, Johnson, Aldrich, & Goodman, 2011; Lee & Cui, 2010; Li, Al-Khalili, Ramosevac, Eaton, & Denson, 2010; Ma et al., 2007; Saito, Nelson, Salkoff, & Lingle, 1997; Shelley, Whitt, Montgomery, & Meredith, 2013; Singh et al., 2013; Tian et al., 2001; Xie, 2008; Yang, Zhang, & Cui, 2015; Zarei et al., 2004).

Four auxiliary β subunits (β1–4) and four auxiliary γ subunits (γ1–4) have been identified to date. These auxiliary subunits are coexpressed with the pore-forming subunit of BK channels to form functional complexes and contribute to the variation in the conductance, open probability, and pharmacological properties of BK channels (Brenner, Jegla, Wickenden, Liu, & Aldrich, 2000; Knaus et al., 1994; Meera, Wallner, & Toro, 2000; Uebele et al., 2000; Wallner, Meera, & Toro, 1999; Yan & Aldrich, 2010, 2012). For example, β subunits regulate the effects of voltage and Ca^{2+} on BK channel opening in the plasma membrane (Dworetzky et al., 1996; McManus et al., 1995; Meera, Wallner, Jiang, & Toro, 1996; Tseng-Crank et al., 1996). Additionally, β subunits can alter the trafficking and pharmacological properties of BK channels (Cox, Toro, Pacheco-Otalora, Garrido-Sanabria, & Zarei, 2014; Shruti et al., 2012; Toro et al., 2006; Zarei et al., 2007). For instance, β4 subunits in brain BK channels confer resistance to IbTX (Meera et al., 2000). Leucine-rich repeat (LRR)-containing proteins are regarded as γ subunits of BK channels. LRRC26 can cause a large negative shift (~140 mV) in the voltage dependence of channel activation (Yan & Aldrich, 2010). Other LRRC26-type proteins, such as LRRC52, LRRC55, and LRRC38, also regulate the voltage dependence of BK channels (Yan & Aldrich, 2012). A recent proteomic study showed that LRRC proteins associate with BK α in mitochondria (Singh et al., 2016). Additional studies are necessary to determine the effects of these proteins on mitoBK channels activity in mitochondria. Posttranslational modifications of mitoBK channel proteins may also alter channel properties (Alioua, Li, Wu, Stefani, & Toro, 2011; Chen, Bi, et al., 2013; Jeffries et al., 2010; Zhou et al., 2012). Phosphorylation, oxidation, and acetylation are common mitochondrial protein modifications. Like other proteins in the IMM, mitoBK channels are vulnerable to these modifications. However, the details of these modifications need to be elucidated.

1.2 mitoBK Channel Composition

In addition to their biophysical and pharmacological features, the occurrence and composition of mitoBK channels have been characterized. Using Western blotting with antibodies against BK channel pore-forming α subunit, a 125-kDa protein was detected in isolated cardiac mitochondria (Shi et al., 2007). Noteworthy, a protein with a smaller molecular weight (55 kDa) was detected in the same preparation (Wang, Yin, Xi, & Kukreja, 2004; Xu et al., 2002). Double labeling of cells with antibodies against BK channel proteins and mitochondrial proteins also showed the localization of BK channels in the mitochondria (Douglas et al., 2006; Redel et al., 2008; Singh et al., 2013).

The results of these immunohistochemical studies suggest that mitoBK channels share the same molecular identity as pmBK channels. Indeed, recent findings from Toro and colleagues have provided direct and strong evidence that mitoBK channels are encoded by *Kcnma1*, similar to pmBK channels (Singh et al., 2013). They purified mitochondria from the ventricle and evaluated the size of mitoBK channel proteins by Western blotting and mass spectrometry. A specific band at 140 kDa was detected in samples from wild-type mice, but not in those with the $Kcnma1^{-/-}$ mice (BK α subunit knockout mice), providing direct evidence that mitoBK channels are encoded *Kcnma1*. In an independent study, BK channel current could not be detected in mitochondria of BK α subunit knockout mice, further providing the functional evidence (Soltysinska et al., 2014).

This conclusion was further confirmed by a mass spectrometry analysis, in which consensus peptides were identified in the C-terminus of the BK channel. BK channels in mitochondria were verified by immunostaining. They colocalized with MitoTracker and mitochondrial ion channel

VDAC1 in both regular confocal and superresolution fluorescence microscopy analyses. They were not detected in BK α subunit knockout mice. Based on a protein proximity index algorithm (Wu et al., 2010), the majority of mitoBK channels colocalize with MitoTracker-loaded mitochondria, but not all MitoTracker-loaded mitochondria express BK channels. These findings raise the question of whether mitochondria have different functions depending on the presence of mitoBK channels.

In adult cardiomyocytes, BK channels are expressed in mitochondria, but not in plasma membranes; accordingly, a systematic exon scan was performed by purifying total mRNAs from isolated mouse ventriculocytes. Both constitutive exons and alternative exons, including stress-axis-regulated exon, SV27, and DEC, were found. The DEC exon encodes a 50-aa C-terminal sequence. Quantitative RT-PCR results showed that the DEC-containing isoform is dominant for mitoBK channels. Importantly, further functional studies have shown that when ventriculocytes are transduced with adenoviruses encoding DEC-containing BK channels, the immunocytochemical signals of BK channels are located in mitochondria (colocalized with MitoTracker), but not in the plasma membrane. This was not observed for adenoviruses encoding SV27-containing BK channels. These results indicate that the DEC exon, but not SV27, determines BK channel mitochondrial targeting (Singh et al., 2013).

Of note, when a DEC-containing isoform is overexpressed in CHO cells, BK channels are located in both plasma membranes and mitochondria (Kathiresan, Harvey, Orchard, Sakai, & Sokolowski, 2009). In contrast, when the isoform is overexpressed in COS7, HEK293, or Neuro-2a cells, it is localized in the endoplasmic reticulum (Chen, Bi, et al., 2013; Kim, Ridgway, Zou, Chiu, & Dryer, 2007; Ma et al., 2007). These findings suggest that the impact of the DEC exon on BK channel targeting to mitochondria depends on the cell type. Additional factors, such as the exon combination and auxiliary subunits, may also influence mitochondrial targeting.

As mentioned above, β and γ auxiliary subunits regulate the biophysical and pharmacological properties of BK channels; thus, studying the localization and identity of these auxiliary subunits in mitochondria is important to fully understand mitoBK channel functions.

Different β subunits have been found in mitochondria of myocytes, skeletal muscle, endothelial cells, and neurons (Banerjee, Lee, Campbell, & Mackinnon, 2013; Bednarczyk, Koziel, et al., 2013; Bednarczyk, Wieckowski, et al., 2013; Loot et al., 2012; Ohya et al., 2005; Piwonska,

Wilczek, Szewczyk, & Wilczynski, 2008; Skalska et al., 2009; Valverde et al., 1999). β1 subunits are highly expressed in the mammalian heart. Based on a yeast two-hybrid assay and immunocytochemistry analyses, β1 subunits colocalize with Snapin, cytochrome *c* oxidase subunit I (Cco1), and CAML in mitochondrial IMM of rat ventricular myocytes, suggesting that mitoBK channels are involved in the regulation of mitochondrial functions (Ohya et al., 2005). Indeed, in the same study, 17β-estradiol significantly increased flavoprotein oxidation and decreased the rate of cell death under simulated ischemia. These effects could be blocked using the BK channel blocker paxilline. Estradiol enhances the activity of the pmBK channel only in the presence of the β1 subunit (Valverde et al., 1999), suggesting that the β1 subunit may be an important component of functional mitoBK channels and may be involved in the regulation of mitochondrial function.

In pulmonary artery smooth muscle cells, BK channels are localized in mitochondria. In these cells, 11,12-epoxyeicosatrienoic acid (11,12-EET) elicits a loss of mitochondrial membrane potential (JC-1 fluorescence) and leads to plasma membrane depolarization (Loot et al., 2012). These effects can be blocked by the BK channel blocker IbTX and β1 subunit gene knockout, indicating that mitoBK channels, and especially the β1 subunit, are important for 11,12-EET-induced pulmonary vasoconstriction.

In the mitoplasts of the human endothelial cell line EA.hy926, the combination of the β2 subunit and α subunit of BK channels has been detected by Western blot analyses with antibodies against pmBK channel proteins, confirming the presence of the β2 subunit in the endothelial mitochondrial IMM (Bednarczyk, Koziel, et al., 2013). The β2 subunit, but not β1 or β3, was also found in the rat brain mitochondrial fraction (Piwonska et al., 2008).

The β4 subunit is also localized in mitochondria. Ultrastructural analyses by transmission electron microscopy and immunocytochemical analyses by confocal microscopy have shown that the β4 subunit is localized in both the plasma membrane and mitochondria of freshly isolated rat skeletal muscles (Bednarczyk, Wieckowski, et al., 2013). Immunoreaction with the anti-BK β4 subunit (~26 kDa) was observed in the homogenate as well as in the mitochondrial skeletal muscle protein fraction. The immunoreactivity was blocked with an antigenic peptide (Skalska et al., 2008).

In the rat brain and cultured neurons, a combined approach using Western blotting, high-resolution immunofluorescence, and immunoelectron microscopy with antibodies against β subunits demonstrated their presence in the IMM. Consistent with findings for BK α subunit, the expression of the β4 subunit is restricted to a subpopulation of mitochondria. Interestingly, the β4 subunit is expressed most highly in the thalamus and the brainstem (Piwonska et al., 2008). Immunoreactivity of anti-β4 subunit antibodies with ~26 kDa proteins has also been observed in mitochondria of the rat brain and a human glioblastoma cell line (U-87 MG) in independent studies (Banerjee et al., 2013; Skalska et al., 2009).

It is noteworthy that the specificity of commercially available antibodies raised against BK auxiliary subunits is questionable (Bhattarai et al., 2014; Martinez-Espinosa, Yang, Gonzalez-Perez, Xia, & Lingle, 2014; Wang, Jaffe, & Brenner, 2014). Thus, cautions should be taken when considering evidence that have relied exclusively on immunoreactivity to detect BK β subunits.

The studies summarized in this section focus on the β subunit as part of mitoBK channel complexes. However, additional studies are needed to determine whether β subunits can act independently of α subunits to regulate mitochondrial functions.

1.3 Function of mitoBK Channels

MitoBK channel functions can be easily studied using isolated mitochondria, but the results of these studies cannot be directly applied to intact cells; in vitro systems do not closely reflect in vivo conditions. Adult cardiomyocytes lack pmBK channels but harbor tremendous amounts of mitoBK channels; accordingly, the interpretation of the results of functional studies is far easier and more straightforward.

Given observations that mitochondrial K_{ATP} (mitoK_{ATP}) channels play an important role in protection against ischemic injury, functional studies of mitoBK channels have focused on their protective effects on ischemia and reperfusion injury after the pioneering work by Xu et al. (2002). In this study, perfused hearts were subjected to global ischemia and reperfusion after pretreatment with NS1619. A 5-min preischemic exposure to 30 μM NS1619 significantly improved left ventricular-developed pressure and approximately halved the extent of myocardial infarction. These effects were comparable to those of the mitoK_{ATP} channel opener diazooxide in the same infarction model. These protective effects of NS1619 were completely blocked by paxilline, suggesting that they were mediated by mitoBK channels.

Similar results have been reported in several other independent studies (Bentzen et al., 2009; Borchert, Hlavackova, & Kolar, 2013;

Gaspar et al., 2009; Schmitt, Grunnet, & Olesen, 2014; Singh et al., 2013; Stowe et al., 2006; Testai et al., 2013). Although these studies use different drug concentrations, the opening of the mitoBK channel via NS1619 or NS11021 consistently protected the heart against ischemia and reperfusion injuries, and the protective effects were abolished by the closing of the mitoBK channels via paxilline.

Owing to potential nonspecific drug effects, we should not attribute all the observed effects of these modulators to mitoBK channels. For example, the mitoBK channel antagonist paxilline blocks the protective effect against ischemia/reperfusion injury by the volatile anesthetic isoflurane. However, these effects still exist in BK α subunit knockout mice, indicating that the effects of paxilline are not due to mitoBK channels (Wojtovich et al., 2013, 2011). Thus, it is important to be cautious when interpreting the findings of studies that use only pharmacological methods.

BK α knockout mice have been used to confirm that the protective effects of NS1619 and NS11021 are indeed mediated by mitoBK channels (Singh et al., 2013). Using WT and BK α subunit knockout mice, Singh et al. preconditioned isolated hearts with or without NS1619, followed by ischemia and reperfusion. As expected, WT hearts preconditioned with NS1619 showed an improvement in left ventricular-developed pressure during reperfusion compared with control hearts preconditioned with vehicle alone. In contrast, BK α subunit knockout hearts lacked the protective effects of NS1619. Similar results have been obtained when evaluating the heart rate-pressure product and myocardial infarction size. These findings provide solid evidence that mitoBK channels are responsible for the protective effects of NS1619.

In an independent study (Soltysinska et al., 2014), the role of mitoBK channels in the protective effects of ischemic preconditioning against ischemia/reperfusion injury was evaluated. WT and BK α subunit knockout hearts were preconditioned with two transient periods of ischemia and reperfusion prior to the prolonged ischemia. Infarct size after ischemia/reperfusion was significantly lower in WT hearts than in BK α subunit knockout hearts upon ischemic preconditioning, indicating that mitoBK channels are required for the protective effects of ischemic preconditioning.

Interestingly, Wojtovich et al. found that the protective effects of NS1619 or NS11021 might not be mediated by mitoBK channels in cardiac myocytes (Wojtovich et al., 2013). They found that in an isolated cardiomyocyte model of ischemia/reperfusion injury, protection by NS1619 and NS11021 was insensitive to BK α subunit deletion. They suggested that cardiac neurons and neurotransmission in intact hearts are vital for protection. These findings indicate that the mechanism underlying ischemic protection is not as simple as initially predicted.

mitoBK channels were localized in IMM, which harbors the electron transport system and the ATP synthase complex, thus it was hypothesized that mitoBK channel activity would be important for the regulation of mitochondrial function.

The structure and function of mitochondria were examined in both WT and BK α knockout mice. While mitochondria from BK α subunit knockout ventricular muscle fibers showed normal ultrastructures and matrix dimensions based on transmission electron microscopy, the oxidative phosphorylation capacities at normoxia and upon reoxygenation after anoxia were significantly attenuated. In the absence of BK channels, postanoxic reactive oxygen species (ROS) production from cardiomyocyte mitochondria was significantly elevated, indicating that mitoBK channels regulate the oxidative state at hypoxia and reoxygenation (Soltysinska et al., 2014). Because ROS and the capacity of the myocardium for oxidative metabolism are important for cellular survival, this elevation of ROS production in BK α knockout mice explains the increased myocardial infarction size after ischemia/reperfusion injury.

The regulation of mitochondrial function via mitoBK channels is reminiscent of the findings from pharmacological studies, in which the opening of mitoBK channels regulates respiratory rate, mitochondrial depolarization, and ROS production (Heinen, Aldakkak, et al., 2007; Heinen, Camara, et al., 2007; Hermann, Sitdikova, & Weiger, 2015; Kulawiak, Kudin, Szewczyk, & Kunz, 2008).

mitoBK channels also regulate the mitochondrial Ca^{2+} retention capacity. While NS1619 protects the heart against ischemic injury, it also increases the number of Ca^{2+} pulses needed to cause a massive Ca^{2+} release from isolated mitochondria. This beneficial effect is eliminated in mitochondria from BK α subunit knockout mice, indicating that mitoBK channels are involved in the regulation of the mitochondrial Ca^{2+} retention capacity (Singh et al., 2013).

The mitochondrial Ca^{2+} retention capacity is closely related to the activation of the mitochondrial permeability transition pore (mPTP), which mediates Ca^{2+} release from mitochondria to the cytosol. The abovementioned mitoBK channel regulation of the Ca^{2+} retention capacity suggests that mitoBK channel and mPTP interact. Indeed, the inhibition of mitoBK channels by the proapoptotic protein Bax activates mPTP and induces cytochrome c release. Under hypoxic conditions, in which mitoBK channels are activated, the opening of mPTP is impaired (Cheng et al., 2008, 2011). These findings suggest that mitoBK channels are functionally coupled with mPTP.

Most of the functional studies of mitoBK channels on neuroprotection against ischemic injury are based on pharmacological experiments, but, as mentioned previously, mitoBK channels share similar pharmacological properties with pmBK channels, and BK channels also exist in nuclei and lysosomes in neurons, making it difficult to attribute the observed effects to mitoBK channels. For example, both the BK channel blocker paxilline and the opener NS1619 protect neurons from ischemia injury, but the opposite findings have also been reported (Chen, Sun, et al., 2013; Gaspar et al., 2009; Liao et al., 2010).

In summary, although the molecular identity of the mitoBK channel α subunit has been determined, the subunit composition, interacting proteins, biophysical and pharmacological properties, and regulatory mechanisms of mitoBK channels in different cell types are largely unknown. Studies of mitoBK channel functions have focused on cardiac protection against ischemia injury. It is not clear whether mitoBK channels play different physiological/pathophysiological roles in other cell types.

2. NUCLEAR BK CHANNELS

The nucleus exerts a critical control over cellular functions by regulating gene expression. The nucleus is surrounded by a nuclear envelope (NE), which consists of two membranes: the inner nuclear membrane (INM) and the outer nuclear membrane (ONM). Nuclear pore complexes (NPCs) are located across the NE and mediate the transportation of ions and molecules between the nucleoplasm and cytoplasm. NPCs are thought to be large enough to provide routes for the passive diffusion of ions across the NE, implying that the ion concentration in the nucleus and cytoplasm should be the same. Nevertheless, the segregation of ions in the nucleoplasm was reported decades ago, indicating that there must be ion channels that regulate ion flux in the NE (Century & Horowitz, 1974; Kanno & Loewenstein, 1963; Loewenstein & Kanno, 1963).

2.1 Evidence for BK Channels in the Nucleus

NE ion channel activity was first recorded in murine zygotes by a patch-clamp experiment in 1990 (Mazzanti, DeFelice, Cohn, & Malter, 1990). K^+-selective channel activity with the maximal conductance of 200 pS was successfully recorded. In subsequent studies, voltage- and Ca^{2+}-activated K^+ channels, with a unitary conductance of 200 pS, were also found in the ONM of rat pancreatic acinar cells (Maruyama et al., 1995). The results of these studies suggest the existence of BK channels in the NE. This inference is supported by immunocytochemical analyses using antibodies against BK channel pore-forming α subunit in the isolated nuclei of brain endothelial cells and fibroblast-like synoviocytes (Douglas et al., 2006; Hu et al., 2012). Recently, BK α knockout mice have been used to provide conclusive evidence that BK channels are present in the NE of hippocampal neurons (Li et al., 2014).

We used a monoclonal antibody against the α subunit of the BK channel and detected intracellular ring-like labeling around the nucleus (Li et al., 2014). The labeling was colocalized with the NE marker lamin B, suggesting that BK channels are localized at the NE. Immunoelectron microscopy results further confirmed this finding at a high spatial resolution. In isolated nuclei, a ring-like pattern of BK channel immunofluorescence was detected and colocalized with lamin B. Subcellular fractionation and an immunoblot analysis have provided evidence for BK channels in the nuclear fraction, as well as in the plasma membrane fraction and the whole cell lysate. Importantly, these signals were absent in the neurons of BK α subunit knockout mice, confirming that the BK channels in the nuclei (nBK channel) are encoded by *Kcnma1*. It is noteworthy that the alternative splicing patterns of nBK channels were unclear but may be important for their nuclear localization, as nBK channels were not reported in many other cell types.

Like neuronal pmBK channels, brain-specific auxiliary β4 subunits have also been observed in nucleus, indicating that nBK channels are fully assembled in the NE (Li et al., 2014). Based on single-channel recordings using the patch-clamp technique for isolated nuclei, nBK channels exhibited voltage- and Ca^{2+}-dependent activity with a conductance of approximately 220 pS, comparable to pmBK channels. Paxilline completely blocked the channel activity, indicating that nBK channels also share similar pharmacological properties with pmBK channels. Importantly, channel activity was absent in BK α subunit knockout mice, further confirming that the channel activity was related to BK channels (Li et al., 2014). Moreover, recent proteomic studies indicate that many BK channel-interacting proteins are nuclear proteins, such as nuclear pore complex protein Nup160, prelamin A/C, and many transcription-related proteins (Kathiresan et al., 2009; Singh et al., 2016; Sokolowski, Orchard, Harvey, Sridhar, & Sakai, 2011). Thus, several independent lines of evidence indicate that functional BK channels are present in the NE of hippocampal neurons.

2.2 The Function of nBK Channels

2.2.1 nBK Channels Regulate Nuclear Transmembrane Potential

Nuclei have two components: the perinuclear space, which is the lumen between the ONM and INM, and the nucleoplasm, which is surrounded by the INM. The findings using ion-selective fluorescent indicators have shown that the K^+ concentration is much lower in the perinuclear space than the cytoplasm and nucleoplasm, whereas the Na^+ concentration in the perinuclear space is much higher than that in the cytoplasm and nucleoplasm (Garner, 2002). The differences in ion concentrations across the NE recapitulate those found between the extracellular and intracellular sides of the plasma membrane and suggest that changes in the ion balance across the NE lead to changes in the nuclear transmembrane potential.

Because nBK channels have similar electrophysiological properties to pmBK channels, it was hypothesized that nBK channels could regulate the nuclear transmembrane potential ($\Delta\Psi_n$), as pmBK channels do in the plasma membrane. To address this hypothesis, we loaded the isolated nuclei with the potentiometric probe $DiOC_6(3)$. Paxilline induced a consistent increase in $DiOC_6(3)$ fluorescence, indicating that the perinuclear lumen became more negative. The signal was completely absent in the nuclei of BK α subunit knockout mice (Li et al., 2014).

The regulation of nuclear transmembrane potential by nBK channels was also observed in cerebral endothelial cells. Gobeil et al. loaded the isolated nuclei with the potentiometric probe RH421 and found that the BK channel opener NS1619 affected RH421 fluorescence, which could be blocked by IbTX (Gobeil et al., 2002). Together with observations from hippocampal neurons, these results indicate that the activity of nBK channels does affect nuclear membrane potential.

2.2.2 nBK Channels Regulate Nuclear Ca^{2+} Signals

Recent studies have shown that the NE is an important source of nucleoplasmic Ca^{2+} (Bootman, Fearnley, Smyrnias, MacDonald, & Roderick, 2009; Carafoli, Nicotera, & Santella, 1997; Gerasimenko, Gerasimenko, Tepikin, & Petersen, 1995; Leite et al., 2003; Stehno-Bittel, Luckhoff, & Clapham, 1995). The Ca^{2+} in perinuclear space can be released into the nucleoplasm. Thus, the NE is regarded as a functional perinuclear Ca^{2+} store, independent of the ER and cytoplasm, regulating the nucleoplasmic Ca^{2+} concentration ($[Ca^{2+}]_{nu}$) (Bootman et al., 2009; Carafoli et al., 1997;

Gerasimenko et al., 1995; Leite et al., 2003; Stehno-Bittel et al., 1995). Meanwhile, numerous studies have indicated that nucleoplasmic Ca^{2+} is important in regulating neuronal activity-dependent gene expression and is critical for synapse formation, dendritic complexity, synaptic plasticity, and memory consolidation (Bading, 2013; Hardingham, Arnold, & Bading, 2001; Limback-Stokin, Korzus, Nagaoka-Yasuda, & Mayford, 2004; Mauceri, Freitag, Oliveira, Bengtson, & Bading, 2011; Papadia, Stevenson, Hardingham, Bading, & Hardingham, 2005; Zhang et al., 2009).

To investigate whether nBK channels could regulate $[Ca^{2+}]_{nu}$ in hippocampal neurons, we exposed neurons to BK channel blockers. We found that the membrane-permeable blocker paxilline induced $[Ca^{2+}]_{nu}$ elevation in intact neurons, whereas membrane-impermeant IbTX did not influence $[Ca^{2+}]_{nu}$, suggesting that the effect of paxilline was mainly mediated by intracellular BK channels, and not pmBK channels. To further verify that it nBK channels and perinuclear Ca^{2+} stores mediate the increase in $[Ca^{2+}]_{nu}$, we loaded isolated nuclei with either the Ca^{2+} probe Fluo-4/AM, a membrane-permeant probe that accumulates in the NE, or Fluo-4/dextran, a membrane-impermeant probe distributed in the nucleoplasm. The application of paxilline resulted in a decrease in NE Ca^{2+}, which paralleled a transient Ca^{2+} increase in the nucleoplasm. The paxilline-induced Ca^{2+} transient was abolished in neurons from BK α subunit knockout mice, or when the perinuclear Ca^{2+} store was depleted with thapsigargin (Li et al., 2014).

Numerous studies have demonstrated that inositol 1,4,5-trisphosphate receptors (IP_3R) and ryanodine receptors (RyRs) are located on the IMM of the NE and mediate the release of Ca^{2+} from the perinuclear space into the nucleoplasm (Bootman et al., 2009; Stehno-Bittel et al., 1995). By pharmacological and genetic methods, we found that RyRs, instead of IP_3R, mediated paxilline-induced Ca^{2+} release in isolated nuclei (Li et al., 2014). The mechanism by which RyRs mediate paxilline-induced Ca^{2+} release is unknown. However, the open probability of RyRs is elevated upon changes in voltage (Chen, Li, Ebisawa, & Zhang, 1997; Endo, 2009; Quesada et al., 2002; Zahradnikova & Meszaros, 1998). Thus, it is highly probable that paxilline-induced $\Delta\Psi_n$ affected the RyRs open probability. Similar voltage-induced RyRs-mediated Ca^{2+} release was observed in pancreatic beta cells (Quesada et al., 2002). Nonetheless, the involvement of other voltage-sensitive Ca^{2+} release pathways cannot be excluded.

Potentially reflecting differences in the biological system and experimental conditions, the regulation of nuclear Ca^{2+} signaling by nBK channels is different in studies of cerebral endothelial cells. Gobeil et al. found that if the isolated nuclei are loaded with fura-2/AM in high-potassium buffer, NS1619 induces an increase in nuclear Ca^{2+}, which could be abolished by IbTX (Gobeil et al., 2002).

2.2.3 nBK Channels Regulate Nuclear Ca^{2+} Signaling and Transcription

The nuclear Ca^{2+} concentration is important for the activation of transcription factors and downstream genes. A well-known Ca^{2+}-sensitive transcription factor is cAMP response element-binding protein (CREB), which mediates neuronal development and plasticity, long-term memory formation, and cell survival. The elevation of $[Ca^{2+}]_{nu}$ leads to the phosphorylation of CREB and CREB-mediated transcription (Bito, Deisseroth, & Tsien, 1996; Papadia et al., 2005; Sheng, Thompson, & Greenberg, 1991; Zhang et al., 2009).

Using Western blotting and immunocytochemistry in isolated nuclei and intact neurons, we found that paxilline significantly augments the phosphorylation of CaMKIV and CREB, but not ERK. These results indicate that blocking nBK channels triggers CaMKIV-mediated CREB-dependent gene transcription. This conclusion was further supported by pharmacological and genetic experiments, in which paxilline-induced phosphorylation of CaMKIV and CREB was blocked by nuclear Ca^{2+} buffer, dominant-negative CaMIV, CaMIV shRNA, or the CaMKIV antagonist STO-609, but not by the ERK antagonist U0126. Importantly, the paxilline-induced phosphorylation of CaMIV and CREB was absent in BK α subunit knockout mice, confirming that nBK channels mediate these regulatory effects (Li et al., 2014).

The long-term maintenance of neuronal activity-dependent adaptations requires gene expression. To investigate the role of nBK channels in gene expression, qRT-PCR was used to examine changes in the expression of synaptic activity-evoked genes. We found that blocking nBK channels by paxilline augmented basal neuronal activity-induced gene expression, which was consistent with the activation of CREB phosphorylation. However, if neural activity was elevated by bicuculline, paxilline pretreatment significantly attenuated bicuculline-induced $[Ca^{2+}]_{nu}$ elevation, CREB phosphorylation, and gene expression (Li et al., 2014). This result may seem counterintuitive, but, importantly, bicuculline-induced CREB phosphorylation and gene expression require sustained nuclear Ca^{2+} elevation (Hardingham et al., 2001). We therefore hypothesize that paxilline induced the release of nuclear Ca^{2+} in advance, which significantly reduced the Ca^{2+} store and consequently compromised CREB phosphorylation and gene expression. These findings suggest the key role of nBK channels in maintaining the homeostasis of nuclear Ca^{2+} signaling and gene expression: When the neural activity is low, the closing of nBK channels elevates nuclear Ca^{2+} and maintains the proper level of gene expression that is necessary for cell survival, whereas if neural activity is too high, the closing of nBK channels prevents overshooting of nuclear Ca^{2+} and gene expression that may harm the cell.

2.2.4 nBK Channels Regulate Dendritic Arborization

Dendritic arborization determines neuronal connectivity and is critically important for neuronal development and function. Dendritic geometry itself is highly dependent on neuronal activity and gene expression (Mauceri et al., 2011). We examined the roles of nBK channels in the regulation of dendritic arborization. Based on morphometric analyses, neurons treated with paxilline or transfected with BK α-specific shRNAs show a significant increase in both the complexity of dendrites and total dendritic length, indicating that blocking nBK channels triggers an increase in dendritic arborization. This increase was abolished by the expression of the nuclear calcium/calmodulin signaling inhibitor CaMBP4, or dominant-negative CaMKIV, suggesting that this regulation was dependent on nuclear Ca^{2+} signaling (Li et al., 2014).

Importantly, we used shRNAs to knock down nBK channel target genes to determine the mediator of paxilline-induced morphological changes. shRNA against *Npas4* significantly reduced dendritic arborization and reversed paxilline-induced morphological changes, indicating that *Npas4* participates in shaping dendritic arborization under basal conditions and mediates paxilline-induced dendritic remodeling (Li et al., 2014). This result is consistent with prior findings that the knockdown of *Npas4* significantly inhibits brain-derived neurotrophic factor expression, which is critical for neural development (Bading, 2013; Lin et al., 2008). Because the reduction in dendritic arborization is commonly observed in many pathological processes associated with neurological conditions, such as aging, Alzheimer's disease, and autism (Bading, 2013), the blockade of nBK channels may provide a new strategy for the development of effective therapeutic methods.

3. SUMMARY AND FUTURE PERSPECTIVES

The study of intracellular BK channels is in its infancy. The detailed mechanisms underlying subcellular localization are unknown. The key factors may include α subunit motifs, alternative splicing, interactions with auxiliary subunits (eg, β and γ subunits), posttranslational modifications, and protein interactions.

Similar to pmBK channels, mitoBK and nBK channels are sensitive to both voltage and calcium concentration, but they also exhibit distinct differences. For example, differences in channel structure, posttranslational modifications, auxiliary subunits, and protein interactions can lead to biophysical properties that differ from those of pmBK channels. The transmembrane potentials and Ca^{2+} concentration in organelles are different from those at the plasma membrane; therefore, the activity and regulation of mitoBK and nBK channels are likely different from those of pmBK channels. Therefore, the results of studies on pmBK channels may not be directly applicable to mitoBK and nBK channels.

The molecular characterization of mitoBK and nBK channels provides a basis for future studies to increase our understanding of these channels. The similar biophysical and pharmacological properties among channel types facilitated the discovery of the basic features of these channels. However, the lack of specific drugs targeting BK channels in different organelles limits the potential for functional studies in intact cells. With the increased knowledge of the mitochondrial and nuclear targeting mechanisms, genetic approaches that reexpress BK channels tagged with a nuclear or mitochondrial localization signal on a BK α subunit knockout background will be helpful for the functional studies in intact cells.

As key players in cellular functions, numerous fundamental questions about mitoBK and nBK channels remain, such as how BK channels are targeted to different organelles, how they are regulated in different physiological and pathological conditions, and what other functions these channels may have, in addition to cardiac protection and nuclear Ca signaling, in different cell types. Further studies using new pharmacological, molecular, and genetic (Zemen et al., 2015) tools will help answer these questions.

ACKNOWLEDGMENTS

This work was supported by the National Natural Science Foundation of China (Grants 81329003; U1201225; 31430032 to T.M.G., 30900581; 31571034 to B.L.), Guangdong

Natural Science Foundation (Grant for Distinguished Young Scholar 2015A030306019 to B.L.), Guangzhou Science and Technology Project (Grant 201300000093), and Specialized Research Fund for the Doctoral Program of Higher Education of China (Grant 20134433130002).

REFERENCES

Alioua, A., Li, M., Wu, Y., Stefani, E., & Toro, L. (2011). Unconventional myristoylation of large-conductance Ca(2)(+)-activated K(+) channel (Slo1) via serine/threonine residues regulates channel surface expression. *Proceedings of the National Academy of Sciences of the United States of America*, *108*, 10744–10749.

Bading, H. (2013). Nuclear calcium signalling in the regulation of brain function. *Nature Reviews. Neuroscience*, *14*, 593–608.

Balderas, E., Zhang, J., Stefani, E., & Toro, L. (2015). Mitochondrial BKCa channel. *Frontiers in Physiology*, *6*, 104.

Banerjee, A., Lee, A., Campbell, E., & Mackinnon, R. (2013). Structure of a pore-blocking toxin in complex with a eukaryotic voltage-dependent K(+) channel. *eLife*, *2*, e00594.

Bednarczyk, P., Koziel, A., Jarmuszkiewicz, W., & Szewczyk, A. (2013). Large-conductance Ca(2)(+)-activated potassium channel in mitochondria of endothelial EA.hy926 cells. *American Journal of Physiology. Heart and Circulatory Physiology*, *304*, H1415–H1427.

Bednarczyk, P., Wieckowski, M. R., Broszkiewicz, M., Skowronek, K., Siemen, D., & Szewczyk, A. (2013). Putative structural and functional coupling of the mitochondrial BK channel to the respiratory chain. *PloS One*, *8*, e68125.

Bentzen, B. H., Osadchii, O., Jespersen, T., Hansen, R. S., Olesen, S. P., & Grunnet, M. (2009). Activation ofbig conductance Ca(2+)-activated K (+) channels (BK) protects the heart against ischemia-reperfusion injury. *Pflügers Archiv/European Journal of Physiology*, *457*, 979–988.

Bhattarai, Y., Fernandes, R., Kadrofske, M. M., Lockwood, L. R., Galligan, J. J., & Xu, H. (2014). Western blot analysis of BK channel beta1-subunit expression should be interpreted cautiously when using commercially available antibodies. *Physiological Reports*, *2*, e12189.

Bilmen, J. G., Wootton, L. L., & Michelangeli, F. (2002). The mechanism of inhibition of the sarco/endoplasmic reticulum Ca^{2+} ATPase by paxilline. *Archives of Biochemistry and Biophysics*, *406*, 55–64.

Bito, H., Deisseroth, K., & Tsien, R. W. (1996). CREB phosphorylation and dephosphorylation: A Ca(2+)- and stimulus duration-dependent switch for hippocampal gene expression. *Cell*, *87*, 1203–1214.

Bootman, M. D., Fearnley, C., Smyrnias, I., MacDonald, F., & Roderick, H. L. (2009). An update on nuclear calcium signalling. *Journal of Cell Science*, *122*, 2337–2350.

Borchert, G. H., Hlavackova, M., & Kolar, F. (2013). Pharmacological activation of mitochondrial BK(Ca) channels protects isolated cardiomyocytes against simulated reperfusion-induced injury. *Experimental Biology and Medicine*, *238*, 233–241.

Brenner, R., Chen, Q. H., Vilaythong, A., Toney, G. M., Noebels, J. L., & Aldrich, R. W. (2005). BK channel beta4 subunit reduces dentate gyrus excitability and protects against temporal lobe seizures. *Nature Neuroscience*, *8*, 1752–1759.

Brenner, R., Jegla, T. J., Wickenden, A., Liu, Y., & Aldrich, R. W. (2000). Cloning and functional characterization of novel large conductance calcium-activated potassium channel beta subunits, hKCNMB3 and hKCNMB4. *The Journal of Biological Chemistry*, *275*, 6453–6461.

Cao, Q., Zhong, X. Z., Zou, Y., Zhang, Z., Toro, L., & Dong, X. P. (2015). BK channels alleviate lysosomal storage diseases by providing positive feedback regulation of lysosomal Ca^{2+} release. *Developmental Cell*, *33*, 427–441.

Carafoli, E., Nicotera, P., & Santella, L. (1997). Calcium signalling in the cell nucleus. *Cell Calcium*, *22*, 313–319.
Century, T. J., & Horowitz, S. B. (1974). Sodium exchange in the cytoplasm and nucleus of amphibian oocytes. *Journal of Cell Science*, *16*, 465–471.
Chen, L., Bi, D., Tian, L., McClafferty, H., Steeb, F., Ruth, P., et al. (2013). Palmitoylation of the beta4-subunit regulates surface expression of large conductance calcium-activated potassium channel splice variants. *The Journal of Biological Chemistry*, *288*, 13136–13144.
Chen, S. R., Li, X., Ebisawa, K., & Zhang, L. (1997). Functional characterization of the recombinant type 3 Ca^{2+} release channel (ryanodine receptor) expressed in HEK293 cells. *The Journal of Biological Chemistry*, *272*, 24234–24246.
Chen, M., Sun, H. Y., Hu, P., Wang, C. F., Li, B. X., Li, S. J., et al. (2013). Activation of BKca channels mediates hippocampal neuronal death after reoxygenation and reperfusion. *Molecular Neurobiology*, *48*, 794–807.
Cheng, Y., Gu, X. Q., Bednarczyk, P., Wiedemann, F. R., Haddad, G. G., & Siemen, D. (2008). Hypoxia increases activity of the BK-channel in the inner mitochondrial membrane and reduces activity of the permeability transition pore. *Cellular Physiology and Biochemistry: International Journal of Experimental Cellular Physiology, Biochemistry, and Pharmacology*, *22*, 127–136.
Cheng, Y., Gulbins, E., & Siemen, D. (2011). Activation of the permeability transition pore by Bax via inhibition of the mitochondrial BK channel. *Cellular Physiology and Biochemistry: International Journal of Experimental Cellular Physiology, Biochemistry, and Pharmacology*, *27*, 191–200.
Cox, N., Toro, B., Pacheco-Otalora, L. F., Garrido-Sanabria, E. R., & Zarei, M. M. (2014). An endoplasmic reticulum trafficking signal regulates surface expression of beta4 subunit of a voltage- and Ca(2)(+)-activated K(+) channel. *Brain Research*, *1553*, 12–23.
Douglas, R. M., Lai, J. C., Bian, S., Cummins, L., Moczydlowski, E., & Haddad, G. G. (2006). The calcium-sensitive large-conductance potassium channel (BK/MAXI K) is present in the inner mitochondrial membrane of rat brain. *Neuroscience*, *139*, 1249–1261.
Dufer, M., Neye, Y., Horth, K., Krippeit-Drews, P., Hennige, A., Widmer, H., et al. (2011). BK channels affect glucose homeostasis and cell viability of murine pancreatic beta cells. *Diabetologia*, *54*, 423–432.
Dworetzky, S. I., Boissard, C. G., Lum-Ragan, J. T., McKay, M. C., Post-Munson, D. J., Trojnacki, J. T., et al. (1996). Phenotypic alteration of a human BK (hSlo) channel by hSlobeta subunit coexpression: Changes in blocker sensitivity, activation/relaxation and inactivation kinetics, and protein kinase A modulation. *The Journal of Neuroscience: The Official Journal of the Society for Neuroscience*, *16*, 4543–4550.
Endo, M. (2009). Calcium-induced calcium release in skeletal muscle. *Physiological Reviews*, *89*, 1153–1176.
Faber, E. S., & Sah, P. (2003a). Ca^{2+}-activated K^{+} (BK) channel inactivation contributes to spike broadening during repetitive firing in the rat lateral amygdala. *The Journal of Physiology*, *552*, 483–497.
Faber, E. S., & Sah, P. (2003b). Calcium-activated potassium channels: Multiple contributions to neuronal function. *The Neuroscientist: A Review Journal Bringing Neurobiology, Neurology and Psychiatry*, *9*, 181–194.
Fahanik-Babaei, J., Eliassi, A., Jafari, A., Sauve, R., Salari, S., & Saghiri, R. (2011a). Electropharmacological profile of a mitochondrial inner membrane big-potassium channel from rat brain. *Biochimica et Biophysica Acta*, *1808*, 454–460.
Fahanik-Babaei, J., Eliassi, A., & Saghiri, R. (2011b). How many types of large conductance Ca(+)(2)-activated potassium channels exist in brain mitochondrial inner membrane: Evidence for a new mitochondrial large conductance Ca(2)(+)-activated potassium channel in brain mitochondria. *Neuroscience*, *199*, 125–132.

Filosa, J. A., Bonev, A. D., Straub, S. V., Meredith, A. L., Wilkerson, M. K., Aldrich, R. W., et al. (2006). Local potassium signaling couples neuronal activity to vasodilation in the brain. *Nature Neuroscience, 9*, 1397–1403.

Fodor, A. A., & Aldrich, R. W. (2009). Convergent evolution of alternative splices at domain boundaries of the BK channel. *Annual Review of Physiology, 71*, 19–36.

Garner, M. H. (2002). Na, K-ATPase in the nuclear envelope regulates Na^+: K^+ gradients in hepatocyte nuclei. *The Journal of Membrane Biology, 187*, 97–115.

Gaspar, T., Domoki, F., Lenti, L., Katakam, P. V., Snipes, J. A., Bari, F., et al. (2009). Immediate neuronal preconditioning by NS1619. *Brain Research, 1285*, 196–207.

Gerasimenko, O. V., Gerasimenko, J. V., Tepikin, A. V., & Petersen, O. H. (1995). ATP-dependent accumulation and inositol trisphosphate- or cyclic ADP-ribose-mediated release of Ca^{2+} from the nuclear envelope. *Cell, 80*, 439–444.

Girouard, H., Bonev, A. D., Hannah, R. M., Meredith, A., Aldrich, R. W., & Nelson, M. T. (2010). Astrocytic endfoot Ca^{2+} and BK channels determine both arteriolar dilation and constriction. *Proceedings of the National Academy of Sciences of the United States of America, 107*, 3811–3816.

Glauser, D. A., Johnson, B. E., Aldrich, R. W., & Goodman, M. B. (2011). Intragenic alternative splicing coordination is essential for Caenorhabditis elegans slo-1 gene function. *Proceedings of the National Academy of Sciences of the United States of America, 108*, 20790–20795.

Gobeil, F., Jr., Dumont, I., Marrache, A. M., Vazquez-Tello, A., Bernier, S. G., Abran, D., et al. (2002). Regulation of eNOS expression in brain endothelial cells by perinuclear EP(3) receptors. *Circulation Research, 90*, 682–689.

Gu, X. Q., Pamenter, M. E., Siemen, D., Sun, X., & Haddad, G. G. (2014). Mitochondrial but not plasmalemmal BK channels are hypoxia-sensitive in human glioma. *Glia, 62*, 504–513.

Gu, X. Q., Siemen, D., Parvez, S., Cheng, Y., Xue, J., Zhou, D., et al. (2007). Hypoxia increases BK channel activity in the inner mitochondrial membrane. *Biochemical and Biophysical Research Communications, 358*, 311–316.

Hardingham, G. E., Arnold, F. J., & Bading, H. (2001). Nuclear calcium signaling controls CREB-mediated gene expression triggered by synaptic activity. *Nature Neuroscience, 4*, 261–267.

Heinen, A., Aldakkak, M., Stowe, D. F., Rhodes, S. S., Riess, M. L., Varadarajan, S. G., et al. (2007). Reverse electron flow-induced ROS production is attenuated by activation of mitochondrial Ca^{2+}-sensitive K^+ channels. *American Journal of Physiology. Heart and Circulatory Physiology, 293*, H1400–H1407.

Heinen, A., Camara, A. K., Aldakkak, M., Rhodes, S. S., Riess, M. L., & Stowe, D. F. (2007). Mitochondrial Ca^{2+}-induced K^+ influx increases respiration and enhances ROS production while maintaining membrane potential. *American Journal of Physiology. Cell Physiology, 292*, C148–C156.

Henney, N. C., Li, B., Elford, C., Reviriego, P., Campbell, A. K., Wann, K. T., et al. (2009). A large-conductance (BK) potassium channel subtype affects both growth and mineralization of human osteoblasts. *American Journal of Physiology. Cell Physiology, 297*, C1397–C1408.

Heppner, T. J., Bonev, A. D., & Nelson, M. T. (1997). Ca(2+)-activated K^+ channels regulate action potential repolarization in urinary bladder smooth muscle. *The American Journal of Physiology, 273*, C110–C117.

Hermann, A., Sitdikova, G. F., & Weiger, T. M. (2015). Oxidative stress and maxi calcium-activated potassium (BK) channels. *Biomolecules, 5*, 1870–1911.

Hu, X., Laragione, T., Sun, L., Koshy, S., Jones, K. R., Ismailov, I. I., et al. (2012). KCa1.1 potassium channels regulate key proinflammatory and invasive properties of

fibroblast-like synoviocytes in rheumatoid arthritis. *The Journal of Biological Chemistry, 287*, 4014–4022.
Jeffries, O., Geiger, N., Rowe, I. C., Tian, L., McClafferty, H., Chen, L., et al. (2010). Palmitoylation of the S0-S1 linker regulates cell surface expression of voltage- and calcium-activated potassium (BK) channels. *The Journal of Biological Chemistry, 285*, 33307–33314.
Kang, J., Huguenard, J. R., & Prince, D. A. (1996). Two types of BK channels in immature rat neocortical pyramidal neurons. *Journal of Neurophysiology, 76*, 4194–4197.
Kanno, Y., & Loewenstein, W. R. (1963). A study of the nucleus and cell membranes of oocytes with an intra-cellular electrode. *Experimental Cell Research, 31*, 149–166.
Kathiresan, T., Harvey, M., Orchard, S., Sakai, Y., & Sokolowski, B. (2009). A protein interaction network for the large conductance Ca(2+)-activated K(+) channel in the mouse cochlea. *Molecular & Cellular Proteomics: MCP, 8*, 1972–1987.
Kent, J., & Meredith, A. L. (2008). BK channels regulate spontaneous action potential rhythmicity in the suprachiasmatic nucleus. *PloS One, 3*, e3884.
Kicinska, A., & Szewczyk, A. (2004). Large-conductance potassium cation channel opener NS1619 inhibits cardiac mitochondria respiratory chain. *Toxicology Mechanisms and Methods, 14*, 59–61.
Kim, E. Y., Ridgway, L. D., Zou, S., Chiu, Y. H., & Dryer, S. E. (2007). Alternatively spliced C-terminal domains regulate the surface expression of large conductance calcium-activated potassium channels. *Neuroscience, 146*, 1652–1661.
Knaus, H. G., Folander, K., Garcia-Calvo, M., Garcia, M. L., Kaczorowski, G. J., Smith, M., et al. (1994). Primary sequence and immunological characterization of beta-subunit of high conductance Ca(2+)-activated K^+ channel from smooth muscle. *The Journal of Biological Chemistry, 269*, 17274–17278.
Kulawiak, B., Kudin, A. P., Szewczyk, A., & Kunz, W. S. (2008). BK channel openers inhibit ROS production of isolated rat brain mitochondria. *Experimental Neurology, 212*, 543–547.
Lai, M. H., Wu, Y., Gao, Z., Anderson, M. E., Dalziel, J. E., & Meredith, A. L. (2014). BK channels regulate sinoatrial node firing rate and cardiac pacing in vivo. *American Journal of Physiology. Heart and Circulatory Physiology, 307*, H1327–H1338.
Lee, U. S., & Cui, J. (2010). BK channel activation: Structural and functional insights. *Trends in Neurosciences, 33*, 415–423.
Leite, M. F., Thrower, E. C., Echevarria, W., Koulen, P., Hirata, K., Bennett, A. M., et al. (2003). Nuclear and cytosolic calcium are regulated independently. *Proceedings of the National Academy of Sciences of the United States of America, 100*, 2975–2980.
Li, J., Al-Khalili, O., Ramosevac, S., Eaton, D. C., & Denson, D. D. (2010). Protein-protein interaction between cPLA2 and splice variants of alpha-subunit of BK channels. *American Journal of Physiology. Cell Physiology, 298*, C251–C262.
Li, B., Jie, W., Huang, L., Wei, P., Li, S., Luo, Z., et al. (2014). Nuclear BK channels regulate gene expression via the control of nuclear calcium signaling. *Nature Neuroscience, 17*, 1055–1063.
Li, M., Sun, Y., Simard, J. M., Wang, J. Y., & Chai, T. C. (2009). Augmented bladder urothelial polyamine signaling and block of BK channel in the pathophysiology of overactive bladder syndrome. *American Journal of Physiology. Cell Physiology, 297*, C1445–C1451.
Liao, Y., Kristiansen, A. M., Oksvold, C. P., Tuvnes, F. A., Gu, N., Runden-Pran, E., et al. (2010). Neuronal Ca^{2+}-activated K^+ channels limit brain infarction and promote survival. *PloS One, 5*, e15601.
Limback-Stokin, K., Korzus, E., Nagaoka-Yasuda, R., & Mayford, M. (2004). Nuclear calcium/calmodulin regulates memory consolidation. *The Journal of Neuroscience: The Official Journal of the Society for Neuroscience, 24*, 10858–10867.

Lin, Y., Bloodgood, B. L., Hauser, J. L., Lapan, A. D., Koon, A. C., Kim, T. K., et al. (2008). Activity-dependent regulation of inhibitory synapse development by Npas4. *Nature, 455*, 1198–1204.

Loewenstein, W. R., & Kanno, Y. (1963). Some electrical properties of a nuclear membrane examined with a microelectrode. *The Journal of General Physiology, 46*, 1123–1140.

Longland, C. L., Dyer, J. L., & Michelangeli, F. (2000). The mycotoxin paxilline inhibits the cerebellar inositol 1,4,5-trisphosphate receptor. *European Journal of Pharmacology, 408*, 219–225.

Loot, A. E., Moneke, I., Keseru, B., Oelze, M., Syzonenko, T., Daiber, A., et al. (2012). 11,12-EET stimulates the association of BK channel alpha and beta(1) subunits in mitochondria to induce pulmonary vasoconstriction. *PloS One, 7*, e46065.

Ma, D., Nakata, T., Zhang, G., Hoshi, T., Li, M., & Shikano, S. (2007). Differential trafficking of carboxyl isoforms of Ca^{2+}-gated (Slo1) potassium channels. *FEBS Letters, 581*, 1000–1008.

Martinez-Espinosa, P. L., Yang, C., Gonzalez-Perez, V., Xia, X. M., & Lingle, C. J. (2014). Knockout of the BK beta2 subunit abolishes inactivation of BK currents in mouse adrenal chromaffin cells and results in slow-wave burst activity. *The Journal of General Physiology, 144*, 275–295.

Maruyama, Y., Shimada, H., & Taniguchi, J. (1995). Ca(2+)-activated K(+)-channels in the nuclear envelope isolated from single pancreatic acinar cells. *Pflügers Archiv/European Journal of Physiology, 430*, 148–150.

Mauceri, D., Freitag, H. E., Oliveira, A. M., Bengtson, C. P., & Bading, H. (2011). Nuclear calcium-VEGFD signaling controls maintenance of dendrite arborization necessary for memory formation. *Neuron, 71*, 117–130.

Mazzanti, M., DeFelice, L. J., Cohn, J., & Malter, H. (1990). Ion channels in the nuclear envelope. *Nature, 343*, 764–767.

McCobb, D. P., Fowler, N. L., Featherstone, T., Lingle, C. J., Saito, M., Krause, J. E., et al. (1995). A human calcium-activated potassium channel gene expressed in vascular smooth muscle. *The American Journal of Physiology, 269*, H767–H777.

McManus, O. B., Helms, L. M., Pallanck, L., Ganetzky, B., Swanson, R., & Leonard, R. J. (1995). Functional role of the beta subunit of high conductance calcium-activated potassium channels. *Neuron, 14*, 645–650.

Meera, P., Wallner, M., Jiang, Z., & Toro, L. (1996). A calcium switch for the functional coupling between alpha (hslo) and beta subunits (KV, Ca beta) of maxi K channels. *FEBS Letters, 382*, 84–88.

Meera, P., Wallner, M., & Toro, L. (2000). A neuronal beta subunit (KCNMB4) makes the large conductance, voltage- and Ca^{2+}-activated K^+ channel resistant to charybdotoxin and iberiotoxin. *Proceedings of the National Academy of Sciences of the United States of America, 97*, 5562–5567.

Meredith, A. L., Thorneloe, K. S., Werner, M. E., Nelson, M. T., & Aldrich, R. W. (2004). Overactive bladder and incontinence in the absence of the BK large conductance Ca^{2+}-activated K^+ channel. *The Journal of Biological Chemistry, 279*, 36746–36752.

Meredith, A. L., Wiler, S. W., Miller, B. H., Takahashi, J. S., Fodor, A. A., Ruby, N. F., et al. (2006). BK calcium-activated potassium channels regulate circadian behavioral rhythms and pacemaker output. *Nature Neuroscience, 9*, 1041–1049.

Montgomery, J. R., Whitt, J. P., Wright, B. N., Lai, M. H., & Meredith, A. L. (2013). Misexpression of the BK K(+) channel disrupts suprachiasmatic nucleus circuit rhythmicity and alters clock-controlled behavior. *American Journal of Physiology. Cell Physiology, 304*, C299–C311.

Morita, T., Hanaoka, K., Morales, M. M., Montrose-Rafizadeh, C., & Guggino, W. B. (1997). Cloning and characterization of maxi K^+ channel alpha-subunit in rabbit kidney. *The American Journal of Physiology, 273*, F615–F624.

Nardi, A., & Olesen, S. P. (2008). BK channel modulators: A comprehensive overview. *Current Medicinal Chemistry*, *15*, 1126–1146.

Nelson, M. T., Cheng, H., Rubart, M., Santana, L. F., Bonev, A. D., Knot, H. J., et al. (1995). Relaxation of arterial smooth muscle by calcium sparks. *Science*, *270*, 633–637.

Ohya, S., Kuwata, Y., Sakamoto, K., Muraki, K., & Imaizumi, Y. (2005). Cardioprotective effects of estradiol include the activation of large-conductance Ca(2+)-activated K(+) channels in cardiac mitochondria. *American Journal of Physiology. Heart and Circulatory Physiology*, *289*, H1635–H1642.

O'Rourke, B. (2007). Mitochondrial ion channels. *Annual Review of Physiology*, *69*, 19–49.

Papadia, S., Stevenson, P., Hardingham, N. R., Bading, H., & Hardingham, G. E. (2005). Nuclear Ca^{2+} and the cAMP response element-binding protein family mediate a late phase of activity-dependent neuroprotection. *The Journal of Neuroscience: The Official Journal of the Society for Neuroscience*, *25*, 4279–4287.

Papassotiriou, J., Kohler, R., Prenen, J., Krause, H., Akbar, M., Eggermont, J., et al. (2000). Endothelial K(+) channel lacks the Ca(2+) sensitivity-regulating beta subunit. *FASEB Journal: Official Publication of the Federation of American Societies for Experimental Biology*, *14*, 885–894.

Piwonska, M., Wilczek, E., Szewczyk, A., & Wilczynski, G. M. (2008). Differential distribution of Ca^{2+}-activated potassium channel beta4 subunit in rat brain: Immunolocalization in neuronal mitochondria. *Neuroscience*, *153*, 446–460.

Pyott, S. J., Meredith, A. L., Fodor, A. A., Vazquez, A. E., Yamoah, E. N., & Aldrich, R. W. (2007). Cochlear function in mice lacking the BK channel alpha, beta1, or beta4 subunits. *The Journal of Biological Chemistry*, *282*, 3312–3324.

Quesada, I., Rovira, J. M., Martin, F., Roche, E., Nadal, A., & Soria, B. (2002). Nuclear KATP channels trigger nuclear Ca(2+) transients that modulate nuclear function. *Proceedings of the National Academy of Sciences of the United States of America*, *99*, 9544–9549.

Ramanathan, K., Michael, T. H., Jiang, G. J., Hiel, H., & Fuchs, P. A. (1999). A molecular mechanism for electrical tuning of cochlear hair cells. *Science*, *283*, 215–217.

Ransom, C. B., & Sontheimer, H. (2001). BK channels in human glioma cells. *Journal of Neurophysiology*, *85*, 790–803.

Redel, A., Lange, M., Jazbutyte, V., Lotz, C., Smul, T. M., Roewer, N., et al. (2008). Activation of mitochondrial large-conductance calcium-activated K^{+} channels via protein kinase A mediates desflurane-induced preconditioning. *Anesthesia and Analgesia*, *106*, 384–391. table of contents.

Ruttiger, L., Sausbier, M., Zimmermann, U., Winter, H., Braig, C., Engel, J., et al. (2004). Deletion of the Ca^{2+}-activated potassium (BK) alpha-subunit but not the BKbeta1-subunit leads to progressive hearing loss. *Proceedings of the National Academy of Sciences of the United States of America*, *101*, 12922–12927.

Saito, M., Nelson, C., Salkoff, L., & Lingle, C. J. (1997). A cysteine-rich domain defined by a novel exon in a slo variant in rat adrenal chromaffin cells and PC12 cells. *The Journal of Biological Chemistry*, *272*, 11710–11717.

Sausbier, M., Arntz, C., Bucurenciu, I., Zhao, H., Zhou, X. B., Sausbier, U., et al. (2005). Elevated blood pressure linked to primary hyperaldosteronism and impaired vasodilation in BK channel-deficient mice. *Circulation*, *112*, 60–68.

Sausbier, M., Hu, H., Arntz, C., Feil, S., Kamm, S., Adelsberger, H., et al. (2004). Cerebellar ataxia and Purkinje cell dysfunction caused by Ca^{2+}-activated K^{+} channel deficiency. *Proceedings of the National Academy of Sciences of the United States of America*, *101*, 9474–9478.

Sausbier, M., Zhou, X. B., Beier, C., Sausbier, U., Wolpers, D., Maget, S., et al. (2007). Reduced rather than enhanced cholinergic airway constriction in mice with ablation of the large conductance Ca^{2+}-activated K^{+} channel. *FASEB Journal: Official Publication of the Federation of American Societies for Experimental Biology*, *21*, 812–822.

Schmitt, N., Grunnet, M., & Olesen, S. P. (2014). Cardiac potassium channel subtypes: New roles in repolarization and arrhythmia. *Physiological Reviews, 94*, 609–653.

Scholz, A., Gruss, M., & Vogel, W. (1998). Properties and functions of calcium-activated K^+ channels in small neurones of rat dorsal root ganglion studied in a thin slice preparation. *The Journal of Physiology, 513*(Pt. 1), 55–69.

Shelley, C., Whitt, J. P., Montgomery, J. R., & Meredith, A. L. (2013). Phosphorylation of a constitutive serine inhibits BK channel variants containing the alternate exon "SRKR". *The Journal of General Physiology, 142*, 585–598.

Sheng, M., Thompson, M. A., & Greenberg, M. E. (1991). CREB: A Ca(2+)-regulated transcription factor phosphorylated by calmodulin-dependent kinases. *Science, 252*, 1427–1430.

Shi, Y., Jiang, M. T., Su, J., Hutchins, W., Konorev, E., & Baker, J. E. (2007). Mitochondrial big conductance KCa channel and cardioprotection in infant rabbit heart. *Journal of Cardiovascular Pharmacology, 50*, 497–502.

Shruti, S., Urban-Ciecko, J., Fitzpatrick, J. A., Brenner, R., Bruchez, M. P., & Barth, A. L. (2012). The brain-specific beta4 subunit downregulates BK channel cell surface expression. *PloS One, 7*, e33429.

Siemen, D., Loupatatzis, C., Borecky, J., Gulbins, E., & Lang, F. (1999). Ca^{2+}-activated K channel of the BK-type in the inner mitochondrial membrane of a human glioma cell line. *Biochemical and Biophysical Research Communications, 257*, 549–554.

Singh, H., Li, M., Hall, L., Chen, S., Sukur, S., Lu, R., et al. (2016). MaxiK channel interactome reveals its interaction with GABA transporter 3 and heat shock protein 60 in the mammalian brain. *Neuroscience, 317*, 76–107.

Singh, H., Lu, R., Bopassa, J. C., Meredith, A. L., Stefani, E., & Toro, L. (2013). MitoBK (Ca) is encoded by the Kcnma1 gene, and a splicing sequence defines its mitochondrial location. *Proceedings of the National Academy of Sciences of the United States of America, 110*, 10836–10841.

Singh, H., Stefani, E., & Toro, L. (2012). Intracellular BK(Ca) (iBK(Ca)) channels. *The Journal of Physiology, 590*, 5937–5947.

Skalska, J., Bednarczyk, P., Piwonska, M., Kulawiak, B., Wilczynski, G., Dolowy, K., et al. (2009). Calcium ions regulate K(+) uptake into brain mitochondria: The evidence for a novel potassium channel. *International Journal of Molecular Sciences, 10*, 1104–1120.

Skalska, J., Piwonska, M., Wyroba, E., Surmacz, L., Wieczorek, R., Koszela-Piotrowska, I., et al. (2008). A novel potassium channel in skeletal muscle mitochondria. *Biochimica et Biophysica Acta, 1777*, 651–659.

Smith, M. R., Nelson, A. B., & Du Lac, S. (2002). Regulation of firing response gain by calcium-dependent mechanisms in vestibular nucleus neurons. *Journal of Neurophysiology, 87*, 2031–2042.

Sokolowski, B., Orchard, S., Harvey, M., Sridhar, S., & Sakai, Y. (2011). Conserved BK channel-protein interactions reveal signals relevant to cell death and survival. *PloS One, 6*, e28532.

Soltysinska, E., Bentzen, B. H., Barthmes, M., Hattel, H., Thrush, A. B., Harper, M. E., et al. (2014). KCNMA1 encoded cardiac BK channels afford protection against ischemia-reperfusion injury. *PloS One, 9*, e103402.

Stehno-Bittel, L., Luckhoff, A., & Clapham, D. E. (1995). Calcium release from the nucleus by InsP3 receptor channels. *Neuron, 14*, 163–167.

Stowe, D. F., Aldakkak, M., Camara, A. K., Riess, M. L., Heinen, A., Varadarajan, S. G., et al. (2006). Cardiac mitochondrial preconditioning by Big Ca^{2+}-sensitive K^+ channel opening requires superoxide radical generation. *American Journal of Physiology. Heart and Circulatory Physiology, 290*, H434–H440.

Szabo, I., & Zoratti, M. (2014). Mitochondrial channels: Ion fluxes and more. *Physiological Reviews, 94*, 519–608.

Szewczyk, A., Jarmuszkiewicz, W., & Kunz, W. S. (2009). Mitochondrial potassium channels. *IUBMB Life*, *61*, 134–143.
Szewczyk, A., Skalska, J., Glab, M., Kulawiak, B., Malinska, D., Koszela-Piotrowska, I., et al. (2006). Mitochondrial potassium channels: From pharmacology to function. *Biochimica et Biophysica Acta*, *1757*, 715–720.
Testai, L., Martelli, A., Marino, A., D'Antongiovanni, V., Ciregia, F., Giusti, L., et al. (2013). The activation of mitochondrial BK potassium channels contributes to the protective effects of naringenin against myocardial ischemia/reperfusion injury. *Biochemical Pharmacology*, *85*, 1634–1643.
Tian, L., Duncan, R. R., Hammond, M. S., Coghill, L. S., Wen, H., Rusinova, R., et al. (2001). Alternative splicing switches potassium channel sensitivity to protein phosphorylation. *The Journal of Biological Chemistry*, *276*, 7717–7720.
Toro, B., Cox, N., Wilson, R. J., Garrido-Sanabria, E., Stefani, E., Toro, L., et al. (2006). KCNMB1 regulates surface expression of a voltage and Ca^{2+}-activated K^+ channel via endocytic trafficking signals. *Neuroscience*, *142*, 661–669.
Tseng-Crank, J., Foster, C. D., Krause, J. D., Mertz, R., Godinot, N., DiChiara, T. J., et al. (1994). Cloning, expression, and distribution of functionally distinct Ca(2+)-activated K^+ channel isoforms from human brain. *Neuron*, *13*, 1315–1330.
Tseng-Crank, J., Godinot, N., Johansen, T. E., Ahring, P. K., Strobaek, D., Mertz, R., et al. (1996). Cloning, expression, and distribution of a Ca(2+)-activated K^+ channel beta-subunit from human brain. *Proceedings of the National Academy of Sciences of the United States of America*, *93*, 9200–9205.
Typlt, M., Mirkowski, M., Azzopardi, E., Ruettiger, L., Ruth, P., & Schmid, S. (2013). Mice with deficient BK channel function show impaired prepulse inhibition and spatial learning, but normal working and spatial reference memory. *PloS One*, *8*, e81270.
Uebele, V. N., Lagrutta, A., Wade, T., Figueroa, D. J., Liu, Y., McKenna, E., et al. (2000). Cloning and functional expression of two families of beta-subunits of the large conductance calcium-activated K^+ channel. *The Journal of Biological Chemistry*, *275*, 23211–23218.
Valverde, M. A., Rojas, P., Amigo, J., Cosmelli, D., Orio, P., Bahamonde, M. I., et al. (1999). Acute activation of Maxi-K channels (hSlo) by estradiol binding to the beta subunit. *Science*, *285*, 1929–1931.
Wallner, M., Meera, P., & Toro, L. (1999). Molecular basis of fast inactivation in voltage and Ca^{2+}-activated K^+ channels: A transmembrane beta-subunit homolog. *Proceedings of the National Academy of Sciences of the United States of America*, *96*, 4137–4142.
Wang, B., Jaffe, D. B., & Brenner, R. (2014). Current understanding of iberiotoxin-resistant BK channels in the nervous system. *Frontiers in Physiology*, *5*, 382.
Wang, X., Yin, C., Xi, L., & Kukreja, R. C. (2004). Opening of Ca^{2+}-activated K^+ channels triggers early and delayed preconditioning against I/R injury independent of NOS in mice. *American Journal of Physiology. Heart and Circulatory Physiology*, *287*, H2070–H2077.
Werner, M. E., Zvara, P., Meredith, A. L., Aldrich, R. W., & Nelson, M. T. (2005). Erectile dysfunction in mice lacking the large-conductance calcium-activated potassium (BK) channel. *The Journal of Physiology*, *567*, 545–556.
Wojtovich, A. P., Nadtochiy, S. M., Urciuoli, W. R., Smith, C. O., Grunnet, M., Nehrke, K., et al. (2013). A non-cardiomyocyte autonomous mechanism of cardioprotection involving the SLO1 BK channel. *PeerJ*, *1*, e48.
Wojtovich, A. P., Sherman, T. A., Nadtochiy, S. M., Urciuoli, W. R., Brookes, P. S., & Nehrke, K. (2011). SLO-2 is cytoprotective and contributes to mitochondrial potassium transport. *PloS One*, *6*, e28287.
Wu, Y., Eghbali, M., Ou, J., Lu, R., Toro, L., & Stefani, E. (2010). Quantitative determination of spatial protein-protein correlations in fluorescence confocal microscopy. *Biophysical Journal*, *98*, 493–504.

Xie, J. (2008). Control of alternative pre-mRNA splicing by Ca(++) signals. *Biochimica et Biophysica Acta*, *1779*, 438–452.

Xu, W., Liu, Y., Wang, S., McDonald, T., Van Eyk, J. E., Sidor, A., et al. (2002). Cytoprotective role of Ca^{2+}-activated K^{+} channels in the cardiac inner mitochondrial membrane. *Science*, *298*, 1029–1033.

Yan, J., & Aldrich, R. W. (2010). LRRC26 auxiliary protein allows BK channel activation at resting voltage without calcium. *Nature*, *466*, 513–516.

Yan, J., & Aldrich, R. W. (2012). BK potassium channel modulation by leucine-rich repeat-containing proteins. *Proceedings of the National Academy of Sciences of the United States of America*, *109*, 7917–7922.

Yang, H., Zhang, G., & Cui, J. (2015). BK channels: Multiple sensors, one activation gate. *Frontiers in Physiology*, *6*, 29.

Zahradnikova, A., & Meszaros, L. G. (1998). Voltage change-induced gating transitions of the rabbit skeletal muscle Ca^{2+} release channel. *The Journal of Physiology*, *509*(Pt. 1), 29–38.

Zarei, M. M., Eghbali, M., Alioua, A., Song, M., Knaus, H. G., Stefani, E., et al. (2004). An endoplasmic reticulum trafficking signal prevents surface expression of a voltage- and Ca^{2+}-activated K^{+} channel splice variant. *Proceedings of the National Academy of Sciences of the United States of America*, *101*, 10072–10077.

Zarei, M. M., Song, M., Wilson, R. J., Cox, N., Colom, L. V., Knaus, H. G., et al. (2007). Endocytic trafficking signals in KCNMB2 regulate surface expression of a large conductance voltage and Ca(2+)-activated K^{+} channel. *Neuroscience*, *147*, 80–89.

Zemen, B. G., Lai, M. H., Whitt, J. P., Khan, Z., Zhao, G., & Meredith, A. L. (2015). Generation of Kcnma1fl-tdTomato, a conditional deletion of the BK channel alpha subunit in mouse. *Physiological Reports*, *3*, e12612.

Zhang, S. J., Zou, M., Lu, L., Lau, D., Ditzel, D. A., Delucinge-Vivier, C., et al. (2009). Nuclear calcium signaling controls expression of a large gene pool: Identification of a gene program for acquired neuroprotection induced by synaptic activity. *PLoS Genetics*, *5*, e1000604.

Zhou, X., Wulfsen, I., Korth, M., McClafferty, H., Lukowski, R., Shipston, M. J., et al. (2012). Palmitoylation and membrane association of the stress axis regulated insert (STREX) controls BK channel regulation by protein kinase C. *The Journal of Biological Chemistry*, *287*, 32161–32171.

CHAPTER SIX

Modulation of BK Channels by Small Endogenous Molecules and Pharmaceutical Channel Openers

T. Hoshi*,[1], S.H. Heinemann†
*University of Pennsylvania, Philadelphia, PA, United States
†Friedrich Schiller University Jena & Jena University Hospital, Jena, Germany
[1]Corresponding author: e-mail address: hoshi@hoshi.org

Contents

Abstract

Voltage- and Ca^{2+}-activated K^+ channels of big conductance (BK channels) are abundantly found in various organs and their relevance for smooth muscle tone and neuronal signaling is well documented. Dysfunction of BK channels is implicated in an array of human diseases involving many organs including the nervous, pulmonary, cardiovascular, renal, and urinary systems. In humans a single gene (*KCNMA1*) encodes the

International Review of Neurobiology, Volume 128
ISSN 0074-7742
http://dx.doi.org/10.1016/bs.irn.2016.03.020

pore-forming α subunit (Slo1) of BK channels, but the channel properties are variable because of alternative splicing, tissue- and subcellular-specific auxiliary subunits (β, γ), posttranslational modifications, and a multitude of endogenous signaling molecules directly affecting the channel function. Initiatives to develop drugs capable of activating BK channels (channel openers) therefore need to consider the tissue-specific variability of BK channel structure and the potential interference with endogenously produced regulatory factors. The atomic structural basis of BK channel function is only beginning to be revealed. However, building on detailed knowledge of BK channel function, including its single-channel characteristics, voltage- and Ca^{2+} dependence of channel gating, and modulation by diffusible messengers, a multi-tier allosteric model of BK channel gating (Horrigan and Aldrich (HA) model) has become a valuable tool in studying modulation of the channel. Using the conceptual framework of the HA model, we here review the functional impact of endogenous modulatory factors and select small synthetic compounds that regulate BK channel activity. Furthermore, we devise experimental approaches for studying BK channel—drug interactions with the aim to classify BK-modulating substances according to their molecular mode of action.

1. INTRODUCTION

BK channels participate in numerous physiological functions by coupling transmembrane K^+ flux, changes in membrane potential (V_m), and intracellular Ca^{2+} concentration ($[Ca^{2+}]_i$). For example, in excitable cells, voltage- and Ca^{2+}-dependent activation of BK channels acts to shorten the action potential (AP) duration and underlies the fast after-hyperpolarization, thereby modulating the overall AP frequency and transmitter release (Griguoli, Sgritta, & Cherubini, 2016; Salkoff, Butler, Ferreira, Santi, & Wei, 2006) (see Chapter "BK Channels in Neurons" by Barth and Contet). In tonic vascular smooth muscle cells, BK channels open in response to local increases in $[Ca^{2+}]_i$ and create spontaneous transient outward currents, opposing depolarization and facilitating vessel relaxation (Jaggar, Porter, Lederer, & Nelson, 2000; Nelson et al., 1995) (see Chapter "BK Channels in the Vascular System" by Krishnamoorthy-Natarajan and Koide). In ion-transporting cells such as those in the kidney and probably in neuronal glial cells, BK channels constitute a high-throughput transport pathway for K^+, contributing to the tissue and body electrolyte and fluid homeostasis (Filosa et al., 2006; Welling, 2016; Wu & Marx, 2010).

The physiological importance of BK channels in a wide variety of cells, tissues, and organs is readily inferred from the results of the studies using genetically modified model organisms where the gene encoding for the pore-forming subunit (*KCNMA1* in the human genome, also SLO1) or one of the auxiliary subunits is disrupted (Brenner et al., 2005; Brenner, Perez, et al., 2000; Filosa et al., 2006; Kurt et al., 2012; Liao et al., 2010; Martinez-Espinosa, Yang, Gonzalez-Perez, Xia, & Lingle, 2014; Meredith, Thorneloe, Werner, Nelson, & Aldrich, 2004; Meredith et al., 2006; Plüger et al., 2000; Sachse, Faulhaber, Seniuk, Ehmke, & Pongs, 2014; Sausbier et al., 2005, 2004; Sprossmann et al., 2009; Werner, Zvara, Meredith, Aldrich, & Nelson, 2005). Such gene disruptions, whether organism wide or tissue specific, lead to a variety of disease-like phenotypes, many of which are neuronal, vascular, and renal. Some of the examples from mouse studies include hypertension (Brenner, Perez, et al., 2000; Plüger et al., 2000), overactive bladder and incontinence (Meredith et al., 2004; Sprossmann et al., 2009), erectile dysfunction (Werner et al., 2005), seizures (Brenner et al., 2005), altered hearing (Kurt et al., 2012), altered circadian rhythm (Meredith et al., 2006), and cerebellar ataxia (Sausbier et al., 2004). As with other animal studies, however, the phenotypes observed in the genetically modified animals are influenced by many factors and some divergent observations have been reported (Xu, Garver, Galligan, & Fink, 2011).

Human genetics studies suggest that various mutations in the genes coding for the BK complex appear associated with epilepsy (Du et al., 2005; Lorenz, Heils, Kasper, & Sander, 2007), paroxysmal dyskinesias (Du et al., 2005), hypertension (Tomás et al., 2008), protection against hypertension (Fernández-Fernández et al., 2004; Kelley-Hedgepeth et al., 2009; Nielsen et al., 2008), and severity of asthma (Seibold et al., 2008) in some select cohorts (but not in others; Kokubo et al., 2005; Sentí et al., 2005). Furthermore, the BK channel activity/expression is reported to be altered in multiple human disease states including hypertension (Yang et al., 2013), overactive bladder (Chang et al., 2010; Hristov, Chen, Kellett, Rovner, & Petkov, 2011; Oger et al., 2011), morbid obesity (Jiao et al., 2011), autism (Laumonnier et al., 2006), and potentially adult-onset neuronal ceroid lipofuscinosis (Donnelier et al., 2015).

Thus, the animal and human studies clearly show that the BK channel plays numerous physiologically significant roles. This multifunctional capability is made possible in part because the BK channel complex is endowed with multiple mechanisms for functional versatility (Contreras et al., 2013; Hoshi, Pantazis, & Olcese, 2013; Hou, Heinemann, & Hoshi, 2009; Kyle & Braun, 2014). For example, while only one gene (*KCNMA1*) exists for the pore-forming BK α (or Slo1) protein in mammals, its transcript is extensively spliced to create a large number of variants (Chen et al., 2005; Shipston, 2001; Xie & McCobb, 1998), for example, potentially up to one thousand full-length transcripts in theory in mice (Sakai, Harvey, & Sokolowski, 2011). Tissue- and subcellular-specific coassembly with auxiliary subunits, β1–4 coded by *KCNMBs*, and various γ subunits coded by *LRRCs* (Behrens et al., 2000; Brenner, Jegla, Wickenden, Liu, & Aldrich, 2000; Gonzalez-Perez, Xia, & Lingle, 2015; Uebele et al., 2000; Wallner, Meera, & Toro, 1999; Weiger et al., 2000; Wu & Marx, 2010; Yan & Aldrich, 2010, 2012) (see Chapter "Modulation of BK Channel Function by Auxiliary Beta and Gamma Subunits" by Yan and Li), capable conferring distinct gating and pharmacological characteristics, further increases the diversity. Participation of the BK channel in different macromolecular signaling complexes involving other proteins and lipids also contribute to the functional versatility of the channel (Berkefeld et al., 2006; Cox, 2014; Li et al., 2010; Liu et al., 2004; Peng, Sakai, Kurgan, Sokolowski, & Uversky, 2014; Toro et al., 2014; Williams et al., 2004; Zhang et al., 2014) (see Chapter "Protein Network Interacting with BK Channels" by Kim and Oh).

The BK channel is furthermore subject to various posttranslational modifications (Kyle & Braun, 2014) such as Ser/Thr phosphorylation (Yan et al., 2008) (see Chapter "Posttranscriptional and Posttranslational Regulation of BK Channels" by Shipston and Tian) and also to direct regulation by small endogenous signaling molecules such as H^+, carbon monoxide (CO), heme, steroids, and fatty acids (Dopico & Bukiya, 2014; Hou et al., 2009). Additionally, because of the potential involvement of the channels in multiple disease processes, the BK channel has been a target of therapeutic drug discovery (Bentzen, Olesen, Ronn, & Grunnet, 2014) (see also Chapter "Developing the Molecular Pharmacology of BK Channels for Therapeutic Benefit" by Kaczorowski and Garcia). In particular, many of the pathophysiological phenotypes are associated with a loss of function or downregulation of the BK channel. Thus, BK "openers" are of particular interest for therapeutic drug discovery. Regrettably, however, as of this writing, no therapeutic drug targeting the BK channel has reached market (see Chapter "Developing the Molecular Pharmacology of BK Channels for Therapeutic Benefit" by Kaczorowski and Garcia). The reasons for this slow development are manifold. Frequently, the corresponding compounds are not potent enough or suffer from undesired side effects. Thus, an urgent need exists for a deeper understanding of the tissue-specific pharmacological properties of BK channels and the crosstalk of potential drugs with endogenously produced BK modulators. Moreover, the availability of BK specific or better yet BK α splice variant/auxiliary subunit-specific modulators whose mechanisms of action are well characterized will also tremendously help to identify the roles that BK channels play under physiological conditions, complementing the on-going genetic manipulation studies. Extensive information on functional features of the BK channel under control conditions and the impact of small-molecule modulators on such features continues to offer fundamental insights into the channel's structure–function relationship. While the atomic-scale structural dynamics information is lacking, a continuously growing body of channel function information and the impact of small molecule modulators provide fundamental insight into the channel's structure–function relationship and its mode of operation.

Using a functional description of BK channels as conceptual framework, this chapter discusses the fundamental aspects of short-term modulation of these channels. Further, the discussion will focus on mammalian, especially human, BK channels in plasma membranes; those in other species and those in intracellular organelle membranes (Cao et al., 2015; Singh, Stefani, & Toro, 2012) (see Chapter "Functional Role of Mitochondrial and Nuclear BK Channels" by Gao and Li), which are a very active area of investigation, will not be included. Lastly, blockers and inhibitors of the BK channel are only briefly mentioned. The authors regret that the space constraint does not allow every published article to be included.

2. SALIENT FEATURES OF THE BK CHANNEL

Two distinguishing functional features of the BK channel are their large conductance and multitier allosteric gating (see Chapter "Biophysics of BK Channel Gating" by Olcese and Pantazis). The unitary conductance of the BK channel under a symmetrical 140 mM K^+ condition is typically ≥200 pS. Under physiological conditions with a few mM K^+ outside, the unitary conductance is smaller but remains much greater than those of other K^+ channels. Thus, opening of even a few BK channels could potentially influence cellular electrical properties. For instance, the channels could contribute to neuronal voltage "noise," which has been postulated to influence AP firing (O'Donnell & van Rossum, 2014). The large conductance of the channel is also utilized in various ion-transporting cells to maintain proper electrolyte balance (Wu & Marx, 2010). An in-depth discussion of the biophysical and structural basis of the BK channel's high conductance is found in Diaz-Franulic, Sepulveda, Navarro-Quezada, Gonzalez-Nilo, and Naranjo (2015), Geng, Niu, and Magleby (2011), and Zhang, Niu, Brelidze, and Magleby (2006).

The multitier allosteric gating mechanism is another well-known feature of the BK channel (see Chapter "Biophysics of BK Channel Gating" by Olcese and Pantazis) (Horrigan, 2012), allowing the channel to respond to changes in V_m and $[Ca^{2+}]_i$. in a versatile and complex manner (Cox, Cui, & Aldrich, 1997; Cui & Aldrich, 2000; Horrigan & Aldrich, 2002; Horrigan, Cui, & Aldrich, 1999; Rothberg & Magleby, 2000). One of the important features of this allosteric gating mechanism is that depolarization without Ca^{2+} can increase open probability (P_o) and that Ca^{2+} without depolarization can also increase P_o; neither is strictly *required* for channel activation. Quantitatively, the allosteric nature of BK channel gating (Cox et al., 1997; Cui & Aldrich, 2000; Horrigan et al., 1999; Rothberg & Magleby, 2000) is well represented by the model of Horrigan and Aldrich (HA model) (Fig. 1A) (Horrigan & Aldrich, 2002). The model accounts for the BK channel behavior without any auxiliary subunit; with some quantitative modifications, the model can also describe the channel behavior in the presence of various auxiliary subunits including β1 (Orio & Latorre, 2005), β2 (Orio & Latorre, 2005), β4 (Wang, Rothberg, & Brenner, 2006), and γ1 (Yan & Aldrich, 2010). The HA model is "semimechanistic" in nature (Horrigan & Hoshi, 2015) such that it incorporates important structural information about the channel, such as its tetrameric organization (Shen et al., 1994). As detailed later, the model has served as an invaluable framework to study the mechanisms of many small-molecule modulators of the BK channel complex and allowed for mechanistic conclusions with ample structural connotations. In this regard, the BK channel is one of the exceptions. Many other channel types lack such a quantitative semimechanistic model (Horrigan & Hoshi, 2015) and the accounts of their modulations are often phenomenological.

3. CLASSIFICATION OF MODULATORS BY THEIR MODE OF ACTION

Modulation framework. A modulator binds to the channel complex ("binding") and the information about the binding event is then transmitted ("coupling") to one or more components of the channel, ultimately affecting opening and closing of the ion conduction gate. In general, currents

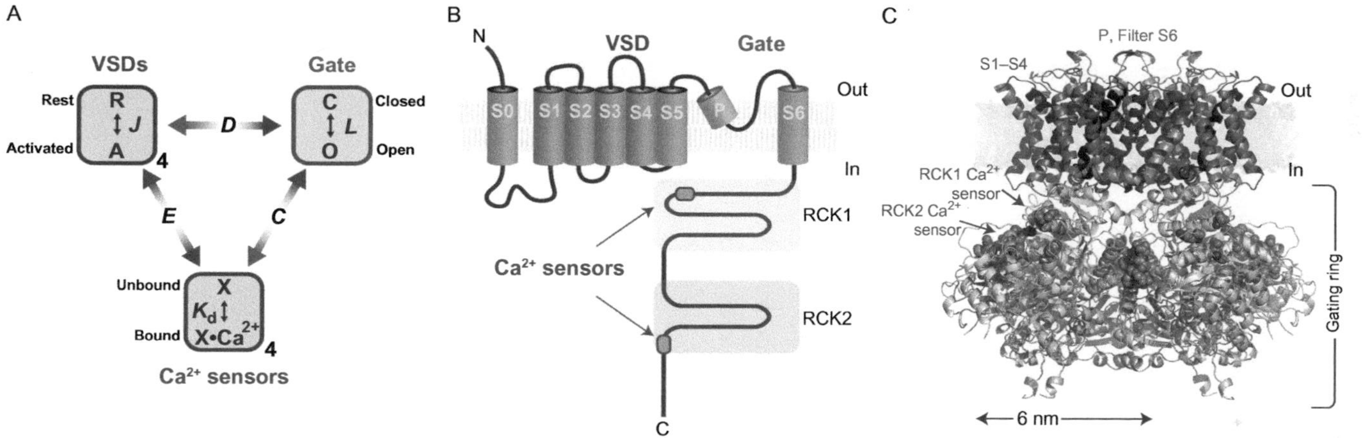

Fig. 1 Functional and structural organization of the BK channel. (A) The equilibrium Horrigan and Aldrich model. (B) Organization of one pore-forming BK α polypeptide containing ~1100 residues. The first 330–340 residues comprise the transmembrane domain (N–S6). (C) Probable organization of one tetrameric BK channel complex. The transmembrane domain is modeled using the structure of a voltage-gated K^+ channel (PDB ID: 2R9R) and the cytoplasmic gating ring is from PDB ID: 3MT5. S0 is not shown. The Ca^{2+}-coordinating residues are shown using *spheres*.

through a single class of ion channels, each with n kinetically distinguishable states, are typically described using the following expression:

$$I(V_m, t) = N \sum_{k=1}^{n} i_k(V_m) P_k(V_m, t) \tag{1}$$

where $I(V_m, t)$ represents the total ionic current as a function of membrane potential V_m and time t, N is the total number of the channels present in the system under consideration (eg, a cell or a membrane patch), $i_k(V_m)$ is the unitary current size of state k at V_m, and $P_k(V_m, t)$ represents the probability of being in state k. Each BK channel contains multiple kinetically distinguishable states as revealed especially well in single-channel studies (summarized in Geng & Magleby, 2014) and many of the states are well conceptualized in the HA model (Horrigan & Aldrich, 2002).

The single-channel current–voltage curve ($i_k(V_m)$) over a wide voltage range can be complex, reflecting the multiple K^+ binding sites in the selectivity filter region of the channel (Zhou, Morais-Cabral, Kaufman, & MacKinnon, 2001) and the influences of local K^+ concentrations in the vicinity of the ion conduction pathway and blocking particles such as H^+ (Brelidze & Magleby, 2004), Na^+ (Yellen, 1984), Ca^{2+} (Geng, Wang, & Magleby, 2013), Mg^{2+} (Geng et al., 2013), and polyamines (Zhang et al., 2006). However, at least near the reversal potential of the channel in a limited voltage range, a simplifying assumption that the unitary current size is a linear function of voltage can be made. With this assumption, the term $i_k(V_m)$ is represented as

$$i_k(V_m) = g_k(V_m - V_{rev}) \tag{2}$$

where g_k signifies the unitary conductance of state k and V_{rev} is the reversal potential. Because the BK channel is highly selective for K^+ (Blatz & Magleby, 1984), V_{rev} may be substituted with E_K, the equilibrium potential for K^+. Furthermore, because each channel complex is considered to be either fully closed or fully open and so-called subconductance levels are typically ignored (Guo et al., 2008), the following familiar expression is often employed.

$$i_k(V_m) = g_k(V_m - E_K) \tag{3}$$

The term $P_k(V_m, t)$ in Eq. (1), representing gating of the channel, is described by the HA model (see Chapter "Biophysics of BK Channel

Gating" by Olcese and Pantazis and also Hoshi, Pantazis, et al., 2013). In essence, the HA model stipulates that the ion conduction gate of the channel is allosterically and reciprocally controlled by its voltage sensor domains (VSDs) and Ca^{2+}-sensor domains, which in turn allosterically and reciprocally interact. The equilibrium energetic version of the HA model is shown in Fig. 1A (Horrigan & Aldrich, 2002; Horrigan et al., 1999; Hoshi, Pantazis, et al., 2013). As explained elsewhere (see Chapter "Biophysics of BK Channel Gating" by Olcese and Pantazis and also Hoshi, Pantazis, et al., 2013), structural connotations may be given to the HA model. The ion conduction gate, whose closed–open equilibrium at 0 mV without Ca^{2+} is described by the equilibrium constant L_0, probably involves those residues near the selectivity filter (Fig. 1B and C) (Chen & Aldrich, 2011; Tang, Zhang, Xia, & Lingle, 2010; Thompson & Begenisich, 2012; Wilkens & Aldrich, 2006; Zhou, Xia, & Lingle, 2011), unlike in other voltage-gated K^+ channels that utilize the cytoplasmic ends of their S6 segments (Jensen et al., 2010, 2012). Depending on the subunit composition and other factors, the value of L_0 is about 1×10^{-6}; even without Ca^{2+} and without VSD activation, the channel does open, albeit infrequently. While each channel has a very low probability of opening, the likelihood of observing openings can be appreciable if a large number of channels are present. Each VSD, which can be at rest or activated as specified by J in the model, encompasses S1, S2, S3, and S4 (Fig. 1B and C) (Hoshi, Pantazis, et al., 2013; Ma, Lou, & Horrigan, 2006). Activation of each VSD promotes opening of the ion conduction gate according to the allosteric coupling factor D and vice versa. The greater the value of D, the greater is the influence of VSD activation on the gate. The BK channel complex is only modestly voltage-dependent, corresponding to ~2.6 voltage-sensing equivalent electron charges per channel ($z_L + 4\ z_J$ in Fig. 1A) (Ma et al., 2006) compared with 14 or more in other voltage-gated K^+ channels (Aggarwal & MacKinnon, 1996; Schoppa & Sigworth, 1998; Zagotta, Hoshi, & Aldrich, 1994). The conformational changes in the BK VSD may be different from those in other voltage-gated K^+ channels (for summary, see Hoshi, Pantazis, et al., 2013). The weak overall voltage sensitivity of the BK channel is required for its wide dynamic voltage range (Cui & Aldrich, 2000). Each pore-forming subunit has at least two distinct "high-affinity" (μM level K_d values) Ca^{2+} sensor sites: one in the RCK1 domain (herein referred to as the RCK1 Ca^{2+} sensor) and the other in the RCK2 domain (herein referred to as the RCK2 Ca^{2+} sensor)

(Fig. 1B and C). The RCK2 Ca^{2+} sensor is also called the Ca^{2+} bowl sensor (Schreiber & Salkoff, 1997). In the original HA model, the two sites are lumped together for simplicity and the Ca^{2+} bound–unbound equilibrium is specified by the constant K ($=[Ca^{2+}]_i/K_d$). The RCK1 and RCK2 sites can be incorporated separately into the model at the cost of increasing the number of parameters (Savalli, Pantazis, Yusifov, Sigg, & Olcese, 2012). The K_d values of the Ca^{2+} sensors are regulated by the closed-open status of the ion conduction gate as specified the factor C and also by the VSDs according to the factor E. Because $C > 1$, the Ca^{2+} sensors are readily influenced by the ion conduction gate such the K_d values are approximately three- to fivefold lower ("higher affinity") when the ion conduction gate is open. The RCK1 and RCK2 Ca^{2+} sensors differ in their functional properties as summarized in Hoshi, Pantazis, et al. (2013) and in Chapter "Biophysics of BK Channel Gating" by Olcese and Pantazis. While the two Ca^{2+}-sensor sites are distant in the amino acid sequence, in the three-dimensional structure the two sites are located relatively close to each other, facing toward the transmembrane area (Fig. 1C). The cytoplasmic area containing the RCK1 and RCK2 domains, together called the gating ring domain, places itself very close to the transmembrane domain such that divalent cations like Mg^{2+} ions, whose dehydrated ionic *diameter* is about 1.3–1.4 Å, may be coordinated between them (Chen, Geng, & Magleby, 2011; Yang et al., 2007; Zhang, Solaro, & Lingle, 2001) (not depicted explicitly in the HA model). The close juxtaposition of the cytoplasmic gating ring domain and the transmembrane domain undoubtedly contribute to their allosteric interactions (C and E in the HA model). This structural arrangement may also have some implications as to how K^+ ions permeate through the channel and how various blockers including the inactivation particles of the auxiliary β2 and select β3 subunits occlude the ion conduction pore (Wallner et al., 1999; Xia, Ding, Zeng, Duan, & Lingle, 2000).

The HA model diagram in Fig. 1A represents only the steady-state energetic interactions of the BK channel. The rate constants underlying the equilibrium constants are also specified in the HA model so that the kinetics of the channel, especially without Ca^{2+}, are well reproduced (Horrigan et al., 1999). The values of the model parameters can be estimated in an experimentally constrained manner (Horrigan & Hoshi, 2015) with a few reasonable assumptions (Horrigan et al., 1999).

A modulator can alter any one or more of the terms in Eq. (1): N, $i_k(V_m)$, and $P_k(V_m, t)$. Further, those compounds affecting $P_k(V_m, t)$ can do so in multiple ways, affecting various components of the HA model. We will describe modulatory phenomena of the BK channel complex according to their underlying mechanisms using Eq. (1) and the HA model.

4. MODULATION OF "*N*"

A change in the number of BK channels available to open can manifest as a simple up- or downsizing of ionic currents without any effect on the time course. For example, this could occur if additional channels are rapidly inserted into the plasma membrane or some of the channels are removed by vesicular translocation. Such a mechanism occurring in seconds to minutes has been documented for some glutamate receptor channels (Malinow & Malenka, 2002) and transient receptor potential channels (Bezzerides, Ramsey, Kotecha, Greka, & Clapham, 2004). Multiple processes that regulate the number of BK channels available in the membrane are known: myristoylation (Alioua, Li, Wu, Stefani, & Toro, 2011) and cholesterol depletion (Shmygol, Noble, & Wray, 2007), for example. How rapidly these processes occur, for instance whether it is possible within a typical electrophysiological experiment, is unknown. Removal/addition of a cofactor required for ion channel function may be another way to alter the number of channels available to open. For instance, phosphatidylinositol 4,5-bisphosphate (PIP_2) is required for many channel types to function properly, acting as a functionally essential cofactor (Suh & Hille, 2008). While PIP_2 does modulate the BK channel complex in multiple ways (Tang, Zhang, Meng, Cui, & Logothetis, 2014; Tian et al., 2015; Vaithianathan et al., 2008) (see later), the lipid does not appear absolutely required for channel function. Overall, rapid dynamic insertion and removal of BK complexes is a plausible way to regulate their contributions to cell function; however, a clear case is yet to be documented. As cell biological techniques to detect specific ion channel complexes in live cells improve, new BK modulators that regulate N rapidly and reversibly may be found. Leo et al. presents an interesting case where N may be modulated (Leo et al., 2014).

Experimentally, it is feasible in principle to determine whether a stimulatory modulator increases N if the measurements are conducted under the conditions in which $P_k(V_m, t)$ is saturated (eg, very positive voltages and high $[Ca^{2+}]_i$). In contrast, it may be exceedingly difficult to establish whether an inhibitory modulator dramatically and stably decreases $P_k(V_m, t)$ or decreases N.

5. MODULATION OF "$I(V_M)$"

Open-channel properties of the channel ($i(V_m)$) include the conductance (g) and the ion selectivity (V_{rev}) (Eq. 2). These biophysical characteristics are structurally conferred by the pore domain (S5–S6) of the channel (Fig. 1B). The information obtained from various K^+ channels collectively suggests that the local ion concentrations near the internal and external mouths of the channel, the access resistance of K^+ ions to the selectivity filter region from the cytoplasmic side through the cavity or vestibule, and the selectivity filter ion occupancy are among the important factors controlling the conductance (Diaz-Franulic et al., 2015; Sack & Tilley, 2015).

The access resistance of K^+ from the cytoplasmic side to the selectivity filter region in the BK channel is kept low to allow for the high conductance (Diaz-Franulic et al., 2015) and if any, a modulator is expected to only increase the resistance, thus decreasing the conductance. Similarly, the local concentration of K^+ ions near the internal mouth of the channel is also high owing to the presence of multiple negatively charged residues (Brelidze, Niu, & Magleby, 2003), any modulation is expected to diminish the conductance.

A single-particle cryoEM study of the Na^+-activated Slo2 K^+ channel indicates that K^+ ions traverse through the water-filled central hole of the torus-shaped gating ring domain as necessitated by the very close apposition of the transmembrane domain and the cytoplasmic gating ring domain (Hite et al., 2015). A similar arrangement probably exists in BK channels. According to the atomic structures of the BK gating ring domain (Wu, Yang, Ye, & Jiang, 2010; Yuan, Leonetti, Hsiung, & MacKinnon, 2012; Yuan, Leonetti, Pico, Hsiung, & Mackinnon, 2010), the central hole of the torus-shaped BK gating ring domain is 30–40 Å in diameter, large enough to accommodate multiple hydrated K^+ ions (~8 Å in diameter with the first hydration shell). Considering this structural arrangement, modulators of the BK channel may alter the single-channel conductance in multiple ways. The compounds may directly alter the local K^+ concentrations, perhaps within the central pore of the torus gating ring domain; for example, electrostatic repulsion may be observed between positively charged compounds and K^+ ions. Binding of a modulator to the channel complex away from the gating ring central pore may indirectly affect the conductance.

Some compounds have been reported to alter the single-channel conductance of the BK channel. Tamoxifen, which also increases P_o (see later), decreases the unitary conductance by 10% (Dick, Rossow, Smirnov, Horowitz, & Sanders, 2001). The "slow" membrane potential dye DiBAC$_4$(3) increases P_o (see later) but also decreases the unitary conductance by ~20–25% (Scornik et al., 2013). How tamoxifen and DiBAC$_4$(3) alter the single-channel conductance remains obscure. It is probable that other modulators that alter $P_k(V_m, t)$ also change $i(V_m)$. In particular, this may be expected for BK channel openers that bind near the inner vestibule to keep the gate open; the flow of K^+ ions to/from the selectivity filter may be impeded. Such phenomena were observed, for example, for voltage-gated Na^+ channels in which veratridine and also some drugs akin to local anesthetics bind to the channel to keep the gate open but, simultaneously, strongly affect the ion flow (Garber & Miller, 1987). For BK channel openers, such an interference with $i(V_m)$ is expected to diminish the overall potency to increase the BK channel current. However, data on how drugs affect the $i(V_m)$ properties are often lacking.

Besides drugs binding to the channel, various ions present near the channel complex may modulate the single-channel conductance. Prolonged depolarization, for example by a train of APs especially under pathophysiological conditions, may drastically alter the intracellular and extracellular ion concentrations (Raimondo, Burman, Katz, & Akerman, 2015). While not typically recognized as "modulators" of the BK channels, physiological ions such as intracellular Na^+ (Yellen, 1984), intracellular Mg^{2+} (Geng et al., 2013), intracellular Ca^{2+} (Geng et al., 2013), and intracellular H^+ (Brelidze & Magleby, 2004) are capable of diminishing the overall time-averaged single-channel conductance of the BK channel, often by the voltage-dependent "fast" or "flickery" block mechanism in which the compounds associate and dissociate from the ion conduction pathway beyond the frequency limit of the electrophysiological apparatus (Yellen, 1984). For instance, 20 mM Na^+, through its ultrafast association and dissociation kinetics, decreases the time-averaged single-channel current size down to ~20% of that without Na^+ at 40 mV (Yellen, 1984). Thus, influx of Na^+ ions during APs may have a functional impact in a dynamic manner albeit this possibility has not been thoroughly investigated. The conductance-diminishing effects of intracellular H^+ and Mg^{2+} are more modest than that of intracellular Na^+ and it is unlikely that H^+ and Mg^{2+} act as biologically significant inhibitory modulators of the BK channel through their effects on the channel conductance. But instead, H^+ and Mg^{2+} increase open probability of the BK channel (see later), thus acting as endogenous activators (Avdonin, Tang, & Hoshi, 2003; Hou, Xu, Heinemann, & Hoshi,

2008b; Shi & Cui, 2001; Shi et al., 2002; Xia, Zeng, & Lingle, 2002; Zeng, Xia, & Lingle, 2005; Zhang et al., 2001).

The extracellular milieu surrounding native BK channels is often restricted in volume especially in the central nervous system and the extracellular K^+ concentration ($[K^+]_o$) is only a few mM. Under such conditions, efflux of K^+ ions through K^+-permeant channels including BK channels has potential to markedly increase $[K^+]_o$ (Filosa et al., 2006; Koide, Bonev, Nelson, & Wellman, 2012; Raimondo et al., 2015). The $[K^+]_o$ increase in turn changes E_K and the single-channel conductance of the BK channel by altering the ion occupancy in the selectivity filter. The conductance of the channel is noticeably greater when $[K^+]_o$ is greater (Yellen, 1984). Experimentally, high concentrations of K^+ in both the intracellular and extracellular solutions are typically employed to avoid undesired current size-dependent accumulation of K^+ and subsequent changes in $i(V_m)$.

Inclusion of β3 subunits in the BK complex alters the single-channel conductance characteristics (Zeng, Xia, & Lingle, 2003). With β3 included, the time-averaged conductance, especially that of inward openings, is diminished. The reducing agent dithiothreitol largely removes the conductance-diminishing effect of β3, presumably affecting one or more of the disulfide bridges formed by the Cys residues in the extracellular linker region of β3. This observation raises the possibility that extracellular reducing agents could alter the single-channel conductance. Physiologically, however, the extracellular environment is expected to remain oxidizing and the physiological significance of the altered conductance by β3 is to be established.

The ion selectivity of a given ion channel is considered to be a static or stable characteristic. For example, K^+ ions permeate through BK channels much more frequently than Na^+ (Blatz & Magleby, 1984). Only under some extreme experimental conditions, Na^+ and *N*-methyl-D-glucamine may permeate through select K^+ channels (Loboda, Melishchuk, & Armstrong, 2001; Starkus, Kuschel, Rayner, & Heinemann, 1997). But, no firm evidence is available that appreciable and dynamic changes in ion selectivity of ion channels take place under physiological conditions. Some reports of drastic and rapid selectivity changes, within seconds to minutes, exist for other channel types (Khakh, Bao, Labarca, & Lester, 1999; Virginio, MacKenzie, Rassendren, North, & Surprenant, 1999); however, it is doubtful these postulates will survive experimental scrutiny (Li, Toombes, Silberberg, & Swartz, 2015). The ion selectivity of the BK channel is unlikely to be reversibly regulated.

6. MODULATION OF OPEN PROBABILITY

Changes in $P_k(V_m, t)$ are clearly the primary way by which modulators alter currents through BK channel complexes. Ideally, changes in $P_k(V_m, t)$ could be expressed as changes in the values of the HA model parameters in an experimentally constrained manner so that semimechanistic and structurally relevant interpretations of the modulator action could be obtained and experimentally verifiable predictions could be made. Such parameter estimation requires systematic and careful manipulations of V_m and $[Ca^{2+}]_i$ and different measurements such as those of single-channel ionic currents, macroscopic ionic currents, and macroscopic gating currents.

The results of electrophysiological measurements at an intermediate concentration of Ca^{2+} (eg, a few μM) and at a moderate voltage (eg, 0 mV) may shed light on the physiological relevance. However, such measurements are not particularly useful in implicating or excluding a mechanism; the observed change in current size could be caused by a change in any one or more of the parameters in the HA model. For example, activation of VSDs can influence the ion conduction gate but also influences the binding affinities of the Ca^{2+} sensors, which then may influence the ion conduction gate. Elucidation of the mechanism of the action of a modulator is facilitated by functionally isolating a subset of the multitier allosteric gating mechanism of the BK channel. One or more of the allosteric components of the HA model in Fig. 1A could be eliminated from consideration by performing measurements under extreme conditions: 0 Ca^{2+}, very high $[Ca^{2+}]_i$ ($\geq$100 μM), very negative voltages, and/or very positive voltages. For example, the effect of a modulator on the intrinsic behavior of the ion conduction gate (L_0, z_L) without any allosteric influence from activation of the VSDs or the Ca^{2+} sensors could be assessed by examining single-channel openings at very negative voltages where the VSDs are at rest in the absence of Ca^{2+}. The values of K_d and C associated with the Ca^{2+} sensors may be in turn estimated by examining L_0 at different $[Ca^{2+}]_i$ in the absence of any influence of VSD activation. Further, these experiments may be complemented with the measurements of mutant channels that are impaired in specific aspects of the channel function (Budelli, Geng, Butler, Magleby, & Salkoff, 2013; Hoshi, Pantazis, et al., 2013). For instance, the use of the divalent cation-insensitive BK mutant without the cytoplasmic gating ring domain (Budelli et al., 2013) may be useful to focus on the ion conduction gate and the VSD domains. For more details of HA model parameter estimation, interested readers are referred to Horrigan and Aldrich (2002) and Horrigan and Hoshi (2015). Some modulators of the BK channel have been analyzed using the HA model framework and a classification scheme based on the HA model has been proposed to more easily indicate the mechanism of action (Gessner et al., 2012). For example, a modulator that increases P_o primarily by increasing L_0 is called an L_0^+ modulator.

Voltage-dependent activation of currents through BK complexes is typically analyzed using a Boltzmann-type function:

$$P_o(V_m, \infty) \propto \frac{G(V_m, \infty)}{G_{max}(\infty)} \propto \frac{1}{1 + e^{\left(-(V_m - V_{0.5})Q_{app}/k_B T\right)}} \tag{4}$$

where P_o represents the open probability, G represents the macroscopic conductance, $V_{0.5}$ is the half-activation voltage, Q_{app} is the apparent number of voltage-sensing particles (or steepness factor), k_B is the Boltzmann constant, and T is temperature in kelvin. It is important to note that especially for the BK channel, the parameters $V_{0.5}$ and Q_{app} are simply data descriptors and do not possess clear mechanistic meanings. Multiple aspects of the channel gating (Fig. 1) contribute to $V_{0.5}$ and Q_{app}, and any experimentally observed changes in $V_{0.5}$ ($\Delta V_{0.5}$) and Q_{app} are difficult to interpret. Nonetheless, changes in $V_{0.5}$ and Q_{app} (frequently $V_{0.5}$ only) have been utilized to characterize voltage- and Ca^{2+}-dependent activation of the channel (Fig. 2).

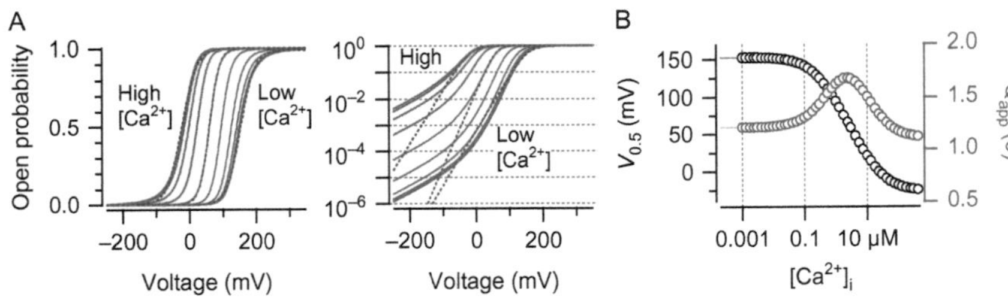

Fig. 2 Model-dependent parameterization of steady-state BK channel currents generated with the HA model. (A) Using an HA model parameter set that best fits electrophysiological data obtained for human BK channels without any auxiliary subunit heterologously expressed in HEK293 cells, steady-state channel open probability was calculated as a function of voltage for various $[Ca^{2+}]_i$ (0 nM, 1 nM, 10 nM, 30 nM, 100 nM, etc., until 300 μM) in linear (*left*) and logarithmic presentation (*right*). Brief "flicker" closed states are not considered. The *dashed red* (*gray* in the print version) *curves* represent simple Boltzmann fits to the results at three $[Ca^{2+}]_i$. (B) Boltzmann functions (Eq. 4) were fit to the data in (A) to yield the operational data descriptors $V_{0.5}$ and Q_{app}, here plotted as a function of $[Ca^{2+}]_i$. When the divalent cation concentration is $\geq\sim$1 mM, a distinct low-affinity divalent cation-dependent activation process, often called the Mg^{2+}-dependent activation, is observed (Shi & Cui, 2001; Shi et al., 2002; Xia et al., 2002; Zeng et al., 2005; Zhang et al., 2001) but this is not depicted here.

Increasing $[Ca^{2+}]_i$ shifts $V_{0.5}$ to the negative direction and alters Q_{app} in a complex way (Fig. 2B). Inspection of the $P_o(V_m)$ curves (Fig. 2A), which are proportional to the $G(V_m)$ curves, also shows that the curves deviate markedly from the predictions of the simple Boltzmann function above (dashed curves in Fig. 2A), especially at very negative voltages and also when $[Ca^{2+}]_i$ is high ($\geq$10 μM). Thus, it is very important to keep in mind that shifts in $V_{0.5}$ alone may not accurately describe the functional impacts of modulators. This issue may be important when comparing different modulators and estimating their potencies in more physiological conditions, in which different cells may preferentially utilize different "operating" windows of the BK channel—not necessarily near the $V_{0.5}$ range. It is unfortunate that many modulators have been characterized only by $V_{0.5}$ as the primary dependent variable. To infer the mechanism, as explored below, at the minimum, $G(V_m)$ changes by a modulator must be tested at multiple $[Ca^{2+}]_i$, from 0 to 100 μM.

7. $G(V_M)$ CHANGES BY MODULATORS WITH DIFFERENT MODES OF ACTION

Fig. 3 presents the results of HA model simulations in which each parameter in the model was increased or decreased, in most cases by a factor of two, one at a time. Changes in $V_{0.5}$ and Q_{app} are plotted as a function of $[Ca^{2+}]_i$ as frequently performed. The simulated results may serve an initial guide as to what to expect from drugs affecting isolated aspects of BK channel function, ideally one that is expressed as a single parameter of the HA model. In practice, however, a single modulator may alter multiple parameters and the results may not quite resemble the simulation results in Fig. 3.

8. INTRINSIC BEHAVIOR OF THE ION CONDUCTION GATE

The ion conduction gate of the BK channel has a low but nonzero probability of opening in the absence of allosteric influences of activation of the four VSDs and Ca^{2+} sensors. The equilibrium constant L_0 (ie, at 0 mV) is typically about 1×10^{-6} without any auxiliary subunit, and with β1 or β4 present, the value is >17-fold smaller (Orio & Latorre, 2005; Wang et al., 2006), corresponding to the free energy difference between the closed and open conformations of about 34–37 kJ/mol. The ion conduction gate has weak but noticeable voltage dependence such that the L_0 value increases with depolarization, equivalent to 0.38 voltage-sensing charges per channel (z_L). This voltage dependence may be too weak to definitively identify a

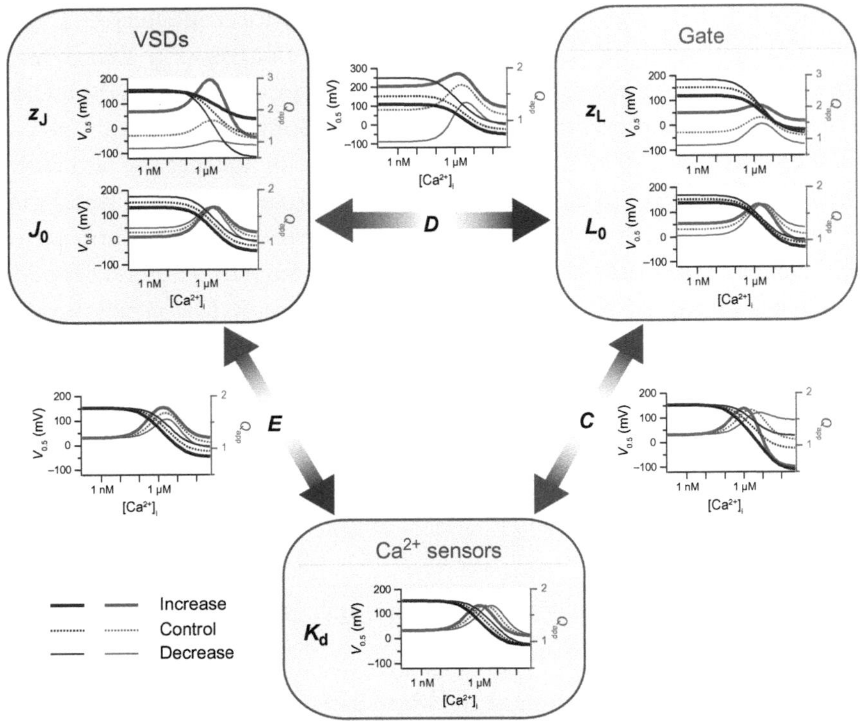

Fig. 3 Impact of variations in equilibrium constants describing the HA model on the operational data descriptors $V_{0.5}$ and Q_{app}. The individual data panels present $V_{0.5}$ (*black*) and Q_{app} (*purple* (*gray* in the print version)) estimated from simulated data in which individual model parameters were altered with respect to a control value, ie, the best fit to BK channel data under standard conditions. Control data are shown with *dashed intermediate line thickness*, increased parameter with *thick*, decreased parameter with *thin lines*. Specifically, the model parameters were (low, control, and high): z_J: 0.275, 0.55, 1.1; α/β (J_0); 176, 144, 113; z_L: 0.19, 0.38, 0.76; L_0: 25×10^{-6}, 50×10^{-6}, 100×10^{-6}; K_D (Ca^{2+}): 5.5, 11, 22 μM; *C*: 4, 8, 16; *D*: 5.8, 11.6, 23.2; *E*: 1.2, 2.4, 4.8. The results describe BK channels without any auxiliary subunit. Those with β1, β2 without inactivation, and β4 will be qualitatively similar. The parameter values have not been estimated for β3. For γ1, see Yan and Aldrich (2010).

single structural basis. Binding of Ca^{2+} ions ($K = [Ca^{2+}]_i/K_d$) and changes in membrane potential working through the voltage-sensing charges in the VSDs (4 z_L), in addition to the gate's intrinsic voltage-sensing charges (z_L), via the reciprocal allosteric interactions as specified by the allosteric interaction factors *C* and *D*, promote opening of the ion conduction gate ($P_o \propto L_o$). The physicochemical and atomic structural mechanisms are mostly obscure (see Horrigan, 2012; Wu et al., 2009). Using the expression

$\Delta G = -RT\ln(K_{eq})$ where ΔG is the free energy difference, R is gas constant, T is temperature, and K_{eq} is an equilibrium constant, a 10-fold change in L_0 requires ~5.71 kJ/mol of energy. This may be contrasted with the free energy of a hydrogen bond, several kJ/mol. Because the binding affinity of a modulator is related to the energy added to the channel complex, high-affinity modulators have potential to more noticeably alter the ion conduction gate. A comparable action by low-affinity modulators requires binding of multiple molecules.

Increases in L_0 (ie, L_0^+ modulators) enhance P_o under every condition (except when P_o is saturated). For instance, $V_{0.5}$ shifts to the negative direction at every $[Ca^{2+}]_i$ although the shift is somewhat diminished at intermediate concentrations (eg, a few μM without any auxiliary subunit; Fig. 3, top right). Thus, if a modulator shifts $V_{0.5}$ in this manner and if it increases P_o at very negative voltages when the VSDs are at rest, its effect on L_0 may be strongly suspected. The voltage dependence of P_o at such negative voltages where the VSDs are at rest may be also used to assess if z_L is altered.

The selectivity filter and S6 segments play critical roles in gating (Chen & Aldrich, 2011; Chen, Yan, & Aldrich, 2014; Tang et al., 2010; Thompson & Begenisich, 2012; Wilkens & Aldrich, 2006; Zhou et al., 2011). In particular, the opening and closing processes may be accompanied by reorientation of the side chains of select residues within S6 with the side-chain hydrophilicity acting as an important factor (Chen & Aldrich, 2011; Chen et al., 2014). Thus, those compounds that regulate the environment around the side chains of the S6 residues directly or indirectly could function as L_0 modulators.

9. VOLTAGE-SENSOR DOMAIN FUNCTION

Two major ways for modulators to affect the VSD function exist. First, the relative stabilities of the resting and activated states of each VSD, as represented by the equilibrium constant at 0 mV J_0, could be altered. The greater the value of J_0, the more stable the activated state is and a comparable level of VSD activation is achieved with smaller depolarization. An increase in J_0 (ie, J_0^+ modulator) shifts $V_{0.5}$ at every $[Ca^{2+}]_i$ (Fig. 3, top left), in a manner similar to that seen with an increase in L_0 (Fig. 3, top right). The two mechanisms, J_0 vs L_0, cannot be distinguished by the $\Delta V_{0.5}$ information alone. Single-channel measurements at very negative voltages where the VSDs are at rest are required. In theory, increases in J_0 and L_0 predict differential changes in Q_{app} or steepness of $G(V_m)$ curves; however, the changes may be too small to be detected reliably in experimental settings. Those modulators that preferentially stabilize the VSDs in the resting or activated state are expected to alter J_0. A clear example of a modulator affecting the BK channel in this manner has not been reported; however, synthetic compounds that stabilize the Na^+ channel VSDs in the activated position by interacting with the positively charged residues in S4, thereby trapping the channel in the inactivated state, are now known (Ahuja et al., 2015). Second, modulators can affect the voltage dependence of J specified by z_L (the number of voltage-sensing charges in the VSD). While changes in J_0 or z_L alter $V_{0.5}$ and Q_{app}, the changes are better assayed by directly measuring gating charge movement and constructing charge movement-voltage ($Q(V_m)$) curves (Horrigan & Aldrich, 1999). An increase in J_0 shifts the $Q(V_m)$ curve to the negative direction and an increase in z_L increases the maximal amount of charge.

10. COUPLING BETWEEN THE ION CONDUCTION GATE AND VSDs

Activation of VSDs is coupled to the ion conduction gate through the coupling factor D. When $D=1$, the VSDs and ion conduction gate are functionally independent and when $D>1$, VSD activation stabilizes the open conformation of the gate. Consequently, increasing the value of D (ie, D^+ modulator; stronger coupling between the VSDs and the gate) shifts $V_{0.5}$ to the negative direction and increases Q_{app}; the $G(V_m)$ curves are left-shifted and steeper at every $[Ca^{2+}]_i$. These $G(V_m)$ changes are relatively intuitive. Steeper slopes (greater Q_{app}) of the $G(V_m)$ curves by strengthening the coupling between the VSDs and the gate also illustrate that Q_{app} is not a good estimate of the number of voltage-sensing charges involved. Conversely, decreases in D shift $V_{0.5}$ to the positive direction and decrease Q_{app}.

Changes in D and L_0 produce relatively similar changes in $V_{0.5}$ and Q_{app}. These two mechanisms could be, however, readily distinguished by functionally removing the contribution of the VSDs. At very negative voltages without Ca^{2+}, changes in L_0 would be observable in the single-channel measurements. Whether a modulator regulates the VSD function (J_0) or the coupling strength to the gate (D) is often more difficult to assess experimentally (see Gessner et al., 2012).

The coupling factor D is the principal gating component regulated by γ1, a single-transmembrane auxiliary subunit (Yan & Aldrich, 2010) (see Chapter "Modulation of BK Channel Function by Auxiliary Beta and

Gamma Subunits" by Yan and Li). Coassembly with γ1 shifts $V_{0.5}$ of the channel complex to the negative direction by 140 mV. This tremendous shift is caused by a ~20-fold increase in D (Yan & Aldrich, 2010).

It is easy to rationalize how transmembrane subunits like γ1 regulate the interaction between the VSDs and the ion conduction gate. Similarly, one can easily imagine a modulator binding to a transmembrane area altering the VSD-gate coupling strength. As noted below, some modulators appear to interact with the cytoplasmic gating ring domain and alter the coupling factor D. It may be speculated that the conformation of the gating ring changes upon modulator binding and then this information is transmitted to the transmembrane domain because of their close juxtaposition. This postulate needs an experimental validation.

11. CA²⁺ SENSORS

Activation of the Ca^{2+} sensors in the BK channel increases P_o even at very negative voltages where the VSDs are at rest (Fig. 1A right). The two Ca^{2+} sensors types, the RCK1 and RCK2 Ca^{2+} sensors, are functionally different as summarized in Hoshi, Pantazis, et al. (2013). For example, the Ca^{2+} affinity K_d values of the RCK2 Ca^{2+} sensor, whether the ion conduction gate is closed or open, are lower (ie, higher affinities) than those of the RCK1 Ca^{2+} sensor (Horrigan & Aldrich, 2002; Sweet & Cox, 2008, 2009). As noted, the two Ca^{2+} sensors are lumped together in the original HA model for simplicity (but see Savalli et al., 2012). Changes in the Ca^{2+} sensor K_d produce expected changes in $G(V_m)$; only at intermediate concentrations of Ca^{2+}, the values of $V_{0.5}$ and Q_{app} are altered. No change in the two parameters is observed when the Ca^{2+} sensors are totally unoccupied or when fully occupied (≥100 μM). As trivial as they seem, measurements at the two extreme Ca^{2+} concentrations are a very valuable first step in elucidating the mechanism of a modulator action.

In theory, a modulator may regulate the Ca^{2+} affinity by altering the Ca^{2+} coordination mechanisms of the RCK1 and RCK2 sensors involving six to eight Ca^{2+} ligands such as the side-chain oxygens, backbone oxygens, and water (Yuan et al., 2012, 2010) (summarized in Hoshi, Pantazis, et al., 2013). However, one serious shortcoming in examining whether the properties of the Ca^{2+} sensors are altered is that full BK channel proteins have been difficult to isolate and purify, and quantitative Ca^{2+} binding assays to infer the K_d values have not been successful. Radioactive Ca^{2+} binding assays have been performed using short (~200 residues) fragments of the gating ring domain (Bao, Kaldany, Holmstrand, & Cox, 2004; Braun & Sy, 2001) and the conformational changes of the isolated and purified gating ring domain proteins caused by Ca^{2+} have been detected using tryptophan fluorescence-based methods (Javaherian et al., 2011; Yusifov, Javaherian, Pantazis, Gandhi, & Olcese, 2010; Yusifov, Javaherian, Gandhi, et al., 2010; Yusifov, Savalli, Gandhi, Ottolia, & Olcese, 2008). While very encouraging, the radioactive binding and fluorescence methods using isolated gating ring domain proteins may not have enough sensitivity to detect small changes, eg, a few fold, in the Ca^{2+} sensor affinity that a modulator may induce. The lack of the transmembrane domain containing the ion conduction gate and VSDs in these assays is also a concern, in part because the sensor K_d values are influenced by the gate and VSDs (Fig. 1).

12. COUPLING BETWEEN THE ION CONDUCTION GATE AND CA^{2+} SENSORS

The ratio of the Ca^{2+} sensor K_d value when the ion conduction gate is closed ($K_{d,\ c}$) over that when the gate is open ($K_{d,\ o}$) defines the allosteric interaction factor $C = K_{d,\ c}/K_{d,\ o}$. The ion conduction gate equilibrium constant L_0 is stabilized by C fold for each Ca^{2+} sensor activated and that the open conformation of the gate is stabilized by $\Delta G = -RT \ln C$ where ΔG is the free energy of stabilization per mole. When $C = 1$, the Ca^{2+} sensors and the gate are functionally independent. The value of C is $>\sim 4$, depending on the subunit composition (Sweet & Cox, 2008, 2009), and thus the Ca^{2+} sensor K_d value is lower (ie, higher affinity) when the ion conduction gate is open ($K_{d,\ c} > K_{d,\ o}$). Increasing the value of C or strengthening the coupling between the gate and Ca^{2+} sensors (ie, K_d^- modulators) facilitates opening of the gate and moves $V_{0.5}$ to the negative direction, as expected only when Ca^{2+} ions are present and the shift becomes greater at higher $[Ca^{2+}]_i$. In electrophysiological experiments, the value of $K_{d,\ c}$, that of $K_{d,\ o}$, and thus that of the factor C may be inferred by fitting logarithm of P_o at negative voltages (ie, VSDs are at rest) at a variety of $[Ca^{2+}]_i$ (Horrigan & Aldrich, 2002).

Structurally, the RCK1 Ca^{2+} sensor probably has a more intimate association with the transmembrane domain than the RCK2 Ca^{2+} sensor (Fig. 1C). Perhaps consistent with this idea, the RCK1 Ca^{2+} sensor, whose $K_{d,\ c}$ and $K_{d,\ o}$ are greater than those of the RCK2 Ca^{2+} sensor, is more strongly coupled to the ion conduction gate (ie, greater C value) and also slightly more voltage dependent than the RCK2 Ca^{2+} sensor (Sweet & Cox, 2008).

13. COUPLING BETWEEN THE VSDs AND CA^{2+} SENSORS

In the original HA model formulation, the interaction factor E between the VSDs and Ca^{2+} sensors (Fig. 1A) is estimated to be ~2 (Horrigan & Aldrich, 2002). With $E > 1$, activation of the VSDs should be easier with higher $[Ca^{2+}]_i$. The interaction of $E = \sim 2$ is modest but provides a way for the electrical information transduced by the transmembrane VSDs to be communicated to the cytoplasmic chemical (ie, Ca^{2+}) sensors and vice versa. The idea is consistent with the probable structural arrangement of the transmembrane domain and the cytoplasmic gating ring domain (Fig. 1C) and some evidence exists to suggest the nature of the interaction (Savalli et al., 2012). Increasing the value of E, strengthening coupling between the Ca^{2+} sensors and VSDs, shifts $G(V_m)$ to the negative direction only when Ca^{2+} ions are present (Fig. 3). However, a more rigorous method to test whether a modulator alters E is to measure gating charge and construct $Q(V_m)$ curves at multiple $[Ca^{2+}]_i$ (Horrigan & Aldrich, 1999, 2002).

14. MULTISTEP INTERACTIONS

In the simulations in Fig. 3, the model parameters were changed one at a time. It is extremely important to remember that the allosteric interactions are all reciprocal. For example, a decrease in the Ca^{2+} sensor K_d facilitates opening of the ion conduction gate as specified by the interaction factor C. The K_d decrease also promotes activation of the VSDs according to the interaction factor E. But activation of the VSD in turn promotes opening of the gate as stipulated by the factor D. Thus, a decrease in the Ca^{2+} sensor K_d promotes opening of the ion conduction gate via two different pathways.

15. MODULATORS WITH WELL-CHARACTERIZED MECHANISMS OF ACTION

The mechanisms of action of some endogenous signaling molecules and synthetic compounds have been analyzed using the HA model framework.

Heme and heme-breakdown products. Heme is well known as a stable protein prosthetic group; however, increasing evidence suggests a new role as a nongenomic signaling molecule (Hou, Reynolds, Horrigan, Heinemann, & Hoshi, 2006). One such example is its modulation of the BK channel. Free hemin (Fe^{3+} protoporphyrin IV) markedly decreases currents through BK channels when the control basal P_o is >0.001 (>0 mV without Ca^{2+}) and this inhibitory effect is accompanied by marked slowing of the activation kinetics of ionic currents (Jaggar et al., 2005; Tang et al., 2003). At functionally saturating concentrations (≥300 nM), hemin causes a ~80 mV shift in $V_{0.5}$ and decreases Q_{app} by 40% (Horrigan, Heinemann, & Hoshi, 2005). At very negative voltages without Ca^{2+}, hemin actually *increases* P_o by ~10-fold and increases the mean open duration by ~30% (Horrigan et al., 2005). These results and others including single-channel ionic current and macroscopic ionic and gating current measurements showed that hemin has multiple effects. First, hemin either removes some channels from the active ready-to-open pool or dramatically slows down the latency to first opening, thereby decreasing the maximal macroscopic conductance level. This is evidenced as an increase in the number of depolarization epochs during which no opening is observed even at positive voltages. Second, hemin increases the frequency of subconductance openings whose $i(V_m)$ characteristics are indistinguishable from those of the full conductance state. Third, hemin decreases the coupling strength between the VSD and the gate (D) to 27% of the control level—this largely accounts for the considerable positive shift in $V_{0.5}$ and the smaller Q_{app}. Fourth, hemin renders the open conformation of the ion conduction gate more stable, causing a 10-fold increase in L_0 and experimentally constrained changes in the opening and closing rate constants underlying L_0. The decreased D and increased L_0 explain the gating changes by hemin, classifying hemin as a $D^-L_0^+$ modulator.

Hemin and heme (Fe^{2+} protoporphyrin IX) bind to the intracellular segment linking the RCK1 and RCK2 domain (Jaggar et al., 2005; Tang et al., 2003; Yi, Morgan, & Ragsdale, 2010; Yusifov, Javaherian, Gandhi, et al., 2010), not well resolved in the crystallographic structures (Wu et al., 2010; Yuan et al., 2012, 2010), and induces conformational changes in the gating ring domain distinct from those induced by Ca^{2+} (Yusifov, Javaherian, Heinemann, Hoshi, & Olcese, 2012; Yusifov, Javaherian, Gandhi, et al., 2010). The altered conformation of the gating ring domain may then influence the transmembrane domain to decrease D and to increase L_0. Alternatively and/or additionally, hemin bound to the structurally disordered RCK1–RCK2 linker segment may directly interact with the transmembrane domain. Select mutations in the RCK1–RCK2 linker segment mostly eliminate binding and the inhibitory effect of hemin (Jaggar et al., 2005; Tang et al., 2003; Yi et al., 2010). Whether these mutations abolish other effects of hemin, such as the increased L_0, remain to be established; hemin may bind to multiple loci of the BK channel.

The physiological significance of the modulatory effect of heme/hemin may be related to sensing of biologically relevant gases such as oxygen and CO (Jaggar et al., 2005; Wang, Wu, & Wang, 1997; Williams et al., 2004). Heme may exert a basal inhibitory influence on the BK channel and binding of CO to heme may relieve this inhibitory influence of heme (Jaggar et al., 2005). CO is in turn produced by oxygen-dependent catabolism of heme by heme oxygenase. Application of CO and CO-releasing molecules increases P_o of the BK channel but the physicochemical mechanisms are only very poorly understood (Hou, Xu, Heinemann, & Hoshi, 2008a; Telezhkin et al., 2011; Williams et al., 2008). The modulation of the BK channel by CO may contribute to cellular hypoxia sensing (Williams et al., 2004) although this is not the main cellular hypoxia-sensing mechanism (Fernandez-Aguera et al., 2015).

Ions other than Ca^{2+}. Ca^{2+} is the principal intracellular ion that promotes opening of the ion conduction gate. P_o of the channel can be increased by other intracellular ions via different mechanisms. Mg^{2+} (and other divalent cations including Ca^{2+}) at ≥ 1 mM increases P_o but the mechanism is distinct from that by activation of the RCK1 and RCK2 Ca^{2+} sensors in the gating ring domain. Mg^{2+} can be coordinated in the space between the transmembrane domain and the cytoplasmic gating ring domain: the residues from the intracellular S0–S1 loop, S2–S3 loop, and two residues near the RCK1 Ca^{2+} sensor in the gating ring domain (Shi et al., 2002; Xia et al., 2002; Yang, Hu, Shi, & Cui, 2006; Yang et al., 2007), in addition to a couple of water molecules, contribute to the coordination of Mg^{2+}. Activation of the channel by Mg^{2+} depends on activation of the VSD; at negative voltages where the VSDs are at rest, Mg^{2+} has no effect (Horrigan & Ma, 2008; Hu et al., 2003; Yang et al., 2007). This is in contrast with the Ca^{2+}-dependent activation of the channel, which does not require activation of the VSD; even at negative voltages, Ca^{2+} increases L_0 and P_o. Because coordination of Mg^{2+} stabilizes the activated state of the VSD when the ion conduction gate is open via the main voltage-sensing Arg residue in S4, intracellular Mg^{2+} increases the coupling between the VSD and the ion conduction gate (Horrigan & Ma, 2008). Mg^{2+} increases the value of D and it is a D^+ modulator.

Intracellular H^+ and Zn^{2+} also activate the BK channel. Their stimulatory effects are abolished by mutation of select His residues near/in the RCK1 Ca^{2+} sensor (Hou et al., 2008b, 2010). H^+ and Zn^{2+} essentially act as Ca^{2+} surrogates or RCK1 Ca^{2+} sensor agonists. Functional impacts of these Ca^{2+} surrogates will be qualitatively similar to those of Ca^{2+}, increasing L_0 and P_o without the requirement for VSD activation (Fig. 2). However, as these Ca^{2+} surrogates and Ca^{2+} compete for the same sensor site, their effects diminish with increasing $[Ca^{2+}]_i$ (Hou et al., 2008b, 2010). The RCK1 and RCK2 Ca^{2+} sensors are differentially activated by many other cations albeit often at nonphysiological high concentrations (Zhou, Zeng, & Lingle, 2012).

Free long-chain polyunsaturated fatty acids. Free long-chain polyunsaturated fatty acids such as docosahexaenoic acid (DHA) found in oily fish may lower blood pressure in humans (Miller, Van Elswyk, & Alexander, 2014). Acute injection of DHA into mice lowers their blood pressure in a BK channel-dependent manner (Hoshi, Wissuwa, et al., 2013). Excised-patch electrophysiological measurements show that DHA with an EC_{50} of ~500 nM reversibly shifts the $V_{0.5}$ of β1- or β4-containing BK channels to the negative direction by ~60 mV without Ca^{2+} and the stimulatory effect is maintained at every $[Ca^{2+}]_i$ (Hoshi, Tian, Xu, Heinemann, & Hou, 2013; Hoshi, Wissuwa, et al., 2013). While HA model parameter estimation has not been reported, the observation that DHA increases P_o at negative voltages in the absence of Ca^{2+} by 20-fold or more without markedly altering the mean open duration and other results collectively suggest that DHA is an L_0^+ modulator.

The structural and physicochemical bases of the modulation by DHA remain obscure. A Tyr residue near the cytoplasmic end of S6 is critical in the stimulatory effect of DHA; however, how this position is involved in the modulatory phenomenon is unknown (Hoshi, Xu, Hou, Heinemann, & Tian, 2013). Fatty acids other than long-chain polyunsaturated fatty acids have a variety of stimulatory and inhibitory effects on the BK channel but the underlying mechanisms are not clear (Clarke, Petrou, Walsh, & Singer, 2002, 2003).

Cym04 and NS1619. While for many BK channel openers, the opening characteristics have largely been described by the amount of the phenotypical $V_{0.5}$ shift, only a few synthetic low-molecular-weight compounds have been studied by also considering other aspects of BK channel function. The dehydroabietic acid derivate Cym04 activates BK channels (Cui et al., 2008) more potently than the benzimidazolone derivative NS1619

(Olesen, Munch, Moldt, & Drejer, 1994), which can be considered a prototypic BK opener as it has been used in many experimental studies. At 10 μM, Cym04 shifts $V_{0.5}$ by about −30 mV, while an about threefold higher concentration of NS1619 is needed to obtain the same effect. In either case, the Q_{app}, estimated with single Boltzmann function fits to $G(V_m)$ data, is not altered appreciably.

In an attempt to classify Cym04's mode of action, Gessner et al. (2012) studied various functional aspects. Single-channel recordings revealed no significant impact on the unitary conductance, so the drug most likely only affects the channel's gating characteristics. Gating currents recorded in the whole-cell configuration were not altered by Cym04 application arguing that the drug does not affect the total gating charge and its equilibrium voltage. The shift in $V_{0.5}$ was also observed at high intracellular Ca^{2+} and Mg^{2+} concentrations, thus ruling out a strong influence of the coupling factors connecting the Ca^{2+}-sensor domains to VSD and gate. However, $\Delta V_{0.5}$ was noticeably diminished at intermediate Ca^{2+} concentrations. As discussed earlier (see Fig. 3), in particular at intermediate Ca^{2+} concentrations the HA model predicts significant changes in Q_{app}, which will compromise a clear mechanistic conclusion. Therefore, all available data need to be considered for finding the best set of model parameters accounting for the action of Cym04. Following such an approach, Cym04 was characterized as a $J_0^+ L_0^+$ modulator, ie, a drug that stabilizes the gate (L_0 increase by a factor of 1.5) and the voltage sensors (J_0) in the open and activated position, respectively (Gessner et al., 2012). There is some ambiguity regarding J_0 because part of the effect is possibly accounted for by changes in the coupling between VSD and the ion-conducting gate (factor D). Although belonging to a different structural class, NS1619 was also functionally characterized as a $J_0^+ L_0^+$ modulator.

The study by Gessner et al. (2012) additionally provided mechanistic insight into the action of Cym04 and NS1619 because they also investigated a splice variant in which an alternative exon 9 (Slo1_9a), which encodes the end of S6 and the beginning of the S6/RCK1 linker, is utilized. This variant was previously shown to be expressed in the brain and to show a high P_o even in the absence of Ca^{2+}, ie, exhibiting a larger L_0 value (Soom, Gessner, Heuer, Hoshi, & Heinemann, 2008). While Cym04 and NS1619 shifted $V_{0.5}$, they were ineffective for Slo1_9a. Interestingly, the naturally occurring BK opener mallotoxin (Zakharov, Morrow, Liu, Yang, & Marx, 2005) (see later) shifted $V_{0.5}$ by about the same extent in Slo1 and Slo1_9a, thus indicating a different mode of action. The functional interaction of Cym04 with the S6/RCK1 linker was further analyzed by site-directed mutagenesis, identifying K330 as an important residue. A mechanism by which Cym04, and probably also NS1619, could affect BK channel gating originates from experiments in which the length of the S6/RCK1 linker was varied; this was motivated by a previous study that has shown that this linker acts like a passive mechanical gating spring (Niu, Qian, & Magleby, 2004), where linker shortening facilitates channel opening (increase in L_0). Using BK α constructs with S6/RCK1 linkers extended (LL+2) and shortened (LL−2) by two residues, it was shown that Cym04 loses its ability to activate BK channels in LL−2 variants (Gessner et al., 2012). This may be interpreted to suggest that the interaction of Cym04 with the BK channel complex induces a structural change equivalent to a shortening of the S6/RCK1 linker. Although it is still elusive where and how Cym04 and NS1619 directly *bind* to the channel protein, it can be summarized that in particular Cym04 is a prototypical $J_0^+ L_0^+$ modulator, which, in addition, discriminates between Slo1 and Slo1_9a variants.

GoSlo-SR synthetic openers. GoSlo-SR compounds are synthetic anthraquinone analogs that activate the BK channel (Roy et al., 2012). One of the compounds, GoSlo-SR-5-6, at saturating concentrations (≥10 μM) shifts $V_{0.5}$ to the negative direction by ~120 mV without any requirement for auxiliary subunits, Ca^{2+}, Mg^{2+}, or the cytoplasmic gating ring domain (Webb et al., 2015). The P_o-increasing effect of GoSlo-SR-5-6 without Ca^{2+} was particularly effective at negative voltages, increasing P_o by up to 1700-fold. The HA model-based analysis showed that the action of GoSlo-SR-5-6 is complex, affecting multiple components: a large, ≥650-fold increase in L_0, an eightfold increase in J_0, and a decrease in D to 40% of the control level. GoSlo-SR-5-6 is thus an $L_0^+ J_0^+ D^-$ modulator. Additionally, the equivalent number of voltage-sensing charges associated with the ion conduction gate opening (z_L) may be reduced. These changes together account for the overall effects of this opener such as $\Delta V_{0.5} = -120$ mV and the large increase in P_o at negative voltages.

The observation that the effect of GoSlo-SR-5-6 is smaller in Slo1_9a (see earlier) combined with measurements from various point mutants suggests that the S4/S5 linker, the cytoplasmic end of S6, and the S6/RCK1 linker are involved in the P_o-enhancing effect of GoSlo-SR-5-6. The lack of appropriate binding assays preclude any precise structural interpretations but GoSlo-SR-5-6 has been suggested to enter a hydrophobic pocket between the S4/S5 linker and the S6 area, leading to the observed alterations in the VSD equilibrium (J_0), its coupling to the gate (D) and the gate closed–open equilibrium (L_0) (Webb et al., 2015).

16. MODULATORS WHOSE MECHANISMS CAN BE INFERRED

PIP$_2$. Like many other membrane proteins, the BK channel is regulated by PIP_2. While it is clear that exogenous PIP_2 modulates the BK channel, divergent observations and mechanisms of its action have been reported (Tang et al., 2014; Tian et al., 2015; Vaithianathan et al., 2008). Vaithianathan et al. (2008) postulated that PIP_2 enhances currents through BK channels, especially those with β1, by increasing P_o in a Ca^{2+}-dependent manner such that the effect is negligible without Ca^{2+} and saturates at <10 μM. The Ca^{2+} dependence reported does not match any of the simulated results shown in Fig. 3. The sequence RKK in the S6/RCK1 linker segment, which may interact with the negative charges of PIP_2, was implicated as an important mediator. Tang et al. (2014) also reported a Ca^{2+}-dependent enhancement of the BK channel but via a different mechanism. According to their study, PIP_2 interacts with two positively charged residues near the RCK1 Ca^{2+} sensor and in the absence of Ca^{2+} the RCK1 domain impairs the influence of PIP_2 on the channel. Ca^{2+} relieves this inhibitory influence, leading to PIP_2-mediated enhancement of the channel activity. It is noteworthy that in their study PIP_2 application following scavenging endogenous PIP_2 with poly-lysine and an antibody against PIP/PIP_2 increased G_{max} in excised patch experiments by up to ~10-fold; PIP_2 may increase the number of channels available to open (N) through an unknown mechanism. Tian et al. (2015) documented multiple effects of PIP_2 depending on the channel subunit composition. Their results suggest that the subunit-invariant common effect of PIP_2 is to increase P_o at negative voltages without Ca^{2+} by ~20-fold; PIP_2 increases L_0. Furthermore, a robust increase in P_o was observed in the truncated BK channel with the sequence RKK but without the cytoplasmic gating ring. Neutralization of the aforementioned sequence RKK eliminated the P_o increase. These results collectively show that PIP_2 can exert multiple effects on the BK channel. It is unclear what factors account for the divergent observations reported.

17. MODULATORS WITH AN UNKNOWN MODE OF ACTION

Many compounds that increase currents through BK channels have been described; however, their mechanisms often remain to be investigated fully.

Rottlerin (mallotoxin). Multiple effects of *rottlerin*, a polyphenol compound, are known. Rottlerin inhibits protein kinase C (Gschwendt et al., 1994) and stimulates hERG1 (Zeng et al., 2006) and BK currents (Zakharov et al., 2005). In the BK channel, in the absence of Ca^{2+}, rottlerin (500 nM) shifted $V_{0.5}$ to the negative direction by 120 mV without any change in Q_{app}. This large $\Delta V_{0.5}$ is similar to what is observed with full activation of the Ca^{2+} sensors by $\geq$100 μM $[Ca^{2+}]_i$ (see Fig. 2). The biophysical mechanism underlying the effect of rottlerin has not been fully investigated but the rottlerin effect does not depend on intact Ca^{2+} sensors (Zakharov et al., 2005) or the alternative splicing covering the S6/RCK1 linker region (Gessner et al., 2012). The therapeutic use of rottlerin has been explored (Goldklang et al., 2013).

BMS-204352. This fluoro-oxindole compound reversibly enhanced currents through BK channels with an $EC_{50} = \sim$350 nM and shifted $V_{0.5}$ to the negative direction by ~50 mV when $[Ca^{2+}]_i$ was $\geq$1 μM and the effect was smaller at lower $[Ca^{2+}]_i$ (Gribkoff et al., 2001). This Ca^{2+} dependence suggests that BMS-204352 may be a C^+, K_d^-, E^+, and/or possibly z_J^+ modulator. In an animal model of ischemic stroke, BMS-204352 decreased the brain infarct volume by ~20% without affecting the blood pressure or cerebral blood flow (Gribkoff et al., 2001). Note that at the time of this study, BK $\alpha^{-/-}$ knockout mice (Meredith et al., 2004; Sausbier et al., 2004) had not been developed. Despite the promising preclinical results, BMS-204352 failed to show efficacy in stroke patients in a Phase III clinical trial (Jensen, 2002). Subsequent studies showed that BMS-204352 is not very selective; it can activate KCNQ channels (Schroder et al., 2001; Schroder, Strobaek, Olesen, & Christophersen, 2003) and inhibit L-type voltage-gated Ca^{2+} channels (Son et al., 2011).

Other BK openers. Numerous BK openers have been described, most of whose mechanisms of action remain to be investigated using the HA model framework: NS11021 (Bentzen et al., 2007), NS19504 (Nausch et al., 2014), dihydrosoyasaponin-1 (Giangiacomo, Kamassah, Harris, & McManus, 1998), 17β-estradiol (Valverde et al., 1999), tamoxifen (Dick et al., 2001), lithocholate and its derivatives (Bukiya, Liu, Toro, & Dopico, 2007; Bukiya, Singh, Parrill, & Dopico, 2011; Bukiya, Vaithianathan, Toro, & Dopico, 2009; Dopico & Bukiya, 2014), DiBAC/oxonol "slow" voltage-sensitive dyes (Morimoto et al., 2007; Scornik et al., 2013), resveratrol (Calderone, Martelli, Testai, Martinotti, & Breschi, 2007), tetrahydroquinoline derivatives (Ponte et al., 2012), DS201 (Tan et al., 2014), ethanol (Dopico, Bukiya, & Martin, 2014) (also see Chapter "Modulation of BK Channels by Ethanol" by Dopico), and human β defensin 2 (HBD2) peptide (Liu et al., 2013). Most of these compounds are relatively hydrophobic in nature and often have marked effects on other channels. The observation that some oxonol "slow" voltage-sensitive dyes are BK channel openers has practical implications. The BK opening effect of tamoxifen may also constitute a confounding factor in tamoxifen-mediated conditional expression systems.

HBD2 is an unusual case—it is a 64-residue Cys-rich antimicrobial peptide. From the extracellular side, HBD2 rapidly increases currents through β1-containing BK channels at 100 nM (Liu et al., 2013) when 1 μM Ca^{2+} is present. The stimulatory effect is accompanied by a shift of $V_{0.5}$ to the negative direction by ~25 mV, a relatively modest effect. Yet, intravenous injection of HBD2 decreases both diastolic and systolic blood pressures of chair-restrained monkeys by 15–20 mmHg (Liu et al., 2013). Whether the hypotensive effect is mediated entirely by the effect of HBD2 on the BK channel needs to be established.

18. STRUCTURAL BASES OF MODULATORY ACTION

Atomic structures of the BK channel are only beginning to be elucidated (Wang & Sigworth, 2009; Wu et al., 2010; Yuan et al., 2012, 2010) and, except for Ca^{2+} (Yuan et al., 2012, 2010), how different modulators interact with the channel complex and how the binding events influence the ion conduction gate remain essentially unknown. Fortunately, however, it is expected that emerging atomic structural techniques such as the cryoEM single-particle reconstruction method will rectify the knowledge gap. For some of the modulators mentioned, mutagenesis coupled with functional assays has identified the residues critical for their effects (see earlier). However, in particular because of the heavily allosteric nature of BK channel gating, mutagenesis studies are of limited use for the identification of physical binding sites. Consequently, concerted application of physical interaction studies involving fragments or the whole BK channel complex, improved structural data of the channel—ideally in various conformations—and in silico docking of modulators to protein models will be necessary to obtain a detailed insight into the molecular mechanism of modulator-induced BK channel opening. At present, however, sophisticated functional analysis based on electrophysiological data obtained under an array of different experimental conditions is a powerful approach for the functional classification of BK channel openers.

19. IMPORTANCE OF MECHANISTIC INFORMATION

The multitier allosteric gating mechanism operative in wide ranges of $[Ca^{2+}]_i$ and V_m allows the BK channel to participate in numerous physiological processes. These diverse physiological phenomena may preferentially utilize different aspects of the channel gating. For example, the Ca^{2+}-dependent activation of the channel may be particularly important in tonic smooth muscle cells. In spiking neurons, the voltage-dependent activation aspect may be crucial for shaping the AP duration while, at negative resting potentials, L_0 may play an important role. It is also possible that the wide modulatory repertoire of the BK channel is necessitated to selectively regulate only certain gating components in a cell- or even a subcellular-specific manner. Using this line of reasoning, a synthetic modulator focusing on the Ca^{2+}-dependent components of the channel may be more useful in tonic smooth muscle cells and a compound primarily targeting the voltage-dependent components of the gating machinery may be more effective in shaping the AP. It must be remembered, however, that the gating components are tightly coupled and it is difficult to make clear-cut separations. As reasonable as the above postulates may be, however, firm supporting evidence is largely absent. Acute pharmacological studies employing modulators with clear modes of action as well as studies utilizing genetically engineered organisms with BK channels altered in specific aspects of gating are expected to provide a better insight into the potential physiological outcome associated with activation of BK channels.

An illustrative example of the importance of mechanistic information may be drawn from the properties of paxilline, a fungal alkaloid inhibitor, which has been used to assess the contributions of BK channels in different tissues. A mechanistic study shows that paxilline stabilizes the closed conformation of the channel such that the IC_{50} for the inhibitor is only 10 nM when the channel is closed; however, it is 10 μM, a thousand fold greater, when P_o is high (Zhou & Lingle, 2014). Without the knowledge of the mechanism, the use of paxilline to assess the contribution of the BK channel in different physiological processes would easily lead to erroneous conclusions. A lack of effect may indicate the absence of BK channels or that the channels are very active.

20. SUMMARY AND FUTURE OUTLOOK

Owing to the ubiquitous expression of BK channels in various cell and tissue types, the pathophysiology associated with BK channel dysfunction is extremely diverse. Unlike for many other ion channel types, however, there are numerous endogenous modulators and synthetic compounds that are known to increase P_o of the channels, so-called BK channel openers. Initially, this has triggered high-flying hopes regarding the development of BK-specific drugs that possibly could correct for loss-of-function phenotypes. However, problems in drug stability and, more importantly, severe side effects had a sobering influence on drug development. In a way this did not come as a surprise because the occurrence of BK channels in many organs makes undesired side effects very likely. Thus, if local drug delivery systems, which are currently only conceivable for the treatment of some diseases such as asthma, for example, cannot be applied, the only hope is to identify tissue-specific properties of BK channels that are druggable. Among them may be the presence of specific auxiliary subunits, the splicing of specific exons, posttranslational modifications, or the presence of site-specific endogenous modulators. Since in most cases these additional factors only play a modulatory role, clear-cut results in tissue-specific targeting are not expected. Nevertheless, some experimental studies still fuel the expectation that there might be solutions in at least some cases. This, however, requires precise knowledge of the molecular mode of action by which drugs under consideration modulate BK channel function.

Here we reviewed how a classification of drugs with respect to their impact on functional parameters of the channel might be very helpful. The multitier model introduced by Horrigan and Aldrich (2002) (Figs. 1 and 3) served as a conceptual framework. However, it will not be possible to study drug effects on BK channels in all cases in enough detail as to be able to actually fit the HA model to the data. Instead, often one will have to make do with more general data descriptors such as $V_{0.5}$ and Q_{app} as shown in Fig. 2. Although limited, such parameters can yield valuable information if determined at both, very low and high Ca^{2+} concentrations. In addition, faithful estimations of P_o at negative voltages are very helpful if inferences about drug effects on L_0, ie, the equilibrium position of the ion conduction gate, are to be derived.

The BK channel represents a unique system where a quantitative model with ample structural connotations is available. With anticipated advances in biochemical and atomic structural methods, we will gain insight into how natural and synthetic modulators govern individual rate constants of the channel gating mechanism and their atomic structural basis.

ACKNOWLEDGMENTS

The work in the authors' laboratories was supported in part through the National Institutes of Health (T.H., R01GM057654), the German Research Foundation (S.H.H., T.H., DFG FOR1738), and German Federal Ministry of Education and Research (S.H.H., BMBF Competence Cluster nutriCARD). We thank Dr. Frank Horrigan for discussion and Dr. Franziska Jahns for assistance in figure preparation.

REFERENCES

Aggarwal, S. K., & MacKinnon, R. (1996). Contribution of the S4 segment to gating charge in the *Shaker* K^+ channel. *Neuron, 16,* 1169–1177.

Ahuja, S., Mukund, S., Deng, L., Khakh, K., Chang, E., Ho, H., et al. (2015). Structural basis of Nav1.7 inhibition by an isoform-selective small-molecule antagonist. *Science, 350,* aac5464.

Alioua, A., Li, M., Wu, Y., Stefani, E., & Toro, L. (2011). Unconventional myristoylation of large-conductance Ca^{2+}-activated K^+ channel (Slo1) via serine/threonine residues regulates channel surface expression. *Proceedings of the National Academy of Sciences of the United States of America, 108,* 10744–10749.

Avdonin, V., Tang, X. D., & Hoshi, T. (2003). Stimulatory action of internal protons on Slo1 BK channels. *Biophysical Journal, 84,* 2969–2980.

Bao, L., Kaldany, C., Holmstrand, E. C., & Cox, D. H. (2004). Mapping the BKCa channel's "Ca^{2+} bowl": Side-chains essential for Ca^{2+} sensing. *Journal of General Physiology, 123,* 475–489.

Behrens, R., Nolting, A., Reimann, F., Schwarz, M., Waldschutz, R., & Pongs, O. (2000). hKCNMB3 and hKCNMB4, cloning and characterization of two members of the large-conductance calcium-activated potassium channel beta subunit family. *FEBS Letters, 474,* 99–106.

Bentzen, B. H., Nardi, A., Calloe, K., Madsen, L. S., Olesen, S. P., & Grunnet, M. (2007). The small molecule NS11021 is a potent and specific activator of Ca^{2+}-activated big-conductance K^+ channels. *Molecular Pharmacology, 72,* 1033–1044.

Bentzen, B. H., Olesen, S. P., Ronn, L. C., & Grunnet, M. (2014). BK channel activators and their therapeutic perspectives. *Frontiers in Physiology, 5,* 389.

Berkefeld, H., Sailer, C. A., Bildl, W., Rohde, V., Thumfart, J.-O., Eble, S., et al. (2006). BK_{Ca}-Cav channel complexes mediate rapid and localized Ca^{2+}-activated K^+ signaling. *Science, 314,* 615–620.

Bezzerides, V. J., Ramsey, I. S., Kotecha, S., Greka, A., & Clapham, D. E. (2004). Rapid vesicular translocation and insertion of TRP channels. *Nature Cell Biology, 6,* 709–720.

Blatz, A. L., & Magleby, K. L. (1984). Ion conductance and selectivity of single calcium-activated potassium channels in cultured rat muscle. *Journal of General Physiology, 84*, 1–23.

Braun, A. F., & Sy, L. (2001). Contribution of potential EF hand motifs to the calcium-dependent gating of a mouse brain large conductance, calcium-sensitive K^+ channel. *Journal of Physiology (London), 533*, 681–695.

Brelidze, T. I., & Magleby, K. L. (2004). Protons block BK channels by competitive inhibition with K^+ and contribute to the limits of unitary currents at high voltages. *Journal of General Physiology, 123*, 305–319.

Brelidze, T. I., Niu, X., & Magleby, K. L. (2003). A ring of eight conserved negatively charged amino acids doubles the conductance of BK channels and prevents inward rectification. *Proceedings of the National Academy of Sciences of the United States of America, 100*, 9017–9022.

Brenner, R., Chen, Q. H., Vilaythong, A., Toney, G. M., Noebels, J. L., & Aldrich, R. W. (2005). BK channel beta4 subunit reduces dentate gyrus excitability and protects against temporal lobe seizures. *Nature Neuroscience, 8*, 1752–1759.

Brenner, R., Jegla, T. J., Wickenden, A., Liu, Y., & Aldrich, R. W. (2000). Cloning and functional characterization of novel large conductance calcium-activated potassium channel beta subunits, hKCNMB3 and hKCNMB4. *Journal of Biological Chemistry, 275*, 6453–6461.

Brenner, R., Perez, G. J., Bonev, A. D., Eckman, D. M., Kosek, J. C., Wiler, S. W., et al. (2000). Vasoregulation by the beta 1 subunit of the calcium-activated potassium channel. *Nature, 407*, 870–876.

Budelli, G., Geng, Y., Butler, A., Magleby, K. L., & Salkoff, L. (2013). Properties of Slo1 K^+ channels with and without the gating ring. *Proceedings of the National Academy of Sciences of the United States of America, 110*, 16657–16662.

Bukiya, A. N., Liu, J., Toro, L., & Dopico, A. M. (2007). β1 (KCNMB1) subunits mediate lithocholate activation of large-conductance Ca^{2+}-activated K^+ channels and dilation in small, resistance-size arteries. *Molecular Pharmacology, 72*, 359–369.

Bukiya, A. N., Singh, A. K., Parrill, A. L., & Dopico, A. M. (2011). The steroid interaction site in transmembrane domain 2 of the large conductance, voltage- and calcium-gated potassium (BK) channel accessory beta1 subunit. *Proceedings of the National Academy of Sciences of the United States of America, 108*, 20207–20212.

Bukiya, A. N., Vaithianathan, T., Toro, L., & Dopico, A. M. (2009). Channel β2-4 subunits fail to substitute for beta1 in sensitizing BK channels to lithocholate. *Biochemical and Biophysical Research Communications, 390*, 995–1000.

Calderone, V., Martelli, A., Testai, L., Martinotti, E., & Breschi, M. C. (2007). Functional contribution of the endothelial component to the vasorelaxing effect of resveratrol and NS 1619, activators of the large-conductance calcium-activated potassium channels. *Naunyn-Schmiedeberg's Archives of Pharmacology, 375*, 73–80.

Cao, Q., Zhong, X. Z., Zou, Y., Zhang, Z., Toro, L., & Dong, X. P. (2015). BK channels alleviate lysosomal storage diseases by providing positive feedback regulation of lysosomal Ca^{2+} release. *Developmental Cell, 33*, 427–441.

Chang, S., Gomes, C. M., Hypolite, J. A., Marx, J., Alanzi, J., Zderic, S. A., et al. (2010). Detrusor overactivity is associated with downregulation of large-conductance calcium- and voltage-activated potassium channel protein. *American Journal of Physiology. Renal Physiology, 298*, F1416–F1423.

Chen, X., & Aldrich, R. W. (2011). Charge substitution for a deep-pore residue reveals structural dynamics during BK channel gating. *Journal of General Physiology, 138*, 137–154.

Chen, R. S., Geng, Y., & Magleby, K. L. (2011). Mg^{2+} binding to open and closed states can activate BK channels provided that the voltage sensors are elevated. *Journal of General Physiology, 138*, 593–607.

Chen, L., Tian, L., MacDonald, S. H., McClafferty, H., Hammond, M. S., Huibant, J. M., et al. (2005). Functionally diverse complement of large conductance calcium- and voltage-activated potassium channel (BK) α-subunits generated from a single site of splicing. *Journal of Biological Chemistry*, *280*, 33599–33609.

Chen, X., Yan, J., & Aldrich, R. W. (2014). BK channel opening involves side-chain reorientation of multiple deep-pore residues. *Proceedings of the National Academy of Sciences of the United States of America*, *111*, E79–E88.

Clarke, A. L., Petrou, S., Walsh, J. V., Jr., & Singer, J. J. (2002). Modulation of BK_{Ca} channel activity by fatty acids: Structural requirements and mechanism of action. *American Journal of Physiology. Cell Physiology*, *283*, C1441–C1453.

Clarke, A. L., Petrou, S., Walsh, J. V., Jr., & Singer, J. J. (2003). Site of action of fatty acids and other charged lipids on BK_{Ca} channels from arterial smooth muscle cells. *American Journal of Physiology. Cell Physiology*, *284*, C607–C619.

Contreras, G. F., Castillo, K., Enrique, N., Carrasquel-Ursulaez, W., Castillo, J. P., Milesi, V., et al. (2013). A BK (Slo1) channel journey from molecule to physiology. *Channels (Austin, Tex.)*, *7*, 442–458.

Cox, D. H. (2014). Modeling a Ca^{2+} channel/BK_{Ca} channel complex at the single-complex level. *Biophysical Journal*, *107*, 2797–2814.

Cox, D. H., Cui, J., & Aldrich, R. W. (1997). Allosteric gating of a large conductance Ca-activated K^+ channel. *Journal of General Physiology*, *110*, 257–281.

Cui, J., & Aldrich, R. W. (2000). Allosteric linkage between voltage and Ca^{2+}-dependent activation of BK-type mslo1 K^+ channels. *Biochemistry*, *39*, 15612–15619.

Cui, Y. M., Yasutomi, E., Otani, Y., Yoshinaga, T., Ido, K., Sawada, K., et al. (2008). Novel oxime and oxime ether derivatives of 12,14-dichlorodehydroabietic acid: Design, synthesis, and BK channel-opening activity. *Bioorganic & Medicinal Chemistry Letters*, *18*, 6386–6389.

Diaz-Franulic, I., Sepulveda, R. V., Navarro-Quezada, N., Gonzalez-Nilo, F., & Naranjo, D. (2015). Pore dimensions and the role of occupancy in unitary conductance of Shaker K channels. *Journal of General Physiology*, *146*, 133–146.

Dick, G. M., Rossow, C. F., Smirnov, S., Horowitz, B., & Sanders, K. M. (2001). Tamoxifen activates smooth muscle BK channels through the regulatory β1 subunit. *Journal of Biological Chemistry*, *276*, 34594–34599.

Donnelier, J., Braun, S. T., Dolzhanskaya, N., Ahrendt, E., Braun, A. P., Velinov, M., et al. (2015). Increased expression of the large conductance, calcium-activated K^+ (BK) channel in adult-onset neuronal ceroid lipofuscinosis. *PloS One*, *10*, e0125205.

Dopico, A. M., & Bukiya, A. N. (2014). Lipid regulation of BK channel function. *Frontiers in Physiology*, *5*, 312.

Dopico, A. M., Bukiya, A. N., & Martin, G. E. (2014). Ethanol modulation of mammalian BK channels in excitable tissues: Molecular targets and their possible contribution to alcohol-induced altered behavior. *Frontiers in Physiology*, *5*, 466.

Du, W., Bautista, J. F., Yang, H., Diez-Sampedro, A., You, S. A., Wang, L., et al. (2005). Calcium-sensitive potassium channelopathy in human epilepsy and paroxysmal movement disorder. *Nature Genetics*, *37*, 733–738.

Fernandez-Aguera, M. C., Gao, L., Gonzalez-Rodriguez, P., Pintado, C. O., Arias-Mayenco, I., Garcia-Flores, P., et al. (2015). Oxygen sensing by arterial chemoreceptors depends on mitochondrial complex I signaling. *Cell Metabolism*, *22*, 825–837.

Fernández-Fernández, J. M., Tomás, M., Vázquez, E., Orio, P., Latorre, R., Sentí, M., et al. (2004). Gain-of-function mutation in the KCNMB1 potassium channel subunit is associated with low prevalence of diastolic hypertension. *Journal of Clinical Investigation*, *113*, 1032–1039.

Filosa, J. A., Bonev, A. D., Straub, S. V., Meredith, A. L., Wilkerson, M. K., Aldrich, R. W., et al. (2006). Local potassium signaling couples neuronal activity to vasodilation in the brain. *Nature Neuroscience*, *9*, 1397–1403.

Garber, S. S., & Miller, C. (1987). Single Na^+ channels activated by veratridine and batrachotoxin. *Journal of General Physiology, 89*, 459–480.
Geng, Y., & Magleby, K. L. (2014). Single-channel kinetics of BK (Slo1) channels. *Frontiers in Physiology, 5*, 532.
Geng, Y., Niu, X., & Magleby, K. L. (2011). Low resistance, large dimension entrance to the inner cavity of BK channels determined by changing side-chain volume. *Journal of General Physiology, 137*, 533–548.
Geng, Y., Wang, X., & Magleby, K. L. (2013). Lack of negative slope in I-V plots for BK channels at positive potentials in the absence of intracellular blockers. *Journal of General Physiology, 141*, 493–497.
Gessner, G., Cui, Y. M., Otani, Y., Ohwada, T., Soom, M., Hoshi, T., et al. (2012). Molecular mechanism of pharmacological activation of BK channels. *Proceedings of the National Academy of Sciences of the United States of America, 109*, 3552–3557.
Giangiacomo, K. M., Kamassah, A., Harris, G., & McManus, O. B. (1998). Mechanism of maxi-K channel activation by dehydrosoyasaponin-I. *Journal of General Physiology, 112*, 485–501.
Goldklang, M. P., Perez-Zoghbi, J. F., Trischler, J., Nkyimbeng, T., Zakharov, S. I., Shiomi, T., et al. (2013). Treatment of experimental asthma using a single small molecule with anti-inflammatory and BK channel-activating properties. *FASEB Journal, 27*, 4975–4986.
Gonzalez-Perez, V., Xia, X. M., & Lingle, C. J. (2015). Two classes of regulatory subunits coassemble in the same BK channel and independently regulate gating. *Nature Communications, 6*, 8341.
Gribkoff, V. K., Starrett, J. E., Jr., Dworetzky, S. I., Hewawasam, P., Boissard, C. G., Cook, D. A., et al. (2001). Targeting acute ischemic stroke with a calcium-sensitive opener of maxi-K potassium channels. *Nature Medicine, 7*, 471–477.
Griguoli, M., Sgritta, M., & Cherubini, E. (2016). Presynaptic BK channels control transmitter release: Physiological relevance and potential therapeutic implications. *Journal of Physiology (London)*. http://dx.doi.org/10.1113/JP271841. [Epub ahead of print]. PMID: 26969302.
Gschwendt, M., Muller, H. J., Kielbassa, K., Zang, R., Kittstein, W., Rincke, G., et al. (1994). Rottlerin, a novel protein kinase inhibitor. *Biochemical and Biophysical Research Communications, 199*, 93–98.
Guo, Z., Lv, C., Yi, H., Xiong, Y., Wu, Y., Li, W., et al. (2008). A residue at the cytoplasmic entrance of BK-type channels regulating single-channel opening by its hydrophobicity. *Biophysical Journal, 94*, 3714–3725.
Hite, R. K., Yuan, P., Li, Z., Hsuing, Y., Walz, T., & MacKinnon, R. (2015). Cryo-electron-microscopy structure of the Slo2.2 Na^+-activated K^+ channel. *Nature, 527*, 198–203.
Horrigan, F. T. (2012). Conformational coupling in BK potassium channels. *Journal of General Physiology, 140*, 625–634.
Horrigan, F. T., & Aldrich, R. W. (1999). Allosteric voltage gating of potassium channels II. Mslo channel gating charge movement in the absence of Ca^{2+}. *Journal of General Physiology, 114*, 305–336.
Horrigan, F. T., & Aldrich, R. W. (2002). Coupling between voltage sensor activation, Ca^{2+} binding and channel opening in large conductance (BK) potassium channels. *Journal of General Physiology, 120*, 267–305.
Horrigan, F. T., Cui, J., & Aldrich, R. W. (1999). Allosteric voltage gating of potassium channels I. Mslo ionic currents in the absence of Ca^{2+}. *Journal of General Physiology, 114*, 277–304.
Horrigan, F. T., Heinemann, S. H., & Hoshi, T. (2005). Heme regulates allosteric activation of the Slo1 BK channel. *Journal of General Physiology, 126*, 7–21.
Horrigan, F. T., & Hoshi, T. (2015). Models of ion channel gating. In J. Zheng & M. C. Trudeau (Eds.), *Handbook of ion channels* (pp. 81–99). Boca Raton, FL: CRC Press.

Horrigan, F. T., & Ma, Z. (2008). Mg^{2+} enhances voltage sensor/gate coupling in BK channels. *Journal of General Physiology, 131*, 13–32.

Hoshi, T., Pantazis, A., & Olcese, R. (2013). Transduction of voltage and Ca^{2+} signals by Slo1 BK channels. *Physiology (Bethesda), 28*, 172–189.

Hoshi, T., Tian, Y., Xu, R., Heinemann, S. H., & Hou, S. (2013). Mechanism of the modulation of BK potassium channel complexes with different auxiliary subunit compositions by the omega-3 fatty acid DHA. *Proceedings of the National Academy of Sciences of the United States of America, 110*, 4822–4827.

Hoshi, T., Wissuwa, B., Tian, Y., Tajima, N., Xu, R., Bauer, M., et al. (2013). Omega-3 fatty acids lower blood pressure by directly activating large-conductance Ca^{2+}-dependent K^{+} channels. *Proceedings of the National Academy of Sciences of the United States of America, 110*, 4816–4821.

Hoshi, T., Xu, R., Hou, S., Heinemann, S. H., & Tian, Y. (2013). A point mutation in the human Slo1 channel that impairs its sensitivity to omega-3 docosahexaenoic acid. *Journal of General Physiology, 142*, 507–522.

Hou, S. W., Heinemann, S. H., & Hoshi, T. (2009). Modulation of BKCa channel gating by endogenous signaling molecules. *Physiology, 24*, 26–35.

Hou, S. W., Reynolds, M. F., Horrigan, F. T., Heinemann, S. H., & Hoshi, T. (2006). Reversible binding of heme to proteins in cellular signal transduction. *Accounts of Chemical Research, 39*, 918–924.

Hou, S., Vigeland, L. E., Zhang, G., Xu, R., Li, M., Heinemann, S. H., et al. (2010). Zn^{2+} activates large-conductance Ca^{2+}-activated K^{+} channel via an intracellular domain. *Journal of Biological Chemistry, 285*, 6434–6442.

Hou, S., Xu, R., Heinemann, S. H., & Hoshi, T. (2008a). The RCK1 high-affinity Ca^{2+} sensor confers carbon monoxide sensitivity to Slo1 BK channels. *Proceedings of the National Academy of Sciences of the United States of America, 105*, 4039–4043.

Hou, S., Xu, R., Heinemann, S. H., & Hoshi, T. (2008b). Reciprocal regulation of the Ca^{2+} and H^{+} sensitivity in the SLO1 BK channel conferred by the RCK1 domain. *Nature Structural & Molecular Biology, 15*, 403–410.

Hristov, K. L., Chen, M., Kellett, W. F., Rovner, E. S., & Petkov, G. V. (2011). Large-conductance voltage- and Ca^{2+}-activated K^{+} channels regulate human detrusor smooth muscle function. *American Journal of Physiology. Cell Physiology, 301*, C903–C912.

Hu, L., Shi, J., Ma, Z., Krishnamoorthy, G., Sieling, F., Zhang, G., et al. (2003). Participation of the S4 voltage sensor in the Mg^{2+}-dependent activation of large conductance (BK) K^{+} channels. *Proceedings of the National Academy of Sciences of the United States of America, 100*, 10488–10493.

Jaggar, J. H., Li, A., Parfenova, H., Liu, J., Umstot, E. S., Dopico, A. M., et al. (2005). Heme is a carbon monoxide receptor for large-conductance Ca^{2+}-activated K^{+} channels. *Circulation Research, 97*, 805–812.

Jaggar, J. H., Porter, V. A., Lederer, W. J., & Nelson, M. T. (2000). Calcium sparks in smooth muscle. *American Journal of Physiology. Cell Physiology, 278*, C235–C256.

Javaherian, A. D., Yusifov, T., Pantazis, A., Franklin, S., Gandhi, C. S., & Olcese, R. (2011). Metal-driven operation of the human large-conductance voltage- and Ca^{2+}-dependent potassium channel (BK) gating ring apparatus. *Journal of Biological Chemistry, 286*, 20701–20709.

Jensen, B. S. (2002). BMS-204352: A potassium channel opener developed for the treatment of stroke. *CNS Drug Reviews, 8*, 353–360.

Jensen, M. O., Borhani, D. W., Lindorff-Larsen, K., Maragakis, P., Jogini, V., Eastwood, M. P., et al. (2010). Principles of conduction and hydrophobic gating in K^{+} channels. *Proceedings of the National Academy of Sciences of the United States of America, 107*, 5833–5838.

Jensen, M. Ø., Jogini, V., Borhani, D. W., Leffler, A. E., Dror, R. O., & Shaw, D. E. (2012). Mechanism of voltage gating in potassium channels. *Science, 336*, 229–233.

Jiao, H., Arner, P., Hoffstedt, J., Brodin, D., Dubern, B., Czernichow, S., et al. (2011). Genome wide association study identifies KCNMA1 contributing to human obesity. *BMC Medical Genomics, 4*, 51.

Kelley-Hedgepeth, A., Peter, I., Montefusco, M. C., Levy, D., Benjamin, E. J., Vasan, R. S., et al. (2009). The KCNMB1 E65K variant is associated with reduced central pulse pressure in the community-based Framingham Offspring Cohort. *Journal of Hypertension, 27*, 55–60.

Khakh, B. S., Bao, X. R., Labarca, C., & Lester, H. A. (1999). Neuronal P2X transmitter-gated cation channels change their ion selectivity in seconds. *Nature Neuroscience, 2*, 322–330.

Koide, M., Bonev, A. D., Nelson, M. T., & Wellman, G. C. (2012). Inversion of neurovascular coupling by subarachnoid blood depends on large-conductance Ca^{2+}-activated K^+ (BK) channels. *Proceedings of the National Academy of Sciences of the United States of America, 109*, E1387–E1395.

Kokubo, Y., Iwai, N., Tago, N., Inamoto, N., Okayama, A., Yamawaki, H., et al. (2005). Association analysis between hypertension and CYBA, CLCNKB, and KCNMB1 functional polymorphisms in the Japanese population—The Suita Study. *Circulation Journal, 69*, 138–142.

Kurt, S., Sausbier, M., Ruttiger, L., Brandt, N., Moeller, C. K., Kindler, J., et al. (2012). Critical role for cochlear hair cell BK channels for coding the temporal structure and dynamic range of auditory information for central auditory processing. *FASEB Journal, 26*, 3834–3843.

Kyle, B. D., & Braun, A. P. (2014). The regulation of BK channel activity by pre- and post-translational modifications. *Frontiers in Physiology, 5*, 316.

Laumonnier, F., Roger, S., Guerin, P., Molinari, F., M'Rad, R., Cahard, D., et al. (2006). Association of a functional deficit of the BK_{Ca} channel, a synaptic regulator of neuronal excitability, with autism and mental retardation. *The American Journal of Psychiatry, 163*, 1622–1629.

Leo, M. D., Bannister, J. P., Narayanan, D., Nair, A., Grubbs, J. E., Gabrick, K. S., et al. (2014). Dynamic regulation of β1 subunit trafficking controls vascular contractility. *Proceedings of the National Academy of Sciences of the United States of America, 111*, 2361–2366.

Li, M., Tanaka, Y., Alioua, A., Wu, Y., Lu, R., Kundu, P., et al. (2010). Thromboxane A2 receptor and MaxiK-channel intimate interaction supports channel trans-inhibition independent of G-protein activation. *Proceedings of the National Academy of Sciences of the United States of America, 107*, 19096–19101.

Li, M., Toombes, G. E., Silberberg, S. D., & Swartz, K. J. (2015). Physical basis of apparent pore dilation of ATP-activated P2X receptor channels. *Nature Neuroscience, 18*, 1577–1583.

Liao, Y., Kristiansen, A. M., Oksvold, C. P., Tuvnes, F. A., Gu, N., Runden-Pran, E., et al. (2010). Neuronal Ca^{2+}-activated K^+ channels limit brain infarction and promote survival. *PloS One, 5*, e15601.

Liu, G., Shi, J., Yang, L., Cao, L., Park, S. M., Cui, J., et al. (2004). Assembly of a Ca^{2+}-dependent BK channel signaling complex by binding to β2 adrenergic receptor. *EMBO Journal, 23*, 2196–2205.

Liu, R., Zhang, Z., Liu, H., Hou, P., Lang, J., Wang, S., et al. (2013). Human β-defensin 2 is a novel opener of Ca^{2+}-activated potassium channels and induces vasodilation and hypotension in monkeys. *Hypertension, 62*, 415–425.

Loboda, A., Melishchuk, A., & Armstrong, C. (2001). Dilated and defunct K channels in the absence of K^+. *Biophysical Journal, 80*, 2704–2714.

Lorenz, S., Heils, A., Kasper, J. M., & Sander, T. (2007). Allelic association of a truncation mutation of the KCNMB3 gene with idiopathic generalized epilepsy. *American Journal of Medical Genetics. Part B, Neuropsychiatric Genetics*, *144B*, 10–13.

Ma, Z., Lou, X. J., & Horrigan, F. T. (2006). Role of charged residues in the S1-S4 voltage sensor of BK channels. *Journal of General Physiology*, *127*, 309–328.

Malinow, R., & Malenka, R. C. (2002). AMPA receptor trafficking and synaptic plasticity. *Annual Review of Neuroscience*, *25*, 103–126.

Martinez-Espinosa, P. L., Yang, C., Gonzalez-Perez, V., Xia, X. M., & Lingle, C. J. (2014). Knockout of the BK β2 subunit abolishes inactivation of BK currents in mouse adrenal chromaffin cells and results in slow-wave burst activity. *Journal of General Physiology*, *144*, 275–295.

Meredith, A. L., Thorneloe, K. S., Werner, M. E., Nelson, M. T., & Aldrich, R. W. (2004). Overactive bladder and incontinence in the absence of the BK large conductance Ca^{2+}-activated K^+ channel. *Journal of Biological Chemistry*, *279*, 36746–36752.

Meredith, A. L., Wiler, S. W., Miller, B. H., Takahashi, J. S., Fodor, A. A., Ruby, N. F., et al. (2006). BK calcium-activated potassium channels regulate circadian behavioral rhythms and pacemaker output. *Nature Neuroscience*, *9*, 1041–1049.

Miller, P. E., Van Elswyk, M., & Alexander, D. D. (2014). Long-chain omega-3 fatty acids eicosapentaenoic acid and docosahexaenoic acid and blood pressure: A meta-analysis of randomized controlled trials. *American Journal of Hypertension*, *27*, 885–896.

Morimoto, T., Sakamoto, K., Sade, H., Ohya, S., Muraki, K., & Imaizumi, Y. (2007). Voltage-sensitive oxonol dyes are novel BK channel activators selective for β1 and β4 but not for β2 subunits. *Molecular Pharmacology*, *71*, 1075–1088.

Nausch, B., Rode, F., Jorgensen, S., Nardi, A., Korsgaard, M. P., Hougaard, C., et al. (2014). NS19504: A novel BK channel activator with relaxing effect on bladder smooth muscle spontaneous phasic contractions. *Journal of Pharmacology and Experimental Therapeutics*, *350*, 520–530.

Nelson, M. T., Cheng, H., Rubart, M., Santana, L. F., Bonev, A. D., Knot, H. J., et al. (1995). Relaxation of arterial smooth muscle by calcium sparks. *Science*, *270*, 633–637.

Nielsen, T., Burgdorf, K. S., Grarup, N., Borch-Johnsen, K., Hansen, T., Jorgensen, T., et al. (2008). The KCNMB1 Glu65Lys polymorphism associates with reduced systolic and diastolic blood pressure in the Inter99 study of 5729 Danes. *Journal of Hypertension*, *26*, 2142–2146.

Niu, X., Qian, X., & Magleby, K. L. (2004). Linker-gating ring complex as passive spring and Ca^{2+}-dependent machine for a voltage- and Ca^{2+}-activated potassium channel. *Neuron*, *42*, 745–756.

O'Donnell, C., & van Rossum, M. C. (2014). Systematic analysis of the contributions of stochastic voltage gated channels to neuronal noise. *Frontiers in Computational Neuroscience*, *8*, 105.

Oger, S., Behr-Roussel, D., Gorny, D., Bernabe, J., Comperat, E., Chartier-Kastler, E., et al. (2011). Effects of potassium channel modulators on myogenic spontaneous phasic contractile activity in human detrusor from neurogenic patients. *BJU International*, *108*, 604–611.

Olesen, S. P., Munch, E., Moldt, P., & Drejer, J. (1994). Selective activation of Ca^{2+}-dependent K^+ channels by novel benzimidazolone. *European Journal of Pharmacology*, *251*, 53–59.

Orio, P., & Latorre, R. (2005). Differential effects of β1 and β2 subunits on BK channel activity. *Journal of General Physiology*, *125*, 395–411.

Peng, Z., Sakai, Y., Kurgan, L., Sokolowski, B., & Uversky, V. (2014). Intrinsic disorder in the BK channel and its interactome. *PloS One*, *9*, e94331.

Plüger, S., Faulhaber, J., Furstenau, M., Löhn, M., Waldschutz, R., Gollasch, M., et al. (2000). Mice with disrupted BK channel β1 subunit gene feature abnormal Ca^{2+} spark/STOC coupling and elevated blood pressure. *Circulation Research*, *87*, E53–E60.

Ponte, C. G., McManus, O. B., Schmalhofer, W. A., Shen, D. M., Dai, G., Stevenson, A., et al. (2012). Selective, direct activation of high-conductance, calcium-activated potassium channels causes smooth muscle relaxation. *Molecular Pharmacology*, *81*, 567–577.

Raimondo, J. V., Burman, R. J., Katz, A. A., & Akerman, C. J. (2015). Ion dynamics during seizures. *Frontiers in Cellular Neuroscience*, *9*, 419.

Rothberg, B. S., & Magleby, K. L. (2000). Voltage and Ca^{2+} activation of single large-conductance Ca^{2+}-activated K^+ channels described by a two-tiered allosteric gating mechanism. *Journal of General Physiology*, *116*, 75–99.

Roy, S., Morayo Akande, A., Large, R. J., Webb, T. I., Camarasu, C., Sergeant, G. P., et al. (2012). Structure-activity relationships of a novel group of large-conductance Ca^{2+}-activated K^+ (BK) channel modulators: The GoSlo-SR family. *ChemMedChem*, *7*, 1763–1769.

Sachse, G., Faulhaber, J., Seniuk, A., Ehmke, H., & Pongs, O. (2014). Smooth muscle BK channel activity influences blood pressure independent of vascular tone in mice. *Journal of Physiology (London)*, *592*, 2563–2574. http://dx.doi.org/10.1113/jphysiol.2014.272880.

Sack, J. T., & Tilley, D. C. (2015). What keeps Kv channels small? The molecular physiology of modesty. *Journal of General Physiology*, *146*, 123–127.

Sakai, Y., Harvey, M., & Sokolowski, B. (2011). Identification and quantification of full-length BK channel variants in the developing mouse cochlea. *Journal of Neuroscience Research*, *89*, 1747–1760.

Salkoff, L., Butler, A., Ferreira, G., Santi, C., & Wei, A. (2006). High-conductance potassium channels of the SLO family. *Nature Reviews. Neuroscience*, *7*, 921–931.

Sausbier, M., Arntz, C., Bucurenciu, I., Zhao, H., Zhou, X. B., Sausbier, U., et al. (2005). Elevated blood pressure linked to primary hyperaldosteronism and impaired vasodilation in BK channel-deficient mice. *Circulation*, *112*, 60–68.

Sausbier, M., Hu, H., Arntz, C., Feil, S., Kamm, S., Adelsberger, H., et al. (2004). Cerebellar ataxia and Purkinje cell dysfunction caused by Ca^{2+}-activated K^+ channel deficiency. *Proceedings of the National Academy of Sciences of the United States of America*, *101*, 9474–9478.

Savalli, N., Pantazis, A., Yusifov, T., Sigg, D., & Olcese, R. (2012). The contribution of RCK domains to human BK channel allosteric activation. *Journal of Biological Chemistry*, *287*, 21741–21750.

Schoppa, N. E., & Sigworth, F. J. (1998). Activation of Shaker potassium channels III—An activation gating model for wild-type and V2 mutant channels. *Journal of General Physiology*, *111*, 313–342.

Schreiber, M., & Salkoff, L. (1997). A novel calcium-sensing domain in the BK channel. *Biophysical Journal*, *73*, 1355–1363.

Schroder, R. L., Jespersen, T., Christophersen, P., Strobaek, D., Jensen, B. S., & Olesen, S. P. (2001). KCNQ4 channel activation by BMS-204352 and retigabine. *Neuropharmacology*, *40*, 888–898.

Schroder, R. L., Strobaek, D., Olesen, S. P., & Christophersen, P. (2003). Voltage-independent KCNQ4 currents induced by (+/−)BMS-204352. *Pflügers Archiv*, *446*, 607–616.

Scornik, F. S., Bucciero, R. S., Wu, Y., Selga, E., Bosch Calero, C., Brugada, R., et al. (2013). $DiBAC_4(3)$ hits a "sweet spot" for the activation of arterial large-conductance Ca^{2+}-activated potassium channels independently of the β_1-subunit. *American Journal of Physiology Heart and Circulatory Physiology*, *304*, H1471–H1482.

Seibold, M. A., Wang, B., Eng, C., Kumar, G., Beckman, K. B., Sen, S., et al. (2008). An African-specific functional polymorphism in KCNMB1 shows sex-specific association with asthma severity. *Human Molecular Genetics*, *17*, 2681–2690.

Sentí, M., Fernandez-Fernandez, J. M., Tomas, M., Vazquez, E., Elosua, R., Marrugat, J., et al. (2005). Protective effect of the KCNMB1 E65K genetic polymorphism against diastolic hypertension in aging women and its relevance to cardiovascular risk. *Circulation Research*, *97*, 1360–1365.

Shen, K. Z., Lagrutta, A., Davies, N. W., Standen, N. B., Adelman, J. P., & North, R. A. (1994). Tetraethylammonium block of Slowpoke calcium-activated potassium channels expressed in *Xenopus* oocytes: Evidence for tetrameric channel formation. *Pflügers Archiv*, *426*, 440–445.

Shi, J., & Cui, J. (2001). Intracellular Mg^{2+} enhances the function of BK-type Ca^{2+}-activated K^{+} channels. *Journal of General Physiology*, *118*, 589–606.

Shi, J., Krishnamoorthy, G., Yang, Y., Hu, L., Chaturvedi, N., Harilal, D., et al. (2002). Mechanism of magnesium activation of calcium-activated potassium channels. *Nature*, *418*, 876–880.

Shipston, M. J. (2001). Alternative splicing of potassium channels: A dynamic switch of cellular excitability. *Trends in Cell Biology*, *11*, 353–358.

Shmygol, A., Noble, K., & Wray, S. (2007). Depletion of membrane cholesterol eliminates the Ca^{2+}-activated component of outward potassium current and decreases membrane capacitance in rat uterine myocytes. *Journal of Physiology (London)*, *581*, 445–456.

Singh, H., Stefani, E., & Toro, L. (2012). Intracellular BK_{Ca} (iBK_{Ca}) channels. *Journal of Physiology (London)*, *590*, 5937–5947.

Son, Y. K., Choi, S. W., Jung, W. K., Jo, S. H., Jung, I. D., Park, Y. M., et al. (2011). The inhibitory effect of Ca^{2+}-activated K^{+} channel activator, BMS on L-type Ca^{2+} channels in rat ventricular myocytes. *Life Sciences*, *89*, 331–336.

Soom, M., Gessner, G., Heuer, H., Hoshi, T., & Heinemann, S. H. (2008). A mutually exclusive alternative exon of slo1 codes for a neuronal BK channel with altered function. *Channels (Austin, Tex.)*, *2*, 278–282.

Sprossmann, F., Pankert, P., Sausbier, U., Wirth, A., Zhou, X. B., Madlung, J., et al. (2009). Inducible knockout mutagenesis reveals compensatory mechanisms elicited by constitutive BK channel deficiency in overactive murine bladder. *FEBS Journal*, *276*, 1680–1697.

Starkus, J. G., Kuschel, L., Rayner, M. D., & Heinemann, S. H. (1997). Ion conduction through C-type inactivated Shaker channels. *Journal of General Physiology*, *110*, 539–550.

Suh, B. C., & Hille, B. (2008). PIP_2 is a necessary cofactor for ion channel function: How and why? *Annual Review of Biophysics*, *37*, 175–195.

Sweet, T. B., & Cox, D. H. (2008). Measurements of the BK_{Ca} channel's high-affinity Ca^{2+} binding constants: Effects of membrane voltage. *Journal of General Physiology*, *132*, 491–505.

Sweet, T. B., & Cox, D. H. (2009). Measuring the influence of the BK_{Ca} β1 subunit on Ca^{2+} binding to the BK_{Ca} channel. *Journal of General Physiology*, *133*, 139–150.

Tan, X. Q., Cheng, X. L., Yang, Y., Yan, L., Gu, J. L., Li, H., et al. (2014). Tanshinone II-A sodium sulfonate (DS-201) enhances human BK_{Ca} channel activity by selectively targeting the pore-forming α subunit. *Acta Pharmacologica Sinica*, *35*, 1351–1363.

Tang, X. D., Xu, R., Reynolds, M. F., Garcia, M. L., Heinemann, S. H., & Hoshi, T. (2003). Haem can bind to and inhibit mammalian calcium-dependent Slo1 BK channels. *Nature*, *425*, 531–535.

Tang, Q. Y., Zhang, Z., Meng, X. Y., Cui, M., & Logothetis, D. E. (2014). Structural determinants of phosphatidylinositol 4,5-bisphosphate (PIP2) regulation of BK channel activity through the RCK1 Ca^{2+} coordination site. *Journal of Biological Chemistry*, *289*, 18860–18872.

Tang, Q. Y., Zhang, Z., Xia, X. M., & Lingle, C. J. (2010). Block of mouse Slo1 and Slo3 K^+ channels by CTX, IbTX, TEA, 4-AP and quinidine. *Channels (Austin, Tex.)*, *4*, 22–41.

Telezhkin, V., Brazier, S. P., Mears, R., Muller, C. T., Riccardi, D., & Kemp, P. J. (2011). Cysteine residue 911 in C-terminal tail of human BK_{Ca} alpha channel subunit is crucial for its activation by carbon monoxide. *Pflügers Archiv*, *461*, 665–675.

Thompson, J., & Begenisich, T. (2012). Selectivity filter gating in large-conductance Ca^{2+}-activated K^+ channels. *Journal of General Physiology*, *139*, 235–244.

Tian, Y., Ullrich, F., Xu, R., Heinemann, S. H., Hou, S., & Hoshi, T. (2015). Two distinct effects of PIP_2 underlies auxiliary subunit-dependent modulation of Slo1 BK channels. *Journal of General Physiology*, *145*, 331–343.

Tomás, M., Vazquez, E., Fernandez-Fernandez, J. M., Subirana, I., Plata, C., Heras, M., et al. (2008). Genetic variation in the KCNMA1 potassium channel α subunit as risk factor for severe essential hypertension and myocardial infarction. *Journal of Hypertension*, *26*, 2147–2153.

Toro, L., Li, M., Zhang, Z., Singh, H., Wu, Y., & Stefani, E. (2014). MaxiK channel and cell signalling. *Pflügers Archiv*, *466*, 875–886.

Uebele, V. N., Lagrutta, A., Wade, T., Figueroa, D. J., Liu, Y., McKenna, E., et al. (2000). Cloning and functional expression of two families of β-subunits of the large conductance calcium-activated K^+ channel. *Journal of Biological Chemistry*, *275*, 23211–23218.

Vaithianathan, T., Bukiya, A., Liu, J., Liu, P., Asuncion-Chin, M., Fan, Z., et al. (2008). Direct regulation of BK channels by phosphatidylinositol 4,5-bisphosphate as a novel signaling pathway. *Journal of General Physiology*, *132*, 13–28.

Valverde, M. A., Rojas, P., Amigo, J., Cosmelli, D., Orio, P., Bahamonde, M. I., et al. (1999). Acute activation of Maxi-K channels (hSlo) by estradiol binding to the β subunit. *Science*, *285*, 1929–1931.

Virginio, C., MacKenzie, A., Rassendren, F. A., North, R. A., & Surprenant, A. (1999). Pore dilation of neuronal P2X receptor channels. *Nature Neuroscience*, *2*, 315–321.

Wallner, M., Meera, P., & Toro, L. (1999). Molecular basis of fast inactivation in voltage and Ca^{2+}-activated K^+ channels: A transmembrane β-subunit homolog. *Proceedings of the National Academy of Sciences of the United States of America*, *96*, 4137–4142.

Wang, B., Rothberg, B. S., & Brenner, R. (2006). Mechanism of β4 subunit modulation of BK channels. *Journal of General Physiology*, *127*, 449–465.

Wang, L., & Sigworth, F. J. (2009). Structure of the BK potassium channel in a lipid membrane from electron cryomicroscopy. *Nature*, *461*, 292–295.

Wang, R., Wu, L., & Wang, Z. (1997). The direct effect of carbon monoxide on KCa channels in vascular smooth muscle cells. *Pflügers Archiv*, *434*, 285–291.

Webb, T. I., Kshatri, A. S., Large, R. J., Akande, A. M., Roy, S., Sergeant, G. P., et al. (2015). Molecular mechanisms underlying the effect of the novel BK channel opener GoSlo: Involvement of the S4/S5 linker and the S6 segment. *Proceedings of the National Academy of Sciences of the United States of America*, *112*, 2064–2069.

Weiger, T. M., Holmqvist, M. H., Levitan, I. B., Clark, F. T., Sprague, S., Huang, W. J., et al. (2000). A novel nervous system beta subunit that downregulates human large conductance calcium-dependent potassium channels. *Journal of Neuroscience*, *20*, 3563–3570.

Welling, P. A. (2016). Roles and regulation of renal K channels. *Annual Review of Physiology*, *78*, 415–435.

Werner, M. E., Zvara, P., Meredith, A. L., Aldrich, R. W., & Nelson, M. T. (2005). Erectile dysfunction in mice lacking the large-conductance calcium-activated potassium (BK) channel. *Journal of Physiology (London)*, *567*, 545–556.

Wilkens, C. M., & Aldrich, R. W. (2006). State-independent block of BK channels by an intracellular quaternary ammonium. *Journal of General Physiology*, *128*, 347–364.

Williams, S. E., Brazier, S. P., Baban, N., Telezhkin, V., Muller, C. T., Riccardi, D., et al. (2008). A structural motif in the C-terminal tail of slo1 confers carbon monoxide sensitivity to human BK_{Ca} channels. *Pflügers Archiv*, *456*, 561–572.

Williams, S. E., Wootton, P., Mason, H. S., Bould, J., Iles, D. E., Riccardi, D., et al. (2004). Hemoxygenase-2 is an oxygen sensor for a calcium-sensitive potassium channel. *Science*, *306*, 2093–2097.

Wu, R. S., & Marx, S. O. (2010). The BK potassium channel in the vascular smooth muscle and kidney: α- and β-subunits. *Kidney International*, *78*, 963–974.

Wu, Y., Xiong, Y., Wang, S., Yi, H., Li, H., Pan, N., et al. (2009). Intersubunit coupling in the pore of BK channels. *Journal of Biological Chemistry*, *284*, 23353–23363.

Wu, Y., Yang, Y., Ye, S., & Jiang, Y. (2010). Structure of the gating ring from the human large-conductance Ca^{2+}-gated K^+ channel. *Nature*, *466*, 393–397.

Xia, X. M., Ding, J. P., Zeng, X. H., Duan, K. L., & Lingle, C. J. (2000). Rectification and rapid activation at low Ca^{2+} of Ca^{2+}-activated, voltage-dependent BK currents: Consequences of rapid inactivation by a novel β subunit. *Journal of Neuroscience*, *20*, 4890–4903.

Xia, X. M., Zeng, X., & Lingle, C. J. (2002). Multiple regulatory sites in large-conductance calcium-activated potassium channels. *Nature*, *418*, 880–884.

Xie, J., & McCobb, D. P. (1998). Control of alternative splicing of potassium channels by stress hormones. *Science*, *280*, 443–446.

Xu, H., Garver, H., Galligan, J. J., & Fink, G. D. (2011). Large-conductance Ca^{2+}-activated K^+ channel β1-subunit knockout mice are not hypertensive. *American Journal of Physiology. Heart and Circulatory Physiology*, *300*, H476–H485.

Yan, J., & Aldrich, R. W. (2010). LRRC26 auxiliary protein allows BK channel activation at resting voltage without calcium. *Nature*, *466*, 513–516.

Yan, J., & Aldrich, R. W. (2012). BK potassium channel modulation by leucine-rich repeat-containing proteins. *Proceedings of the National Academy of Sciences of the United States of America*, *109*, 7917–7922.

Yan, J., Olsen, J. V., Park, K. S., Li, W., Bildl, W., Schulte, U., et al. (2008). Profiling the phospho-status of the BKCa channel alpha subunit in rat brain reveals unexpected patterns and complexity. *Molecular & Cellular Proteomics*, 7, 2188–2198.

Yang, H., Hu, L., Shi, J., & Cui, J. (2006). Tuning magnesium sensitivity of BK channels by mutations. *Biophysical Journal*, *91*, 2892–2900.

Yang, H., Hu, L., Shi, J., Delaloye, K., Horrigan, F. T., & Cui, J. (2007). Mg^{2+} mediates interaction between the voltage sensor and cytosolic domain to activate BK channels. *Proceedings of the National Academy of Sciences of the United States of America*, *104*, 18270–18275.

Yang, Y., Li, P. Y., Cheng, J., Mao, L., Wen, J., Tan, X. Q., et al. (2013). Function of BKCa channels is reduced in human vascular smooth muscle cells from Han Chinese patients with hypertension. *Hypertension*, *61*, 519–525.

Yellen, G. (1984). Ionic permeation and blockade in Ca^{2+}-activated K^+ channels of bovine chromaffin cells. *Journal of General Physiology*, *84*, 157–186.

Yi, L., Morgan, J. T., & Ragsdale, S. W. (2010). Identification of a thiol/disulfide redox switch in the human BK channel that controls its affinity for heme and CO. *Journal of Biological Chemistry*, *285*, 20117–20127.

Yuan, P., Leonetti, M. D., Hsiung, Y., & MacKinnon, R. (2012). Open structure of the Ca^{2+} gating ring in the high-conductance Ca^{2+}-activated K^+ channel. *Nature*, *481*, 94–97.

Yuan, P., Leonetti, M. D., Pico, A. R., Hsiung, Y., & Mackinnon, R. (2010). Structure of the human BK channel Ca^{2+}-activation apparatus at 3.0 Å resolution. *Science*, *329*, 182–186.

Yusifov, T., Javaherian, A., Gandhi, C., Hou, S., Heinemann, S. H., Hoshi, T., et al. (2010a). Heme-driven conformational changes in the human Slo1 BKCa channel gating ring. *Biophysical Journal*, *98*, 126a.

Yusifov, T., Javaherian, A. D., Heinemann, S. H., Hoshi, T., & Olcese, R. (2012). Calcium and heme induce distinct conformational states of the human BK (Slo1) channel gating ring. *Biophysical Journal, 102*, 687a–688a.

Yusifov, T., Javaherian, A. D., Pantazis, A., Gandhi, C. S., & Olcese, R. (2010b). The RCK1 domain of the human BK_{Ca} channel transduces Ca^{2+} binding into structural rearrangements. *Journal of General Physiology, 136*, 189–202.

Yusifov, T., Savalli, N., Gandhi, C. S., Ottolia, M., & Olcese, R. (2008). The RCK2 domain of the human BK_{Ca} channel is a calcium sensor. *Proceedings of the National Academy of Sciences of the United States of America, 105*, 376–381.

Zagotta, W. N., Hoshi, T., & Aldrich, R. W. (1994). Shaker potassium channel gating. III: Evaluation of kinetic models for activation. *Journal of General Physiology, 103*, 321–362.

Zakharov, S. I., Morrow, J. P., Liu, G., Yang, L., & Marx, S. O. (2005). Activation of the BK (SLO1) potassium channel by mallotoxin. *Journal of Biological Chemistry, 280*, 30882–30887.

Zeng, H., Lozinskaya, I. M., Lin, Z., Willette, R. N., Brooks, D. P., & Xu, X. (2006). Mallotoxin is a novel human *ether-a-go-go*-related gene (hERG) potassium channel activator. *Journal of Pharmacology and Experimental Therapeutics, 319*, 957–962.

Zeng, X. H., Xia, X. M., & Lingle, C. J. (2003). Redox-sensitive extracellular gates formed by auxiliary β subunits of calcium-activated potassium channels. *Nature Structural Biology, 10*, 448–454.

Zeng, X. H., Xia, X. M., & Lingle, C. J. (2005). Divalent cation sensitivity of BK channel activation supports the existence of three distinct binding sites. *Journal of General Physiology, 125*, 273–286.

Zhang, Z., Li, M., Lu, R., Alioua, A., Stefani, E., & Toro, L. (2014). The angiotensin II type 1 receptor (AT1R) closely interacts with large conductance voltage- and Ca^{2+}-activated K^+ (BK) channels and inhibits their activity independent of G-protein activation. *Journal of Biological Chemistry, 289*, 25678–25689.

Zhang, Y., Niu, X., Brelidze, T. I., & Magleby, K. L. (2006). Ring of negative charge in BK channels facilitates block by intracellular Mg^{2+} and polyamines through electrostatics. *Journal of General Physiology, 128*, 185–202.

Zhang, X., Solaro, C. R., & Lingle, C. J. (2001). Allosteric regulation of BK channel gating by Ca^{2+} and Mg^{2+} through a nonselective, low affinity divalent cation site. *Journal of General Physiology, 118*, 607–635.

Zhou, Y., & Lingle, C. J. (2014). Paxilline inhibits BK channels by an almost exclusively closed-channel block mechanism. *Journal of General Physiology, 144*, 415–440.

Zhou, Y., Morais-Cabral, J. H., Kaufman, A., & MacKinnon, R. (2001). Chemistry of ion coordination and hydration revealed by a K^+ channel-Fab complex at 2.0 Å resolution. *Nature, 414*, 43–48.

Zhou, Y., Xia, X. M., & Lingle, C. J. (2011). Cysteine scanning and modification reveal major differences between BK channels and Kv channels in the inner pore region. *Proceedings of the National Academy of Sciences of the United States of America, 108*, 12161–12166.

Zhou, Y., Zeng, X. H., & Lingle, C. J. (2012). Barium ions selectively activate BK channels via the Ca^{2+}-bowl site. *Proceedings of the National Academy of Sciences of the United States of America, 109*, 11413–11418.

CHAPTER SEVEN

Modulation of BK Channels by Ethanol

A.M. Dopico[1], A.N. Bukiya, G. Kuntamallappanavar, J. Liu
College of Medicine, The University of Tennessee Health Science Center, Memphis, TN, United States
[1]Corresponding author: e-mail address: adopico@uthsc.edu

Contents

Abstract

In alcohol-naïve systems, ethanol (<100 mM) exposure of calcium-gated BK channels perturbs physiology and behavior. Brief (several minutes) ethanol exposure usually leads to increased BK current, which results from ethanol interaction with a pocket mapped to the BK channel-forming slo1 protein cytosolic tail domain. The importance of this region in ethanol-induced intoxication has been independently supported by an unbiased screen of *Caenorhabditis elegans* slo1 mutants. However, ethanol-induced BK activation is not universal as refractoriness and inhibition have been reported. The final effect depends on many factors, including intracellular calcium levels, slo1 isoform, BK beta subunit composition, posttranslational modification of BK proteins, channel lipid microenvironment, and type of ethanol administration. Studies in *Drosophila melanogaster*, *C. elegans*, and rodents show that protracted/repeated ethanol administration leads to tolerance to ethanol-induced modification of BK-driven physiology and behavior. Unveiling the mechanisms underlying tolerance is of major importance, as tolerance to ethanol has been proposed as predictor of risk for alcoholism.

International Review of Neurobiology, Volume 128
ISSN 0074-7742
http://dx.doi.org/10.1016/bs.irn.2016.03.019

1. INTRODUCTION: IS THE BK CHANNEL AN ETHANOL RECEPTOR?

Ion channel proteins serve fundamental roles in the physiology of excitable tissues, including initiation and propagation of action potentials (APs), generation of postsynaptic potentials, and electrochemical coupling leading to neurotransmitter and hormone release (see Chapters "BK Channels in Neurons" by Barth and Contet and "BK Channels in the Pituitary and Hypothalamus" by Duncan and Shipston of the present volume). It has been widely documented that alcohol (ethyl alcohol, ethanol) alters the physiology of excitable tissues, including central nervous system (CNS) neurons and behavior. Moreover, ethanol has pharmacological effects common to those of drugs that act primarily by modifying ion channel protein function (eg, anxiolytics, barbiturates, and volatile anesthetics; reviewed in Fleming, Harris, & Mihic, 2001; Masters, 2009). Thus, it is not surprising that the literature on interactions between ethanol and ion channel proteins is truly massive. With the goal of narrowing down the number of ethanol receptors based on clinical relevance, Harris, Trudell, and Mihic (2008) advanced several criteria to be fulfilled by an "ethanol receptor," which we now slightly modify and adapt to *ion channel proteins*: (1) The function of the ion channel protein must be altered by in vitro exposure to $\leq$150 mM ethanol (Harris et al., 2008). In most states of the Union, blood ethanol levels (BALs) that constitute legal intoxication correspond to 17.4 mM (0.08 g/dL) while 100–150 mM BALs are usually lethal to most ethanol-naïve mammals, including humans (Diamond, 1992; Thombs, Olds, & Snyder, 2003). Ideally, alteration of channel function by ethanol should be observed in recombinant and native ion channel proteins under physiological conditions; (2) Manipulation of specific amino acids in the channel-forming protein should provide evidence of an ethanol-recognition site of discrete dimensions, as well as information on the chemical bonds participating in ethanol–ion channel protein site interaction(s); (3) Manipulation of the channel-forming protein in vivo (using sRNAi or epi/genetic manipulations) should alter ethanol modification of physiological and/or behavioral end points controlled by the ion channel protein under examination; (4) Ideally, Structural Biology methods (crystallography, NMR, etc.) should demonstrate the presence of ethanol molecule(s) at the proposed site(s) within the ion channel protein.

Since the first report on selective and concentration-dependent enhancement of ionic currents mediated by large conductance, voltage- and $Ca^{2+}{}_i$-gated K^+ (BK) channels in response to a brief exposure (1–2 min) of rat neurohypophysial nerve endings to 10–100 mM ethanol (Dopico, Lemos, & Treistman, 1996), functional interactions between ethanol and BK channels have been documented in a wide variety of systems (Table 1), which fairly fulfills criterion (1) mentioned earlier. More recent studies using engineered BK channel-forming slo1 proteins (Bukiya et al., 2014) and genetic screening or mutation of the BK channel-coding gene in *Caenorhabditis elegans* (*Slo1*; Davis, Scott, Hu, & Pierce-Shimomura, 2014; Davis et al., 2015) have buttressed the idea of an ethanol-recognition site in the BK channel-forming protein (also termed BK α subunit or slo1 channel) by satisfying criteria (2) and (3). While criterion (4) has yet to be met, it should be noted that the molecular modeling (MM) and molecular dynamics (MD) data from Bukiya et al. (2014) were based on crystallographic data of the BK channel cytosolic tail domain (CTD; wherein the ethanol binding site has been mapped) in both "open" ($Ca^{2+}{}_i$-bound; Yuan, Leonetti, Pico, Hsiung, & MacKinnon, 2010) and "closed" ($Ca^{2+}{}_i$-free; Wu, Yang, Ye, & Jiang, 2010) states of the BK channel. The model of ethanol docking onto a site of discrete dimensions in the mslo1 channel (cloned from mouse brain; Bukiya et al., 2014) explains previously described features of modulation of BK current by ethanol and related *n*-alkanols: (1) ethanol action's dependence on $Ca^{2+}{}_i$ presence (Liu, Bukiya, Kuntamallappanavar, Singh, & Dopico, 2013; Liu, Vaithianathan, Manivannan, Parrill, & Dopico, 2008) (Fig. 1) and (2) the refractoriness of slo1 channels to *n*-alkanols longer than heptanol (Chu & Treistman, 1997) (Fig. 2). Finally, the coupling of ethanol action to Ca^{2+}-driven gating is supported by in vitro electrophysiology on mslo1 proteins with substitutions in the channel's Ca^{2+} sensors, single-channel gating models (Liu et al., 2008), and modification of ethanol action following slo1 gene mutations in *C. elegans*, using both behavioral and electrophysiological assays as end points (Davis et al., 2015; see later). The role of specific residues and bonds that participate in the model of ethanol docking onto mslo1 proteins is shown in Fig. 1. Additionally, Thr 381 in *C. elegans* slo1 (equivalent to hslo1 Thr 352) determines the overall ethanol sensitivity of BK channel-forming subunits and slo1-driven behavioral responses in *C. elegans* (Davis et al., 2014). A structural role for this conserved residue within the proposed ethanol-docking discrete site is discussed elsewhere (Dopico, Bukiya, &

Table 1 Effects of Brief Exposure (Typically <5 min) to Clinically Relevant Concentrations of Ethanol (≤ 100 mM) on Native BK Channel Function in Ethanol-Naïve Systems[a]

System	V-Clamp Method	Ethanol (mM)	Effect on Channel Activity/Current	Likely Consequence on Tissue Physiology	References
Neuroendocrine cells and neurons					
Rat GH3 and GH4/C1 pituitary cell lines	O/O, C/A patches	1–300	Potentiation	Inhibition of growth hormone release	Jakab, Weiger, and Hermann (1997) and Handlechner, Hermann, Fuchs, and Weiger (2013)
Rat supraoptic/paraventricular neurons (nerve terminals)	I/O patch, conventional WCR	10–100	Potentiation	Inhibition of vasopressin or oxytocin release	Dopico, Lemos, and Treistman (1995), Dopico et al. (1996), Knott, Dopico, Dayanithi, Lemos, and Treistman (2002), Pietrzykowski et al. (2004), and Wynne, Puig, Martin, and Treistman (2009)
Rat supraoptic/paraventricular neurons (somata)	I/O patch	25–100	Refractoriness	N/A	Dopico, Chu, Lemos, and Treistman (1999), Dopico, Widmer, Wang, Lemos, and Treistman (1999), and Wynne et al. (2009)
Rat primary sensory dorsal root ganglia neurons	O/O patch	40–80	Potentiation	AP shortening, increase in refractory period with decreased firing rate	Gruss et al. (2001)

Rat NAcc neurons (somata)	I/O patch	1–100	Potentiation	AP shortening; reduced neuronal firing	Martin et al. (2004)
Rat NAcc neurons (dendrites)	I/O patch	10, 50	Refractoriness	N/A	Martin et al. (2004)
Mouse striatal MSN	C/A single-cell patch, WCR	50	Potentiation	AP shortening?; reduced neuronal firing	Martin et al. (2008)
C. elegans type IV DA CEP neurons; also interneurons	WCR, I/O patch	20	Potentiation	Disruption of locomotion, egg laying[b]	Davies et al. (2003)
Other cell types					
Human umbilical vein endothelial cells in culture	I/O patch	10–50	Potentiation	Increased endothelial proliferation and $NO^{\bullet}$ production	Kuhlmann et al. (2004)
Bovine aortic smooth muscle	Planar lipid bilayer	5–80	Inhibition	Aortic constriction	Walters, Covarrubias, and Ellingson (2000)
Rat cerebral artery myocytes	I/O, O/O patches, PP WCR	3–200	Inhibition	Cerebral artery constriction	Liu, Xi, Ahmed, Jaggar, and Dopico (2004) and Bukiya, Vaithianathan, Kuntamallappanavar, Asuncion-Chin, and Dopico (2011)

Continued

Table 1 Effects of Brief Exposure (Typically <5 min) to Clinically Relevant Concentrations of Ethanol (≤100 mM) on Native BK Channel Function in Ethanol-Naïve Systems—cont'd

System	V-Clamp Method	Ethanol (mM)	Effect on Channel Activity/ Current	Likely Consequence on Tissue Physiology	References
Mouse cerebral artery myocytes	I/O patch	50 mM	Inhibition	Cerebral artery constriction	Bukiya, Liu, and Dopico (2009), Bukiya, Vaithianathan, Toro, and Dopico (2009), Bukiya, Singh, Parrill, and Dopico (2011), and Bukiya, Vaithianathan, et al. (2011)

[a]The listing *excludes* cell/animals where genes/transcripts encoding for BK channel subunits, whether channel-forming or regulatory, have been deleted, overexpressed, or engineered (see main text).

[b]In this case, worms were exposed to ethanol-containing bath for 22 min, ie, more than what is considered a typical "acute/brief" application. However, pharmacokinetic considerations should be taken into account as ethanol does not seem to readily cross the *C. elegans* cuticle. This is reflected by the very different external, ie, bath, and corresponding internal, ie, from worm lysates, ethanol concentrations: 400/500 and 22/29, respectively.

AP, action potential; *C/A*, cell-attached; *DA CEP*, dopaminergic cephalic sensilla; *I/O*, inside-out; *MSN*, medium spiny neurons; *NAcc*, nucleus accumbens; *O/O*, outside-out; *PP*, perforated patch, macroscopic current recording; *WCR*, standard whole-cell, macroscopic current recording.

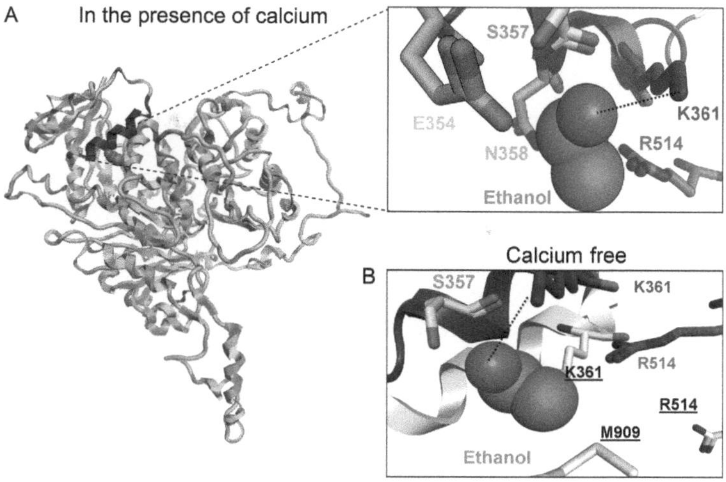

Fig. 1 Ethanol-sensing site in the BK channel-forming slo1 protein cytosolic tail domain (CTD). (A) Location of the ethanol-sensing site in the slo1 CTD; the CTD is modeled based on the crystallographic structure obtained in *Ca^{2+}-containing* solution (see main text). The ethanol carbon chain is shown in bright *pink* (*gray* in the print version) and the oxygen is in *red* (*dark gray* in the print version). *Right top insert* shows the ethanol-sensing site provided by residues 354–367 where Lys361 hydrogen bonds (*dotted line*) with ethanol. Nitrogen is in *blue* (*black* in the print version); hydrogens are hidden to improve display. (B) The ethanol–slo1 CTD model based on the crystallographic structure obtained in the *absence of Ca^{2+}* (see main text) shows substantial perturbations within the ethanol-sensing site: repositioning of Lys361-containing helix; obstructed ethanol's access to the site due to the movement of Met909; displacement of Arg514 away from the ethanol-sensing area. Key elements in ethanol-sensing pocket are *color* (*black-gray-white* in the print version) coded as in (A). Lys361, Met909, and Arg514 experience the bigger perturbations by the absence of Ca^{2+}; their location under this condition is shown in *black* and *underlined. With modifications from Bukiya, A. N., Kuntamallappanavar, G., Edwards, J., Singh, A. K., Shivakumar, B., & Dopico, A. M. (2014). An alcohol-sensing site in the calcium- and voltage-gated, large conductance potassium (BK) channel. Proceedings of the National Academy of Sciences of the United States of America, 111(25), 9313–9318. http://dx.doi.org/10.1073/pnas.1317363111.*

Martin, 2014). Collectively, the facts that (a) ethanol modulates native BK and recombinant slo1 channels at concentrations below 100 mM; (b) point mutants of slo1 confer resistance to ethanol; (c) a pocket of discrete dimensions mapped to the slo1 protein where ethanol may dock, and chemical bonds between this site and the ligand have been identified (see Section 3.2); (d) the ethanol–slo1 pocket interaction model by MM is based on high-resolution crystallography of the slo1 protein cytosolic end in both Ca^{2+}-bound and Ca^{2+}-unbound states; (e) docking models for ethanol in these two states explain phenomenology of ethanol interaction

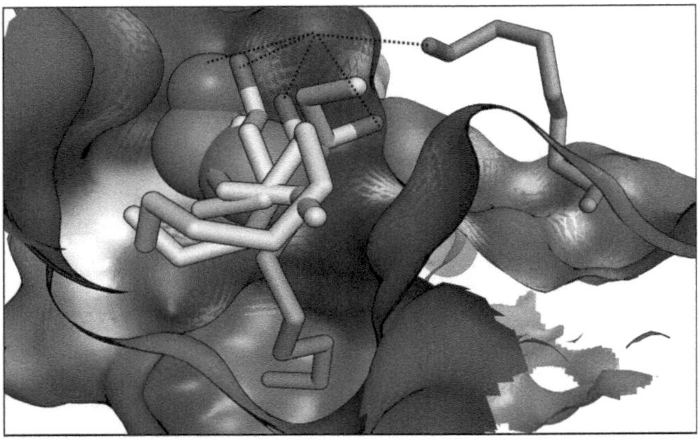

Fig. 2 Role of the ethanol-docking slo1 CTD site in long-chain alkanol's effects on slo1 currents. Poses of *n*-alkanols on the site surface map show that slo1-activating *n*-alkanols, such as ethanol, propanol, butanol, heptanol, and hexanol, fit within the ethanol-sensing pocket while ineffective *n*-alkanols, such as octanol and nonanol, do not (see main text). Ethanol carbon chain is in *bright pink* (*gray* in the print version); other activating *n*-alkanols are shown in *yellow* (*light gray* in the print version), whereas ineffective *n*-alkanols are in *orange* (*gray* in the print version). In all cases, oxygen is in *red* (*dark gray* in the print version), nitrogen is in *blue* (*darker gray-black* in the print version). Hydrogen bonds between hydroxyl groups of alkanols and Lys361 are presented as *dotted lines*. This figure is based on previously published material by the authors (Bukiya, A. N., Kuntamallappanavar, G., Edwards, J., Singh, A. K., Shivakumar, B., & Dopico, A. M. (2014). An alcohol-sensing site in the calcium- and voltage-gated, large conductance potassium (BK) channel. *Proceedings of the National Academy of Sciences of the United States of America, 111(25), 9313–9318. http://dx.doi.org/10.1073/pnas.1317363111).*

with BK channels, including the dependence of ethanol effect on the presence of $Ca^{2+}{}_i$ and the existence of "cut-off" between heptanol and octanol (Bukiya et al., 2014), and (f) manipulation of the *Slo/Slo1* genes in vivo alters ethanol-driven behaviors in *Drosophila and C. elegans* (see Sections 3.2 and 4), it is highly likely that the BK channel-forming slo1 protein constitutes an ethanol receptor.

While a direct slo1 protein–ethanol interaction in the presence of activating $Ca^{2+}{}_i$ (Crowley, Treistman, & Dopico, 2003; Liu et al., 2013, 2008) seems sufficient for slo1 channels to fulfill criterion (1) for a functional ethanol receptor, the final ethanol effect on native BK channel-mediated ionic current is highly variable, with potentiation, inhibition, and refractoriness all being reported (reviewed in Dopico et al., 2014). This heterogeneous profile is the result of ethanol targeting a wide variety of molecular entities (ethanol receptors additional to slo1 proteins) and mechanisms that may regulate slo1 channel function itself. Thus, we will succinctly present to the reader the many molecular factors that determine the final response of BK current to clinically relevant concentrations of ethanol. Moreover, demonstration of BK channel involvement in ethanol-altered behavior and/or animal models of alcoholism requires some form of "protracted" (longer than a couple of minutes) or repeated exposure of slo1 channel proteins/coding genes to ethanol. As with other drugs of misuse, the response of a biological system to protracted/repeated ethanol exposure may include receptors and mechanisms that were perturbed by ethanol when ethanol was applied to the system for the first time (Dopico & Lovinger, 2009). In Section 2, we summarize and analyze ethanol actions on BK currents provided that ethanol was applied: (a) to an ethanol-naïve system, (b) for a brief interval (ie, a few minutes), and (c) just once. Section 3 concentrates on the genetic, epigenetic, and post-translational mechanisms mediating modifications in the BK channel's ethanol response that are driven by protracted and/or repeated ethanol exposure. Finally, Section 4 analyzes involvement of BK channels/coding genes in modification of behavior by ethanol.

2. RESPONSES OF NATIVE BK CHANNELS TO CLINICALLY RELEVANT ETHANOL CONCENTRATIONS IN ETHANOL-NAÏVE SYSTEMS

Ionic current mediated by *native* BK channels renders all possible outcomes (increase, decrease, or refractoriness) when these channels are briefly (<5 min) exposed to clinically relevant ethanol concentrations (10–100 mM) for the first time, as summarized in Table 1. It should be underscored that such heterogeneity can be observed even when native channels are studied within a same neuronal type (albeit within different neuronal domains; eg, somata vs axon terminals; Dopico et al., 1996; Dopico, Widmer, et al., 1999; Wynne et al., 2009; or *nucleus accumbens* (NAcc) neuronal somata vs dendrites; Martin et al., 2004) underscoring that ethanol–BK channel results obtained in a presynaptic domain cannot be simply extrapolated to those in a postsynaptic environment.

In a more general context, the consequences of ethanol action on BK channels in a given excitable tissue (eg, neurons) may completely differ from another. For instance, ethanol potentiation of BK current reduces AP firing rate and thus, decreases the excitability of medium spiny neurons (MSN) in rat and mouse (Martin et al., 2008, 2004). This action has been linked to ethanol-induced perturbation of motor behavior and ethanol preference (reviewed by Bettinger & Davies, 2014; Treistman & Martin, 2009; see Section 4). A decrease in AP frequency in response to 40 mM ethanol has also been reported in dorsal root ganglia (DRG) neurons that show positive staining for isolectin B4, a marker for nociceptive neurons. Ethanol also shortens AP duration and increases AP mean threshold, these actions being blunted by selective block of BK channels with the peptide iberiotoxin (Ibtx; Gruss et al., 2001). Thus, authors of this study proposed that ethanol actions leading to reduced firing and excitability of distinct DRG neurons might contribute to ethanol's analgesic effect in the peripheral nervous system. In contrast to the studies on NAcc MSN and DRG neurons, ethanol-induced slo1 expression and the drug interaction with BK channels actually *increases* motor neuron excitability within the *Drosophila melanogaster* giant fiber pathway. This effect is related to a substantial decrease in the neuronal refractory period, which might provide an explanation for the long-standing counter-adaptive theory of "drug tolerance" (Cowmeadow, Krishnan, & Atkinson, 2005; Ghezzi, Pohl, Wang, & Atkinson, 2010). Finally, pharmacological block of presynaptic BK channels prevents ethanol-induced enhancement of GABAergic *inhibitory* neurotransmission in the rat central nucleus of the amygdala (Li, Madison, & Moore, 2014).

Heterogeneity of ethanol responses across different neuronal types and domains can be largely explained by heterogeneity in the subunit composition and distribution of BK channels themselves (see later). However, disruption of physiology by ethanol's targeting of BK channels must be analyzed having in mind that ethanol is a very simple molecule, a feature that allows this alkanol to "fit" in many putative docking areas and readily cross-membranes. Thus, ethanol has the capability to simultaneously target many receptors within a single-cell domain. In some cases, ethanol's targeting of more than one ion channel/receptor population results in synergistic modulation of a given cell process by ethanol. For example, in rat neurohypophysial terminals, ethanol-induced potentiation of BK current in concert with ethanol inhibition of voltage-dependent Ca^{2+} channels (Wang, Wang, Lemos, & Treistman, 1994) has been recognized as a central mechanism leading to inhibition of vasopressin and oxytocin release by such nerve terminals (Dopico et al., 1996; Knott et al., 2002). Likewise, ethanol inhibition of smooth muscle BK channels in concert with ethanol inhibition of local Ca^{2+} sparks driven by ryanodine receptors of type 2 (RyR2) synergistically contributes to blunt spontaneous transient outward currents (STOCs) eventually leading to endothelium-independent cerebral artery constriction (Bukiya, Liu, et al., 2009; Liu et al., 2004; Ye, Jian, Jaggar,

Bukiya, & Dopico, 2014). In other cases, ethanol functional targeting of BK current is antagonized, and can be overridden, by ethanol actions on other cell targets. For example, ethanol-induced BK channel activation in GH3 and GH4/C1 rat pituitary tumor cells would likely lead to inhibition of hormone release (Jakab et al., 2006, 1997). In spite of BK current potentiation, ethanol actually increases growth hormone secretion in these cells, which has been attributed to ethanol-induced cell swelling and increase in $Ca^{2+}{}_i$ (Jakab et al., 2006; Stojilkovic, Zemkova, & Van Goor, 2005; Strbak, 2006).

Heterogeneity in the impact of ethanol–BK channel interactions on cell function is not limited to neurons/neuroendocrine systems. As mentioned earlier, such interaction has been demonstrated to evoke cerebral artery constriction in both rat and mouse (Bukiya, Liu, et al., 2009; Bukiya, Vaithianathan, et al., 2011; Liu et al., 2004). Likewise, this interaction has been hypothesized to contribute to ethanol-induced aortic constriction (Walters et al., 2000). In both cases, ethanol interaction on smooth muscle BK channels results in ionic current inhibition. In contrast, ethanol-induced BK channel activation of either or both neuronal and cerebral microvascular endothelial cells has been proposed as a key mechanism for the neuroprotective effect of ethanol preconditioning against postischemic neuronal injury in mice (Wang, Kalogeris, Wang, Jones, & Korthuis, 2010). Targeting of myocyte BK channels by ethanol is not limited to vascular smooth muscle: a recent study shows that BK channel activation plays a critical role in ethanol-induced relaxation of guinea pig urinary bladder smooth muscle (Malysz, Afeli, Provence, & Petkov, 2014).

Collectively, reported functional consequences of an ethanol–BK channel interaction have been largely limited to neurons, neuroendocrine, and smooth muscle cells. A single study, however, has documented that 10–50 mM ethanol potentiates BK currents in human endothelial umbilical vein cells, an ethanol action that leads to increased nitric oxide ($NO^{\bullet}$) production and cell proliferation with eventual bolstering of endothelial function (Kuhlmann et al., 2004). The functional consequences of ethanol action on native BK channels in ethanol-naïve systems are also summarized in Table 1. The behavioral consequences of the interactions between neuronal BK channels and ethanol have been recently discussed in detail elsewhere (Bettinger & Davies, 2014) and briefly described in Section 4.

In spite of the heterogeneities in ethanol–BK channel interactions mentioned in the previous paragraphs, some generalizations can be made. Brief exposure of native/recombinant BK channels to intoxicating levels of ethanol in ethanol-naïve systems is: (1) In most cases, rather selective, with this type of ethanol exposure perturbing BK current without altering other voltage-gated K^+ channels (eg, other K^+ outward rectifiers of the TM6 superfamily) in the same nerve terminal (Dopico et al., 1996), or even the slo1-related, slo2 and slo3 channels after expression in the same system where slo1 channels have been expressed (Liu et al., 2013); (2) Unexplained by a simple osmotic challenge to the channel/membrane as a result of exposure to high mM levels of ethanol (reviewed in Brodie, Scholz, Weiger, & Dopico, 2007). Indeed, both hypotonicity and 30 mM ethanol lead to cell swelling and BK channel activation in GH4/C1 pituitary tumor cells, yet the former and not the later increases $Ca^{2+}{}_i$-dependent Cl^- currents (Jakab et al., 2006). In general, multiple studies favor the idea that ethanol effect on BK channel function is rather specific and does not simply reflect changes caused by an osmotic challenge (reviewed in Brodie et al., 2007; Mulholland et al., 2009); (3) Due to ethanol itself. Data obtained from isolated membrane patches long after patch excision and from channels reconstituted into artificial lipid bilayers, supported by the mutagenesis and MM–MD data mentioned earlier, advance the notion that ethanol action on BK channels is due to ethanol molecules rather than ethanol metabolites. It should be noted however, that acetaldehyde, a major metabolite of ethanol, antagonizes ethanol-induced BK channel activation in excised patches from GH3 cells (Handlechner et al., 2013). Considering that BK channels in vivo are likely to be exposed to a mixture of ethanol and acetaldehyde, it is conceivable that some of the in vivo effects of ethanol on pituitary cells (and on organ systems in general) result from the contribution of acetaldehyde on top of ethanol itself; (4) Concentration-dependent, with $EC_{50}/IC_{50}=22$–24 mM and $E_{max} \sim 100$ mM (Dopico, 2003; Dopico, Anantharam, & Treistman, 1998; Dopico et al., 1996); (5) Independent of changes in unitary current amplitude (reviewed in Dopico et al., 2014; see also Davis et al., 2014; Malysz et al., 2014; Marrero, Treistman, & Lemos, 2015) or change in K^+ selectively over other monovalents (Dopico et al., 1998). Given the conservation of ion channel permeation properties in the presence of ethanol, potentiation and inhibition of BK current by this alcohol have been attributed to ethanol-induced increase and decrease in BK channel steady-state activity, respectively. In all studies dealing with ethanol–BK channel interactions, channel "steady-state activity" has been routinely determined from the product of the number of functional channels in the membrane patch under study (N) and the individual open channel probability (P_o; Dopico et al., 1996). The degree of ethanol modification of NP_o in channels of unknown N has been found identical to that in membrane patches containing a single BK channel (ie, $N=1$; Dopico et al., 1998). This parallelism and other considerations that rule out any possible increase in N by a short exposure of a cell-free patch or artificial bilayer containing BK channels to ethanol leads to the conclusion that change in BK current to brief, low concentrations of intoxicating ethanol levels is solely dependent on drug-induced changes in P_o. These changes are consequence of modifications in both open- and closed-times distributions (discussed in more detail in Dopico et al., 2014). A gating analysis of ethanol action on BK channels will be summarized later, framed within the ethanol dependence on activating $Ca^{2+}{}_i$ to modify BK function. Changes in N in the presence of ethanol seem to only occur when the system is continuously exposed for at least 10 min to ethanol and/or to repeated, brief ethanol exposures, paradigms that are discussed in Section 4 of this chapter.

3. MOLECULAR DETERMINANTS OF CHANGES IN BK CHANNEL STEADY-STATE ACTIVITY EVOKED BY CLINICALLY RELEVANT ETHANOL CONCENTRATIONS IN ETHANOL-NAÏVE SYSTEMS

Table 1 shows that some of the studies demonstrating heterogeneity of BK current response to a first and brief ethanol exposure were conducted in cell-free membrane patches (reviewed in Brodie et al., 2007; Dopico et al., 2014). The experimental conditions of these studies underscore that the different effects of ethanol on native BK channels are largely independent of Ca^{2+} release/buffering from internal organelles, continuous presence of freely diffusible cytosolic signals, and cell/systemic metabolism of ethanol. Moreover, comparison of ethanol responses between BK channels in a cell-free, native membrane environment, recombinant slo1 channels in cell-free membrane patches following heterologous expression, and recombinant or native BK channels following channel-forming protein reconstitution into artificial bilayers of controlled lipid composition (reviewed in Brodie et al., 2007; Dopico et al., 2014) indicates that the molecular factors determining the final BK current response to ethanol are inherent to the BK channel protein complex itself (participating channel subunits and their posttranslational modification), the lipid microenvironment of the BK channel complex, and, fundamentally, dependent on the presence and amount of $Ca^{2+}{}_i$ nearby the BK channel Ca^{2+} sensors. These factors have been discussed in detail elsewhere (Dopico et al., 2014) and summarized later.

3.1 Activating Ions

Overwhelming amount of data show that brief ethanol exposure of homomeric slo1 channels in ethanol-naïve systems results in increased channel steady-state activity/macroscopic current provided that $Ca^{2+}{}_i$ is present at low μM levels (reviewed in Brodie et al., 2007; Dopico et al., 2014; Mulholland et al., 2009). *In the absence of activating* $Ca^{2+}{}_i$, that is, when the slo1 channel is gated by voltage/$Mg^{2+}{}_i$, slo1 channels become ethanol resistant even when probed with fully effective activating ethanol concentrations (100 mM; Liu et al., 2008). Moreover, combination of amino acid substitutions (5D5N) in the mslo1 CTD "Ca^{2+}-bowl" and in the high-affinity regulator of conductance for K^+ (RCK) 1 domain (D362A, D367A), which render both high-affinity $Ca^{2+}{}_i$-recognition sites nonfunctional, results in a channel that is ethanol resistant (Liu et al., 2008). In contrast, mutations that render the low-affinity, divalent $\left(Mg^{2+}{}_i\right)$ -recognition site nonfunctional do not have any impact on the ethanol responsiveness of mslo1 channels (Liu et al., 2008). Consistently with the in vitro results on Ca^{2+}-bowl and high-affinity RCK1 mslo1 mutants, in vivo studies on transgenic *C. elegans* show that WT human slo gene insertion into slo1-null worm mutants rescues the egg-laying and locomotion responses to ethanol (which are blunted in the slo1-null worms) whereas insertion in the same genetic background of hslo-containing mutations that render *both* Ca^{2+}-bowl and high-affinity RCK1 nonfunctional did not (Davis et al., 2015). It is noteworthy that insertion of equivalent cslo1 mutants on a slo1-null background renders a different outcome (see Davis et al., 2015 for discussion). In vitro data from mslo1 channels also demonstrate that amino acid substitutions that hamper *either* the Ca^{2+}-bowl *or* the high-affinity RCK1 site do not modify the ethanol sensitivity of BK channels. Thus, as far as $Ca^{2+}{}_i$ is able to interact with one of its high-affinity, physiological recognition sites present in the slo1 protein CTD, BK channels are ethanol-sensitive (Liu et al., 2008). Consistent with the electrophysiological results from mutated mslo1 channels mentioned earlier and with earlier data demonstrating that ethanol fails to modify the rate-limiting slope of the $V - \log NP_o$ relationship (a crude reader of the BK channel voltage dependence; Dopico et al., 1998, 1996), a 10-state allosteric gating model underscored that ethanol modified only Ca^{2+}-dependent parameters, such as the channel Ca^{2+} dissociation constants for both open (K_O) and closed (K_C) BK channel conformations. In contrast, Ca^{2+}-independent parameters, such as the equivalent gating charge associated with the BK closed-to-open equilibrium (Q) and the closed-to-open equilibrium constant in the absence of $Ca^{2+}{}_i$ and transmembrane (TM) voltage (L_o, an index of "intrinsic gating"), remained unchanged (Liu et al., 2008).

Ethanol action on slo1 channels is not only dependent on the presence of $Ca^{2+}{}_i$ but also modulated by the amount of this divalent sensed by the channel. Ethanol potentiation of P_o and macroscopic current progressively diminishes as concentration of $Ca^{2+}{}_i$ increases from high nM to low μM levels, finally turning to inhibition of slo1 activity when the divalent reaches several tens of μM to close to 1 mM. For homomeric slo1 channels, whether mslo1 (slo1 cloned from mouse brain; mbr5 variant) or cbv1 (slo1 cloned from rat cerebral artery myocytes), the ethanol-induced activation-to-inhibition "crossover" occurs at ~20 μM $Ca^{2+}{}_i$ (Bukiya, Liu, et al., 2009; Liu et al., 2008), this crossover being heavily modulated by BK regulatory β_1 subunits (see later). An empiric single-channel kinetic model revealed that ethanol-induced inhibition of slo1 activity was related to ethanol-induced facilitation of channel dwelling into $Ca^{2+}{}_i$-driven low P_o modes (Liu et al., 2008), which can be interpreted as ethanol facilitating $Ca^{2+}{}_i$-driven slo1 channel desensitization (Dopico & Lovinger, 2009). A very simple scheme summarizing our current understanding of ethanol and the basic processes that gate the slo1 channel is given in Fig. 3. The gating of BK channels, however, is rather complex: while many parameters of voltage-, intrinsic-, and Ca^{2+}-driven gating can be obtained, full understanding of the slo1 gating behavior requires incorporation of allosteric factors that couple the different equilibria between channel closed–open states, inactive–active voltage sensors, and Ca^{2+} bound–unbound Ca^{2+}-sensing sites on slo1 CTDs of the channel-forming tetramer (Cox & Aldrich, 2000; Horrigan & Aldrich, 1999, 2002). Which specific processes are disrupted by ethanol resulting in modification of channel function and BK current is under current investigation in our laboratory. The structural basis of ethanol activation of slo1 channels and its relation to $Ca^{2+}{}_i$ is provided under Section 3.2.

Ethanol is able to perturb Ca^{2+}-driven gating of slo1 and *Methanobacterium thermoautotrophicum* (MthK) channels. However, it neither modifies the gating of slo1-related channels (slo2, slo3) by Na^+/Cl^- or H^+/OH^- (Liu et al., 2013) nor the gating of slo1 itself by $Mg^{2+}{}_i$ (ie, in the absence of Ca^{2+}) (Liu et al., 2008). A recent report, however, has shown that *in the presence of* Ca^{2+}, $Mg^{2+}{}_i$ modulates ethanol action on BK currents in cultured hippocampal neurons: this modulation is bimodal, as at $Mg^{2+}{}_i \leq 200\,\mu M$ ethanol decreases BK activity whereas activity increases as $Mg^{2+}{}_i$ reaches

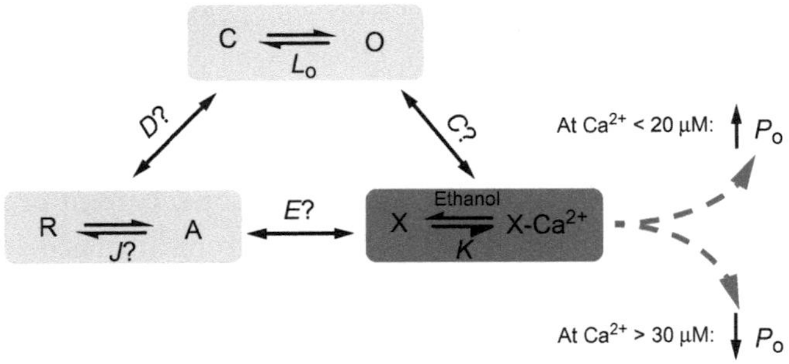

Fig. 3 Cartoon illustrating our current interpretation of ethanol modulation of slo1 channel steady-state activity (P_o). Slo1 channel gating is basically presented as the integration of three allosterically coupled equilibria: channel close (C)–open (O) transitions in the absence of voltage and $Ca^{2+}{}_i$ with "L_o" being the equilibrium constant; transition between resting (R)-activated (A) stages of the voltage sensor, with *J* as the equilibrium constant, and $Ca^{2+}{}_i$-free–$Ca^{2+}{}_i$-bound channel, with *K* being the equilibrium constant. Allosteric coupling factors between equilibria are given by *C*, *D*, and *E*, following Horrigan and Aldrich (2002). Ethanol cannot modify slo1 P_o in the absence of $Ca^{2+}{}_i$ or activating voltage (ie, ethanol does not modify L_o; Liu et al., 2008). While the ethanol sensitivity of *J* remains undetermined, crude parameters of voltage sensing seem to indicate that voltage sensing is not altered by ethanol. In contrast, ethanol modifies *K*, leading to increase or decrease in P_o dependent upon $Ca^{2+}{}_i$ levels (Bukiya, Liu, et al., 2009; Bukiya, Vaithianathan, et al., 2009; Liu et al., 2008). Ethanol-sensitive and -resistant equilibria are given in *red* (*dark gray* in the print version) and *light green* (*light gray* in the print version) boxes, respectively. *Questions marks* underscore that the ethanol sensitivity of key gating parameters in this allosteric model remains unexplored.

1 mM (Marrero et al., 2015). The study seems to underscore that Mg^{2+} direct interaction with slo1 channels does not influence ethanol action, as the reported bimodal effect of ethanol as function of $Mg^{2+}{}_i$ levels only occurs in the presence of a preserved intracellular actin: destabilization of actin at the cytoplasmic side of the patch results in $Mg^{2+}{}_i$-independent activation of BK currents by ethanol. In contrast, presence of an actin stabilizer leads to ethanol-induced BK current inhibition even at 1 mM $Mg^{2+}{}_i$ (Marrero et al., 2015). $Mg^{2+}{}_i$ modulation of ethanol action is of particular relevance as $Mg^{2+}{}_i$ has been reported to be reduced in both experimental systems exposed to ethanol and in intracellular compartments and serum of alcoholics (Altura & Altura, 1999; Babu, Cheng, Zhang, Altura, & Altura, 1999; Brown, Ilg, Chen, & Ren, 2002; Li et al., 2001; Poikolainen & Alho, 2008; Romani, 2008).

3.2 Slo1 Protein Isoforms

Ethanol modulation of BK channel activity has been consistently reported in vitro using highly simplified model systems (reviewed in Brodie et al., 2007; Dopico et al., 2014; Mulholland et al., 2009). In a most reductionist approach, ethanol-induced channel activation has been documented after reconstituting hslo1 (cloned from human brain) proteins in an artificial planar bilayer made of two phospholipid species: phosphoglyceride, ie, 1-palmitoyl-2-oleoyl-*sn*-glycero-3-phosphoethanolamine (POPE) and 1-palmitoyl-2-oleoyl phosphatidylserine (POPS; Crowley et al., 2003), indicating that this very simple proteolipid system must contain an ethanol sensor(s) that leads to modification of BK current. Indeed, a combination of electrophysiological studies on slo1 and related channels, including its prokaryotic analog MthK and chimeric slo channels (Liu et al., 2013), followed by ethanol probing of point mutants, and MM data based on crystallography of the slo1 CTD led to the identification of a protein pocket of discrete dimensions in the slo1 protein that may dock ethanol leading to increased channel activity (Bukiya et al., 2014). The ethanol pocket (Fig. 1) involves: (1) Lys 361 forming hydrogen bond with ethanol; (2) net positive charge provided by R514 near the ethanol–Lys 361 interaction; (3) Glu 354, Ser 357, and Asn 358 allowing ethanol access to its Lys 361 bonding partner. Interestingly, the T352I substitution in cslo1 or hslo1 (introduced into *C. elegans*) abolishes both ethanol-induced channel activation and introduces resistance to ethanol modification of slo1-mediated behavior in *C. elegans* (Davis et al., 2014). MM shows that Thr 352 is near the identified ethanol pocket, yet it points away from the pocket itself (A. N. Bukiya & A. M. Dopico, unpublished). A possible mechanism for this residue to control the ethanol response of BK channels has been recently discussed elsewhere (Dopico et al., 2014).

The ethanol site briefly described in the previous paragraph contains some residues (most notably, Thr352) that are highly conserved among species, from *C. elegans* and *D. melanogaster* to humans (Davis et al., 2014). However, not all BK channel-forming slo1 isoforms respond similarly to a first exposure to clinically relevant concentrations of ethanol. For instance, ALCOREX-slo1 channels, ie, a slo1 splice variant that contains exon 29 (ALCOhol-Regulated EXon) showed greatest ethanol sensitivity in both magnitude and persistence of potentiation by ethanol. However, INSERTLESS-slo1 channels, ie, completely spliced-out variant (thus, having neither ALCOREX nor STREX exons), were potentiated by ethanol in a reduced and briefer fashion. Finally, STREX-slo1 channels, ie, a splice variant that contains the so-called "STress axis-Regulated EXon," were completely insensitive to ethanol (Pietrzykowski et al., 2008). In addition, bslo channels (cloned from bovine aortic smooth muscle) are inhibited rather than activated by 10–100 mM ethanol (Dopico, 2003). In the latter, however, the presence of a Thr residue targeted by phosphorylation in the S0–S1 cytosolic loop seems to determine this BK channel's response to ethanol (see Section 3.4).

3.3 Channel Regulatory β Subunits

In most mammalian tissues, BK channel-forming slo1 proteins are tightly associated with regulatory proteins termed BK β subunits. These subunits are much smaller than slo1 proteins as they consist of two short intracellular ends and two TM segments joined by an extracellular (EC) loop. Four isoforms have been characterized, each coded by a different gene (*KCNMB1-4*). BK β subunits are characterized by rather specific tissue expression, with β1 and β4 prevailing in smooth muscle cells and CNS neurons, respectively (Orio, Rojas, Ferreira, & Latorre, 2002; see Chapter "Modulation of BK Channel Function by Auxiliary Beta and Gamma Subunits" by Yan and Li). Remarkably, BK β expression introduces drastic changes in the BK channel phenotype, some of these changes suited to serve physiology in a cell-specific manner. For example, association of β1 with slo1 subunits drastically increases the apparent Ca^{2+} sensitivity of the channel complex, an action not exerted by β4 (Behrens et al., 2000; Brenner, Jegla, Wickenden, Liu, & Aldrich, 2000; Kuntamallappanavar, Toro, & Dopico, 2014). This β1 action allows smooth muscle BK channels to activate in response to physiological levels of $Ca^{2+}{}_i$ reached near the BK channel during myocyte contraction (Perez, Bonev, & Nelson, 2001; see Chapter "BK Channels in the Vascular System" by Krishnamoorthy-Natarajan and Koide). Thus, BK channel activation renders STOCs, which negatively feedback onto depolarization-driven $Ca^{2+}{}_i$ influx, limiting cell contraction and favoring smooth muscle relaxation. Indeed, physiological and pharmacological activators of BK channels that act via recognition by distinct site mapped to β1 TM2 serve as effective vasodilators both in vitro and in vivo (Bukiya, Singh, et al., 2011; Bukiya, Vaithianathan, et al., 2009). Considering that β1 drastically modifies Ca^{2+}-driven gating parameters of the BK complex and ethanol modulation of BK channels is coupled to Ca^{2+}-driven gating (see earlier), it is not surprising that coexpression of β1 subunits exerts a powerful influence on ethanol modulation of BK current. Ethanol studies on slo1 (cbv1) and β1 subunits cloned from rat cerebrovascular myocytes demonstrated that β1 subunits shift the ethanol-induced activation-to-inhibition "crossover" toward lower $Ca^{2+}{}_i$ (≤3 μM; Bukiya, Liu, et al., 2009). Consistent with its lack of modulation on the channel apparent Ca^{2+} sensitivity, β4 subunits fail to modify such activation-to-inhibition crossover in mslo1 (Liu et al., 2008). Thus, β1 modification of ethanol action results in ethanol reduction of BK channel activity when evaluated at low μM $Ca^{2+}{}_i$ (Bukiya, Liu, et al., 2009), an ethanol action reported with the native cerebrovascular channel (Liu et al., 2004). Eventually, intoxicating levels of ethanol (50 mM) constrict both rat and WT mouse cerebral arteries while failing to do so in arteries from *KCNMB1* knockout (K/O) mice (Bukiya, Liu, et al., 2009; Bukiya, Vaithianathan, et al., 2011).

Characterization of BK channel in nerve terminals vs somata of rat supraoptic neurons revealed two distinct phenotypes displaying ethanol sensitivity/charybdotoxin (Cbtx) and Ibtx resistance and ethanol resistance/ Cbtx and Ibtx sensitivity, respectively (Dopico, Chu, et al., 1999). This pharmacological profile could be attributed to differential expression of BK β4 subunits between the different neuronal compartments, as heteromeric BK channels containing the neuronally abundant β4 subunit are resistant to the BK channel blockers Cbtx and Ibtx (Meera, Wallner, & Toro, 2000). Indeed, BK channels from supraoptic nerve terminals are Cbtx/Ibtx resistant (up to 360 nM Cbtx; Wynne et al., 2009). In contrast, somatic channels are blocked by ~100 nM Cbtx/Ibtx. These channels are insensitive to ethanol whereas their nerve terminal counterparts are potentiated by ethanol (Dopico, Chu, et al., 1999; Dopico, Widmer, et al., 1999; Wynne et al., 2009). Finally, somatic BK channel kinetics is fast ("type I") whereas their nerve terminal counterparts are slow ("type II"; Dopico, Chu, et al., 1999; Dopico, Widmer, et al., 1999). Based on the biophysical and pharmacological properties of the somatic and neuroterminal BK channels and by immunehistochemical examination, Wynne et al. (2009) confirmed the expression of β1 and β4 subunits in somatic and nerve terminal compartments, respectively. Thus, the differential ethanol response of BK channels in different domains of supraoptic neurons has been attributed to this selective distribution of β1 and β4 subunits in different neuronal compartments (Wynne et al., 2009).

In rat NAcc MSN, single-channel recordings obtained from BK channels in the somata showed a phenotype similar to that of slo1 channels coexpressed with β4 subunits while those in the dendritic domain of these neurons showed a phenotype similar to that of slo1 channels coexpressed with β1. Remarkably, ethanol (10–50 mM) increased BK P_o in MSN somata while failing to do so in their dendritic counterparts (Martin et al., 2008). Collectively, results from different excitable tissues (vascular smooth muscle, neurons, and neuronal domains) point at β subunits as critical modulators of BK channel pharmacology, and acute ethanol sensitivity in particular. The role of BK β subunits in adaptive responses of BK channels to protracted/repeated ethanol exposure is discussed in the next section.

3.4 Posttranslational Modifications of BK Subunits

Abundant literature documents that ethanol modulation of BK currents is regulated by phosphorylation or dephosphorylation, with the channel slo1 subunits being the likely targets of such posttranslational modifications. The possible contribution of phosphorylation process in ethanol action was first reported in GH3 cells, where ethanol-induced potentiation of BK channel activity was abolished by protein kinase C (PKC) inhibition and bolstered by phosphatase (PP) inhibitors (Jakab et al., 1997). In GH4/C1 and GH4/C1-STREX cells, ethanol-induced increase in BK steady-state activity was observed only in some membrane patches; authors attributed this variability to posttranslational modification of slo1 proteins. Indeed, PKC blockers diminished ethanol potentiation of BK channel activity in GH4/C1 cells. However, PKC blockers had no effect on GH4/C1-STREX cells (Jakab et al., 2006, 1997). Further investigation demonstrated that ethanol-mediated inhibition of STREX-containing BK channels is PKA dependent. In the absence of PKA inhibitor (14–22 Amide), ethanol exposure decreased activity of BK-STREX channels. Block of BK-STREX phosphorylation by PKA caused the potentiation of the channel by ethanol (Brodie et al., 2007; Pietrzykowski et al., 2008).

As mentioned under "*Slo1 protein isoforms*," bslo1 channels are inhibited by 10–100 mM ethanol when studied in the same expression system and under the same conditions in which the drug evokes activation of mslo1 channels (Dopico, 2003; Liu, Liu, Huang, Li, & Dopico, 2003). Bslo1 protein distinctly includes a Thr residue in the S0–S1 cytosolic loop (bslo position 107). Incremental phosphorylation of channel subunits on this residue by Ca^{2+}/calmodulin-dependent protein kinase II (CaMKII) progressively increases channel P_o and also gradually switches the channel's ethanol responses from robust activation to inhibition (Liu, Asuncion-Chin, Liu, & Dopico, 2006). Thus, CaMKII phosphorylation of bslo1 Thr107 seems to override ethanol allosteric coupling to channel activation by physiological levels of $Ca^{2+}{}_i$. It is important to note that Thr107 is not conserved in mslo1, hslo1, or cbv1; all these isoforms contain nonphosphorylatable residues and are activated by ethanol at nM to low μM $Ca^{2+}{}_i$ (reviewed in Brodie et al., 2007; Dopico et al., 2014).

A recent investigation examining the influence of PKA, CaMKII, and PP on ethanol actions on slo1 ± β4 channels in HEK293 cells and native BK channels in NAcc MSN has revealed that β4 subunits drastically alter the effects of PKA, CaMKII, and PP (Velazquez-Marrero, Seale, Treistman, & Martin, 2014). This study clearly demonstrates that PKA potentiates BK steady-state activity, an action that is completely blocked by β4 subunit expression. In addition, in the presence of this subunit, PKA modulation of the BK channel is not necessary for potentiation by ethanol. In contrast, homomeric BK channels are tonically inhibited by CaMKII, an effect that is not blocked by β4 subunits. The study also shows that in the presence of CaMKIIN, a specific CaMKII inhibitor, slo1 + β4 channels develop ethanol tolerance; this type of tolerance was reported in *KCNMB4* K/O mouse but absent in WT (β4 subunit-containing) BK channels (Martin et al., 2008). However, Velazquez-Marrero et al. (2014) document that coexpression of β4 subunits does not modulate ethanol action on homomeric BK channels, confirming previous data from recombinant mslo1 channels (Liu et al., 2008). On the other hand, phosphatases inhibit homomeric slo1 channels, an action that is reversed by β4 subunit coexpression. In turn, slo1 channel's rapid tolerance to ethanol is reversed by phosphatase inhibition (Velazquez-Marrero et al., 2014). The molecular mechanisms by which BK channel β1 and β4 participate in phosphorylation/dephosphorylation regulation of ethanol action on BK channels remain to be elucidated. The role of phosphorylation/dephosphorylation processes in BK channel adaptation to protracted/repeated ethanol exposure is discussed in Section 4.

3.5 Lipid Microenvironment of the BK Channel Complex

The role of membrane lipids in modification of ethanol's final effect on BK steady-state activity has been reported in many studies (reviewed by Bettinger & Davies, 2014; Dopico et al., 2014; Dopico, Bukiya, & Singh, 2012). Overall, bilayer thickness, lipid molecule "effective shape," and the presence of distinct lipid species in the bilayer all play critical roles in modulating ethanol action on BK P_o.

Modification of bilayer thickness by changing the lipid acyl chain length affects the time course of the acute response to ethanol and can turn ethanol-induced activation into inhibition: while ethanol strongly potentiated hslo1 channels embedded in a thin bilayer, this alcohol inhibited these channels when probed in a thicker bilayer (Yuan, O'Connell, Wilson, Pietrzykowski, & Treistman, 2008). Insights on the mechanisms by which bilayer thickness affects ethanol modulation of BK function will become more accessible from our growing understanding of the structural bases of slo1 channel gating. In slo1 proteins, the linker that connects the S6 gate to the RCK domains forms a passive spring connected to the gating ring that is involved in $Ca^{2+}{}_{i}$-dependent activation (Niu, Qian, & Magleby, 2004). A simple mechanical model to explain ethanol modulation of channel function by bilayer thickness has been hypothesized: lateral stress (which tends to decrease P_o) is counteracted by local hydrophobic mismatch between membrane lipids and the slo1 protein and by forces generated by Ca^{2+} binding to the gating ring and communicated to the gate itself via the spring, which leads to increased P_o in thick bilayers (Yuan, O'Connell, Jacob, Mason, & Treistman, 2007). In the presence of ethanol, *Ca^{2+} activation* of BK channels appears reduced, as determined from the $V_{0.5}$–Ca^{2+} plots (Bukiya, Liu, et al., 2009; Liu et al., 2008). It is possible to speculate that ethanol-reduced action in thicker bilayers results from ethanol-induced perturbation of Ca^{2+} binding (in particular at tens of μM Ca^{2+}) and/or of coupling between Ca^{2+} binding and the gate, probably via perturbation of S6-gate spring function.

Modulation of ethanol action on BK channel activity by the lipid "effective shape" is independent of lipid head net charge. Thus "type I" lipids (those with a polar head cross-sectional area larger than the tail area), such as phosphatidylserine, introduce "positive monolayer curvature" and favor ethanol-induced activation of hslo1 channels following their reconstitution into planar lipid bilayers. In contrast, ethanol-induced potentiation of hslo1 channels is blunted by "type II" lipids (those with a polar head cross-sectional area smaller than the hydrophobic tails/rings), such as phosphatidylglycerol and cholesterol (CLR; Crowley et al., 2003; Crowley, Treistman, & Dopico, 2005). Remarkably, ethanol and CLR change channel dwell times in opposite ways: while ethanol increases mean open time and decreases mean closed time (both actions leading to increased BK P_o) CLR decreases mean open time and increases mean closed time. Although dwell times represent an aggregate reflection of multiple processes that underlie channel gating, the most straightforward explanation for CLR functional antagonism on ethanol-induced BK current potentiation is that CLR and ethanol share a same target(s) that they modulate in opposite directions. Considering that ethanol can be conceptualized as a "type I" molecule, authors have speculated that the reduced ethanol activation of BK current in the presence of CLR in a POPE/POPS bilayer was due to opposing effects of "type I" ethanol molecules and "type II" CLR molecules on bilayer physical properties (Crowley et al., 2003, 2005). Other possibilities cannot be excluded: CLR antagonism of slo1 channel potentiation by ethanol may be attributed to CLR making it more difficult for ethanol molecules to partition into the bilayer and thus reaching eventual channel subunit targets. This partitioning-lipid phase hypothesis, modification of bilayer physical properties by CLR and ethanol, and possible CLR–ethanol interaction at common BK protein target sites are discussed in detail elsewhere (Dopico et al., 2012).

Remarkably, BK channels made of channel-forming cbv1 and β1 subunits in CLR-free bilayers show resistance to 50 mM ethanol. Inclusion of 23 mol% CLR into the lipid mixture results in ethanol-induced BK channel inhibition (Bukiya, Vaithianathan, et al., 2011). Collectively, studies from homomeric and heteromeric BK channels in lipid bilayers where CLR can be titrated show that CLR blunts ethanol-induced activation (homomeric slo1 channels) or turns channel refractoriness to inhibition by ethanol (slo1 + β1 channels). Thus, ethanol seems to require CLR presence to inhibit β1 subunit-containing BK channel. This conclusion obtained from experiments in artificial lipid bilayers is supported by data obtained in patches from rat and mouse cerebral artery myocytes that contained native (cbv1 + β1) BK channel complexes. Ethanol-induced cerebral artery constriction that results from ethanol-induced BK channel inhibition is also ablated in rat and mouse arteries following CLR depletion (Bukiya, Vaithianathan, et al., 2011).

The molecular mechanism(s) underlying ethanol–CLR interactions on BK channels remain unclear. Remarkably, ethanol–CLR antagonism on slo1 channels could not be observed when CLR was substituted by its enatiomer, *ent*-CLR (Yuan, Chen, Covey, Johnston, & Treistman, 2011), suggesting that CLR modulation of ethanol action on BK channels involves specific CLR–protein interactions. Both ethanol-recognition (see earlier) and CLR-sensing sites have been mapped to the CTD of slo1 proteins, yet they do not overlap. CLR-sensing areas include seven CLR recognition amino acid consensus (CRAC) domains, with CRAC4 (the domain adjacent to the inner membrane leaflet where CLR is particularly abundant) playing a major role in the channel's CLR sensitivity (Singh et al., 2012). How CLR-sensing sites and the ethanol-docking site described earlier may interact with each other upon respective ligand recognition remains unknown. Finally, it should be underscored that BK channels cluster in CLR-enriched membrane domains, eg, "rafts" (reviewed in Dopico et al., 2012). It has been speculated that BK channels may redistribute in cell membranes upon chronic ethanol exposure (Crowley et al., 2003; Yuan et al., 2008). The role of lipids in BK channel plasticity in response to ethanol is reviewed in the next section.

4. ADAPTATIONS IN BK CHANNEL/CODING GENES TO PROTRACTED OR REPEATED ETHANOL EXPOSURE: FROM MOLECULAR MECHANISMS TO MODIFICATION OF ETHANOL-INDUCED PERTURBATION OF BEHAVIOR

Adaptation(s) in the response of a biological system that results in reduced response to a given stimulus is referred to as "tolerance." When the stimulus is a drug (eg, ethanol exposure) tolerance is usually associated with repeated exposure to that drug (Chandler, Harris, & Crews, 1998). Some individuals, however, can be relatively resistant or tolerant to ethanol compared to others when first exposed to this alcohol. Thus, from a time frame point of view, several types of tolerance to ethanol have been distinguished: "innate/intrinsic" tolerance, also termed "low level of response to ethanol" (Schuckit, 2000), can be described as a reduced response to ethanol from an ethanol-naïve system, ie, the response of the individual is significantly smaller than that of the normal population. The terms "acute" or "within-session" tolerance have also been used to describe a reduced response to ethanol that develops within a single exposure or single episode of intoxication (in case of behavioral evaluation) and typically occurs within minutes of the initial exposure (Le, 1990; Tabakoff, Cornell, & Hoffman, 1986). In turn, development of tolerance during a second (or more) ethanol exposure 8–24 h after the previous exposure has been traditionally defined as "rapid tolerance" (Khanna, Chau, & Shah, 1996). Finally, "chronic tolerance" has been applied to a reduced response to ethanol following repeated or protracted exposure to ethanol over a period of days or weeks (Khanna et al., 1996). Disregarding time frame, the two major "branches" of pharmacology, pharmacokinetics and pharmacodynamics, offer several explanations for developing tolerance to ethanol.

The former involves modifications in absorption, distribution, metabolism, and/or excretion (ADME) of ethanol by the organism. To our knowledge, there are no studies documenting that specific changes in ADME at the organism level contribute to reduced response of BK channels to ethanol. On the other hand, "pharmacodynamics-driven" tolerance involves adaptation(s) occurring in one or (usually for a ligand with simple chemical structure such as ethanol) many "sites of action" whether at genetic, epigenetic, or posttranslational levels. In this section, we will briefly summarize the literature dealing with "pharmacodynamics-based" tolerance in the BK channel–ethanol interaction, expanding from subcellular to behavioral studies. The latter have been recently described in detail elsewhere (Bettinger & Davies, 2014). It should be pointed out from the beginning that the contribution of memory, operant, associative, or other forms of learning (Kalant, 1998) to tolerance in ethanol–BK channel interactions remains to be established. In addition to the time-based terminology that categorizes tolerance, we will use the term "molecular tolerance" (Treistman & Martin, 2009) when plasticity in the interactions between BK channel proteins/coding gene(s) and ethanol has been resolved at the subcellular level but several macromolecules seem to participate in such adaptation and/or the role of a unique ethanol–macromolecule interaction has not yet been determined. Within the concept of "molecular tolerance," if a given macromolecule that had responded to ethanol first application was unequivocally identified as mediator or final effector in the ethanol adaptive response, the term "desensitization" should be applied (Dopico & Lovinger, 2009). Whichever the time frame and level of resolution, identification of the mechanisms that underlie tolerance/resistance to ethanol exposure acquires particular relevance as tolerance to ethanol in humans has been advanced as a predictor for the likelihood of developing alcoholism (Schuckit, 2000; Schuckit, Klein, Twitchell, & Smith, 1994).

Critical advances in the identification of targets and mechanisms that contribute to development and/or maintenance of behavioral tolerance to ethanol have been initially limited to invertebrate model systems, such as *D. melanogaster* and *C. elegans*; these species are amenable to wide genetic manipulation, characterized by short generation times, and present neural circuits that are simpler and better characterized than those of mammals. Rapid tolerance, measured as a reduction in the duration of the sedative phase that follows a single exposure to ethanol vapor, has been documented in WT *D. melanogaster* but not in *Slowpoke* (*Slo*; ortholog to *Slo1* or *KCNMA1* in mammals) null flies (Cowmeadow et al., 2005). In these experiments, flies were sedated by ethanol vapor within 20 min of vapor introduction. After full recovery in fresh air, vapor was delivered again. After switching ethanol vapor to fresh air again, the percentage of flies recovering from sedation every minute was used to characterize the tolerance. Remarkably, block of *Slo* expression in the nervous system prevented development of rapid tolerance (Cowmeadow et al., 2005). On the other hand, ethanol-induced sedation increased Slo transcripts when using neuronal promoters but not when using muscle or tracheal cell promoters (Cowmeadow et al., 2006). Increased neuronal Slo expression and thus, BK channel function allows for shorter neuron refractory periods, which results in enhanced neuronal excitability and faster recovery from ethanol sedation (Ghezzi et al., 2010). More recent work in *Drosophila* unveiled DNA elements responsible for modulation of BK channel expression following alcohol (benzyl alcohol, in this case) exposure (Li, Ghezzi, Pohl, Bohm, & Atkinson, 2013). In this set of experiments, flies were exposed to benzyl alcohol-coated vials until fully sedated within minutes. Following recovery, flies were again exposed to benzyl alcohol 24 h later. Histone profile modification in response to benzyl alcohol suggests that increased BK channel expression in neurons of benzyl alcohol-exposed flies is related to the 6b element (60 nt). Moreover, genomic deletion of 6b DNA results in a phenotype with prolonged tolerance and increased symptoms of withdrawal. However, this deletion does not alter other *Slo*-dependent behaviors (Li et al., 2013). Moreover, cyclic AMP response element-binding (CREB) proteins may be involved in *Drosophila*'s tolerance to benzyl alcohol. Indeed, sedation by this alcohol alters expression of dCREB-A and dCREB2-b genes, which results in increased production of positively acting CREB isoforms while reducing the presence of negatively acting CREB isoforms. Furthermore, a loss-of-function allele of dCREB2 ablates the increase in BK channel expression and eventual tolerance following exposure to benzyl alcohol (Wang, Ghezzi, Yin, & Atkinson, 2009).

In mammals (rats), tolerance in response to 20- to 40-day-long ethanol intake (rendering 30 mM BAL) has been measured as a reduction in both activation of BK channels and inhibition of vasopressin and oxytocin release by dissociated neurohypophysial terminals in response to acute ethanol challenge (Knott et al., 2000, 2002). Authors attributed the plasticity in the BK channel response associated with chronic ethanol exposure to a decrease in BK current density (Knott et al., 2002). Also in the rat species, combination of patch-clamp electrophysiology and immunolabeling against BK channel-forming slo1 subunits in hypothalamic–neurohypophysial explants exposed to ethanol in vitro shows two components of tolerance. The first is given by a decreased potentiation of BK steady-state current by ethanol. This component of tolerance was evident at 30 min of prolonged ethanol exposure and persisted throughout 24 h of exposure. The second component of tolerance consisted of decreased current density, this component having a delayed time course: it was first seen several hours after the initiation of prolonged exposure to ethanol. Regarding this second component, confocal microscopy showed reduction in BK channel protein clustering in the plasma membrane during ethanol exposure (Pietrzykowski et al., 2004). A recent study in primary cultures of rat hippocampal neurons unveiled a complex pattern of BK channel membrane expression in response to 25 mM ethanol applied for 10 min, 3, 6, and 24 h: total internal fluorescence (TIRF) microscopy showed that perimembrane channel expression increased during 10 min and 3 h applications, decreased after 6 h, and returned to preethanol levels after a day of ethanol exposure. In contrast, this recovery was not observed when BK perimembrane expression was determined after a 6-h-long ethanol exposure followed by an 18-h-long ethanol withdrawal period. On the other hand, electrophysiological data did not parallel the TIRF data at every time point, as current density was actually decreased following a 3-h-long ethanol exposure. Therefore, differences in BK channel membrane clustering (Palacio et al., 2015) and/or receptor desensitization likely contribute to the overall adaptive response of BK current density to protracted ethanol exposure.

In addition to internalization of membrane BK channels as a mechanism underlying rapid tolerance to ethanol, ethanol-driven changes in BK channel splice variant profile have been reported in hypothalamic–neurohypophysial explants and cultured striatal neurons exposed to ethanol in vitro. This rearrangement is controlled by microRNAs. Specifically, miR-9 mediates destabilization of BK mRNAs that contain 3′UTRs with a miR-9 recognition element, which leads to a switch between alternatively spliced BK mRNA variants. Thus, drop in the highly ethanol-sensitive ALCOREX/INSERTLESS and increase in the ethanol-resistant INSERTLESS/STREX variant are detected (Pietrzykowski et al., 2008). Moreover, data from cultured or acutely dissociated striatal neurons from mice revealed that longer exposure times to ethanol required longer periods for the BK channel to recover ethanol sensitivity. BK channel ethanol sensitivity after 6 h exposure to 20 mM ethanol was diminished beyond 24 h after withdrawal (Velazquez-Marrero et al., 2014). In contrast, after a 1- or 3-h-long exposure to ethanol, the ethanol sensitivity of BK channels had significantly recovered 4 h after the exposure. This time-controlled tolerance to ethanol exposure was paralleled by changes in the BK mRNA profile, with increased levels of the ethanol-insensitive STREX splice variant (Velazquez-Marrero et al., 2011).

Subunit composition of BK channel heteromers is another important determinant of the two components of ethanol tolerance identified by Pietrzykowski et al. (2004) (see earlier). Patch-clamp data from hslo1 channels expressed in HEK cells indicate that ethanol-driven downregulation of BK current density is found in homomeric hslo1 and heteromeric hslo1 + β4 but not in hslo1 + β1 channel complexes. Thus, β1 subunits decrease the "second component" of BK channel tolerance to acute challenge by ethanol. In contrast, "the first component" of ethanol tolerance has been observed in channels made of hslo1 homomers, hslo1 + β1, and hslo1 + β4 heteromers. Homomeric slo1 channels, however, displayed the greatest tolerance. Thus, β1 and β4 subunits seem to be similarly efficacious in blunting the first component of acute tolerance (Feinberg-Zadek, Martin, & Treistman, 2008). In a different experimental setting of protracted, but not repeated, challenge with ethanol (50 mM × 8 min), β4 subunit-containing BK complexes exhibit less tolerance to acute ethanol challenge than homomeric channels (consisting of α subunits only; Martin et al., 2008). Consistently, the BK channel in freshly isolated mouse striatal MSNs, which contains β4 subunits, exhibits little tolerance to acute ethanol challenge. In contrast, striatal MSNs neurons from *KCNMB4* K/O mouse do not exhibit ethanol tolerance (Martin et al., 2008). A recent study using similar experimental patch-clamp paradigm points at the ability of β4 subunit to influence regulation of BK channels by PKA and CaMKII as a mechanism for β4 subunit-mediated modulation of tolerance. Interestingly, slo1 channel's rapid tolerance to ethanol is reversed following PP inhibition. In addition, slo1 + β4 channels under pharmacological CaMKII inhibition develop ethanol tolerance (Velazquez-Marrero et al., 2014). The specific mechanisms engaged in the interplay between slo1, β1 and β4 subunits, kinases, phosphatases, and ethanol remain speculative. Considering that β4 subunit contains several phosphorylation substrate motifs, it is highly likely that phosphorylation of β4 subunits participates in ethanol tolerance of BK channels. It is also possible that β4 modulates adaptive responses of slo1 to ethanol via protein–protein interaction independently of phosphorylation processes.

The data described earlier from subcellular resolution experiments have been complemented by behavioral studies in rodents. Constitutive deficiency in BK β1 or β4 subunits in mice does not alter BAL nor sensitivity to ataxia, sedation, and hypothermia evoked by acute ethanol administration. Acute (ie, within-session) tolerance is not altered either. BK β1 deletion, however, reduces chronic tolerance (after repeated ethanol exposure) to sedation and hypothermia while BK β4 deletion does not affect these adaptations to chronic ethanol exposure (Kreifeldt, Cates-Gatto, Roberts, & Contet, 2015). The lack of effect of BK β4 deletion in this work is somewhat surprising in light of the work by Martin et al. (2008) described earlier. Authors speculate on two possibilities: (1) the intoxication conditions used in their study (7 cycles of 16-h-long vapor exposure) could have induced maximal tolerance and thus failed to capture the intermediate points of tolerance development, at which a role of β4 subunits could have been most prominent; (2) the role of β4 subunits in the development of tolerance in MSNs may be different from the role of these subunits in tolerance to sedation and hypothermia (Kreifeldt et al., 2015). Interestingly, the hyperexcitable state (determined by handling-induced convulsions) associated with ethanol withdrawal appears earlier in the absence of BK β1 subunits and lasts longer in the absence of BK β4 (Kreifeldt et al., 2015). These findings bolster the case for a critical role of BK channel subunit composition in behavioral adaptations to protracted ethanol exposure.

Genetic ablation of *KCNMB1* or *KCNMB4* not only affects neurological perturbations linked to protracted or repeated ethanol administration but also modulates escalation of ethanol drinking in mice. Remarkably, genetic deletion of β1 or β4 regulatory subunits (on a C57BL/6J background) altered voluntary ethanol drinking in ethanol-dependent mice while failing to do so in nondependent animals (Table 2). Indeed, whether mice were subjected to a continuous two-bottle choice followed by intermittent two-bottle choice paradigm, or trained to drink ethanol in a limited-access two-bottle choice paradigm, baseline ethanol drinking in WT, *KCNMB1* K/O, or *KCNMB4* K/O was all similar. In contrast, *KCNMB1* K/O and *KCNMB4* K/O, respectively, displayed an increased and decreased escalation of ethanol drinking during withdrawal from chronic intermittent ethanol administration (Kreifeldt, Le, Treistman, Koob, & Contet, 2013). Following the authors' arguments, it is conceivable that both β1-induced blunting of neuronal BK channel's responses to ethanol and β4-induced suppression of BK channel's acute tolerance to protracted ethanol exposure (see earlier) lead to the aforementioned changes in ethanol escalation in dependent mice. The mechanisms linking the in vitro observations to changes in drinking escalation, however, need to be established.

Table 2 Factors That Contribute to Changes in the Ethanol Response of BK Channels/BK Subunit-Coding Genes Following Protracted or Repeated Ethanol Administration

System	Type of Ethanol Administration	Factor(s), Variable(s), or Parameter(s)	Consequence on BK Current and/or Tissue Response	References
BK in HEK293 cells; lipid bilayer	50 mM ethanol for 15 min	Changes in bilayer bulk lipid composition	Loss of BK channel activation by acute ethanol	Yuan et al. (2008)
HEK293 cells	Two 2-min-long exposures to 25 mM ethanol separated by 10 min wash. Also: 25 mM ethanol for 24 h	Steady-state activity and current density (amplitude/capacitance) in BK channel heteromers	Rapid tolerance is robust in homomeric slo1 and in slo1 + β4. Both constructs and slo1 + β1 are all ethanol tolerant after 24 h ethanol.	Feinberg-Zádek et al. (2008)
Rat cultured striatal neurons	Vapor on cell culture dish (20 mM ethanol)	Measurement (PCR) of slo1 splice variants; BK steady-state activity	Major drops in ALCOREX and major increase in STREX; ethanol potentiation of current is greatly reduced	Pietrzykowski et al. (2008)
Striatal neurons of C57BL/6J mice	In vitro exposure to 20 mM ethanol for 1, 3, or 6 h	Slo1 transcripts	6-h-long ethanol exposure lead to increased STREX and delayed recovery from tolerance	Velazquez-Marrero et al. (2011)
Rat primary hippocampal neuronal cultures	25 mM ethanol applied for 10 min, 3, 6, and 24 h	BK channel current, BK membrane expression and clustering (TIRF microscopy)	Successive increases and decreases in BK membrane expression and reduction in active channels after 3 and 6 h of ethanol exposure	Palacio et al. (2015)

Hypothalamic–neurohypophysial explants (rat)	Ethanol vapor on culture (20 mM). Exposure times of 0.5, 1, 3, 6, and 24 h	Ion current response and channel membrane distribution and internalization	Reduced BK potentiation by ethanol within several minutes of exposure; BK internalization during hours-long exposure	Pietrzykowski et al. (2004)
Isolated rat neurohypophysial terminals	Ethanol-rich diet for 20–40 days (30 mM BAL)	Peptide release (radioimmunoassay)	Blunting of vasopressin release by 10 mM ethanol is reduced in rats on ethanol-rich diet	Knott et al. (2000)
Isolated rat neurohypophysial terminals	Ethanol-rich diet for 3–4 weeks (30–35 mM BAL by day 24 on diet)	BK current density (current amplitude/capacitance) and peptide release (radioimmunoassay)	Current potentiation and oxytocin release inhibition by ethanol are blunted in the chronic alcohol-diet group	Knott et al. (2002)
Mouse striatal medium spiny neurons	In vitro: 50 mM ethanol on neuron/slice for 8 min. In vivo drinking in the dark (DID): 10% ethanol bottles for 2 h	BK current in isolated neurons, spike patterning in brain slices, motor activity, and ethanol consumption (restricted access)	BK channels from WT mice do not exhibit tolerance, while β4 KO do. PKA, CAMKII, PP are involved. Rapid tolerance to ethanol locomotor effects develops in β4 KO but not in WT. β4 KO drink more than WT	Martin et al. (2008) and Velazquez-Marrero et al. (2014)

Continued

Table 2 Factors That Contribute to Changes in the Ethanol Response of BK Channels/BK Subunit-Coding Genes Following Protracted or Repeated Ethanol Administration—cont'd

System	Type of Ethanol Administration	Factor(s), Variable(s), or Parameter(s)	Consequence on BK Current and/or Tissue Response	References
BK β1-, BK β4-deficient mice	4 routes (16 h ethanol vapor/4 hr air); before each route: ethanol load (1.5 g/kg, i.p.); 43 mM BAL maintained with ethanol vapor	Motor coordination, righting reflex, body temperature, handling-induced convulsions after ethanol withdrawal	Appearance of hyperexcitable state driven by ethanol withdrawal is accelerated in BK β1 KO while delayed in BK β4 KO	Kreifeldt et al. (2015)
Drosophila melanogaster	Ethanol vapor twice; 4 h between sedations	*Slowpoke* gene expression in the nervous system	Decreased time for ethanol-induced sedation	Cowmeadow et al. (2005)
D. melanogaster	Double exposure to benzyl alcohol-coated vials; 24 h in-between	CREB transcription factor-stimulated gene expression/reduced CREB repressor	Rapid tolerance to sedation-induced *Slo* gene expression	Wang et al. (2009)
D. melanogaster	Benzyl alcohol-coated vials given 1, 4, 7, 14, 21, or 28 days after first exposure	Replacement of 6b element with loxP site in *Slo* transcription control area	Tolerance to benzyl alcohol even 28 days after first sedation episode	Li et al. (2013)
Caenorhabditis elegans	400 mM ethanol in media for 30 min (<25 mM in worm tissue)	CLR starvation, manipulation of triacylglycerol levels; locomotion	Decreased ability to develop fast tolerance as measured by locomotion assays	Bettinger, Leung, Bolling, Goldsmith, and Davies (2012)

BAL, blood ethanol level; *CaMKII*, Ca^{2+}-calmodulin kinase II; *PKA*, protein kinase A; *PP*, phosphatase; *TIRF*, total internal reflection fluorescence.

In summary, while details of β subunit-type modulation of ethanol tolerance are still under investigation, there is little doubt that BK subunits play a critical role in both acute and chronic behavioral tolerance to ethanol. Considering that BK β4 subunits are the prevalent β subunit type in the brain (Brenner et al., 2000) and that they control development of ethanol tolerance at molecular, cellular, and organ levels, BK β4 subunit may be postulated as a novel therapeutic target for alcoholism.

Data discussed earlier seem to indicate that a major component underlying acute tolerance to ethanol is limited to the BK protein complex, including channel subunit ability to disrupt modulation of channel function by kinases and phosphatases. However, this type of tolerance has been shown to be conditioned by membrane lipids as well. In *C. elegans*, regulation of acute tolerance by lipid environment has been studied in vivo. Worms that have been starved of exogenous CLR (which is necessary for their normal physiology) do not exhibit acute tolerance when placed into media containing 400 mM ethanol, which renders 22 mM ethanol from the worm's lysate (Davies et al., 2003). Thus, ethanol tolerance seems to require CLR presence (Bettinger et al., 2012). In addition, lips-7 (ok3110) worms, which carry a mutation in the triacylglycerol pathway lipase LIPS-7, are resistant to high dosage of ethanol in the environment, and are fast developers of acute (within 30 min) tolerance (Bettinger et al., 2012).

Data obtained in *C. elegans* are in agreement with in vitro data from hslo1 channel incorporated into the planar lipid bilayers. In this simplified setting, BK channels exhibit acute (within 10 min) tolerance to channel activation in response to 50 mM ethanol. An initial ethanol-induced BK channel activation (observed within 5 minutes) is favored by thin lipid bilayers, yet this activation vanishes and BK channel inhibition appears as bilayers consist of lipid species that form thicker bilayers (Yuan et al., 2008). Moreover, a sustained response to ethanol ("ethanol tolerance") is also modified by bilayer thickness: in thicker bilayers, a protracted ethanol exposure of 5–8 min leads to channel inhibition, the latter not being observed in thin bilayers (Yuan et al., 2008). In addition to possible pharmacodynamic explanations, authors hypothesized that modification of the channel's lipid environment might selectively alter the access of ethanol molecules to its site(s) in the channel protein, leading to differential ethanol effects on BK P_o (Yuan et al., 2008). Collectively, in vivo findings from *C. elegans* and in vitro findings with BK channels from human brain reconstituted into lipid bilayers of controlled proteolipid composition indicate that membrane lipids play an important role in shaping up the adaptive responses of slo1 channels to protracted ethanol administration.

5. CONCLUSIONS

During the last two decades, several studies have demonstrated that in vitro exposure of BK channels to ethanol concentrations reached in blood during alcohol intoxication in humans disrupts physiology. Recently, an ethanol-recognition pocket of discrete dimensions has been mapped to the BK channel-forming slo1 protein cloned from mouse brain, this region being highly conserved across species. The importance of this region in acute ethanol intoxication has been determined by evaluating cslo1 and hslo1 mutants in *C. elegans*. It is also clearly established that the final response of BK channels to ethanol action from ethanol-naïve systems, however, is determined by a wide variety of molecules that impact on BK channels. Moreover, the role of some of these factors in adaptive responses of BK channel proteins/coding genes to protracted and/or repeated ethanol exposure is beginning to emerge. Thus, tolerance to ethanol exposure seems to involve modifications at posttranslational, epigenetic, and genetic levels. Expression of *Slo* in *D. melanogaster* is linked to development of tolerance to ethanol intoxication. On the other hand, expression of its ortholog cSlo1 mediates ethanol-induced intoxication of *C. elegans*. While the different forms of ethanol administration might contribute to the differential role of *Slo* genes in ethanol-induced behavior, it is more likely that posttranslational, epigenetic, and/or genetic mechanisms play primary role in the ethanol response of the two species. Adding to this complexity, very recent studies in rodents underscore the impact of genetic manipulation of BK channel regulatory *β subunits* in ethanol-driven behaviors, including ethanol drinking in ethanol-dependent mice. The relative contribution of *KCNMA1* vs *KCNMB(1–4)* expression to ethanol-induced modification of behavior in mammals, however, remains undetermined. In addition, whether memory or other mental processes involved in learning contribute to tolerance in ethanol–BK channel interactions remains unknown. Finally, the contribution of specific pharmacokinetic processes to individual tolerance to ethanol effects on BK channels also remains to be explored. While major unknowns in the ethanol–BK channel field lie ahead, the importance of this area of investigation cannot be overstated. On one hand, identification of specific ethanol-recognition sites in effectors of ethanol action on BK channels is a critical first step for designing agents that could counteract modification of physiology elicited by ethanol action on BK channels. On the other hand, identification of mechanisms determining or, at least, contributing to BK channel–ethanol tolerance is of paramount importance as tolerance to ethanol has been proposed as a predictor of risk for developing alcoholism

ACKNOWLEDGMENT

This work has been supported by the National Institute of Alcohol Abuse and Alcoholism through Grants R37-AA11560 (A.M.D.) and R01-AA023764 (A.N.B.).

REFERENCES

Altura, B. M., & Altura, B. T. (1999). Association of alcohol in brain injury, headaches, and stroke with brain-tissue and serum levels of ionized magnesium: A review of recent findings and mechanisms of action. *Alcohol*, *19*(2), 119–130.

Babu, A. N., Cheng, T. P., Zhang, A., Altura, B. T., & Altura, B. M. (1999). Low concentrations of ethanol deplete type-2 astrocytes of intracellular free magnesium. *Brain Research Bulletin*, *50*(1), 59–62.

Behrens, R., Nolting, A., Reimann, F., Schwarz, M., Waldschutz, R., & Pongs, O. (2000). hKCNMB3 and hKCNMB4, cloning and characterization of two members of the large-conductance calcium-activated potassium channel beta subunit family. *FEBS Letters*, *474*(1), 99–106. Retrieved from, http://www.ncbi.nlm.nih.gov/pubmed/10828459.

Bettinger, J. C., & Davies, A. G. (2014). The role of the BK channel in ethanol response behaviors: Evidence from model organism and human studies. *Frontiers in Physiology*, *5*, 346. http://dx.doi.org/10.3389/fphys.2014.00346.

Bettinger, J. C., Leung, K., Bolling, M. H., Goldsmith, A. D., & Davies, A. G. (2012). Lipid environment modulates the development of acute tolerance to ethanol in Caenorhabditis elegans. *PloS One*. 7(5). e35192. http://dx.doi.org/10.1371/journal.pone.0035192.

Brenner, R., Jegla, T. J., Wickenden, A., Liu, Y., & Aldrich, R. W. (2000). Cloning and functional characterization of novel large conductance calcium-activated potassium channel beta subunits, hKCNMB3 and hKCNMB4. *The Journal of Biological Chemistry*, *275*(9), 6453–6461.

Brodie, M. S., Scholz, A., Weiger, T. M., & Dopico, A. M. (2007). Ethanol interactions with calcium-dependent potassium channels. *Alcoholism, Clinical and Experimental Research*, *31*(10), 1625–1632. http://dx.doi.org/10.1111/j.1530-0277.2007.00469.x.

Brown, R. A., Ilg, K. J., Chen, A. F., & Ren, J. (2002). Dietary Mg^{2+} supplementation restores impaired vasoactive responses in isolated rat aorta induced by chronic ethanol consumption. *European Journal of Pharmacology*, *442*(3), 241–250.

Bukiya, A. N., Kuntamallappanavar, G., Edwards, J., Singh, A. K., Shivakumar, B., & Dopico, A. M. (2014). An alcohol-sensing site in the calcium- and voltage-gated, large conductance potassium (BK) channel. *Proceedings of the National Academy of Sciences of the United States of America*, *111*(25), 9313–9318. http://dx.doi.org/10.1073/pnas.1317363111.

Bukiya, A. N., Liu, J., & Dopico, A. M. (2009). The BK channel accessory beta1 subunit determines alcohol-induced cerebrovascular constriction. *FEBS Letters*, *583*(17), 2779–2784. http://dx.doi.org/10.1016/j.febslet.2009.07.019.

Bukiya, A. N., Singh, A. K., Parrill, A. L., & Dopico, A. M. (2011). The steroid interaction site in transmembrane domain 2 of the large conductance, voltage- and calcium-gated potassium (BK) channel accessory beta1 subunit. *Proceedings of the National Academy of Sciences of the United States of America*, *108*(50), 20207–20212. http://dx.doi.org/10.1073/pnas.1112901108.

Bukiya, A. N., Vaithianathan, T., Kuntamallappanavar, G., Asuncion-Chin, M., & Dopico, A. M. (2011). Smooth muscle cholesterol enables BK beta1 subunit-mediated channel inhibition and subsequent vasoconstriction evoked by alcohol. *Arteriosclerosis, Thrombosis, and Vascular Biology*, *31*(11), 2410–2423. http://dx.doi.org/10.1161/atvbaha.111.233965.

Bukiya, A. N., Vaithianathan, T., Toro, L., & Dopico, A. M. (2009). Channel beta2-4 subunits fail to substitute for beta1 in sensitizing BK channels to lithocholate. *Biochemical and Biophysical Research Communications*, *390*(3), 995–1000. http://dx.doi.org/10.1016/j.bbrc.2009.10.091.

Chandler, L. J., Harris, R. A., & Crews, F. T. (1998). Ethanol tolerance and synaptic plasticity. *Trends in Pharmacological Sciences*, *19*(12), 491–495.

Chu, B., & Treistman, S. N. (1997). Modulation of two cloned potassium channels by 1-alkanols demonstrates different cutoffs. *Alcoholism, Clinical and Experimental Research*, *21*(6), 1103–1107.

Cowmeadow, R. B., Krishnan, H. R., & Atkinson, N. S. (2005). The slowpoke gene is necessary for rapid ethanol tolerance in Drosophila. *Alcoholism, Clinical and Experimental Research*, *29*(10), 1777–1786.

Cowmeadow, R. B., Krishnan, H. R., Ghezzi, A., Al'Hasan, Y. M., Wang, Y. Z., & Atkinson, N. S. (2006). Ethanol tolerance caused by slowpoke induction in Drosophila. *Alcoholism, Clinical and Experimental Research*, *30*(5), 745–753.

Cox, D. H., & Aldrich, R. W. (2000). Role of the beta1 subunit in large-conductance Ca^{2+}-activated K^+ channel gating energetics. Mechanisms of enhanced Ca^{2+} sensitivity. *The Journal of General Physiology*, *116*(3), 411–432.

Crowley, J. J., Treistman, S. N., & Dopico, A. M. (2003). Cholesterol antagonizes ethanol potentiation of human brain BKCa channels reconstituted into phospholipid bilayers. *Molecular Pharmacology*, *64*(2), 365–372. http://dx.doi.org/10.1124/mol.64.2.365.

Crowley, J. J., Treistman, S. N., & Dopico, A. M. (2005). Distinct structural features of phospholipids differentially determine ethanol sensitivity and basal function of BK channels. *Molecular Pharmacology*, *68*(1), 4–10. http://dx.doi.org/10.1124/mol.105.012971.

Davies, A. G., Pierce-Shimomura, J. T., Kim, H., VanHoven, M. K., Thiele, T. R., Bonci, A., … McIntire, S. L. (2003). A central role of the BK potassium channel in behavioral responses to ethanol in C. elegans. *Cell*, *115*(6), 655–666. Retrieved from, http://www.ncbi.nlm.nih.gov/pubmed/14675531.

Davis, S. J., Scott, L. L., Hu, K., & Pierce-Shimomura, J. T. (2014). Conserved single residue in the BK potassium channel required for activation by alcohol and intoxication in C. elegans. *The Journal of Neuroscience*, *34*(29), 9562–9573. http://dx.doi.org/10.1523/jneurosci.0838-14.2014.

Davis, S. J., Scott, L. L., Ordemann, G., Philpo, A., Cohn, J., & Pierce-Shimomura, J. T. (2015). Putative calcium-binding domains of the Caenorhabditis elegans BK channel are dispensable for intoxication and ethanol activation. *Genes, Brain, and Behavior*, *14*(6), 454–465. http://dx.doi.org/10.1111/gbb.12229.

Diamond, I. (1992). *Cecil textbook of medicine* (pp. 44–47). Philadelphia, PA: WB Saunders.

Dopico, A. M. (2003). Ethanol sensitivity of BK_{Ca} channels from arterial smooth muscle does not require the presence of the beta 1-subunit. *American Journal of Physiology. Cell Physiology*, *284*(6), C1468–C1480. http://dx.doi.org/10.1152/ajpcell.00421.2002.

Dopico, A. M., Anantharam, V., & Treistman, S. N. (1998). Ethanol increases the activity of Ca^{++}-dependent K^+ (mslo) channels: Functional interaction with cytosolic Ca^{++}. *The Journal of Pharmacology and Experimental Therapeutics*, *284*(1), 258–268.

Dopico, A. M., Bukiya, A. N., & Martin, G. E. (2014). Ethanol modulation of mammalian BK channels in excitable tissues: Molecular targets and their possible contribution to alcohol-induced altered behavior. *Frontiers in Physiology*, *5*, 466. http://dx.doi.org/10.3389/fphys.2014.00466.

Dopico, A. M., Bukiya, A. N., & Singh, A. K. (2012). Large conductance, calcium- and voltage-gated potassium (BK) channels: Regulation by cholesterol. *Pharmacology & Therapeutics*, *135*(2), 133–150. http://dx.doi.org/10.1016/j.pharmthera.2012.05.002.

Dopico, A. M., Chu, B., Lemos, J. R., & Treistman, S. N. (1999). Alcohol modulation of calcium-activated potassium channels. *Neurochemistry International*, *35*(2), 103–106.

Dopico, A. M., Lemos, J. R., & Treistman, S. N. (1995). Alcohol and the release of vasopressin and oxytocin. In R. R. Watson (Ed.), *Drug and alcohol abuse reviews, Vol. 6: Alcohol and hormones* (pp. 209–226). Totowa, NJ: Humana Press.

Dopico, A. M., Lemos, J. R., & Treistman, S. N. (1996). Ethanol increases the activity of large conductance, Ca^{2+}-activated K^+ channels in isolated neurohypophysial terminals. *Molecular Pharmacology*, *49*(1), 40–48.

Dopico, A. M., & Lovinger, D. M. (2009). Acute alcohol action and desensitization of ligand-gated ion channels. *Pharmacological Reviews*, *61*(1), 98–114. http://dx.doi.org/10.1124/pr.108.000430.

Dopico, A. M., Widmer, H., Wang, G., Lemos, J. R., & Treistman, S. N. (1999). Rat supraoptic magnocellular neurones show distinct large conductance, Ca^{2+}-activated K^+ channel subtypes in cell bodies versus nerve endings. *The Journal of Physiology*, *519*(Pt. 1), 101–114.

Feinberg-Zadek, P. L., Martin, G., & Treistman, S. N. (2008). BK channel subunit composition modulates molecular tolerance to ethanol. *Alcoholism, Clinical and Experimental Research*, *32*(7), 1207–1216. http://dx.doi.org/10.1111/j.1530-0277.2008.00704.x.

Fleming, M., Harris, R. A., & Mihic, S. J. (2001). Ethanol. In J. G. Hardman, L. E. Limbird, & G. A. Gilman (Eds.), *The pharmacological basis of therapeutics* (10th ed., pp. 429–445). New York: McGraw-Hill.

Ghezzi, A., Pohl, J. B., Wang, Y., & Atkinson, N. S. (2010). BK channels play a counter-adaptive role in drug tolerance and dependence. *Proceedings of the National Academy of Sciences of the United States of America*, *107*(37), 16360–16365. http://dx.doi.org/10.1073/pnas.1005439107.

Gruss, M., Henrich, M., Konig, P., Hempelmann, G., Vogel, W., & Scholz, A. (2001). Ethanol reduces excitability in a subgroup of primary sensory neurons by activation of BK(Ca) channels. *The European Journal of Neuroscience*, *14*(8), 1246–1256.

Handlechner, A. G., Hermann, A., Fuchs, R., & Weiger, T. M. (2013). Acetaldehyde--ethanol interactions on calcium-activated potassium (BK) channels in pituitary tumor (GH3) cells. *Frontiers in Behavioral Neuroscience*, 7, 58. http://dx.doi.org/10.3389/fnbeh.2013.00058.

Harris, R. A., Trudell, J. R., & Mihic, S. J. (2008). Ethanol's molecular targets. *Science Signaling*, *1*(28), re7. http://dx.doi.org/10.1126/scisignal.128re7.

Horrigan, F. T., & Aldrich, R. W. (1999). Allosteric voltage gating of potassium channels II. Mslo channel gating charge movement in the absence of Ca^{2+}. *The Journal of General Physiology*, *114*(2), 305–336.

Horrigan, F. T., & Aldrich, R. W. (2002). Coupling between voltage sensor activation, Ca^{2+} binding and channel opening in large conductance (BK) potassium channels. *The Journal of General Physiology*, *120*(3), 267–305.

Jakab, M., Schmidt, S., Grundbichler, M., Paulmichl, M., Hermann, A., Weiger, T., & Ritter, M. (2006). Hypotonicity and ethanol modulate BK channel activity and chloride currents in GH4/C1 pituitary tumour cells. *Acta Physiologica (Oxford, England)*, *187*(1–2), 51–59. http://dx.doi.org/10.1111/j.1748-1716.2006.01544.x.

Jakab, M., Weiger, T. M., & Hermann, A. (1997). Ethanol activates maxi Ca^{2+}-activated K^+ channels of clonal pituitary (GH3) cells. *The Journal of Membrane Biology*, *157*(3), 237–245.

Kalant, H. (1998). Research on tolerance: What can we learn from history? *Alcoholism, Clinical and Experimental Research*, *22*(1), 67–76.

Khanna, J. M., Chau, A., & Shah, G. (1996). Characterization of the phenomenon of rapid tolerance to ethanol. *Alcohol*, *13*(6), 621–628.

Knott, T. K., Dayanithi, G., Coccia, V., Custer, E. E., Lemos, J. R., & Treistman, S. N. (2000). Tolerance to acute ethanol inhibition of peptide hormone release in the isolated neurohypophysis. *Alcoholism, Clinical and Experimental Research*, *24*(7), 1077–1083.

Knott, T. K., Dopico, A. M., Dayanithi, G., Lemos, J., & Treistman, S. N. (2002). Integrated channel plasticity contributes to alcohol tolerance in neurohypophysial terminals. *Molecular Pharmacology*, *62*(1), 135–142.

Kreifeldt, M., Cates-Gatto, C., Roberts, A. J., & Contet, C. (2015). BK channel beta1 subunit contributes to behavioral adaptations elicited by chronic intermittent ethanol exposure. *Alcoholism, Clinical and Experimental Research*, *39*, 2394–2402. http://dx.doi.org/10.1111/acer.12911.

Kreifeldt, M., Le, D., Treistman, S. N., Koob, G. F., & Contet, C. (2013). BK channel beta1 and beta4 auxiliary subunits exert opposite influences on escalated ethanol drinking in dependent mice. *Frontiers in Integrative Neuroscience*, 7, 105. http://dx.doi.org/10.3389/fnint.2013.00105.

Kuhlmann, C. R., Li, F., Ludders, D. W., Schaefer, C. A., Most, A. K., Backenkohler, U., … Wiecha, J. (2004). Dose-dependent activation of Ca^{2+}-activated K^{+} channels by ethanol contributes to improved endothelial cell functions. *Alcoholism, Clinical and Experimental Research*, *28*(7), 1005–1011.

Kuntamallappanavar, G., Toro, L., & Dopico, A. M. (2014). Both transmembrane domains of BK beta1 subunits are essential to confer the normal phenotype of beta1-containing BK channels. *PloS One*, *9*(10), e109306. http://dx.doi.org/10.1371/journal.pone.0109306.

Le, A. D. (1990). Factors regulating ethanol tolerance. *Annals of Medicine*, *22*(4), 265–268.

Li, X., Ghezzi, A., Pohl, J. B., Bohm, A. Y., & Atkinson, N. S. (2013). A DNA element regulates drug tolerance and withdrawal in Drosophila. *PloS One*, *8*(9), e75549. http://dx.doi.org/10.1371/journal.pone.0075549.

Li, Q., Madison, R., & Moore, S. D. (2014). Presynaptic BK channels modulate ethanol-induced enhancement of GABAergic transmission in the rat central amygdala nucleus. *The Journal of Neuroscience*, *34*(41), 13714–13724. http://dx.doi.org/10.1523/jneurosci.5284-13.2014.

Li, W., Zheng, T., Babu, A. N., Altura, B. T., Gupta, R. K., & Altura, B. M. (2001). Importance of magnesium ions in development of tolerance to ethanol: Studies on cultured cerebral vascular smooth muscle cells, type-2 astrocytes and intact rat brain. *Brain Research Bulletin*, *56*(2), 153–158.

Liu, J., Asuncion-Chin, M., Liu, P., & Dopico, A. M. (2006). CaM kinase II phosphorylation of slo Thr107 regulates activity and ethanol responses of BK channels. *Nature Neuroscience*, *9*(1), 41–49. http://dx.doi.org/10.1038/nn1602.

Liu, J., Bukiya, A. N., Kuntamallappanavar, G., Singh, A. K., & Dopico, A. M. (2013). Distinct sensitivity of slo1 channel proteins to ethanol. *Molecular Pharmacology*, *83*(1), 235–244. http://dx.doi.org/10.1124/mol.112.081240.

Liu, P., Liu, J., Huang, W., Li, M. D., & Dopico, A. M. (2003). Distinct regions of the slo subunit determine differential BK_{Ca} channel responses to ethanol. *Alcoholism, Clinical and Experimental Research*, *27*(10), 1640–1644. http://dx.doi.org/10.1097/01.alc.0000094756.41638.5d.

Liu, J., Vaithianathan, T., Manivannan, K., Parrill, A., & Dopico, A. M. (2008). Ethanol modulates BKCa channels by acting as an adjuvant of calcium. *Molecular Pharmacology*, *74*(3), 628–640. http://dx.doi.org/10.1124/mol.108.048694.

Liu, P., Xi, Q., Ahmed, A., Jaggar, J. H., & Dopico, A. M. (2004). Essential role for smooth muscle BK channels in alcohol-induced cerebrovascular constriction. *Proceedings of the National Academy of Sciences of the United States of America*, *101*(52), 18217–18222. http://dx.doi.org/10.1073/pnas.0406096102.

Malysz, J., Afeli, S. A., Provence, A., & Petkov, G. V. (2014). Ethanol-mediated relaxation of guinea pig urinary bladder smooth muscle: Involvement of BK and L-type Ca^{2+} channels. *American Journal of Physiology. Cell Physiology, 306*(1), C45–C58. http://dx.doi.org/10.1152/ajpcell.00047.2013.

Marrero, H. G., Treistman, S. N., & Lemos, J. R. (2015). Ethanol effect on BK channels is modulated by magnesium. *Alcoholism, Clinical and Experimental Research, 39*(9), 1671–1679. http://dx.doi.org/10.1111/acer.12821.

Martin, G. E., Hendrickson, L. M., Penta, K. L., Friesen, R. M., Pietrzykowski, A. Z., Tapper, A. R., & Treistman, S. N. (2008). Identification of a BK channel auxiliary protein controlling molecular and behavioral tolerance to alcohol. *Proceedings of the National Academy of Sciences of the United States of America, 105*(45), 17543–17548. http://dx.doi.org/10.1073/pnas.0801068105.

Martin, G., Puig, S., Pietrzykowski, A., Zadek, P., Emery, P., & Treistman, S. (2004). Somatic localization of a specific large-conductance calcium-activated potassium channel subtype controls compartmentalized ethanol sensitivity in the nucleus accumbens. *The Journal of Neuroscience, 24*(29), 6563–6572. http://dx.doi.org/10.1523/jneurosci.0684-04.2004.

Masters, S. B. (2009). The alcohols. In B. G. Katzung, S. B. Masters, & A. J. Trevor (Eds.), *Basic and clinical pharmacology* (11th ed., pp. 387–398). New York: McGraw-Hill.

Meera, P., Wallner, M., & Toro, L. (2000). A neuronal beta subunit (KCNMB4) makes the large conductance, voltage- and Ca^{2+}-activated K^+ channel resistant to charybdotoxin and iberiotoxin. *Proceedings of the National Academy of Sciences of the United States of America, 97*(10), 5562–5567. http://dx.doi.org/10.1073/pnas.100118597.

Mulholland, P. J., Hopf, F. W., Bukiya, A. N., Martin, G. E., Liu, J., Dopico, A. M., … Chandler, L. J. (2009). Sizing up ethanol-induced plasticity: The role of small and large conductance calcium-activated potassium channels. *Alcoholism, Clinical and Experimental Research, 33*(7), 1125–1135. http://dx.doi.org/10.1111/j.1530-0277.2009.00936.x.

Niu, X., Qian, X., & Magleby, K. L. (2004). Linker-gating ring complex as passive spring and Ca(2+)-dependent machine for a voltage- and Ca(2+)-activated potassium channel. *Neuron, 42*(5), 745–756. http://dx.doi.org/10.1016/j.neuron.2004.05.001.

Orio, P., Rojas, P., Ferreira, G., & Latorre, R. (2002). New disguises for an old channel: MaxiK channel beta-subunits. *News in Physiological Sciences, 17*, 156–161.

Palacio, S., Velazquez-Marrero, C., Marrero, H. G., Seale, G. E., Yudowski, G. A., & Treistman, S. N. (2015). Time-dependent effects of ethanol on BK channel expression and trafficking in hippocampal neurons. *Alcoholism, Clinical and Experimental Research, 39*(9), 1619–1631. http://dx.doi.org/10.1111/acer.12808.

Perez, G. J., Bonev, A. D., & Nelson, M. T. (2001). Micromolar Ca^{2+} from sparks activates Ca^{2+}-sensitive K^+ channels in rat cerebral artery smooth muscle. *American Journal of Physiology. Cell Physiology, 28*(6), C1769–C1775. Retrieved from, http://www.ncbi.nlm.nih.gov/pubmed/11698234.

Pietrzykowski, A. Z., Friesen, R. M., Martin, G. E., Puig, S. I., Nowak, C. L., Wynne, P. M., … Treistman, S. N. (2008). Posttranscriptional regulation of BK channel splice variant stability by miR-9 underlies neuroadaptation to alcohol. *Neuron, 59*(2), 274–287. http://dx.doi.org/10.1016/j.neuron.2008.05.032.

Pietrzykowski, A. Z., Martin, G. E., Puig, S. I., Knott, T. K., Lemos, J. R., & Treistman, S. N. (2004). Alcohol tolerance in large-conductance, calcium-activated potassium channels of CNS terminals is intrinsic and includes two components: Decreased ethanol potentiation and decreased channel density. *The Journal of Neuroscience, 24*(38), 8322–8332. http://dx.doi.org/10.1523/jneurosci.1536-04.2004.

Poikolainen, K., & Alho, H. (2008). Magnesium treatment in alcoholics: A randomized clinical trial. *Substance Abuse Treatment, Prevention, and Policy, 3*, 1. http://dx.doi.org/10.1186/1747-597x-3-1.

Romani, A. M. (2008). Magnesium homeostasis and alcohol consumption. *Magnesium Research, 21*(4), 197–204.

Schuckit, M. A. (2000). Biological phenotypes associated with individuals at high risk for developing alcohol-related disorders. Part 2. *Addiction Biology, 5*(1), 23–36. http://dx.doi.org/10.1080/13556210071243.

Schuckit, M. A., Klein, J., Twitchell, G., & Smith, T. (1994). Personality test scores as predictors of alcoholism almost a decade later. *The American Journal of Psychiatry, 151*(7), 1038–1042. http://dx.doi.org/10.1176/ajp.151.7.1038.

Singh, A. K., McMillan, J., Bukiya, A. N., Burton, B., Parrill, A. L., & Dopico, A. M. (2012). Multiple cholesterol recognition/interaction amino acid consensus (CRAC) motifs in cytosolic C tail of Slo1 subunit determine cholesterol sensitivity of Ca^{2+}- and voltage-gated K^{+} (BK) channels. *The Journal of Biological Chemistry, 287*(24), 20509–20521. http://dx.doi.org/10.1074/jbc.M112.356261.

Stojilkovic, S. S., Zemkova, H., & Van Goor, F. (2005). Biophysical basis of pituitary cell type-specific Ca^{2+} signaling-secretion coupling. *Trends in Endocrinology and Metabolism, 16*(4), 152–159. http://dx.doi.org/10.1016/j.tem.2005.03.003.

Strbak, V. (2006). Cell volume and peptide hormone secretion. *Contributions to Nephrology, 152*, 210–220. http://dx.doi.org/10.1159/000096325.

Tabakoff, B., Cornell, N., & Hoffman, P. L. (1986). Alcohol tolerance. *Annals of Emergency Medicine, 15*(9), 1005–1012.

Thombs, D. L., Olds, R. S., & Snyder, B. M. (2003). Field assessment of BAC data to study late-night college drinking. *Journal of Studies on Alcohol, 64*(3), 322–330.

Treistman, S. N., & Martin, G. E. (2009). BK channels: Mediators and models for alcohol tolerance. *Trends in Neurosciences, 32*(12), 629–637. http://dx.doi.org/10.1016/j.tins.2009.08.001.

Velazquez-Marrero, C., Seale, G. E., Treistman, S. N., & Martin, G. E. (2014). Large conductance voltage- and Ca^{2+}-gated potassium (BK) channel beta4 subunit influences sensitivity and tolerance to alcohol by altering its response to kinases. *The Journal of Biological Chemistry, 289*(42), 29261–29272. http://dx.doi.org/10.1074/jbc.M114.604306.

Velazquez-Marrero, C., Wynne, P., Bernardo, A., Palacio, S., Martin, G., & Treistman, S. N. (2011). The relationship between duration of initial alcohol exposure and persistence of molecular tolerance is markedly nonlinear. *The Journal of Neuroscience, 31*(7), 2436–2446. http://dx.doi.org/10.1523/jneurosci.5429-10.2011.

Walters, F. S., Covarrubias, M., & Ellingson, J. S. (2000). Potent inhibition of the aortic smooth muscle maxi-K channel by clinical doses of ethanol. *American Journal of Physiology Cell Physiology, 279*(4), C1107–C1115.

Wang, Y., Ghezzi, A., Yin, J. C., & Atkinson, N. S. (2009). CREB regulation of BK channel gene expression underlies rapid drug tolerance. *Genes, Brain, and Behavior, 8*(4), 369–376. http://dx.doi.org/10.1111/j.1601-183X.2009.00479.x.

Wang, Q., Kalogeris, T. J., Wang, M., Jones, A. W., & Korthuis, R. J. (2010). Antecedent ethanol attenuates cerebral ischemia/reperfusion-induced leukocyte-endothelial adhesive interactions and delayed neuronal death: Role of large conductance, Ca^{2+}-activated K^{+} channels. *Microcirculation, 17*(6), 427–438. http://dx.doi.org/10.1111/j.1549-8719.2010.00041.x.

Wang, X., Wang, G., Lemos, J. R., & Treistman, S. N. (1994). Ethanol directly modulates gating of a dihydropyridine-sensitive Ca^{2+} channel in neurohypophysial terminals. *The Journal of Neuroscience, 14*(9), 5453–5460.

Wu, Y., Yang, Y., Ye, S., & Jiang, Y. (2010). Structure of the gating ring from the human large-conductance Ca^{2+}-gated K^{+} channel. *Nature, 466*(7304), 393–397. http://dx.doi.org/10.1038/nature09252.

Wynne, P. M., Puig, S. I., Martin, G. E., & Treistman, S. N. (2009). Compartmentalized beta subunit distribution determines characteristics and ethanol sensitivity of somatic, dendritic, and terminal large-conductance calcium-activated potassium channels in the rat central nervous system. *The Journal of Pharmacology and Experimental Therapeutics*, *329*(3), 978–986. http://dx.doi.org/10.1124/jpet.108.146175.

Ye, Y., Jian, K., Jaggar, J. H., Bukiya, A. N., & Dopico, A. M. (2014). Type 2 ryanodine receptors are highly sensitive to alcohol. *FEBS Letters*, *588*(9), 1659–1665. http://dx.doi.org/10.1016/j.febslet.2014.03.005.

Yuan, C., Chen, M., Covey, D. F., Johnston, L. J., & Treistman, S. N. (2011). Cholesterol tuning of BK ethanol response is enantioselective, and is a function of accompanying lipids. *PloS One*, *6*(11), e27572. http://dx.doi.org/10.1371/journal.pone.0027572.

Yuan, P., Leonetti, M. D., Pico, A. R., Hsiung, Y., & MacKinnon, R. (2010). Structure of the human BK channel Ca^{2+}-activation apparatus at 3.0 A resolution. *Science*, *329*(5988), 182–186. http://dx.doi.org/10.1126/science.1190414.

Yuan, C., O'Connell, R. J., Jacob, R. F., Mason, R. P., & Treistman, S. N. (2007). Regulation of the gating of BKCa channel by lipid bilayer thickness. *The Journal of Biological Chemistry*, *282*(10), 7276–7286. http://dx.doi.org/10.1074/jbc.M607593200.

Yuan, C., O'Connell, R. J., Wilson, A., Pietrzykowski, A. Z., & Treistman, S. N. (2008). Acute alcohol tolerance is intrinsic to the BKCa protein, but is modulated by the lipid environment. *The Journal of Biological Chemistry*, *283*(8), 5090–5098. http://dx.doi.org/10.1074/jbc.M708214200.

CHAPTER EIGHT

BK Channels in the Central Nervous System

C. Contet*,[1], S.P. Goulding*, D.A. Kuljis†, A.L. Barth†
*Committee on the Neurobiology of Addictive Disorders, The Scripps Research Institute, La Jolla, CA, United States
†Carnegie Mellon University, Pittsburgh, PA, United States
[1]Corresponding author: e-mail address: contet@scripps.edu

Contents

Abstract

Large conductance Ca^{2+}- and voltage-activated K^+ (BK) channels are widely distributed in the postnatal central nervous system (CNS). BK channels play a pleiotropic role in regulating the activity of brain and spinal cord neural circuits by providing a negative feedback mechanism for local increases in intracellular Ca^{2+} concentrations. In

ISSN 0074-7742
http://dx.doi.org/10.1016/bs.irn.2016.04.001

neurons, they regulate the timing and duration of K^+ influx such that they can either increase or decrease firing depending on the cellular context, and they can suppress neurotransmitter release from presynaptic terminals. In addition, BK channels located in astrocytes and arterial myocytes modulate cerebral blood flow. Not surprisingly, both loss and gain of BK channel function have been associated with CNS disorders such as epilepsy, ataxia, mental retardation, and chronic pain. On the other hand, the neuroprotective role played by BK channels in a number of pathological situations could potentially be leveraged to correct neurological dysfunction.

1. OVERVIEW

Large conductance Ca^{2+}-, and voltage-activated K^+ channels (also known as BK [for Big Potassium], Maxi-K, or Slo1) are expressed in various tissues. As their name indicates, opening of BK channels leads to a massive efflux of K^+ ions that hyperpolarizes cellular membrane potential. This property, combined with their ability to act as a coincidence detector for concomitant elevation in intracellular Ca^{2+} concentration ($[Ca^{2+}]_i$) and membrane depolarization, puts BK channels in a unique position to control the activity of excitable cells via negative feedback regulation of Ca^{2+} influx through voltage-activated Ca^{2+} (Ca_V) channels. In the central nervous system (CNS), BK channels located in the plasma membrane of neurons influence the shape, frequency, and propagation of action potentials (APs), as well as neurotransmitter release from presynaptic terminals. BK channels located in the nuclear envelope of neurons can also directly influence gene transcription and neuronal morphology (a topic addressed in Chapter "Functional Role of Mitochondrial and Nuclear BK Channels" by Gao and Li of the present volume). Furthermore, BK channels expressed in non-neuronal cell populations, such as astrocytes or vascular smooth muscle cells, can regulate cerebral blood flow, thereby influencing brain activity. In the present chapter, we review the expression patterns of BK channel subunits in the CNS, the mechanisms by which BK channels shape neuronal excitability, neurotransmission, and neurovascular coupling, as well as the neurological functions and pathologies in which BK channels are implicated. Aspects relevant to BK channels expressed in the pituitary gland, inner ear, and cerebral blood vessels are addressed in Chapters "BK Channels in the Pituitary and Hypothalamus" by Duncan and Shipston, "BK Channels in the Inner Ear" by Pyott and Duncan, and "BK Channels in the Vascular System" by Krishnamoorthy-Natarajan and Koide, respectively.

2. EXPRESSION PATTERNS OF BK CHANNEL SUBUNITS IN THE CNS

2.1 Pore-Forming Subunit (α)

2.1.1 Localization

In the mature mammalian CNS, the pore-forming α subunit of BK channels (encoded by the *KCNMA1* gene in humans) displays a widespread distribution, as assessed by Northern blot, in situ hybridization, autoradiography, or immunohistochemistry. Prominent levels of expression are observed in the olfactory bulb, cortex, basal ganglia, hippocampus, thalamus, habenulo-interpeduncular tract, cerebellum, vestibular nuclei, and spinal cord (Chang, Dworetzky, Wang, & Goldstein, 1997; Grunnet & Kaufmann, 2004; Hu et al., 2001; Knaus et al., 1996; Misonou et al., 2006; Sailer et al., 2006; Sausbier et al., 2006; Tseng-Crank et al., 1994; Wanner et al., 1999). Although BK α immunoreactivity is macroscopically enriched in fiber tracts and terminal fields, it is also present in the somatodendritic compartment of neurons (Knaus et al., 1996; Misonou et al., 2006; Sausbier et al., 2006).

The distribution of BK α subunits at the ultrastructural level varies between neuronal types, as revealed by electron microscopy. In hippocampal pyramidal cells, BK α immunoreactivity is present in the presynaptic membrane facing the synaptic cleft, as well as in the head of dendritic spines, in close proximity to the postsynaptic specialization of glutamatergic synapses (Hu et al., 2001; Sailer et al., 2006). In contrast, in cerebellar Purkinje cells, BK α subunits are found in the soma and at extrasynaptic sites in dendritic shafts and spines, but they are absent from the vicinity of postsynaptic densities (Kaufmann et al., 2009).

Interestingly, Purkinje cell BK α subunits are segregated into two distinct pools: some are clustered in patches of plasma membrane overlying subsurface cisterns (see below) while others are scattered; the clustered pool is restricted to the soma and proximal dendrites while the scattered pool is present throughout the somatodendritic compartment (Kaufmann et al., 2009). This nonhomogenous distribution of somatic BK α subunits is also observed in other populations of projections neurons, both inhibitory (central amygdala medium spiny neurons) and excitatory (layer V pyramidal neurons of the somatosensory cortex, hippocampal CA3 pyramidal neurons, dentate gyrus granule cells, and basolateral amygdala principal neurons), albeit at lower densities than in Purkinje cells (Kaufmann, Kasugai, Ferraguti, & Storm, 2010). In dentate gyrus granule cells, approximately twice as many

BK α subunits belong to the scattered pool as to the clustered pool. It is currently unknown whether the scattered and clustered pools of BK α subunits are independent from each other, whether they are molecularly distinct, or whether individual BK channels can be exchanged between each pool, as was previously described for neurotransmitter receptors that are rapidly trafficked between synaptic and extrasynaptic plasma membrane domains (Triller & Choquet, 2005).

Subsurface cisterns are present underneath BK α clusters in both Purkinje cells and dentate gyrus granule cells (Kaufmann et al., 2009, 2010). These cisterns are apical extensions of smooth (ie, ribosome free) endoplasmic reticulum closely apposed against the plasma membrane that may serve as a source of local Ca^{2+} sparks for coordinated opening of clustered BK channels (Petersen, Tepikin, & Park, 2001). BK channels located in these domains may become activated during AP-stimulated release of Ca^{2+} from internal stores and thereby exert negative feedback on Ca^{2+} entry through Ca_V channels, in a manner reminiscent of their role in smooth muscle cells in response to Ca^{2+} sparks elicited from sarcoplasmic reticulum stores (Brenner, Perez, et al., 2000; Burdyga & Wray, 2005; Ledoux, Werner, Brayden, & Nelson, 2006; see Chapter "BK Channels in the Vascular System" by Krishnamoorthy-Natarajan and Koide).

2.1.2 Ontogeny

Some insight into the role BK channels play in the CNS can be gained by considering when they are expressed. BK α expression is upregulated during late embryonic and early postnatal development throughout the CNS (Hafidi, Beurg, & Dulon, 2005; Higgins et al., 2010; MacDonald, Ruth, Knaus, & Shipston, 2006). Consistent with this observation, BK currents undergo an abrupt increase during the first 2 weeks after birth in neocortical pyramidal neurons, substantia nigra dopamine neurons, and cochlear inner hair cells (Kang, Huguenard, & Prince, 1996; Kros, Ruppersberg, & Rusch, 1998; Langer, Grunder, & Rusch, 2003; Marcotti, Johnson, Holley, & Kros, 2003; Ramirez-Latorre, 2012). Interestingly, the emergence of BK currents in these cells coincides with their functional maturation, as characterized by increased excitability in the neocortex, high-frequency firing in response to salient stimuli in the substantia nigra, and auditory signal transduction in the cochlea. This developmental pattern suggests that BK channels shape the properties of mature neuronal firing and contribute to experience-dependent plasticity. Interestingly, BK currents are transiently reduced in medial prefrontal cortex pyramidal neurons during adolescence, which may participate in the strong plasticity of prefrontal circuits during this phase of development (Benhassine & Berger, 2009; Ksiazek, Ladno, Szulczyk, Grzelka, & Szulczyk, 2013; Selemon, 2013).

BK α precursor mRNA undergoes extensive alternative splicing and the resulting isoforms exhibit different biophysical and trafficking properties (see Chapter "Posttranscriptional and Posttranslational Regulation of BK Channels" by Shipston and Tian). In particular, BK α isoforms containing the stress-regulated exon (STREX) diversify BK channel function during development. These isoforms possess a C-terminal insert that speeds up activation, slows down deactivation, and increases open probabilities at negative-voltage potentials (Xie & McCobb, 1998). STREX-containing isoforms predominate in the embryonic mouse CNS but are drastically repressed after birth, so that the vast majority of BK α transcripts lack STREX in adulthood (Higgins et al., 2010; MacDonald et al., 2006). Accordingly, fast-gating BK currents consistent with STREX inclusion disappear during the first postnatal week, as observed in neocortical pyramidal neurons (Kang et al., 1996). STREX-containing isoforms remain detectable in a limited number of adult neurons, including dentate gyrus granule cells, CA3 pyramidal neurons, and cerebellar Purkinje cells (Petrik & Brenner, 2007).

The regulation of STREX splicing during development has been proposed as a mechanism whereby BK channels facilitate the generation of spontaneous activity necessary for neural circuit formation during embryogenesis, and then later protect against hyperexcitability in the postnatal CNS (MacDonald et al., 2006). Exclusion of the STREX exon can be triggered by cellular depolarization via a mechanism involving a Ca^{2+}/calmodulin-dependent protein kinase IV-responsive RNA element within the STREX 3′ splice site (Xie & Black, 2001; Xie, Jan, Stoilov, Park, & Black, 2005). Whether this mechanism drives the switch in STREX splicing following birth remains to be determined. Hippocampal neurons also contain a partially processed intron-containing BK α mRNA that promotes the distribution of BK channels to dendritic spines and alters membrane excitability (Bell et al., 2008).

2.2 Auxiliary Subunits (β and γ)

2.2.1 Localization

The transcripts encoding BK channel auxiliary subunits are heterogeneously represented in the CNS. BK β4 (encoded by the *KCNMB4* gene) is commonly referred as the "neuronal auxiliary subunit" since it is the most abundant of the four β subunits in the brain and spinal cord, and is expressed at low levels in other tissues (Behrens et al., 2000; Brenner, Jegla, Wickenden, Liu, & Aldrich, 2000; Weiger et al., 2000). The distribution of BK β4 in the brain mostly overlaps with that of BK α, with prominent expression in the olfactory bulb mitral and granule layers, neocortical pyramidal cells, paraventricular nucleus of the hypothalamus, supraoptic nucleus, CA3 pyramidal layer and dentate gyrus granule cell layer, cerebellar Purkinje layer, and facial and trigeminal motor nuclei (Brenner et al., 2005; http://mouse.brain-map.org, Lein et al., 2007; Petrik & Brenner, 2007; Shruti et al., 2012; Weiger et al., 2000).

BK β2 (*KCNMB2*) is also expressed throughout the brain, although at lower levels than BK β4 (Behrens et al., 2000; Brenner, Jegla, et al., 2000; Weiger et al., 2000). In the mouse forebrain, BK β4 and β2 transcripts are enriched almost fourfold in neurons compared to both oligodendrocytes and astrocytes (Cahoy et al., 2008).

BK β1 (*KCNMB1*) is the canonical "smooth muscle auxiliary subunit." Accordingly, it is highly abundant in cerebral artery myocytes but is otherwise undetectable in the brain, with the exception of several hypothalamic subregions, cerebellar Purkinje cells, and several brainstem nuclei (Behrens et al., 2000; Brenner, Jegla, et al., 2000; Chang et al., 1997; Jiang, Wallner, Meera, & Toro, 1999; Salzmann et al., 2010; Tseng-Crank et al., 1994; Weiger et al., 2000). BK β1 is prominently expressed in ependymal cells lining the ventricles and is particularly enriched in tanycytes of the median eminence (Salzmann et al., 2010). Tanycytes are a subtype of ependymal cells with long processes that extend into the hypothalamic parenchyma, thereby coming in close contact with neurons and reaching the fenestrated capillary network, while their tight junctions prevent the diffusion of blood-borne molecules into the adjacent cerebrospinal fluid (Bolborea & Dale, 2013; Langlet, Mullier, Bouret, Prevot, & Dehouck, 2013). It remains to be determined whether functional BK channels are present in tanycytes, and, if so, what role they might play in the physiology of these cells. It is also possible that BK β1 subunits expressed in tanycytes associate with the pore-forming subunit of another (non-BK) ion channel. Such nonconventional functional interactions between ion channel subunits was previously described between the β1 subunit of L-type Ca_V channels and BK α, between BK β4 and the pore-forming subunit of pH-regulated Slo3 K^+ channels, as well as between BK β4 and the Nav1.6 pore-forming subunit of voltage-gated Na^+ channels (Seidel et al., 2011; Yang, Zeng, Xia, & Lingle, 2009; Zou, Jha, Kim, & Dryer, 2008b).

BK β3 (*KCNMB3*) shows particularly weak expression in the CNS (Behrens et al., 2000; Brenner, Jegla, et al., 2000; Poulsen et al., 2009; Uebele et al., 2000).

Another family of BK auxiliary subunits, named γ1–γ4 subunits, was recently identified. Association of these leucine-rich repeat-containing proteins (encoded by *LRRC26*, *LRRC52*, *LRRC55*, and *LRRC38*, respectively) with BK α tetramers shifts the voltage dependence of BK channel activation toward hyperpolarizing voltages (see Chapter "Modulation of BK Channel Function by Auxiliary Beta and Gamma Subunits" by Yan and Li). Comparative analysis of the distribution of BK γ subunits indicated that γ3 is selectively enriched in the brain, while γ1 and γ4 are expressed at lower levels, and γ2 is essentially absent from this tissue (Yan & Aldrich, 2012). BK γ1 subunits located in cerebral arterial myocytes elevate the voltage and Ca^{2+} sensitivity of BK channels, which contributes to lowering myogenic tone and promoting vasodilation (Evanson, Bannister, Leo, & Jaggar, 2014). The role of γ3 in modulating neuronal physiology still remains to be determined.

Unlike the BK α subunit for which a monoclonal antibody has been thoroughly validated in knockout tissue (Misonou et al., 2006), the specificity of commercially available antibodies raised against BK auxiliary subunits has been questionable (see Bhattarai et al., 2014 for anti-β1 antibodies; Martinez-Espinosa, Yang, Gonzalez-Perez, Xia, & Lingle, 2014 for anti-β2 antibodies; comment in Wang, Jaffe, & Brenner, 2014 for anti-β4 antibodies; and our personal experience with a number of commercial anti-β1 and anti-β4 antibodies). Immunohistochemical characterization of these subunits should therefore be interpreted with caution when used as the sole analytical method (eg, Piwonska, Wilczek, Szewczyk, & Wilczynski, 2008 reporting BK β4 immunoreactivity in neuronal mitochondria and BK β2 immunoreactivity in astrocytes; and Seidel et al., 2011 reporting BK β1 and β4 immunoreactivity in astrocytes).

2.2.2 Functional Signatures

Auxiliary subunits alter BK channel activation and inactivation, subcellular trafficking, and sensitivity to channel activators or blockers (see Chapter "Modulation of BK Channel Function by Auxiliary Beta and Gamma Subunits" by Yan and Li). These features have been exploited to identify the subunits that contribute to BK currents in different neuronal populations.

Early studies conducted in brain synaptosomal membranes identified a subtype of BK channels (called "type II") that shows slow gating kinetics and resistance to the BK channel blocker iberiotoxin (Reinhart, Chung, & Levitan, 1989). These properties were later found to be conferred by the association of auxiliary β4 subunits (see Wang et al., 2014 for review). In the presence of β4, peak currents are not observed until >25 ms after depolarization, which is in contrast to currents from BK α (or BK α with other β subunits) that reach steady state in 2–10 ms, and tail currents are significantly prolonged (Behrens et al., 2000; Brenner, Jegla, et al., 2000; Ha, Heo, & Park, 2004). This is particularly relevant for neurons, where the participation of β4-containing BK channels may broaden APs and increase the afterhyperpolarization (AHP) (Brenner et al., 2005). In addition, β4-containing BK channels exhibit unique pharmacological properties, as they are resistant to blockade by charybdotoxin and iberiotoxin (Meera, Wallner, & Toro, 2000).

BK channels showing the characteristic properties of β4-containing channels have been observed in a variety of neuronal populations, including neurohypophysial terminals, dentate gyrus granule cells, cerebellar Purkinje cells, and trigeminal caudal nuclei (TCN) terminals (Benton, Lewis, Bant, & Raman, 2013; Bielefeldt, Rotter, & Jackson, 1992; Brenner et al., 2005; Deng et al., 2013; Dopico, Widmer, Wang, Lemos, & Treistman, 1999; Samengo, Curro, Barrese, Taglialatela, & Martire, 2014; Shruti et al., 2012). It is somewhat puzzling that a large fraction of CNS BK channels is sensitive to iberiotoxin despite the prominent expression of β4 in this tissue. One possible explanation advanced by Wang and colleagues relates to the stoichiometry of β4 subunits assembling with α tetramers, with a full complement of β4 subunits (ie, 1:1 stoichiometry) being possibly required to confer iberiotoxin resistance (Wang et al., 2014). Alternatively, it is possible that β4 subunits control BK channel function through their ability to restrict plasma membrane localization of BK α tetramers (see Section 2.3); in this scenario, BK β4 may be highly expressed but retained within the endoplasmic reticulum and not present in plasma membrane BK channels (Shruti et al., 2012). Martentoxin and conopeptide Vt3.1 were reported to preferentially inhibit β4-containing BK channels and could help provide novel insights into this question, but, as for iberiotoxin, the ratio of β4 subunits required to drive increased sensitivity to these inhibitors remains to be determined (Li, Chang, et al., 2014; Shi et al., 2008).

Another subtype of BK channels, which was first reported in chromaffin cells of the adrenal gland, exhibits rapid inactivation (Solaro & Lingle, 1992). This fast inactivation was later shown to be conferred by the N-terminal residues of BK β2 subunit (Wallner, Meera, & Toro, 1999; Xia, Ding, & Lingle, 1999). Using this unique inactivating property, β2-containing BK channels have been identified in hippocampus, neocortex, and lateral amygdala pyramidal neurons, as well as in cerebellar Purkinje cells and dorsal root ganglion (DRG) neurons (Faber & Sah, 2003; Haghdoost-Yazdi, Janahmadi, & Behzadi, 2008; Hicks & Marrion, 1998; Li et al., 2007; McLarnon, 1995; Sun, Gu, & Haddad, 2003). There is, to the best of our knowledge, no ligand that selectively discriminates β2-containing BK channels.

Alcohol can increase BK currents in neurons, but the association of β1 renders BK channels insensitive to alcohol-induced potentiation (Feinberg-Zadek & Treistman, 2007; Martin et al., 2004; see Chapter "Modulation of BK Channels by Ethanol" by Dopico). This pharmacological signature has been used to identify β1-containing BK channels in the soma and dendrites of nucleus accumbens medium spiny neurons and magnocellular neurons of the supraoptic nucleus (Dopico et al., 1999; Martin et al., 2004; Wynne, Puig, Martin, & Treistman, 2009).

Finally, the association of γ1 confers resistance to mallotoxin (Almassy & Begenisich, 2012), but this feature has not been exploited yet to identify γ1-containing BK channels in the CNS.

2.3 Subcellular Trafficking

A number of factors regulate the subcellular trafficking of BK channels, including pre-mRNA splicing, posttranslational modifications, and interaction with BK auxiliary subunits and other proteins (reviewed in Chapters "Modulation of BK Channel Function by Auxiliary Beta and Gamma Subunits" by Yan and Li, "Posttranscriptional and Posttranslational Regulation of BK Channels" by Shipston and Tian, and "Protein Network Interacting with BK Channels" by Kim and Oh). Examples of parameters promoting the cell surface localization of BK α subunits include incorporation of the alternatively spliced C-terminus ending with the "QEERL" amino acid sequence, palmitoylation of cysteine residues within the S0–S1 linker or the alternatively spliced STREX insert, and interaction with actin (Jeffries et al., 2010; Kim, Ridgway, Zou, Chiu, & Dryer, 2007; Ma et al., 2007; Tian et al., 2008; Zou, Jha, Kim, & Dryer, 2008a). Conversely, inclusion of the SV1 insert or VEDEC C-terminus, myristoylation, ubiquitination, phosphorylation by cyclin-dependent kinase 5 (CDK5), adenylyl cyclase activation, and association with caveolin-1 or cereblon promote BK α endocytosis or retention in the endoplasmic reticulum, thereby reducing surface expression of functional BK channels (Alioua, Li, Wu, Stefani, & Toro, 2011; Alioua et al., 2008; Bai, Surguchev, Joshi, Gross, & Navaratnam, 2012; Jo, Lee, Song, Jung, & Park, 2005;

Kim, Ridgway, et al., 2007; Liu et al., 2014; Ma et al., 2007; Pratt, He, Wang, Barth, & Bruchez, 2015; Tian et al., 2008; Zarei et al., 2004).

A large number of proteins physically interacting with BK α subunits are known to play a role in trafficking/scaffolding and therefore likely dictate the subcellular localization of BK channels in neurons (Gorini et al., 2010; Kathiresan, Harvey, Orchard, Sakai, & Sokolowski, 2009; Kim, Jo, Song, Park, & Park, 2007; Singh et al., 2016; Sokolowski, Orchard, Harvey, Sridhar, & Sakai, 2011; Williams et al., 2004). To add another layer of complexity, there is an intricate cross talk between all parameters known to influence the subcellular trafficking of BK channel subunits (see Chapter "Posttranscriptional and Posttranslational Regulation of BK Channels" by Shipston and Tian). For instance, the influence of BK β1 and β4 subunits depends on the C-terminal variant of BK α: both β1 and β4 enhance the cell surface expression of the VEDEC isoform, but β1 targets the QEERL isoform to endosomes, while β4 promotes the retention of the EMVYR isoform in the endoplasmic reticulum (Chen et al., 2013; Kim, Zou, Ridgway, & Dryer, 2007; Shruti et al., 2012; Toro et al., 2006).

Importantly, most studies examining the regulation of BK channel trafficking have been conducted in transfected cells. Findings obtained in these heterologous expression systems may not necessarily reflect what happens in the CNS, where the expression and posttranslational modifications of BK α isoforms and BK auxiliary subunits are cell type-dependent and dynamically regulated. We therefore highlight the few studies that have examined endogenous mechanisms of BK channel subcellular trafficking in neurons.

In mouse hippocampus CA3 pyramidal cells, despite prominent expression of the β4 subunit, β4-containing BK channels may not participate in whole-cell BK currents, as suggested by an equivalent effect of paxilline and iberiotoxin (Shruti et al., 2012). However, genetic deletion of β4 strongly increases whole-cell BK currents, which suggests, in accordance with observations made in transfected cells, that β4 represses the forward trafficking of BK α in these neurons (Shruti et al., 2012).

Notably, some studies have identified proteins implicated in the addressing of BK α to specific neuronal compartments. In *Caenorhabditis elegans* cholinergic motor neurons, the localization of BK channels to presynaptic terminals is dependent upon α-catulin and dystrobrevin, two proteins that interact with cortical cytoskeletal proteins (Oh et al., 2015). In mouse and rat myelinated cerebellar Purkinje cell axons, the cell adhesion molecule Caspr promotes the localization of BK channels to paranodal junctions, where they support axonal AP conduction during high-frequency firing and ensure synaptic output in the deep cerebellar nuclei (Hirono et al., 2015).

Finally, a recent study provided evidence for activation-induced endocytosis of neuronal BK channels. In primary cultures of hippocampal neurons, alcohol acutely potentiates macroscopic BK currents, but prolonged (3–6 h) exposure to alcohol decreases cell surface expression of functional BK channels without altering total levels of BK α expression, indicating their trafficking to intracellular compartments (Palacio et al., 2015).

3. ROLE OF BK CHANNELS IN CNS CELLULAR PHYSIOLOGY

3.1 Neurons

The synergistic effect of depolarization and Ca^{2+} binding on BK channel opening enables BK channels to act as a coincidence detector (see Chapter "Biophysics of BK Channel Gating" by Pantazis and Olcese), providing a hyperpolarizing drive that contributes to AP repolarization, mediates the fast phase of AHP, and regulates neurotransmitter release and dendritic excitability.

3.1.1 Calcium Nanodomains

Neurons regulate $[Ca^{2+}]_i$ using specialized Ca^{2+}-binding proteins, which ensure that increases in neuronal $[Ca^{2+}]_i$ are tightly restricted in time and space (Augustine, Santamaria, & Tanaka, 2003; Neher, 1998). At resting Ca^{2+} concentrations and within the physiological voltage range found in neurons, BK channel open probability is close to null. Furthermore, the relatively low affinity of BK channels for Ca^{2+} (as compared to other Ca^{2+}-activated channels, such as SK channels) implies that only BK channels located in the immediate vicinity of active Ca^{2+} influx sources can become activated. The $[Ca^{2+}]_i$ required for BK channel activation (~10 μM) is largely restricted to local Ca^{2+}-signaling nanodomains centered around Ca_V channels (Augustine et al., 2003; Brenner, Jegla, et al., 2000; Fakler & Adelman, 2008). Accordingly, the experimental findings reviewed below have revealed the spatial proximity and even direct physical interaction between BK and Ca_V channels.

Early functional evidence for the tight spatial coupling between BK and Ca_V channels in neurons came from the ability of BAPTA, a Ca^{2+}-chelator with fast-binding kinetics, to disrupt BK–Ca_V interactions and suppress BK currents, unlike the slow buffer EGTA (Edgerton & Reinhart, 2003;

Lancaster & Nicoll, 1987; Muller, Kukley, Uebachs, Beck, & Dietrich, 2007; Robitaille, Garcia, Kaczorowski, & Charlton, 1993; Storm, 1987b; Sun et al., 2003). In some instances, the coupling between BK and Ca_V channels is even completely resistant to buffering by BAPTA (Loane, Lima, & Marrion, 2007). Altogether, these experimental data suggest a ~10–50 nm distance between the two channel pores.

A more detailed understanding of the functional coupling between Ca_V and BK channels was obtained using subtype-specific Ca_V channel blockers. In sum, virtually all subtypes of Ca_V channels have been shown to influence BK channel opening (except R-type), with the specific subtype involved depending upon the subcellular location of the BK channel. For instance, Ca^{2+} entry via fast-activating P/Q- and N-type channels in presynaptic terminals can indirectly, via a BK channel-dependent shortening of AP duration, reduce neurotransmitter release (Castillo, Weisskopf, & Nicoll, 1994; Kulik et al., 2004; Pelkey, Topolnik, Lacaille, & McBain, 2006; Wu, Westenbroek, Borst, Catterall, & Sakmann, 1999). Alternatively, the more slowly activating L- and T-type channels located primarily in the dendrites and cell body have been shown to gate BK channels and thereby regulate their role in synaptic summation and AP propagation (Dobremez et al., 2005; Engbers, Anderson, Zamponi, & Turner, 2013; Obermair, Szabo, Bourinet, & Flucher, 2004; Rehak et al., 2013).

The coupling of BK channels with specific Ca_V subtypes may also vary across different brain regions. Near-coincident activation of N-type Ca_V and BK channels in hippocampal CA1 pyramidal neurons suggests a very close spatial relationship between these channels, which is not seen for L-type channels in this cell type (Marrion & Tavalin, 1998). In cortical pyramidal neurons, both N- and L-type channels provide Ca^{2+} to gate BK channel opening (Sun et al., 2003), while in chromaffin cells, L- and P/Q-type channels play a greater role than N-type channels in gating BK channels (Prakriya & Lingle, 1999). Additionally, low-voltage-activated T-type channels in cerebellar neurons complex with BK channels to reduce synaptically evoked neural firing (Engbers et al., 2013). By associating with a diverse set of Ca_V partners, BK channels can adapt to display diverse temporal responses that may enable them to serve different roles in regulating neural activity (Berkefeld & Fakler, 2008).

Although early electrophysiological studies provided important evidence for a functional coupling between BK and Ca_V channels, direct physical association between channel subunits was not demonstrated until fairly recently (Bao, de Jong, Alevra, & Schild, 2015; Berkefeld et al., 2006;

Grunnet & Kaufmann, 2004; Indriati et al., 2013; Kaufmann et al., 2009; Loane et al., 2007; Samaranayake, Saunders, Greene, & Navaratnam, 2004). A first coimmunoprecipitation study using a rat brain membrane preparation found that BK channels predominantly complex with L-type Ca_V channels, and to a limited extent with P/Q-type, but not with N-type channels (Grunnet & Kaufmann, 2004). A subsequent coimmunoprecipitation study however showed that BK channels assemble with N-type channels in the rat whole brain, and in the hippocampus in particular, to a greater extent than with L-type channels (Loane et al., 2007). The discrepancy between these two studies possibly results from the use of different antibodies against Ca_V channels. When mass spectrometry was used to identify proteins interacting with BK α subunit affinity-purified from rat brain plasma membranes, it was found that BK channels complex with L-, P/Q-, and N-type channels, with P/Q-type associations being most abundant (Berkefeld et al., 2006). Consistent with pharmacological observations, R-type Ca_V channels were not specifically copurified with BK α in the latter study.

High-resolution imaging further corroborated the physical colocalization of BK and Ca_V channels in neurons. Specifically, SDS-digested, freeze-fracture replica, immunogold double-labeling experiments showed that BK channel clusters on cerebellar Purkinje cell soma and primary dendrites virtually always colocalize with P/Q-type Ca_V channels (Indriati et al., 2013; Kaufmann et al., 2009).

In contrast to BK channels tightly associated with Ca_V channels, BK channels located further away from Ca^{2+} sources may play a role in sensing global $[Ca^{2+}]_i$ and provide an "emergency brake" under excitotoxic conditions where $[Ca^{2+}]_i$ rises above the physiological range (Hu et al., 2001; Runden-Pran, Haug, Storm, & Ottersen, 2002).

3.1.2 Regulation of Neural Firing

BK channels can shape the AP waveform and influence firing frequency in complex ways. In modeling studies, BK channel currents can slow down or speed up firing, depending on the relative magnitude of other ion channels. For example, when hyperpolarization-activated cyclic nucleotide-gated (HCN) channels are present, slowly activating BK channels (a feature of β4-containing channels; see Section 2.2.2) increase firing rates in cells that are normally relatively silent. This effect is reversed in neurons with high basal firing, where the same slow-activating BK channels reduce overall firing rates (Ly, Melman, Barth, & Ermentrout, 2011). This mixed effect suggests that BK channels, together with HCN channels, act as a gain control mechanism to flatten F–I curves (Ly et al., 2011).

In CNS neurons, pharmacological or genetic inhibition of BK channels can *increase* or *decrease* evoked or spontaneous firing activity (Bielefeldt & Jackson, 1993; Brenner et al., 2005; Du et al., 2005; Gu, Vervaeke, & Storm, 2007; Jin, Sugaya, Tsuda, Ohguchi, & Sugaya, 2000; Li et al., 2007; Matthews, Weible, Shah, & Disterhoft, 2008; Meredith et al., 2006; Nelson, Gittis, & du Lac, 2005; Nelson, Krispel, Sekirnjak, & du Lac, 2003; Pitts, Ohta, & McMahon, 2006). These findings are part of a consistent pattern of experimental evidence supporting the notion that BK channels should not be thought of as strictly excitatory or inhibitory, but rather, similar to dynamic range compressors used in sound recording that reduce the volume of loud sounds and amplify quiet ones. In other words, BK channels can turn down the firing of neurons that are too active and turn up the firing of quiet ones.

What is an explanation for the diversity of BK channel effects on neural activity? It may seem counterintuitive that K^+ conductance (ie, efflux) can increase firing output, since its net effect should be membrane hyperpolarization. However, the timing of BK channel activation (during the falling phase of the AP, when Ca^{2+} from Ca_V channels has had a chance to accumulate, or in later spikes during a high-frequency train) and the presence of other channels in the membrane contribute to its net effect on firing output. Understanding the timing of BK channel activation during isolated or sustained firing may prove helpful in explaining these bidirectional effects.

The three major phases of the AP are: (1) a depolarization phase mediated by voltage-gated Na^+ and Ca^{2+} channel currents, (2) a repolarization phase involving fast-activating voltage-gated K^+ currents, and (3) an AHP phase driven by slowly activating/deactivating voltage-gated K^+ currents that sets the spike refractory period. In isolation, BK α homotetramers show fast-activation kinetics, and thus, can contribute to the repolarization phase and the AHP (Faber & Sah, 2002; Gu et al., 2007; Montgomery & Meredith, 2012; Shao, Halvorsrud, Borg-Graham, & Storm, 1999; Shruti, Clem, & Barth, 2008; Storm, 1987a). In neurons, other currents and cell-specific factors regulate whether BK channels contribute to the repolarization or AHP phases of the AP and, as a result, BK currents can either slow down or speed up spontaneous firing depending on the context.

These nuances have been investigated in detail in the vestibular nucleus, where firing properties and long-term changes in intrinsic excitability are dependent on BK currents (Gittis & du Lac, 2007; Nelson et al., 2003). This brain region contains fast-firing projection neurons and slower-firing GABAergic interneurons, and BK channels are differentially implicated in the regulation of firing in these two populations (Gittis & du Lac, 2007; Gittis, Moghadam, & du Lac, 2010; Smith, Nelson, & Du Lac, 2002). In projection neurons, Kv3 currents activate faster than BK currents and dominate the repolarization phase, while in GABA interneurons that lack Kv3, BK currents dominate the repolarization phase (Gittis et al., 2010). In cells with higher basal firing, pharmacological inhibition of fast-activating BK channels using iberiotoxin increases firing rates and reduces the AHP, whereas for a subset of cells with lower firing rates, iberiotoxin slightly decreases firing rates and has less pronounced effects on AHP (Smith et al., 2002). These data support the notion that BK channels contribute preferentially *either* to the repolarization *or* to the AHP phases of the AP, depending on the coexpression of other voltage-activated currents, and will thereby either increase or decrease firing rates.

Notably, the contribution of BK channels to firing response gain was dependent on the relative activity levels of the cells examined (Gittis et al., 2010; Smith et al., 2002). For instance, BK channel contribution to repolarization is greatest in low-frequency firing neurons in an excited state (Gittis et al., 2010). The interplay between neural activity levels and the role of BK channels in firing might be related to activity-dependent second messengers, which can have bidirectional effects on BK channel gating. In cell-excised patches from vestibular neurons, application of activated Ca^{2+}/calmodulin-dependent kinase II or a peptide inhibitor of protein kinase C increases BK open probability, raising the possibility that BK contribution to neural excitability can be dynamically regulated depending upon activation of different signaling cascades (van Welie & du Lac, 2011).

The state-dependent effects of BK channels on neural excitability have also been demonstrated in the suprachiasmatic nucleus (SCN) of the hypothalamus, a GABAergic circuit controlling circadian rhythms where $[Ca^{2+}]_i$, cAMP-dependent second messengers, fast-activating K^+ currents, and BK currents display circadian oscillations (Doi et al., 2011; Granados-Fuentes, Norris, Carrasquillo, Nerbonne, & Herzog, 2012; Hong, Jeong, Min, & Lee, 2012; Ikeda & Ikeda, 2014; Ikeda et al., 2003; Kudo, Loh, Kuljis, Constance, & Colwell, 2011; Meredith et al., 2006; O'Neill, Maywood, Chesham, Takahashi, & Hastings, 2008; Panda et al., 2002; Pitts et al., 2006). Paradoxically, total BK currents are inversely correlated with neuronal Ca^{2+} levels, which positively regulate BK currents (Meredith et al., 2006); however, they are also inversely correlated with fast-activating K^+

currents that mediate fast-firing rates (Granados-Fuentes et al., 2012; Kudo et al., 2011). These findings make it increasingly apparent that disambiguating BK channel contribution to the regulation of neural excitability will likely also require cell-by-cell characterization of other voltage-gated K^+ channels.

Examination of BK channel contribution to neural excitability in cerebellar Purkinje cells raises additional problems in understanding how BK channels contribute to the regulation of neural excitability. Pharmacological experiments indicate that BK currents generally suppress activity in Purkinje cells by increasing the magnitude of the AHP, which supports an inhibitory role of BK channels in these cells (Edgerton & Reinhart, 2003; Swensen & Bean, 2003; Womack & Khodakhah, 2002). Yet, when BK channels are genetically ablated, a substantial decrease in Purkinje cell tonic firing rates is observed, which supports an excitatory role of BK channels in these cells (Sausbier et al., 2004). Reconciling these two experimental observations is difficult, but might be accounted for if one considers that acute pharmacological inhibition of BK currents and constitutive loss of BK channels are fundamentally different interventions. Unlike acute pharmacological inhibition, constitutive knockout of BK α may elicit homeostatic/compensatory changes in other voltage-gated currents to protect against hyperexcitability. Investigating this possibility will require systematic examination of other currents regulating neural excitability and AP generation in the cerebellar circuit of BK α knockout mice.

Experimental studies focused on other neuronal populations have provided evidence for a proexcitatory role of BK channels. For example, genetic ablation of BK β4 subunit causes hyperexcitability of dentate gyrus granule cells in conjunction with faster gating and higher density of BK channels (Brenner et al., 2005). Furthermore, in neocortical pyramidal neurons primed by seizure activity, increased BK current amplitude is linked to both single-cell and network hyperexcitability, and the elevated firing output of postseizure neurons can be normalized by application of the BK channel blocker paxilline (Shruti et al., 2008). In snail serotonergic neurons, silencing synapsin expression also increases intrinsic excitability while concomitantly increasing the magnitude of BK currents (Brenes, Vandael, Carbone, Montarolo, & Ghirardi, 2015). Paxilline reduces firing frequency down to control levels in this system as well, further buttressing the idea that excessive BK channel activity can generate neuronal hyperexcitability.

Some of the differential effects BK channels have on neural excitability may depend on the specific β subunit that is coexpressed with the pore-forming α subunit. For example, when assembled with the fast-inactivating β2 subunit found in adrenal medullary chromaffin cells, the role of the BK channel is limited to the repolarization phase, which acts to increase the spontaneous firing of these cells (Martinez-Espinosa et al., 2014). In contrast, when assembled with the β4 subunit, the role of the BK channel is more critical during the AHP phase, where it lengthens spike refractory periods to increase interspike intervals and decrease overall firing rates (Brenner et al., 2005; Cloues & Sather, 2003; Petrik, Wang, & Brenner, 2011). This effect may be due to the relatively slow activation kinetics of β4-containing channels that prevent BK channels from opening during the AP (Brenner, Jegla, et al., 2000), or from the role of β4 in reducing BK channel current density via intracellular sequestration (Shruti et al., 2012). Overall, it is clear that the different biophysical characteristics and cellular localization properties conferred by the β subunits can profoundly influence firing output in neurons.

3.1.3 Regulation of Neurotransmitter Release

The role of BK channels in regulating neural activity does not end with their role in shaping AP waveform and firing rate. The pattern of BK α expression within terminal areas of prominent fiber tracts is consistent with its presynaptic targeting and involvement in the regulation of neurotransmitter release (Knaus et al., 1996; Misonou et al., 2006; Raffaelli, Saviane, Mohajerani, Pedarzani, & Cherubini, 2004). Determining whether presynaptic BK channels inhibit or facilitate neurotransmitter release has not been inconsequential, since the small size of most synapses makes isolating the contribution of synaptic vs somatic BK currents methodologically challenging. Nevertheless, multiple lines of evidence including neurotransmitter release from synaptosome preparations, direct examination of the brain largest synapses, indirect measures of synaptic function such as neurotransmitter release probability, and the manipulation of synapse-specific proteins have provided important insights into the functional role of presynaptic BK channels. Additionally, recent technological advancements in superresolution scanning and genetically encoded voltage sensors have facilitated glimpses into even small synapses (≤1 μm), where the hallmark properties of BK channel conductances can be visualized (Novak et al., 2013). Based on these findings and the literature discussed below, BK channels appear to be an important regulator of neurotransmitter release at diverse synapses in the brain.

Once an AP propagates to the presynaptic terminal, it stimulates Ca^{2+} entry through Ca_V channels that serves both as a signal to initiate exocytosis of neurotransmitter-containing vesicles (reviewed in Kaeser & Regehr,

2014), and as a requisite cofactor for activating local BK channels. In general, presynaptic BK channels are thought to act as negative regulators of neurotransmitter release by hyperpolarizing the plasma membrane in response to AP-stimulated depolarization and Ca^{2+} entry through Ca_V channels (Wang, 2008). In addition, Ca^{2+} influx in the terminal is modulated by the width of the presynaptic AP, which is in part determined by BK channel activity (see Section 3.1.2). Experimental evidence for an inhibitory role of BK channels on neurotransmitter release was first obtained at neurosecretory terminals and neuromuscular junctions (Obaid, Flores, & Salzberg, 1989; Robitaille & Charlton, 1992; Robitaille et al., 1993; Dopico et al., 1999), and later at central synapses. In organotypic hippocampal slice cultures, BK channel blockade by iberiotoxin or paxilline increases the probability of glutamate release at synapses between CA3 pyramidal neurons, as reflected by increased frequency of spontaneous excitatory postsynaptic currents (EPSCs), increased amplitude of evoked EPSCs, and reduced paired-pulse ratio (Raffaelli et al., 2004). Likewise, iberiotoxin increases the frequency of spontaneous EPSCs in acute spinal cord slices (Furukawa, Takasusuki, Fukushima, & Hori, 2008).

The inhibitory influence of BK channels on excitatory synaptic transmission was also observed in synaptosomes subjected to depolarization-induced release of radioactively labeled neurotransmitter. Using this approach, pharmacological blockade of BK channels enhances neurotransmitter release from glutamatergic terminals in cortex, hippocampus, and TCN, but not in cerebellum (Martire et al., 2010; Samengo et al., 2014). In contrast, BK channel inhibition does not affect GABA release from cortical, hippocampal, cerebellar, and TCN synaptosomes; noradrenaline release from cortical, hippocampal, and hypothalamic synaptosomes; or dopamine release from striatal synaptosomes (Martire et al., 2010; Samengo et al., 2014). Experiments performed on synaptosomes therefore suggested that BK channels selectively regulate excitatory transmission. However, electrophysiological recordings have provided more refined insights into the influence of BK channels on inhibitory transmission at individual synapses. In primary hippocampal cultures, iberiotoxin can indeed enhance the amplitude of evoked inhibitory postsynaptic currents (IPSCs) in a subset of neurons receiving a monosynaptic GABAergic innervation (Martire et al., 2010). Similarly, the probability of GABA release onto neurons of the central amygdala is enhanced by iberiotoxin and paxilline, as reflected by increased amplitude of evoked IPSCs and increased frequency of miniature IPSCs (Li, Madison, & Moore, 2014). Iberiotoxin also augments glycine release in the spinal cord (Shoudai et al., 2007). In addition, BK channel activity mediates the inhibitory influence of presynaptic α1-adrenoreceptors on GABA release onto ventral tegmental area dopaminergic neurons (Velasquez-Martinez, Vazquez-Torres, Rojas, Sanabria, & Jimenez-Rivera, 2015).

The functional implication of BK channels in inhibiting neurotransmitter release plays an important role in preventing overexcitation. In monosynaptically connected *neonatal* rat hippocampal CA3 pyramidal neurons, BK currents decrease the transmitter release probability and increase the synaptic failure rate, even at low-stimulation frequencies (Raffaelli et al., 2004). In *adult* rat CA3 terminals and mossy fiber boutons, similar Ca^{2+}-dependent negative feedback control of transmitter release is also observed, but only under "excited" conditions, when 4-aminopyridine treatment is used to pharmacologically broaden APs (Alle, Kubota, & Geiger, 2011; Hu et al., 2001). Notably, this is in contrast to the strong contribution of CA3 somatic BK channels to AP repolarization even under baseline conditions (Hu et al., 2001). Thus, the ability of BK channels to brake network activity is expressed differently at the soma and at the synapse.

Presynaptic BK channels have also been characterized in invertebrates. Although genetic ablation of BK channels is not thought to alter neurotransmitter release at the *Drosophila* neuromuscular junction (Gho & Ganetzky, 1992; Lee, Ueda, & Wu, 2008), a recent study using archaerhodopsin to image synaptic voltage changes found that during periods of repetitive firing, BK channels inhibit recurrent electrical activity (Ford & Davis, 2014). Taken together, physiological data suggest that BK channels at and around synaptic terminals support high-fidelity neurotransmitter release during periods of repetitive firing by inhibiting neurotransmitter release from overly excited synapses.

BK channel trafficking to the synapse is an important determinant of neurotransmitter release regulation. Experiments manipulating cytoskeletal proteins to affect synaptic BK channel clustering alter the ability of BK channels to act as a brake on neurotransmitter release. In neonatal rat hippocampal neuron cultures, stimulating cytoskeletal rearrangement with leptin and insulin signaling promotes presynaptic BK clustering, which is necessary for BK channels to block recurrent electrical activity (O'Malley & Harvey, 2007; O'Malley, Irving, & Harvey, 2005). Additionally, BK clustering at the synapse appears to be an important mechanism for normal synapse maintenance. Mutations in the presynaptic protein cysteine string protein α (CSPα), which enhance synaptic BK expression, are associated with overall synapse loss and neurodegeneration (Kyle, Ahrendt, Braun, & Braun, 2013). Taken together, these findings suggest that while synaptic BK currents must be large enough to inhibit excessive neurotransmitter release during an excited state, they must also be not so large as to limit normal neurotransmission.

In some cases, presynaptic BK channel activity increases the amplitude of postsynaptic responses by facilitating Ca^{2+} entry in the nerve terminal (Pattillo et al., 2001; Xu & Slaughter, 2005). This is observed in embryonic *Xenopus* nerve-muscle co-cultures, where BK channel blockers (iberiotoxin and charybdotoxin) reduce transmitter release by approximately a third (Pattillo et al., 2001). This observation was hypothesized to result from a precisely timed broadening of the presynaptic AP, which reduces Ca^{2+} currents in varicosities (Pattillo et al., 2001). Importantly, opposite observations were made at adult frog neuromuscular junctions, where the same blockers produce a twofold increase in transmitter release, suggesting that the timing of BK channel contribution to AP repolarization is developmentally regulated (Robitaille & Charlton, 1992; Robitaille et al., 1993). In nonspiking rod photoreceptors of the salamander retina, BK currents can also facilitate neurotransmitter release through a Ca^{2+}-dependent positive feedback loop (Xu & Slaughter, 2005). Specifically, BK channel activity produces K^+ accumulation in the synaptic cleft, which in turn increases Ca^{2+} channel currents and hence synaptic transmission (Xu & Slaughter, 2005). Taken together, these data indicate that BK channels located at the presynaptic terminal generally inhibit neurotransmitter release; however, under rare conditions, they may enhance synaptic transmission.

3.1.4 Regulation of Dendritic Excitability

In addition to their effects on regulating neuronal firing rates and neurotransmitter release, BK channels exert an additional level of control on neurotransmission by modulating dendritic excitability. Dendritic electrogenesis stems from the backpropagation of axonal APs into the dendritic tree, as well as from locally generated Na^+ and Ca^{2+} spikes (Hausser, Spruston, & Stuart, 2000). Backpropagating APs can amplify dendritic APs and facilitate the induction of long-term potentiation (Magee & Johnston, 1997). Likewise, although dendritic spikes do not necessarily trigger axonal AP generation, they contribute to synaptic plasticity (Golding, Staff, & Spruston, 2002; Holthoff, Kovalchuk, Yuste, & Konnerth, 2004; Rancz & Hausser, 2006). In layer 5 pyramidal neurons, BK channels suppress AP backpropagation during high-frequency firing and increase the threshold for dendritic spike generation (Benhassine & Berger, 2009; Grewe, Bonnan, & Frick, 2010). BK channels also restrict dendritic Ca^{2+} spikes both temporally and spatially. In CA1 pyramidal cells, BK channels shorten dendritic Ca^{2+}-dependent spikes, although dendritic BK channels are not activated by backpropagating APs in this neuronal population (Golding, Jung, Mickus, & Spruston, 1999; Poolos & Johnston, 1999). BK channel inhibition was also shown to improve the propagation of dendritic Ca^{2+} spikes in cerebellar Purkinje neurons, thereby facilitating a form of short-term synaptic plasticity at distant spines (Rancz & Hausser, 2006).

The pathophysiological implications of BK-mediated reduction in dendritic excitability were recently highlighted in mice lacking the Fragile X Mental Retardation Protein (FMRP). These mice exhibit reduced BK channel activity (further discussed in Section 4.7) and, consistent with the findings reported above, recordings performed at distal apical dendrites of layer 5 neocortical neurons revealed an increased amplitude of dendritic Ca^{2+} spikes, as well as an increased amplitude of Ca^{2+} transients associated with bursts of backpropagating APs, in the absence of FMRP. Most strikingly, local application of a BK channel activator (BMS-191011) onto distal dendrites was able to normalize these phenotypes (Zhang et al., 2014).

3.2 Astrocytes

BK channels also play an important role for CNS function in nonneuronal cell types. In astrocytes, BK channels are targeted to endfoot processes that wrap parenchymal blood vessels where they play a critical role in the rapid dilation of intracerebral arterioles produced by adjacent neuronal activity (Filosa et al., 2006; Price, Ludwig, Mi, Schwarz, & Ellisman, 2002). Specifically, following neuronal stimulation, astrocytic BK channels are activated by the elevation of endfoot Ca^{2+} concentration and produce a local K^+ efflux in the perivascular space that activates inward rectifier K^+ (K_{ir}) channels located on arteriolar smooth muscle cells (Filosa et al., 2006). Accordingly, the coupling between neuronal activity and vasodilation is suppressed by iberiotoxin and by genetic inactivation of BK α (Filosa et al., 2006). Similarly, BK channels located in the glia limitans, the barrier of astrocytic endfoot processes that surrounds the brain and spinal cord, participate in the dilation of pial arterioles in response to somatosensory stimulation (Paisansathan, Xu, Vetri, Hernandez, & Pelligrino, 2010). However, at higher astrocytic endfoot Ca^{2+} concentration, BK channels mediate the opposite response (ie, vasoconstriction), by raising perivascular K^+

concentration above a crossover threshold that depolarizes, instead of hyperpolarizing, arteriolar smooth muscle cells (Girouard et al., 2010). This situation occurs following aneurysmal subarachnoid hemorrhage, where astrocytic Ca^{2+} concentrations are sufficiently elevated to invert neurovascular coupling, potentially contributing to subsequent ischemia and neurological deficits (Koide, Bonev, Nelson, & Wellman, 2012).

BK channels and other markers of astrocytic endfeet are downregulated in whole brain samples from mouse models of vascular amyloid deposition representing different stages of cerebral amyloid angiopathy, consistent with a general disruption of the neurovascular unit in this disorder associated with the development of Alzheimer's disease (AD) (Wilcock, Vitek, & Colton, 2009). In addition, pharmacological blockade of BK channels by paxilline prevents the blood–brain barrier disruption induced by HIV-infected astrocytes, which suggests that BK channels may be a target for reducing neuroAIDS pathogenesis (Eugenin, Clements, Zink, & Berman, 2011).

3.3 Microglia

BK currents have been identified in microglial cells of the hippocampus, striatum, neocortex, and entorhinal cortex, as well as in the spinal cord following nerve injury (Bordey & Spencer, 2003; Hayashi et al., 2011; Schilling & Eder, 2007, 2015). Intrathecal administration of a BK channel opener produces microglial activation in the dorsal horn of the spinal cord, as reflected by cell body enlargement and positive staining for the Ca^{2+}-binding protein Iba1 (Hayashi et al., 2011). Spinal nerve injury increases BK currents in spinal microglia and S-ketamine suppresses this effect, which led the authors to propose that the blockade of microglial BK currents by S-ketamine inhibits microglial hyperactivation and mediates the antiallodynic properties of this drug in neuropathic pain (Hayashi et al., 2011). However, the pharmacological approach used in this study does not rule out a role of neuronal BK channels in this phenomenon.

Finally, the density of microglial BK current is not different between brain regions, between juvenile and aged brains, or between ramified and dystrophic microglial cells, thereby discounting a role for BK channels in microglia senescence (Schilling & Eder, 2015). K^+ efflux through BK channels has been hypothesized to play a role in the process motility that enables microglial cells to survey their microenvironment, but experimental evidence supporting this premise still needs to be generated (Schilling & Eder, 2015).

3.4 Other Cell Types

There is, to the best of our knowledge, only one report mentioning the presence of BK channels in oligodendrocytes. Specifically, BK currents were found to regulate Ca^{2+} influx in oligodendroglial precursor cells (Buttigieg, Eftekharpour, Karimi-Abdolrezaee, & Fehlings, 2011). BK channel expression and function gradually diminish during the maturation of precursors into oligodendrocytes, which suggests that BK currents could play a role in the differentiation of these cells (Buttigieg et al., 2011).

Furthermore, BK channels are present in cerebral blood vessels and meninges (Poulsen et al., 2009; Wulf et al., 2008). As in other blood vessels, BK channels play a critical role in promoting cerebral artery vasodilation in response to Ca^{2+} sparks from the sarcoplasmic reticulum (reviewed in Chapter "BK Channels in the Vascular System" by Krishnamoorthy-Natarajan and Koide).

4. ROLE OF BK CHANNELS IN CNS FUNCTION AND PATHOLOGIES

Owing to their multifaceted role in shaping neuronal activity, BK channels help maintain a physiological range of circuit output in the brain and spinal cord. Accordingly, both insufficient and excessive BK channel activity can have detrimental effects on the CNS, as illustrated in Fig. 1.

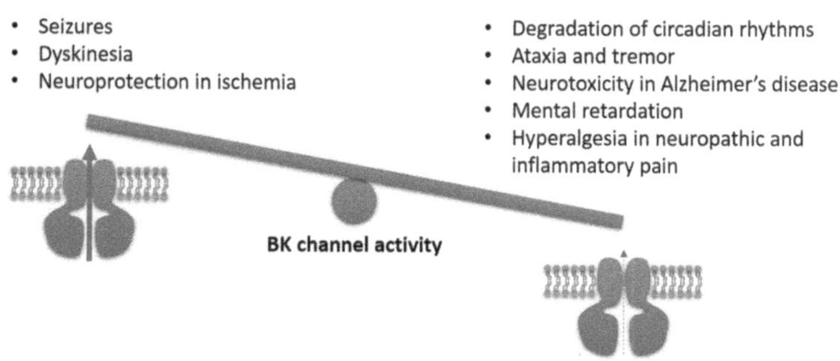

Fig. 1 Summary of the physiological and behavioral outcomes resulting from excessive (*left*) and insufficient (*right*) BK channel activity in the CNS. In some instances, this balance can be tipped by systemic pharmacological manipulation, such that BK channel blockade can normalize abnormalities elicited by BK channel hyperfunction (eg, anticonvulsant effect of paxilline) and BK channel activation can correct abnormalities generated by BK channel hypofunction (eg, analgesic effect of NS-1619). Mechanistic explanations and bibliographic references are provided in the text.

The following section reviews neurological functions and pathologies, in which BK channels have been implicated.

4.1 Control of the Circadian Rhythm

The SCN of the hypothalamus acts as a circadian pacemaker coordinating physiological and behavioral rhythms in mammals (Moore, 2013). Accordingly, SCN neurons exhibit spontaneous rhythms in electrical activity, with high firing rates during daytime and low firing rates or silence at night. This oscillating pattern persists in constant darkness and is independent from synaptic inputs (de Jeu, Hermes, & Pennartz, 1998; Kuhlman & McMahon, 2004; Welsh, Logothetis, Meister, & Reppert, 1995). BK channels contribute to the ionic mechanisms controlling the circadian rhythm of SCN neuronal activity by regulating the magnitude and duration of the AHP (Cloues & Sather, 2003; Montgomery, Whitt, Wright, Lai, & Meredith, 2013). *Kcnma1* was identified as a cycling transcript in the mouse SCN under both entrained (light–dark cycles) and free-running (constant darkness) conditions, with mRNA levels peaking during the dark phase (Meredith et al., 2006; Panda et al., 2002; Pitts et al., 2006). Accordingly, BK current amplitude and density increase during the dark phase in SCN neurons, alongside a decrease in their firing (Montgomery et al., 2013; Pitts et al., 2006).

Consistent with a role of BK channels in silencing SCN neuron activity at night, genetic ablation of BK α selectively increases nighttime firing frequency without affecting daytime frequency (Meredith et al., 2006; Montgomery et al., 2013). In contrast, pharmacological experiments have produced paradoxical results, as iberiotoxin significantly reduces daytime firing in loose patch but not whole-cell recordings, and minimally elevates nighttime firing (Cloues & Sather, 2003; Pitts et al., 2006). These discrepant findings possibly reflect the implication of β4-containing BK channels, which are removed in BK α knockout mice but unaffected by iberiotoxin. The strongly reduced AHP observed in BK α knockout SCN neurons during nighttime is consistent with the hypothesis that slowly deactivating β4-containing BK channels contribute to the suppression of nighttime SCN neuron firing (Montgomery et al., 2013).

At the behavioral level, BK α deletion reduces the amplitude of wheel-running activity, home-cage activity, and body temperature circadian rhythms, but it does not alter the period and phase of these rhythms (Meredith et al., 2006). BK α knockout mice are able to entrain to photic zeitgebers, indicating that BK channels are *not* involved in retinal signaling of ambient light information to the SCN (Meredith et al., 2006). At the molecular level, the absence of BK α does not affect the transcription of core "clock genes"—the set of transcription factors that form autoregulatory feedback loops driving the rhythmic expression of downstream proteins (Meredith et al., 2006). These data imply that the degradation of behavioral and physiological rhythms in BK α knockout mice does not result from a disruption of the transcriptional clockwork.

The role of BK channels in the control of circadian patterning extends beyond mammals. The expression of *Drosophila KCNMA1* orthologue, *slowpoke,* also displays daily oscillations: head mRNA levels peak at the beginning of the dark phase, and protein levels accumulate during the dark phase and drop during the light phase (Ceriani et al., 2002). At the behavioral level, *slowpoke* null mutations weaken the rhythmic behavior of flies by disrupting the typical bursts of activity occurring around the circadian phase transitions (Ceriani et al., 2002). In contrast to the mammalian timekeeping system, fly brain BK channels do not control the activity of core pacemaker cells (small ventral lateral neurons) but rather mediate the rhythmicity of downstream output pathways (dorsal clusters) (Fernandez et al., 2007).

A decay in circadian rhythms is also observed with aging (Monk, 2005). The circadian modulation of BK channel activity in SCN neurons was recently shown to be age-dependent, with BK currents no longer increasing during nighttime in aged SCN neurons (Farajnia, Meijer, & Michel, 2015). The loss of BK current oscillation in aged SCN neurons alters AP waveform and increases $[Ca^{2+}]_i$ at night, changes that could mediate the degradation of behavioral rhythms and other Ca^{2+}-induced disorders in older organisms (Farajnia et al., 2015).

4.2 Epilepsy

Epilepsy is a chronic disorder in which neuronal hyperexcitability and excessive synchronization generate abnormal brain electrical activity (seizures), which can in turn produce absences, loss of consciousness, limb stiffening and/or jerking (convulsions), or atonia. The first evidence for a genetic link between BK channels and epilepsy came from the association between the duplication of a chromosomal segment encompassing the human gene encoding BK β3 (*KCNMB3*) and seizures exhibited by some dup(3q) syndrome patients (Riazi et al., 1999). Gene duplication is hypothesized to increase the expression of β3, but the functional impact of β3 upregulation on BK channel activity is difficult to predict, in particular given the very low abundance of this subunit in the brain (Behrens et al., 2000; Brenner, Jegla, et al., 2000; Poulsen et al., 2009; Uebele et al., 2000).

A truncation mutation in *KCNMB3* (delA750) was later found to confer susceptibility to idiopathic generalized epilepsy (Lorenz, Heils, Kasper, & Sander, 2007). This mutation selectively alters the properties of one of the four β3 splice variants, β3b, by lowering voltage sensitivity and causing rapid inactivation of BK channels containing this isoform (Hu et al., 2003). Whether BK β3 truncation alters AP waveform and neuronal excitability is unknown. Notably, genetic ablation of BK α subunits causes movement disorders (see Section 4.3.), but does not produce an epileptic phenotype in mice and even elevates seizure thresholds in flies, which does not support a causative role for BK channel *hypoactivity* in epileptogenesis (Kuebler, Zhang, Ren, & Tanouye, 2001; Sausbier et al., 2004).

In contrast, a number of studies indicate that BK channel *hyperactivity* increases neuronal excitability. In some instances, BK currents can facilitate high-frequency firing by accelerating spike repolarization and enhancing the AHP amplitude, thereby accelerating the recovery of Na^+ channels from inactivation (see Section 3.1.2). Accordingly, a gain-of-function mutation in the human gene encoding BK α (*KCNMA1*, D434G) was identified in a large family of patients exhibiting generalized epilepsy, paroxysmal nonkinesigenic dyskinesia (ie, involuntary movements of the body), or both (Du et al., 2005). The D434G mutation speeds up BK channel activation and enhances Ca^{2+} sensitivity via an increase in intrinsic gating, Ca^{2+} affinity, and allosteric coupling between Ca^{2+} binding and channel opening (Diez-Sampedro, Silverman, Bautista, & Richerson, 2006; Du et al., 2005; Lee & Cui, 2009; Wang, Rothberg, & Brenner, 2009; Yang et al., 2010). While these effects are preserved when β1, β2, or β4 subunits assemble with BK α, the D434G mutation slows down activation kinetics, decreases voltage dependence, and does not alter Ca^{2+} sensitivity in the presence of β3b (Lee & Cui, 2009). The relevance of this distinction for neuronal physiology remains unclear given the low levels of β3 expression in the CNS.

Further evidence for a causal link between increased BK channel function and seizure activity comes from the manipulation of BK β4. Genetic deletion of BK β4 reduces AP half-width, shortens the AHP, and increases firing rates in dentate gyrus granule cells, which exhibit particularly high levels of β4 expression in wild-type mice (Brenner et al., 2005). At the systemic level, BK β4 knockout mice exhibit spontaneous nonconvulsive seizures that appear to originate in the hippocampus and spread into the whole neocortex (Brenner et al., 2005). Interestingly, a single-nucleotide polymorphism (SNP) located 3′ of the human gene encoding BK β4 (*KCNMB4,* rs398702) confers strong predisposition to mesial temporal lobe epilepsy associated with hippocampal sclerosis in an Irish population (Cavalleri et al., 2007). The impact of this allelic variation on BK β4 expression is currently unknown. In mouse dentate granule cells, the effects of BK β4 deletion on AP waveform are reversed by paxilline, consistent with the idea that the β4 subunit reduces BK currents in wild-type neurons (Brenner et al., 2005). One can therefore speculate that paxilline might suppress temporal lobe seizures in BK β4 knockout mice, and possibly in human carriers of the rs398702 allele. Indeed, the anticonvulsant properties of paxilline have been demonstrated in other models, as reviewed below.

Seizure activity potentiates BK currents in neocortical neurons, enhancing their intrinsic excitability and recurrent activity in the local network. In layer 2/3 pyramidal neurons from the mouse somatosensory cortex, whole-cell BK currents are increased fivefold 24 h after picrotoxin-induced generalized seizures in mice (Shruti et al., 2008). This enhancement of BK channel activity is associated with increased evoked firing rate, which is normalized by pharmacological blockade of BK channels, both in isolated neurons and in acute brain slices (Shruti et al., 2008). At the behavioral level, systemic administration of paxilline 24 h after an initial seizure episode elicited by picrotoxin or pentylenetetrazole virtually abolishes seizures upon second exposure to these chemoconvulsants (Sheehan, Benedetti, & Barth, 2009). Consistent with a direct suppression of neural activity, paxilline pretreatment also blocks the induction of Fos immunoreactivity in the neocortex and hippocampus (Sheehan et al., 2009). Similar findings were obtained in rats injected with pilocarpine. Twenty-four hours after seizure induction, dentate gyrus granule cells exhibit an increased firing rate, with a reduced duration of APs and a larger amplitude of fast AHP, which are consistent with enhanced BK channel activity (Mehranfard, Gholamipour-Badie, Motamedi, Janahmadi, & Naderi, 2014). Again, these changes are normalized by paxilline (Mehranfard et al., 2014). Importantly, BK channel blockade does not exert an anticonvulsant effect in animals that had no previous seizure experience, and even reduces the protective effect of cannabidiol in a first seizure episode (Sheehan et al., 2009; Shirazi-zand, Ahmad-Molaei, Motamedi, & Naderi, 2013).

What is the mechanism mediating the seizure-induced upregulation of BK channel activity? Postseizure BK channel hyperactivity could result from increased transcription and/or translation of BK α. Accordingly, in a rat model of genetic predisposition to audiogenic seizures (Krushinskii–Molodkina rats), exposure to audiogenic stimuli transiently upregulates BK α protein levels in the dentate gyrus, with a significant increase detected 3 days but not 14 days postseizure (Savinaa, Levina, Poletaevab, Fedotovab, & Shchipakinaa, 2014). Alternatively, the increase in neuronal BK currents might result from post-translational modification of existing channels. For example, ubiquitination of BK α subunits by the DNA damage-binding protein 1 (DDB1)/Cullin-4A/RING-box protein 1/Cereblon ubiquitin ligase complex prevents their trafficking from the endoplasmic reticulum to the cell surface and inhibition of this process increases BK currents (Jo et al., 2005; Liu et al., 2014). In accordance with a proepileptogenic effect of BK channel hyperactivity, mice with genetic inactivation of DDB1 in forebrain excitatory neurons develop spontaneous seizures in adulthood, sometimes leading to convulsive death (Liu et al., 2014). In addition, they have an enhanced sensitivity to pentylenetetrazol-induced seizures, which is ameliorated by paxilline pre-treatment (Liu et al., 2014).

BK channel activity has also been implicated in the pathogenesis of alcohol withdrawal seizures. Ablation of neuronal BK channels in *D. melanogaster* completely prevents alcohol-induced enhancement of seizure susceptibility (Ghezzi, Krishnan, & Atkinson, 2014). Conversely, deletion of BK β4 prolongs handling-induced convulsions in mice withdrawn from chronic intermittent alcohol inhalation (Kreifeldt, Cates-Gatto, Roberts, & Contet, 2015). These findings corroborate the notion that enhanced BK channel activity promotes abnormal hyperexcitability and seizures.

In some instances, seizure activity or susceptibility have been correlated with a downregulation of BK channels. In the pilocarpine model of mesial temporal lobe epilepsy, in which pilocarpine precipitates status epilepticus followed by a latent period and later by the appearance of spontaneous recurrent seizures, BK α mRNA and protein expression progressively diminishes in the cortex and hippocampus (Ermolinsky, Arshadmansab, Pacheco Otalora, Zarei, & Garrido-Sanabria, 2008; Pacheco Otalora et al., 2008). In this model of chronic seizure activity, BK channel downregulation may represent a compensatory adaptation to offset sustained neuronal hyperexcitability. Alternatively, if this downregulation affects the presynaptic compartment, as suggested by the prominent decrease in BK α immunoreactivity in the terminal fields of hippocampal granule cells and entorhinal inputs to the hippocampus, it could directly contribute to increased glutamate release and drive hyperactivity of downstream neurons (Pacheco Otalora et al., 2008). Interestingly, BK α downregulation in the dentate gyrus of pilocarpine rats is accompanied by a relative increase in the proportion of STREX-containing BK α isoforms, whose gating kinetics promotes repetitive firing (Ermolinsky et al., 2011; Xie & McCobb, 1998). Altogether, it is possible that a decreased expression of presynaptic BK channels and a preferential expression of STREX-containing variants in the soma of glutamatergic neurons concomitantly drive neuronal hyperexcitability in these chronically epileptic rats.

BK current density was also decreased in inferior colliculus neurons 24 h into withdrawal from subchronic alcohol gavage, a time point at which rats exhibit increased audiogenic seizure susceptibility, and partially returned to baseline 24 h later, when seizure susceptibility is normalized, suggesting that this decrease might play a role in seizure induction, either directly or through a downstream area (N'Gouemo & Morad, 2014). Importantly, recordings were obtained from rats that had not been exposed to audiogenic stimulation, ruling out that BK channel hypoactivity was elicited by seizures. Additional studies are needed to determine whether this decrease in BK currents is indeed mediating neuronal hyperexcitability in the inferior colliculus during alcohol withdrawal (Evans, Li, & Faingold, 2000).

Overall, there is abundant evidence that altered BK channel activity is linked to changes in firing output that can be associated with seizures. Findings that show up- or downregulation of BK currents in different seizure models are consistent with the fact that BK currents can speed up or slow down firing depending on their cellular context (see Section 3.1.2). Critical issues to be resolved are how abnormal activity occurring during seizures drives changes in BK currents, and how changes in BK currents either remedy or exacerbate abnormal excitability.

4.3 Movement Disorders

BK channels have been repeatedly implicated in movement disorders, a phenomenon consistent with their abundant expression in the cerebellum. Ablating BK channel activity leads to tremor and cerebellar ataxia. Early evidence for a tremorgenic effect of BK channel inhibition came from the demonstration that the indole diterpene mycotoxins causing ryegrass staggers in ruminants grazing on fungus-infected pastures (eg, penitrems, lolitrems, paxilline) are selective blockers of BK currents (Knaus et al., 1994). Accordingly, BK α knockout mice exhibit muscular tremor and are insensitive to the tremorgenic effects of lolitrem B and paxilline (Imlach et al., 2008; Typlt et al., 2013). On the other hand, genetic deletion of BK $\beta1$ or $\beta4$ does not produce tremors and does not affect the tremorgenic effect of these mycotoxins (Imlach et al., 2008).

In contrast, a gain-of-function mutation in *KCNMA1* has been associated with paroxysmal nonkinesigenic dyskinesia (D434G; Du et al., 2005). Two additional de novo heterozygous point mutations in *KCNMA1* were later identified in children affected by early-onset paroxysmal nonkinesigenic dyskinesia and developmental delay (E884K and N1053S; Zhang, Tian, Gao, Jiang, & Wu, 2015). The functional impact of these two mutations on BK channel activity is however unknown.

Constitutive deletion of BK α in mice also produces abnormal gait and deficient sensorimotor coordination (Sausbier et al., 2004). These deficits are recapitulated in mice lacking BK α exclusively in cerebellar Purkinje neurons, indicating the critical role of BK channel activity in these cells for normal motor coordination (Chen et al., 2010). Genetic deletion of BK β4, but not BK β1, reduces the ataxic effect of BK channel inhibitors (Imlach et al., 2008). The circuitry mediating the influence of BK channel activity on Purkinje cell output has been elegantly worked out. The absence of BK α depolarizes Purkinje neurons resting membrane potential and decreases their spontaneous activity (Sausbier et al., 2004). In addition, the reduced firing of Purkinje cells perturbs the downstream neuronal circuitry, ultimately resulting in reduced excitatory feedback from the inferior olive (Chen et al., 2010). This disruption of the olivocerebellar feedback loop further contributes to silencing Purkinje neurons, thereby disabling the sole output of the cerebellar cortex.

The ataxic behavior produced by genetic inactivation of BK α is associated with the emergence of a beta oscillatory field potential in the cerebellar cortex, along with a highly rhythmic bursting pattern of Purkinje cells (Cheron et al., 2009). Interestingly, spike rhythmicity of Purkinje cells is increased in other rodent models of ataxia, which suggests that it may represent a general pathophysiological mechanism of cerebellar ataxia (Bearzatto et al., 2006; Cheron et al., 2004; Cheron, Servais, Wagstaff, & Dan, 2005; LeDoux & Lorden, 2002; Servais et al., 2005, 2007).

In a mouse model of spinocerebellar ataxia type 1 (SCA1, ATXN-1 [82Q] mice overexpressing mutant human ataxin-1 in Purkinje neurons), a reduction in BK currents contributes to the depolarization block that abolishes Purkinje cell repetitive firing in juvenile mice and is associated with motor impairment at this age (Dell'Orco et al., 2015). Importantly, these phenotypes can be rescued by virally mediated overexpression of BK α in the cerebellum of ATXN-1[82Q] mice, which normalizes the AHP amplitude and firing frequency of 6-week-old Purkinje neurons and ameliorates motor function in young adult mice (Dell'Orco et al., 2015).

4.4 Cerebral Ischemia

BK channels are well suited to provide an "emergency brake" to limit glutamate release in instances of excessive Ca^{2+} influx, such as during cerebral ischemic stroke (Hu et al., 2001). Accordingly, intravenous administration of a BK channel opener (BMS-204352) 1–2 h after middle cerebral artery occlusion reduces cortical infarct volume in rats (Gribkoff et al., 2001). Consistent with a protective role of BK channels in ischemic stroke, BK α knockout mice display larger infarct volume, more severe neurological deficits, and higher mortality than their wild-type littermates (Liao et al., 2010). Interestingly, these pharmacological and genetic manipulations do not impact cerebral blood flow, ruling out a role of vascular BK channels in their effects (Gribkoff et al., 2001; Liao et al., 2010). Pharmacological blockade or genetic inactivation of BK α also exacerbates NMDA-induced neurotoxicity in vivo and neuronal death in hippocampal or cerebrocortical slice cultures exposed to ischemia-like conditions, which further suggests that neuronal BK channels exert their neuroprotective effect by dampening excitotoxicity (Katsuki, Shinohara, Fujimoto, Kume, & Akaike, 2005; Liao et al., 2010; Runden-Pran et al., 2002). It is also possible that BK channels protect against neuronal death via their interactions with the proapoptotic proteins p53 and fas-associated protein with death domain (Sakai & Sokolowski, 2015). An additional mechanism may involve mitochondrial BK channels, which are known to play a critical cardioprotective role in ischemia–reperfusion injury of the heart (Singh et al., 2013; Xu et al., 2002; see Chapter "Functional Role of Mitochondrial and Nuclear BK Channels" by Gao and Li).

Increased BK channel activity also mediates the neuroprotective effect of the traditional Chinese medicine Baifuzi and the antipsychotic chlorpromazine in rodent models of cerebral ischemia (Chi et al., 2010; Li, Zhang, et al., 2014), as well as the neuroprotective effect of resveratrol and estrogen receptor activators in hippocampal slices deprived from oxygen and glucose (Zhang et al., 2008, 2009). Interestingly, Baifuzi promotes BK channel activation via a direct interaction with the STREX insert of BK α, a domain that also mediates BK channel sensitivity to hypoxia (Chi et al., 2010; McCartney et al., 2005). BK channel activation also contributes to the neuroprotective effect of mitochondrial division inhibitor mdivi-1 in a model of spinal cord ischemia–reperfusion injury (Liu et al., 2015).

Despite these promising preclinical findings, the therapeutic efficacy of BMS-204352 failed to be demonstrated in a phase III clinical trial involving acute stroke patients, possibly due to the intrinsic limitations of that particular compound (Jensen, 2002; see Chapter "Developing the Molecular Pharmacology of BK Channels for Therapeutic Benefit" by Kaczorowski and Garcia). The abundance of evidence pointing to a neuroprotective role of BK channels in cerebral ischemic stroke calls for this therapeutic direction to be further explored, although side effects associated with CNS-wide BK channel activation will be a major hurdle to overcome.

4.5 Alzheimer's Disease

A genome-wide association study found that a SNP in the gene encoding BK α (*KCNMA1,* rs16934131) is nominally associated with late-onset AD risk in humans (Beecham et al., 2009). Although a separate study was unable to replicate this association, the same SNP was found to influence age-at-onset and disease duration between different late-onset AD genotypes (Burns et al., 2011). A more recent genome-wide association study, coupled with an analysis of known genetic risk loci for AD and related dementias, found that a SNP in the gene encoding BK β2 (*KCNMB2,* rs9637454) was strongly associated with hippocampal sclerosis, a comorbid neuropathological feature of AD (Beecham et al., 2014).

While the functional consequence of these point mutations are unknown, rodent models of AD have shed some light on the mechanism through which BK channels may contribute to AD pathogenesis. Intracellular injection of amyloid-β_{1-42} ($A\beta_{1-42}$) peptides into neocortical pyramidal cells suppresses BK channel activity, which broadens evoked AP spike width and indirectly enhances Ca^{2+} influx (Yamamoto et al., 2011). This effect is reversed by prior exposure of the animal to electroconvulsive shock (ECS), a treatment known to facilitate BK channel opening (Sakagami et al., 2005; Yamamoto et al., 2011). Accordingly, in a triple-transgenic mouse model of AD that displays intraneuronal Aβ accumulation prior to the appearance of extracellular Aβ plaques and neurofibrillary tangles (4 months old 3xTg mice; Oddo et al., 2003), spike broadening is observed in neocortical neurons and BK currents cannot be detected unless the mice are first subjected to ECS (Yamamoto et al., 2011). A similar effect of Aβ accumulation on BK channels is also observed in brain mitochondrial inner membranes, as ventricular injection of $A\beta_{1-42}$ in rats decreases the abundance of BK α and the open probability of BK channels in this intracellular compartment (Jafari et al., 2015).

Consistent with these findings, both ventricular injection of isopimaric acid, a BK channel opener, and the use of transcranial magnetic stimulation ameliorate cognitive deficits and cortical hyperexcitability in 3xTg mice (Wang, Kang et al., 2015; Wang, Zhang, et al., 2015). Interestingly, these treatments not only restore BK channel activity but also reduce brain Aβ content, which reveals a two-way relationship between BK currents and Aβ accumulation (Wang, Kang, et al., 2015; Wang, Zhang, et al., 2015).

BK channels seem to play an intricate role in AD pathogenesis: In contrast to the therapeutic effect of BK channel activation discussed above, increased activity of presynaptic BK channels was shown to account for reduced basal excitatory transmission in the hippocampus of another mouse model of preplaque AD (6–9 weeks old TgCRND8 mice; Chishti et al., 2001; Ye, Jalini, Mylvaganam, & Carlen, 2010). As a result, treatments enhancing BK channel activity could potentially aggravate synaptic dysfunction while being beneficial to counteract neurotoxicity in AD.

4.6 Adult-Onset Neuronal Ceroid Lipofuscinosis

BK channel dysfunction has been associated with adult-onset neuronal ceroid lipofuscinosis (ANCL) cases caused by mutations in the gene encoding CSPα. This dominantly inherited neurodegenerative disorder is characterized by the accumulation of an autofluorescent lipopigment (lipofuscin) in neuronal lysosomes and causes seizures, ataxia, psychiatric manifestations, and progressive dementia (Noskova et al., 2011; Velinov et al., 2012). CSPα acts as a presynaptic chaperone for proteins involved in neurotransmitter vesicle trafficking (Chamberlain & Burgoyne, 2000). CSPα knockout mice suffer from a progressive neurodegenerative disorder similar to ANCL (Fernandez-Chacon et al., 2004). Interestingly, CSPα genetic ablation dramatically increases whole brain synaptosomal BK α protein levels without altering BK α mRNA levels (Ahrendt, Kyle, Braun, & Braun, 2014; Kyle et al., 2013). This finding is consistent with the idea that CSPα suppresses BK α trafficking to the presynaptic terminals, and likely enhances neurotransmitter release (see Section 3.1.3). Synaptosomal BK α protein abundance was also significantly elevated in the postmortem cortex of an ANCL patient (Donnelier et al., 2015), but it remains to be determined whether BK channel upregulation actually contributes to the etiology of ANCL.

4.7 Mental Retardation

De novo mutations in *KNCMA1* have been identified in several cases of mental retardation (Laumonnier et al., 2006; Zhang et al., 2015). In a patient affected by severe mental deficiency and autism (but not epilepsy), physical disruption of the *KCNMA1* gene by chromosomal translocation led to haploinsufficiency and decreased BK currents in lymphoblastoid cell lines (Laumonnier et al., 2006). Dysregulation of BK α expression is also hypothesized to contribute to intellectual disability in humans carrying a nonsense mutation (R419X) in *CRBN*, the gene encoding cereblon (Higgins, Hao, Kosofsky, & Rajadhyaksha, 2008). Cereblon is a cytosolic protein that promotes the ubiquitination of BK α and suppresses the surface expression of BK channels in neurons (Jo et al., 2005; Liu et al., 2014). Taken together, these findings suggest that both decreased and increased BK channel activity may lead to mental retardation.

Two SNPs in the gene encoding BK β4 (*KCNMB4*) were also found to be among the strongest predictors of an autism spectrum disorder diagnosis, with a first one (rs968122) conferring vulnerability and the other one (rs12317962) being protective (Skafidas et al., 2014). The impact of these SNPs on BK β4 expression and function is currently unknown.

Mechanistic insights into the link between BK channels and mental retardation have recently been provided by the study of FMRP knockout mice, a mouse model of Fragile X Syndrome (FXS). In contrast with its conventional role in the control of dendritic mRNA translation (Darnell & Klann, 2013), FMRP was shown to regulate glutamate release at CA3-CA1 synapses by interacting with BK β4 in the presynaptic terminals of CA3 pyramidal cells (Deng et al., 2013). Specifically, the loss of FMRP broadens APs, reduces the fast AHP, increases presynaptic Ca^{2+} influx, and facilitates neurotransmitter release during repetitive activity (Deng et al., 2013). These effects result from a reduction in the Ca^{2+} sensitivity of BK channels and are corrected by genetic deletion of BK β4. These findings suggest that in wild-type animals FMRP enhances BK channel activity by sequestering β4 subunits, thereby reducing their influence on the gating kinetics and/or trafficking of BK α tetramers (Deng & Klyachko, 2016; Deng et al., 2013). Furthermore, BK β4 deletion normalizes epileptiform neuronal activity in hippocampal slices from FMRP knockout mice even though it promotes neuronal hyperexcitability in the presence of FMRP (Deng & Klyachko, 2016).

A missense mutation in the gene encoding FMRP (R138Q), which impairs the presynaptic function of FMRP and disrupts the interaction between FMRP and BK β4 without altering its canonical postsynaptic function, was recently identified in an FXS patient, thereby providing a unique opportunity to interrogate the role of presynaptic FMRP at the systemic level (Myrick et al., 2015). This case revealed that presynaptic FMRP–BK channel interaction indeed contributes to some of the neuropathological features of FXS, as this patient exhibits intellectual disability and seizures (Myrick et al., 2015). The behavioral relevance of presynaptic FMRP was further confirmed by the striking ability of a BK channel opener (BMS-204352) to normalize the deficits in social recognition and interaction, nonsocial anxiety, and spatial memory of FMRP knockout mice after a single intraperitoneal injection (Hebert et al., 2014).

BK channels are also implicated in the sensory hypersensitivity associated with the loss of FMRP. In FMRP knockout mice, primary somatosensory cortex pyramidal neurons exhibit an increase in the intrinsic excitability of their dendrites, which results from a reduction in I_h and BK currents in this cellular compartment (Zhang et al., 2014). Specifically, the downregulation of dendritic BK channels in these mice enhances AP backpropagation and facilitates the generation of dendritic spikes (Zhang et al., 2014). Accordingly, pharmacological activation of BK channels rescues the hyperexcitable phenotype of FMRP knockout neurons in slices and corrects hypersensitivity to a moderate acoustic stimulus in vivo (Zhang et al., 2014).

Given the bulk of evidence pointing to a causative role of BK channel hypofunction in mental retardation in humans, the cognitive phenotype of BK α knockout mice appears surprisingly mild. Genetic inactivation of BK α abolishes the experience-dependent increase in prepulse inhibition and slows down spatial learning in the Morris water maze, but leaves working memory and spatial reference memory intact (Typlt et al., 2013). It is possible that constitutive deletion of BK α produces compensatory adaptations, particularly in other ion channels, that may mask the more severe phenotypes resulting from BK channel hypoactivity. Alternatively, the behavioral testing that has been conducted in BK α knockout mice may not have tapped into the cognitive domains most impacted in intellectually disabled humans, or BK channels may fulfill a species-specific function with regard to intellect.

4.8 Alcohol Use Disorders

Alcohol use disorders (AUD), formerly known as alcohol abuse and alcohol dependence, represent a spectrum of problematic patterns of alcoholic beverage consumption. Diagnostic criteria of AUD include loss of control over drinking, craving, and withdrawal symptoms (American Psychiatric Association, 2013). Low level of response to alcohol is a strong predictor of future AUD (Schuckit, 1994; Schuckit, Smith, Anderson, & Brown, 2004). *KCNMA1* has been repeatedly detected by genome-wide linkage and association studies of the level of response to alcohol and alcohol dependence, although the signals were statistically weak (Edenberg et al., 2010; Kendler et al., 2011; Schuckit et al., 2005). In addition, both *KCNMA1* and *KCNMB1* were identified as part of a 39-gene functional network associated with a higher risk of developing alcohol dependence (Han et al., 2013).

BK channels are one of the well-established molecular targets of alcohol (Dopico, Lemos, & Treistman, 1996; see Chapter "Modulation of BK Channels by Ethanol" by Dopico). Residues implicated in alcohol sensing are located between the channel Ca^{2+}-sensing sites and gate (Bukiya et al., 2014; Davis, Scott, Hu, & Pierce-Shimomura, 2014). Clinically relevant concentrations of alcohol (20–25 mM) can lead to either BK current potentiation, refractoriness, or reduction depending on a number of parameters. In particular, the association of BK β1 shifts the activation-to-inhibition crossover to a lower Ca^{2+} concentration, yielding alcohol-resistant channels in neurons and alcohol-inhibited channels in cerebral artery myocytes (Bukiya, Liu, & Dopico, 2009; Martin et al., 2004; Wynne et al., 2009). Lowering the concentration of Mg^{2+} also switches the effect of alcohol from activating to inhibitory, which is relevant to the pathophysiology of AUD as chronic alcohol consumption leads to Mg^{2+} depletion (Marrero, Treistman, & Lemos, 2015; Romani, 2008).

In striatal medium spiny neurons, alcohol-induced potentiation of BK currents rapidly desensitizes, unless auxiliary β4 subunits are present (Martin et al., 2008). An miRNA-mediated BK α isoform switch contributes to this rapid desensitization in supraoptic nucleus and striatal neurons, with miR-9 targeting ALCOREX-containing splice variants to degradation, thereby favoring the expression of alcohol-insensitive STREX-containing isoforms (Pietrzykowski et al., 2008; Velazquez-Marrero et al., 2011). Whether miR-9-driven regulation of BK α splice variants also occurs following in vivo alcohol exposure, possibly in a brain region- or cell type-specific manner, awaits further investigation. Changes in the phosphorylation state of BK α also contribute to rapid desensitization (Velazquez-Marrero, Seale, Treistman, & Martin, 2014). Finally, prolonged exposure of neurons to alcohol (6 h) reduces BK channel cell surface expression (Palacio et al., 2015; Pietrzykowski et al., 2004; Velazquez-Marrero et al., 2011).

The contribution of BK channels to the behavioral effects of alcohol has been investigated by genetically manipulating the pore-forming α subunit in worms and flies, and the auxiliary β subunits in mice.

Studies in invertebrates have yielded conflicting results, showing that BK channels mediate the intoxicating effects of alcohol in *Caenorhabditis elegans* (Davies et al., 2003; Davis et al., 2014) while they mediate rapid tolerance to alcohol-induced sedation and increased seizure susceptibility in *Drosophila melanogaster* (Cowmeadow, Krishnan, & Atkinson, 2005; Cowmeadow et al., 2006; Ghezzi, Al-Hasan, Larios, Bohm, & Atkinson, 2004; Ghezzi et al., 2014; Ghezzi, Pohl, Wang, & Atkinson, 2010). In mice, deletion of BK β4 promotes rapid tolerance to the hypolocomotor effect of alcohol (Martin et al., 2008) and attenuates alcohol drinking escalation in alcohol-dependent mice (Kreifeldt, Le, Treistman, Koob, & Contet, 2013). Conversely, deletion of BK β1 accelerates drinking escalation in dependent mice (Kreifeldt et al., 2013), while reducing chronic tolerance to alcohol-induced sedation and hypothermia (Kreifeldt et al., 2015). BK channel auxiliary subunits therefore seem to play a critical role in the adaptive response to chronic alcohol exposure in mammals. Whether the interaction of alcohol with BK channels directly contributes to the intoxicating effects of alcohol and/or to the adaptive mechanisms generating chronic tolerance or drinking escalation in mammals is still an open question and a hot topic of investigation.

4.9 Pain

In the spinal cord, BK channels are expressed in subsets of small- and medium-diameter DRG neurons, in which they shorten AP duration, drive the fast AHP, reduce firing frequency, and prolong the refractory period (Li et al., 2007; Lu et al., 2014; Sarantopoulos et al., 2007; Scholz, Gruss, & Vogel, 1998; Shieh et al., 2007; Zhang, Gopalakrishnan, & Shieh, 2003; Zhang, Mok, Katz, & Gold, 2010). BK channels are particularly enriched in isolectin-B4 (IB4) positive small-diameter DRG neurons, a subset of nociceptive primary afferents (Gruss et al., 2001; Zhang et al., 2010). In these neurons, clinically relevant concentrations of ethanol and 2,2,2-trichloroethanol, the active metabolite of chloral hydrate, potentiate BK currents and may contribute to the analgesic effects of these compounds (Gruss, Hempelmann, & Scholz, 2002; Gruss et al., 2001). Consistent with this hypothesis, BK channels located in calcitonin gene-related peptide terminals were recently shown to mediate the analgesic effect of a peripherally acting μ opioid receptor agonist in a model of trigeminal nociception (Baillie, Schmidhammer, & Mulligan, 2015). BK channels are also present on presynaptic terminals in the superficial dorsal horn, where they inhibit transmitter release from primary afferents and inhibitory interneurons (Furukawa et al., 2008; Song & Marvizon, 2005).

At the behavioral level, pharmacological blockade of spinal BK channels with iberiotoxin produces mechanical hyperalgesia (Chen, Cai, & Pan, 2009). The effect of spinal BK channel activation in naïve animals may be biphasic and/or modality dependent, as intrathecal administration of a low dose of the BK channel opener NS-1619 (0.072 μg) produces tactile allodynia, but a higher dose (20 μg) does not affect thermal and mechanical nociceptive thresholds (Chen et al., 2009; Hayashi et al., 2011).

Peripheral nerve injury, a preclinical model of neuropathic pain, reduces the expression and activity of BK channels in small and medium DRG neurons, resulting in increased excitatory transmission in the superficial dorsal horn (Cao, Chen, Li, & Pan, 2012; Chen et al., 2009; Furukawa et al., 2008; Sarantopoulos et al., 2007). This downregulation of BK α is mediated by brain-derived neurotrophic factor activation of TrkB receptors (Cao et al., 2012). Glial cell-derived neurotrophic factor (GDNF) also reduces BK currents in IB4-positive nociceptors, an effect hypothesized to mediate GDNF-induced mechanical hyperalgesia (Hendrich, Alvarez, Chen, & Levine, 2012). Consistent with BK channel activity attenuating nociceptor signaling, intrathecal administration of a BK channel opener (NS-1619) reduces tactile allodynia, as well as mechanical and thermal hyperalgesia, in nerve-ligated rats (Chen et al., 2009). Spinal BK channel activation also contributes to the antiallodynic effect of resveratrol (a plant stilbenoid) in neuropathic pain, and to the analgesic effect of MV8612 (a plant pregnane compound) in inflammatory pain (Bermudez-Ocana, Ambriz-Tututi, Perez-Severiano, & Granados-Soto, 2006; Santos, Trentin, Ferreira, Yunes, & Calixto, 2003). In models of chronic inflammatory pain, genetic ablation of BK α in DRG sensory neurons aggravates nociceptive behavior, while systemic administration of a BK channel opener (NS-1619) exerts analgesic properties (Lu et al., 2014).

Surprisingly, DRG-restricted BK α conditional knockout mice display unaltered nociceptive thresholds in models of acute or neuropathic pain (Lu et al., 2014). This lack of phenotype suggests that the effects of intrathecally administered BK channel modulators may be mediated by other spinal cord neuronal populations, such as interneurons (see below). Another paradoxical finding was that charybdotoxin, a BK channel blocker, suppresses nerve injury-induced tactile allodynia (Hayashi et al., 2011). This discrepancy may stem from the promiscuous action of charybdotoxin on intermediate conductance Ca^{2+}-activated K^{+} (IK) channels, as well as on voltage-activated K^+ channels (Garcia et al., 1991; Garcia, Knaus, Munujos, Slaughter, & Kaczorowski, 1995).

BK channels inhibit the release of inhibitory neurotransmitters from spinal cord interneurons. In particular, they reduce spontaneous glycinergic transmission in the sacral dorsal commissural nucleus and mediate the inhibitory effect of NMDA on opioid release in the dorsal horn (Shoudai et al., 2007; Song & Marvizon, 2005). Interestingly, while nerve ligation reduces BK channel expression in DRG neurons (see above), it increases BK α immunoreactivity in dorsal horn interneurons (Chen et al., 2009). The functional impact of this redistribution awaits further investigation, but suppression of inhibitory signaling within the dorsal horn may potentiate nociception in neuropathic pain.

BK channels are also located at juxtaparanodal sites in spinal cord white matter axons, beneath the myelin sheath (Ye et al., 2012). Under normal conditions, juxtaparanodal BK channels do not modulate axonal conduction, but in a model of spinal cord compression injury, resulting demyelination exposes juxtaparanodal BK channels, whose activity then impairs axonal conduction (Ye et al., 2012). In summary, BK channels are implicated in multiple aspects of sensory information flow in the spinal cord, and changes in BK channel activity upon nerve injury or following treatment can have either detrimental or beneficial consequences.

4.10 In the Eye

BK currents can be detected in a large number of retinal cell types, including retinal pigment epithelial cells (Tao & Kelly, 1996; Wimmers, Halsband, Seyler, Milenkovic, & Strauss, 2008), rod photoreceptors (Pelucchi, Grimaldi, & Moriondo, 2008; Xu & Slaughter, 2005), cone photoreceptors (Barnes & Hille, 1989), retinal bipolar cells (Burrone & Lagnado, 1997; Sakaba, Ishikane, & Tachibana, 1997), amacrine cells (Grimes, Li, Chavez, & Diamond, 2009; Mitra & Slaughter, 2002b), and retinal ganglion cells (Henne & Jeserich, 2004; Wang, Robinson, & Chalupa, 1998). They are also present in retinal arterioles (Mori, Suzuki, Sakamoto, Nakahara, & Ishii, 2011) and retinal pericytes (Quignard, Harley, Duhault, Vanhoutte, & Feletou, 2003), which are both involved in the regulation of retinal blood flow, as well as in proliferating Müller (glial) cells (see Bringmann et al., 2000 for review) and in trabecular meshwork, a smooth muscle-like tissue involved in the regulation of aqueous humor outflow (Soto et al., 2004; Stumpff, Strauss, Boxberger, & Wiederholt, 1997).

In a subset of amacrine cells, BK channels mediate spontaneous miniature outward currents, which are hypothesized to serve as synaptic noise suppressors at quiescent glutamatergic synapses (Mitra & Slaughter, 2002a, 2002b). In A17 amacrine cells, BK channels also suppress activation of Ca_V channels to limit GABA release onto rod bipolar cells (Grimes et al., 2009). Accordingly, genetic ablation of BK channels reduces rod bipolar cell activity in vivo, although this modulatory role is restricted to intermediate lighting conditions (Tanimoto et al., 2012). Conversely, at the ribbon synapse of rod photoreceptors, BK channel activity enhances neurotransmitter release (Xu & Slaughter, 2005; see Section 3.1.3). Notably, this facilitatory role is only observed within a physiological range of depolarization, beyond which BK currents hyperpolarize the rod and brake neurotransmitter release. Finally, in the retinal pigment epithelium, BK channels regulate the rhythmic phagocytosis of shed photoreceptor outer segments, a process whose dysfunction leads to blindness (Muller, Mas Gomez, Ruth, & Strauss, 2014; Strauss, 2005).

Although there are no journal publications on the topic, multiple patents cover the use of BK channel *inhibitors* to lower intraocular pressure for the treatment of open-angle ocular hypertension and the prevention of glaucoma (Adorante, WoldeMussie, Ruiz, Kopper, & Moore, 1999; Brnardic et al., 2009; Chen et al., 2004; Gao & Shen, 2006; Goetz, Kaczorowski, Monaghan, Strohl, & Tkacz, 2003). Surprisingly, the drug Rescula (unoprostone isopropyl ophthalmic solution), the only BK channel modulator that has been approved by the U.S. Food and Drug Administration for this indication, is a potent BK channel *activator* (Cuppoletti, Malinowska, Tewari, Chakrabarti, & Ueno, 2007; Yu et al., 2015). It is hypothesized to exert its therapeutic action by increasing the aqueous humor outflow through the trabecular meshwork (Thieme et al., 2001), while inhibitors claimed in the patent literature are hypothesized to act on BK channels of the nonpigmented ciliary epithelium to block humor inflow (see Chapter "Developing the Molecular Pharmacology of BK Channels for Therapeutic Benefit" by Kaczorowski and Garcia).

5. CONCLUSIONS

BK channels have myriad (and sometimes opposite) effects on neural activity, as underscored not only by experimental data but also by modeling studies. Their ability to regulate firing output and synaptic transmission depends upon their coexpression and intracellular association with other ion channels, particularly Ca^{2+} channels, but also HCN and Na^+ channels among others. Overall, it is clear that early studies investigating how specific conditions change total levels of BK α transcript or protein were insufficient to understand how BK channels influence normal neuronal function and confer disease phenotypes. Specific future challenges for experimentalists will be in separating the physiological contribution of intracellular vs plasma membrane BK channels, as well as isolating the contribution of somatic vs presynaptic BK channels. Electrophysiological measures, or biochemical methods that can focus on specific cellular compartments, will be required to fully comprehend how BK channels contribute to firing and synaptic communication.

The regulation of BK α pre-mRNA alternative splicing and isoform-specific mechanisms of posttranscriptional control can have dramatic functional consequences on BK channel activity in vitro. Analyzing the abundance of specific splice variants, their subcellular localization, and interactions with other proteins in different CNS regions and/or cell types will be essential for fully understanding the role BK channels play in diverse pathologies in vivo. Such knowledge may then enable the development of therapeutic approaches targeting critically involved subpopulations of BK channels, in an attempt to circumvent the detrimental effects associated with CNS-wide BK channel activation or inhibition (summarized in Fig. 1).

ACKNOWLEDGMENTS

We are grateful for the support of NIH Grants U01 AA-020913 and R21 AA-024198 to C.C., and R21 MH-100612 to A.L.B.

REFERENCES

Adorante, J. S., WoldeMussie, E., Ruiz, G., Kopper, K., & Moore, A. M. (1999). Method for reducing intraocular pressure in the mammalian eye by administration of potassium channel blockers. Patent No. US5925342 A.

Ahrendt, E., Kyle, B., Braun, A. P., & Braun, J. E. (2014). Cysteine string protein limits expression of the large conductance, calcium-activated K(+) (BK) channel. *PloS One*, *9*(1), e86586.

Alioua, A., Li, M., Wu, Y., Stefani, E., & Toro, L. (2011). Unconventional myristoylation of large-conductance Ca(2)(+)-activated K(+) channel (Slo1) via serine/threonine residues regulates channel surface expression. *Proceedings of the National Academy of Sciences of the United States of America*, *108*(26), 10744–10749.

Alioua, A., Lu, R., Kumar, Y., Eghbali, M., Kundu, P., Toro, L., & Stefani, E. (2008). Slo1 caveolin-binding motif, a mechanism of caveolin-1-Slo1 interaction regulating Slo1 surface expression. *The Journal of Biological Chemistry*, *283*(8), 4808–4817.

Alle, H., Kubota, H., & Geiger, J. R. (2011). Sparse but highly efficient Kv3 outpace BKCa channels in action potential repolarization at hippocampal mossy fiber boutons. *The Journal of Neuroscience*, *31*(22), 8001–8012.

Almassy, J., & Begenisich, T. (2012). The LRRC26 protein selectively alters the efficacy of BK channel activators. *Molecular Pharmacology*, *81*(1), 21–30.

American Psychiatric Association (2013). *Diagnostic and statistical manual of mental disorders* (5th ed.). Arlington, VA: American Psychiatric Publishing.

Augustine, G. J., Santamaria, F., & Tanaka, K. (2003). Local calcium signaling in neurons. *Neuron*, *40*(2), 331–346.

Bai, J. P., Surguchev, A., Joshi, P., Gross, L., & Navaratnam, D. (2012). CDK5 interacts with Slo and affects its surface expression and kinetics through direct phosphorylation. *American Journal of Physiology. Cell Physiology*, *302*(5), C766–C780.

Baillie, L. D., Schmidhammer, H., & Mulligan, S. J. (2015). Peripheral mu-opioid receptor mediated inhibition of calcium signaling and action potential-evoked calcium fluorescent transients in primary afferent CGRP nociceptive terminals. *Neuropharmacology*, *93*, 267–273.

Barnes, S., & Hille, B. (1989). Ionic channels of the inner segment of tiger salamander cone photoreceptors. *The Journal of General Physiology*, *94*(4), 719–743.

Bao, G., de Jong, D., Alevra, M., & Schild, D. (2015). Ca(2+)-BK channel clusters in olfactory receptor neurons and their role in odour coding. *The European Journal of Neuroscience*, *42*(11), 2985–2995.

Bearzatto, B., Servais, L., Roussel, C., Gall, D., Baba-Aissa, F., Schurmans, S., … Schiffmann, S. N. (2006). Targeted calretinin expression in granule cells of calretinin-null mice restores normal cerebellar functions. *FASEB Journal*, *20*(2), 380–382.

Beecham, G. W., Hamilton, K., Naj, A. C., Martin, E. R., Huentelman, M., Myers, A. J., … Montine, T. J. (2014). Genome-wide association meta-analysis of neuropathologic features of Alzheimer's disease and related dementias. *PLoS Genetics*, *10*(9), e1004606.

Beecham, G. W., Martin, E. R., Li, Y. J., Slifer, M. A., Gilbert, J. R., Haines, J. L., & Pericak-Vance, M. A. (2009). Genome-wide association study implicates a chromosome 12 risk locus for late-onset Alzheimer disease. *American Journal of Human Genetics*, *84*(1), 35–43.

Behrens, R., Nolting, A., Reimann, F., Schwarz, M., Waldschutz, R., & Pongs, O. (2000). hKCNMB3 and hKCNMB4, cloning and characterization of two members of the large-conductance calcium-activated potassium channel beta subunit family. *FEBS Letters*, *474*(1), 99–106.

Bell, T. J., Miyashiro, K. Y., Sul, J. Y., McCullough, R., Buckley, P. T., Jochems, J., … Eberwine, J. (2008). Cytoplasmic BK(Ca) channel intron-containing mRNAs contribute to the intrinsic excitability of hippocampal neurons. *Proceedings of the National Academy of Sciences of the United States of America*, *105*(6), 1901–1906.

Benhassine, N., & Berger, T. (2009). Large-conductance calcium-dependent potassium channels prevent dendritic excitability in neocortical pyramidal neurons. *Pflügers Archiv*, *457*(5), 1133–1145.

Benton, M. D., Lewis, A. H., Bant, J. S., & Raman, I. M. (2013). Iberiotoxin-sensitive and -insensitive BK currents in Purkinje neuron somata. *Journal of Neurophysiology*, *109*(10), 2528–2541.

Berkefeld, H., & Fakler, B. (2008). Repolarizing responses of BKCa-Ca_v complexes are distinctly shaped by their Ca_v subunits. *The Journal of Neuroscience*, *28*(33), 8238–8245.

Berkefeld, H., Sailer, C. A., Bildl, W., Rohde, V., Thumfart, J. O., Eble, S., … Fakler, B. (2006). BKCa-Ca_v channel complexes mediate rapid and localized Ca^{2+}-activated K^+ signaling. *Science*, *314*(5799), 615–620.

Bermudez-Ocana, D. Y., Ambriz-Tututi, M., Perez-Severiano, F., & Granados-Soto, V. (2006). Pharmacological evidence for the participation of NO-cyclic GMP-PKG-K^+ channel pathway in the antiallodynic action of resveratrol. *Pharmacology, Biochemistry, and Behavior*, *84*(3), 535–542.

Bhattarai, Y., Fernandes, R., Kadrofske, M. M., Lockwood, L. R., Galligan, J. J., & Xu, H. (2014). Western blot analysis of BK channel beta1-subunit expression should be interpreted cautiously when using commercially available antibodies. *Physiological Reports*, *2*(10), e12189.

Bielefeldt, K., & Jackson, M. B. (1993). A calcium-activated potassium channel causes frequency-dependent action-potential failures in a mammalian nerve terminal. *Journal of Neurophysiology*, *70*(1), 284–298.

Bielefeldt, K., Rotter, J. L., & Jackson, M. B. (1992). Three potassium channels in rat posterior pituitary nerve terminals. *The Journal of Physiology*, *458*, 41–67.

Bolborea, M., & Dale, N. (2013). Hypothalamic tanycytes: Potential roles in the control of feeding and energy balance. *Trends in Neurosciences*, *36*(2), 91–100.

Bordey, A., & Spencer, D. D. (2003). Chemokine modulation of high-conductance Ca(2+)-sensitive K(+) currents in microglia from human hippocampi. *The European Journal of Neuroscience*, *18*(10), 2893–2898.

Brenes, O., Vandael, D. H., Carbone, E., Montarolo, P. G., & Ghirardi, M. (2015). Knock-down of synapsin alters cell excitability and action potential waveform by potentiating BK and voltage-gated Ca(2+) currents in Helix serotonergic neurons. *Neuroscience*, *311*, 430–443.

Brenner, R., Chen, Q. H., Vilaythong, A., Toney, G. M., Noebels, J. L., & Aldrich, R. W. (2005). BK channel beta4 subunit reduces dentate gyrus excitability and protects against temporal lobe seizures. *Nature Neuroscience*, *8*(12), 1752–1759.

Brenner, R., Jegla, T. J., Wickenden, A., Liu, Y., & Aldrich, R. W. (2000). Cloning and functional characterization of novel large conductance calcium-activated potassium channel beta subunits, hKCNMB3 and hKCNMB4. *The Journal of Biological Chemistry*, *275*(9), 6453–6461.

Brenner, R., Perez, G. J., Bonev, A. D., Eckman, D. M., Kosek, J. C., Wiler, S. W., … Aldrich, R. W. (2000). Vasoregulation by the beta1 subunit of the calcium-activated potassium channel. *Nature*, *407*(6806), 870–876.

Bringmann, A., Francke, M., Pannicke, T., Biedermann, B., Kodal, H., Faude, F., … Reichenbach, A. (2000). Role of glial K(+) channels in ontogeny and gliosis: A hypothesis based upon studies on Muller cells. *Glia*, *29*(1), 35–44.

Brnardic, E., Doherty, J. B., Dorsey, J., Ellwood, C., Fillmore, M., Malaska, M., … Soukri, M. (2009). Maxi-k channel blockers and methods of use. Patent No. WO2009048559 A1.

Bukiya, A. N., Kuntamallappanavar, G., Edwards, J., Singh, A. K., Shivakumar, B., & Dopico, A. M. (2014). An alcohol-sensing site in the calcium- and voltage-gated, large conductance potassium (BK) channel. *Proceedings of the National Academy of Sciences of the United States of America*, *111*(25), 9313–9318.

Bukiya, A. N., Liu, J., & Dopico, A. M. (2009). The BK channel accessory beta1 subunit determines alcohol-induced cerebrovascular constriction. *FEBS Letters*, *583*(17), 2779–2784.

Burdyga, T., & Wray, S. (2005). Action potential refractory period in ureter smooth muscle is set by Ca sparks and BK channels. *Nature*, *436*(7050), 559–562.

Burns, L. C., Minster, R. L., Demirci, F. Y., Barmada, M. M., Ganguli, M., Lopez, O. L., … Kamboh, M. I. (2011). Replication study of genome-wide associated SNPs with late-onset Alzheimer's disease. *American Journal of Medical Genetics. Part B, Neuropsychiatric Genetics*, *156*(4), 507–512.

Burrone, J., & Lagnado, L. (1997). Electrical resonance and Ca^{2+} influx in the synaptic terminal of depolarizing bipolar cells from the goldfish retina. *The Journal of Physiology*, *505*(Pt. 3), 571–584.

Buttigieg, J., Eftekharpour, E., Karimi-Abdolrezaee, S., & Fehlings, M. G. (2011). Molecular and electrophysiological evidence for the expression of BK channels in oligodendroglial precursor cells. *The European Journal of Neuroscience*, *34*(4), 538–547.

Cahoy, J. D., Emery, B., Kaushal, A., Foo, L. C., Zamanian, J. L., Christopherson, K. S., … Barres, B. A. (2008). A transcriptome database for astrocytes, neurons, and oligodendrocytes: A new resource for understanding brain development and function. *The Journal of Neuroscience*, *28*(1), 264–278.

Cao, X. H., Chen, S. R., Li, L., & Pan, H. L. (2012). Nerve injury increases brain-derived neurotrophic factor levels to suppress BK channel activity in primary sensory neurons. *Journal of Neurochemistry*, *121*(6), 944–953.

Castillo, P. E., Weisskopf, M. G., & Nicoll, R. A. (1994). The role of Ca^{2+} channels in hippocampal mossy fiber synaptic transmission and long-term potentiation. *Neuron*, *12*(2), 261–269.

Cavalleri, G. L., Weale, M. E., Shianna, K. V., Singh, R., Lynch, J. M., Grinton, B., … Goldstein, D. B. (2007). Multicentre search for genetic susceptibility loci in sporadic epilepsy syndrome and seizure types: A case-control study. *Lancet Neurology*, *6*(11), 970–980.

Ceriani, M. F., Hogenesch, J. B., Yanovsky, M., Panda, S., Straume, M., & Kay, S. A. (2002). Genome-wide expression analysis in Drosophila reveals genes controlling circadian behavior. *The Journal of Neuroscience*, *22*(21), 9305–9319.

Chamberlain, L. H., & Burgoyne, R. D. (2000). Cysteine-string protein: The chaperone at the synapse. *Journal of Neurochemistry*, *74*(5), 1781–1789.

Chang, C. P., Dworetzky, S. I., Wang, J., & Goldstein, M. E. (1997). Differential expression of the alpha and beta subunits of the large-conductance calcium-activated potassium channel: Implication for channel diversity. *Brain Research. Molecular Brain Research*, *45*(1), 33–40.

Chen, L., Bi, D., Tian, L., McClafferty, H., Steeb, F., Ruth, P., … Shipston, M. J. (2013). Palmitoylation of the beta4-subunit regulates surface expression of large conductance calcium-activated potassium channel splice variants. *The Journal of Biological Chemistry*, *288*(18), 13136–13144.

Chen, S. R., Cai, Y. Q., & Pan, H. L. (2009). Plasticity and emerging role of BKCa channels in nociceptive control in neuropathic pain. *Journal of Neurochemistry*, *110*(1), 352–362.

Chen, M. H., Doherty, J. B., Liu, L., Natarajan, S. R., Shen, D. M., & Tynebor, R. M. (2004). Ophthalmic compositions for treating ocular hypertension. Patent No. WO2004043354 A2.

Chen, X., Kovalchuk, Y., Adelsberger, H., Henning, H. A., Sausbier, M., Wietzorrek, G., … Konnerth, A. (2010). Disruption of the olivo-cerebellar circuit by Purkinje neuron-specific ablation of BK channels. *Proceedings of the National Academy of Sciences of the United States of America*, *107*(27), 12323–12328.

Cheron, G., Gall, D., Servais, L., Dan, B., Maex, R., & Schiffmann, S. N. (2004). Inactivation of calcium-binding protein genes induces 160 Hz oscillations in the cerebellar cortex of alert mice. *The Journal of Neuroscience*, *24*(2), 434–441.

Cheron, G., Sausbier, M., Sausbier, U., Neuhuber, W., Ruth, P., Dan, B., & Servais, L. (2009). BK channels control cerebellar Purkinje and Golgi cell rhythmicity in vivo. *PloS One*, *4*(11), e7991.

Cheron, G., Servais, L., Wagstaff, J., & Dan, B. (2005). Fast cerebellar oscillation associated with ataxia in a mouse model of Angelman syndrome. *Neuroscience*, *130*(3), 631–637.

Chi, S., Cai, W., Liu, P., Zhang, Z., Chen, X., Gao, L., … Qi, Z. (2010). Baifuzi reduces transient ischemic brain damage through an interaction with the STREX domain of BKCa channels. *Cell Death & Disease*, *1*, e13.

Chishti, M. A., Yang, D. S., Janus, C., Phinney, A. L., Horne, P., Pearson, J., … Westaway, D. (2001). Early-onset amyloid deposition and cognitive deficits in transgenic mice expressing a double mutant form of amyloid precursor protein 695. *The Journal of Biological Chemistry*, *276*(24), 21562–21570.

Cloues, R. K., & Sather, W. A. (2003). Afterhyperpolarization regulates firing rate in neurons of the suprachiasmatic nucleus. *The Journal of Neuroscience*, *23*(5), 1593–1604.

Cowmeadow, R. B., Krishnan, H. R., & Atkinson, N. S. (2005). The slowpoke gene is necessary for rapid ethanol tolerance in Drosophila. *Alcoholism, Clinical and Experimental Research*, *29*(10), 1777–1786.

Cowmeadow, R. B., Krishnan, H. R., Ghezzi, A., Al'Hasan, Y. M., Wang, Y. Z., & Atkinson, N. S. (2006). Ethanol tolerance caused by slowpoke induction in Drosophila. *Alcoholism, Clinical and Experimental Research, 30*(5), 745–753.

Cuppoletti, J., Malinowska, D. H., Tewari, K. P., Chakrabarti, J., & Ueno, R. (2007). Cellular and molecular effects of unoprostone as a BK channel activator. *Biochimica et Biophysica Acta, 1768*(5), 1083–1092.

Darnell, J. C., & Klann, E. (2013). The translation of translational control by FMRP: Therapeutic targets for FXS. *Nature Neuroscience, 16*(11), 1530–1536.

Davies, A. G., Pierce-Shimomura, J. T., Kim, H., VanHoven, M. K., Thiele, T. R., Bonci, A., … McIntire, S. L. (2003). A central role of the BK potassium channel in behavioral responses to ethanol in C. elegans. *Cell, 115*(6), 655–666.

Davis, S. J., Scott, L. L., Hu, K., & Pierce-Shimomura, J. T. (2014). Conserved single residue in the BK potassium channel required for activation by alcohol and intoxication in C. elegans. *The Journal of Neuroscience, 34*(29), 9562–9573.

de Jeu, M., Hermes, M., & Pennartz, C. (1998). Circadian modulation of membrane properties in slices of rat suprachiasmatic nucleus. *Neuroreport, 9*(16), 3725–3729.

Dell'Orco, J. M., Wasserman, A. H., Chopra, R., Ingram, M. A., Hu, Y. S., Singh, V., … Shakkottai, V. G. (2015). Neuronal atrophy early in degenerative ataxia is a compensatory mechanism to regulate membrane excitability. *The Journal of Neuroscience, 35*(32), 11292–11307.

Deng, P. Y., & Klyachko, V. A. (2016). Genetic upregulation of BK channel activity normalizes multiple synaptic and circuit defects in a mouse model of fragile X syndrome. *The Journal of Physiology, 594*(1), 83–97.

Deng, P. Y., Rotman, Z., Blundon, J. A., Cho, Y., Cui, J., Cavalli, V., … Klyachko, V. A. (2013). FMRP regulates neurotransmitter release and synaptic information transmission by modulating action potential duration via BK channels. *Neuron, 77*(4), 696–711.

Diez-Sampedro, A., Silverman, W. R., Bautista, J. F., & Richerson, G. B. (2006). Mechanism of increased open probability by a mutation of the BK channel. *Journal of Neurophysiology, 96*(3), 1507–1516.

Dobremez, E., Bouali-Benazzouz, R., Fossat, P., Monteils, L., Dulluc, J., Nagy, F., & Landry, M. (2005). Distribution and regulation of L-type calcium channels in deep dorsal horn neurons after sciatic nerve injury in rats. *The European Journal of Neuroscience, 21*(12), 3321–3333.

Doi, M., Ishida, A., Miyake, A., Sato, M., Komatsu, R., Yamazaki, F., … Okamura, H. (2011). Circadian regulation of intracellular G-protein signalling mediates intercellular synchrony and rhythmicity in the suprachiasmatic nucleus. *Nature Communications, 2*, 327.

Donnelier, J., Braun, S. T., Dolzhanskaya, N., Ahrendt, E., Braun, A. P., Velinov, M., & Braun, J. E. (2015). Increased expression of the large conductance, calcium-activated K+ (BK) channel in adult-onset neuronal ceroid lipofuscinosis. *PloS One, 10*(4), e0125205.

Dopico, A. M., Lemos, J. R., & Treistman, S. N. (1996). Ethanol increases the activity of large conductance, Ca(2+)-activated K+ channels in isolated neurohypophysial terminals. *Molecular Pharmacology, 49*(1), 40–48.

Dopico, A. M., Widmer, H., Wang, G., Lemos, J. R., & Treistman, S. N. (1999). Rat supraoptic magnocellular neurones show distinct large conductance, Ca^{2+}-activated K^{+} channel subtypes in cell bodies versus nerve endings. *The Journal of Physiology, 519*(Pt. 1), 101–114.

Du, W., Bautista, J. F., Yang, H., Diez-Sampedro, A., You, S. A., Wang, L., … Wang, Q. K. (2005). Calcium-sensitive potassium channelopathy in human epilepsy and paroxysmal movement disorder. *Nature Genetics, 37*(7), 733–738.

Edenberg, H. J., Koller, D. L., Xuei, X., Wetherill, L., McClintick, J. N., Almasy, L., ... Foroud, T. (2010). Genome-wide association study of alcohol dependence implicates a region on chromosome 11. *Alcoholism, Clinical and Experimental Research, 34*(5), 840–852.

Edgerton, J. R., & Reinhart, P. H. (2003). Distinct contributions of small and large conductance Ca^{2+}-activated K^+ channels to rat Purkinje neuron function. *The Journal of Physiology, 548*(Pt. 1), 53–69.

Engbers, J. D., Anderson, D., Zamponi, G. W., & Turner, R. W. (2013). Signal processing by T-type calcium channel interactions in the cerebellum. *Frontiers in Cellular Neuroscience, 7*, 230.

Ermolinsky, B., Arshadmansab, M. F., Pacheco Otalora, L. F., Zarei, M. M., & Garrido-Sanabria, E. R. (2008). Deficit of Kcnma1 mRNA expression in the dentate gyrus of epileptic rats. *Neuroreport, 19*(13), 1291–1294.

Ermolinsky, B. S., Skinner, F., Garcia, I., Arshadmansab, M. F., Otalora, L. F., Zarei, M. M., & Garrido-Sanabria, E. R. (2011). Upregulation of STREX splice variant of the large conductance Ca^{2+}-activated potassium (BK) channel in a rat model of mesial temporal lobe epilepsy. *Neuroscience Research, 69*(1), 73–80.

Eugenin, E. A., Clements, J. E., Zink, M. C., & Berman, J. W. (2011). Human immunodeficiency virus infection of human astrocytes disrupts blood-brain barrier integrity by a gap junction-dependent mechanism. *The Journal of Neuroscience, 31*(26), 9456–9465.

Evans, M. S., Li, Y., & Faingold, C. (2000). Inferior colliculus intracellular response abnormalities in vitro associated with susceptibility to ethanol withdrawal seizures. *Alcoholism, Clinical and Experimental Research, 24*(8), 1180–1186.

Evanson, K. W., Bannister, J. P., Leo, M. D., & Jaggar, J. H. (2014). LRRC26 is a functional BK channel auxiliary gamma subunit in arterial smooth muscle cells. *Circulation Research, 115*(4), 423–431.

Faber, E. S., & Sah, P. (2002). Physiological role of calcium-activated potassium currents in the rat lateral amygdala. *The Journal of Neuroscience, 22*(5), 1618–1628.

Faber, E. S., & Sah, P. (2003). Ca^{2+}-activated K^+ (BK) channel inactivation contributes to spike broadening during repetitive firing in the rat lateral amygdala. *The Journal of Physiology, 552*(Pt. 2), 483–497.

Fakler, B., & Adelman, J. P. (2008). Control of K(Ca) channels by calcium nano/microdomains. *Neuron, 59*(6), 873–881.

Farajnia, S., Meijer, J. H., & Michel, S. (2015). Age-related changes in large-conductance calcium-activated potassium channels in mammalian circadian clock neurons. *Neurobiology of Aging, 36*(6), 2176–2183.

Feinberg-Zadek, P. L., & Treistman, S. N. (2007). Beta-subunits are important modulators of the acute response to alcohol in human BK channels. *Alcoholism, Clinical and Experimental Research, 31*(5), 737–744.

Fernandez, M. P., Chu, J., Villella, A., Atkinson, N., Kay, S. A., & Ceriani, M. F. (2007). Impaired clock output by altered connectivity in the circadian network. *Proceedings of the National Academy of Sciences of the United States of America, 104*(13), 5650–5655.

Fernandez-Chacon, R., Wolfel, M., Nishimune, H., Tabares, L., Schmitz, F., Castellano-Munoz, M., ... Sudhof, T. C. (2004). The synaptic vesicle protein CSP alpha prevents presynaptic degeneration. *Neuron, 42*(2), 237–251.

Filosa, J. A., Bonev, A. D., Straub, S. V., Meredith, A. L., Wilkerson, M. K., Aldrich, R. W., & Nelson, M. T. (2006). Local potassium signaling couples neuronal activity to vasodilation in the brain. *Nature Neuroscience, 9*(11), 1397–1403.

Ford, K. J., & Davis, G. W. (2014). Archaerhodopsin voltage imaging: Synaptic calcium and BK channels stabilize action potential repolarization at the Drosophila neuromuscular junction. *The Journal of Neuroscience, 34*(44), 14517–14525.

Furukawa, N., Takasusuki, T., Fukushima, T., & Hori, Y. (2008). Presynaptic large-conductance calcium-activated potassium channels control synaptic transmission in the superficial dorsal horn of the mouse. *Neuroscience Letters, 444*(1), 79–82.

Gao, Y. D., & Shen, D. M. (2006). Compositions ophtalmiques servant a traiter l'hypertension oculaire. Patent No. WO2006044232 A1.

Garcia, M. L., Galvez, A., Garcia-Calvo, M., King, V. F., Vazquez, J., & Kaczorowski, G. J. (1991). Use of toxins to study potassium channels. *Journal of Bioenergetics and Biomembranes, 23*(4), 615–646.

Garcia, M. L., Knaus, H. G., Munujos, P., Slaughter, R. S., & Kaczorowski, G. J. (1995). Charybdotoxin and its effects on potassium channels. *The American Journal of Physiology, 269*(1 Pt. 1), C1–C10.

Ghezzi, A., Al-Hasan, Y. M., Larios, L. E., Bohm, R. A., & Atkinson, N. S. (2004). slo K(+) channel gene regulation mediates rapid drug tolerance. *Proceedings of the National Academy of Sciences of the United States of America, 101*(49), 17276–17281.

Ghezzi, A., Krishnan, H. R., & Atkinson, N. S. (2014). Susceptibility to ethanol withdrawal seizures is produced by BK channel gene expression. *Addiction Biology, 19*(3), 332–337.

Ghezzi, A., Pohl, J. B., Wang, Y., & Atkinson, N. S. (2010). BK channels play a counter-adaptive role in drug tolerance and dependence. *Proceedings of the National Academy of Sciences of the United States of America, 107*(37), 16360–16365.

Gho, M., & Ganetzky, B. (1992). Analysis of repolarization of presynaptic motor terminals in Drosophila larvae using potassium-channel-blocking drugs and mutations. *The Journal of Experimental Biology, 170*, 93–111.

Girouard, H., Bonev, A. D., Hannah, R. M., Meredith, A., Aldrich, R. W., & Nelson, M. T. (2010). Astrocytic endfoot Ca^{2+} and BK channels determine both arteriolar dilation and constriction. *Proceedings of the National Academy of Sciences of the United States of America, 107*(8), 3811–3816.

Gittis, A. H., & du Lac, S. (2007). Firing properties of GABAergic versus non-GABAergic vestibular nucleus neurons conferred by a differential balance of potassium currents. *Journal of Neurophysiology, 97*(6), 3986–3996.

Gittis, A. H., Moghadam, S. H., & du Lac, S. (2010). Mechanisms of sustained high firing rates in two classes of vestibular nucleus neurons: Differential contributions of resurgent Na, Kv3, and BK currents. *Journal of Neurophysiology, 104*(3), 1625–1634.

Goetz, M. A., Kaczorowski, G. J., Monaghan, R. L., Strohl, W. R., & Tkacz, J. S. (2003). Novel maxi-k channel blockers, methods of use and process for making the same. Patent No. WO2003105868 A1.

Golding, N. L., Jung, H. Y., Mickus, T., & Spruston, N. (1999). Dendritic calcium spike initiation and repolarization are controlled by distinct potassium channel subtypes in CA1 pyramidal neurons. *The Journal of Neuroscience, 19*(20), 8789–8798.

Golding, N. L., Staff, N. P., & Spruston, N. (2002). Dendritic spikes as a mechanism for cooperative long-term potentiation. *Nature, 418*(6895), 326–331.

Gorini, G., Ponomareva, O., Shores, K. S., Person, M. D., Harris, R. A., & Mayfield, R. D. (2010). Dynamin-1 co-associates with native mouse brain BKCa channels: Proteomics analysis of synaptic protein complexes. *FEBS Letters, 584*(5), 845–851.

Granados-Fuentes, D., Norris, A. J., Carrasquillo, Y., Nerbonne, J. M., & Herzog, E. D. (2012). I(A) channels encoded by Kv1.4 and Kv4.2 regulate neuronal firing in the suprachiasmatic nucleus and circadian rhythms in locomotor activity. *The Journal of Neuroscience, 32*(29), 10045–10052.

Grewe, B. F., Bonnan, A., & Frick, A. (2010). Back-propagation of physiological action potential output in dendrites of slender-tufted L5A pyramidal neurons. *Frontiers in Cellular Neuroscience, 4*, 13.

Gribkoff, V. K., Starrett, J. E., Jr., Dworetzky, S. I., Hewawasam, P., Boissard, C. G., Cook, D. A., … Yeola, S. W. (2001). Targeting acute ischemic stroke with a calcium-sensitive opener of maxi-K potassium channels. *Nature Medicine*, 7(4), 471–477.

Grimes, W. N., Li, W., Chavez, A. E., & Diamond, J. S. (2009). BK channels modulate pre- and postsynaptic signaling at reciprocal synapses in retina. *Nature Neuroscience*, *12*(5), 585–592.

Grunnet, M., & Kaufmann, W. A. (2004). Coassembly of big conductance Ca^{2+}-activated K^{+} channels and L-type voltage-gated Ca^{2+} channels in rat brain. *The Journal of Biological Chemistry*, *279*(35), 36445–36453.

Gruss, M., Hempelmann, G., & Scholz, A. (2002). Trichloroethanol alters action potentials in a subgroup of primary sensory neurones. *Neuroreport*, *13*(6), 853–856.

Gruss, M., Henrich, M., Konig, P., Hempelmann, G., Vogel, W., & Scholz, A. (2001). Ethanol reduces excitability in a subgroup of primary sensory neurons by activation of BK(Ca) channels. *The European Journal of Neuroscience*, *14*(8), 1246–1256.

Gu, N., Vervaeke, K., & Storm, J. F. (2007). BK potassium channels facilitate high-frequency firing and cause early spike frequency adaptation in rat CA1 hippocampal pyramidal cells. *The Journal of Physiology*, *580*(Pt. 3), 859–882.

Ha, T. S., Heo, M. S., & Park, C. S. (2004). Functional effects of auxiliary beta4-subunit on rat large-conductance Ca(2+)-activated K(+) channel. *Biophysical Journal*, *86*(5), 2871–2882.

Hafidi, A., Beurg, M., & Dulon, D. (2005). Localization and developmental expression of BK channels in mammalian cochlear hair cells. *Neuroscience*, *130*(2), 475–484.

Haghdoost-Yazdi, H., Janahmadi, M., & Behzadi, G. (2008). Iberiotoxin-sensitive large conductance Ca^{2+}-dependent K^{+} (BK) channels regulate the spike configuration in the burst firing of cerebellar Purkinje neurons. *Brain Research*, *1212*, 1–8.

Han, S., Yang, B. Z., Kranzler, H. R., Liu, X., Zhao, H., Farrer, L. A., … Gelernter, J. (2013). Integrating GWASs and human protein interaction networks identifies a gene subnetwork underlying alcohol dependence. *American Journal of Human Genetics*, *93*(6), 1027–1034.

Hausser, M., Spruston, N., & Stuart, G. J. (2000). Diversity and dynamics of dendritic signaling. *Science*, *290*(5492), 739–744.

Hayashi, Y., Kawaji, K., Sun, L., Zhang, X., Koyano, K., Yokoyama, T., … Nakanishi, H. (2011). Microglial Ca(2+)-activated K(+) channels are possible molecular targets for the analgesic effects of S-ketamine on neuropathic pain. *The Journal of Neuroscience*, *31*(48), 17370–17382.

Hebert, B., Pietropaolo, S., Meme, S., Laudier, B., Laugeray, A., Doisne, N., … Briault, S. (2014). Rescue of fragile X syndrome phenotypes in Fmr1 KO mice by a BKCa channel opener molecule. *Orphanet Journal of Rare Diseases*, *9*, 124.

Hendrich, J., Alvarez, P., Chen, X., & Levine, J. D. (2012). GDNF induces mechanical hyperalgesia in muscle by reducing I(BK) in isolectin B4-positive nociceptors. *Neuroscience*, *219*, 204–213.

Henne, J., & Jeserich, G. (2004). Maturation of spiking activity in trout retinal ganglion cells coincides with upregulation of Kv3.1- and BK-related potassium channels. *Journal of Neuroscience Research*, *75*(1), 44–54.

Hicks, G. A., & Marrion, N. V. (1998). Ca^{2+}-dependent inactivation of large conductance Ca^{2+}-activated K^{+} (BK) channels in rat hippocampal neurones produced by pore block from an associated particle. *The Journal of Physiology*, *508*(Pt. 3), 721–734.

Higgins, J. J., Hao, J., Kosofsky, B. E., & Rajadhyaksha, A. M. (2008). Dysregulation of large-conductance Ca^{2+}-activated K^{+} channel expression in nonsyndromal mental retardation due to a cereblon p.R419X mutation. *Neurogenetics*, *9*(3), 219–223.

Higgins, J. J., Tal, A. L., Sun, X., Hauck, S. C., Hao, J., Kosofosky, B. E., & Rajadhyaksha, A. M. (2010). Temporal and spatial mouse brain expression of cereblon, an ionic channel regulator involved in human intelligence. *Journal of Neurogenetics*, *24*(1), 18–26.
Hirono, M., Ogawa, Y., Misono, K., Zollinger, D. R., Trimmer, J. S., Rasband, M. N., & Misonou, H. (2015). BK channels localize to the paranodal junction and regulate action potentials in myelinated axons of cerebellar Purkinje cells. *The Journal of Neuroscience*, *35*(18), 7082–7094.
Holthoff, K., Kovalchuk, Y., Yuste, R., & Konnerth, A. (2004). Single-shock LTD by local dendritic spikes in pyramidal neurons of mouse visual cortex. *The Journal of Physiology*, *560*(Pt. 1), 27–36.
Hong, J. H., Jeong, B., Min, C. H., & Lee, K. J. (2012). Circadian waves of cytosolic calcium concentration and long-range network connections in rat suprachiasmatic nucleus. *The European Journal of Neuroscience*, *35*(9), 1417–1425.
Hu, S., Labuda, M. Z., Pandolfo, M., Goss, G. G., McDermid, H. E., & Ali, D. W. (2003). Variants of the KCNMB3 regulatory subunit of maxi BK channels affect channel inactivation. *Physiological Genomics*, *15*(3), 191–198.
Hu, H., Shao, L. R., Chavoshy, S., Gu, N., Trieb, M., Behrens, R., … Storm, J. F. (2001). Presynaptic Ca^{2+}-activated K^+ channels in glutamatergic hippocampal terminals and their role in spike repolarization and regulation of transmitter release. *The Journal of Neuroscience*, *21*(24), 9585–9597.
Ikeda, M., & Ikeda, M. (2014). Bmal1 is an essential regulator for circadian cytosolic Ca(2)(+) rhythms in suprachiasmatic nucleus neurons. *The Journal of Neuroscience*, *34*(36), 12029–12038.
Ikeda, M., Sugiyama, T., Wallace, C. S., Gompf, H. S., Yoshioka, T., Miyawaki, A., & Allen, C. N. (2003). Circadian dynamics of cytosolic and nuclear Ca2+ in single suprachiasmatic nucleus neurons. *Neuron*, *38*(2), 253–263.
Imlach, W. L., Finch, S. C., Dunlop, J., Meredith, A. L., Aldrich, R. W., & Dalziel, J. E. (2008). The molecular mechanism of "ryegrass staggers," a neurological disorder of K^+ channels. *The Journal of Pharmacology and Experimental Therapeutics*, *327*(3), 657–664.
Indriati, D. W., Kamasawa, N., Matsui, K., Meredith, A. L., Watanabe, M., & Shigemoto, R. (2013). Quantitative localization of Cav2.1 (P/Q-type) voltage-dependent calcium channels in Purkinje cells: Somatodendritic gradient and distinct somatic coclustering with calcium-activated potassium channels. *The Journal of Neuroscience*, *33*(8), 3668–3678.
Jafari, A., Noursadeghi, E., Khodagholi, F., Saghiri, R., Sauve, R., Aliaghaei, A., & Eliassi, A. (2015). Brain mitochondrial ATP-insensitive large conductance Ca(+)(2)-activated K (+) channel properties are altered in a rat model of amyloid-beta neurotoxicity. *Experimental Neurology*, *269*, 8–16.
Jeffries, O., Geiger, N., Rowe, I. C., Tian, L., McClafferty, H., Chen, L., … Shipston, M. J. (2010). Palmitoylation of the S0-S1 linker regulates cell surface expression of voltage- and calcium-activated potassium (BK) channels. *The Journal of Biological Chemistry*, *285*(43), 33307–33314.
Jensen, B. S. (2002). BMS-204352: A potassium channel opener developed for the treatment of stroke. *CNS Drug Reviews*, *8*(4), 353–360.
Jiang, Z., Wallner, M., Meera, P., & Toro, L. (1999). Human and rodent MaxiK channel beta-subunit genes: Cloning and characterization. *Genomics*, *55*(1), 57–67.
Jin, W., Sugaya, A., Tsuda, T., Ohguchi, H., & Sugaya, E. (2000). Relationship between large conductance calcium-activated potassium channel and bursting activity. *Brain Research*, *860*(1–2), 21–28.

Jo, S., Lee, K. H., Song, S., Jung, Y. K., & Park, C. S. (2005). Identification and functional characterization of cereblon as a binding protein for large-conductance calcium-activated potassium channel in rat brain. *Journal of Neurochemistry*, *94*(5), 1212–1224.

Kaeser, P. S., & Regehr, W. G. (2014). Molecular mechanisms for synchronous, asynchronous, and spontaneous neurotransmitter release. *Annual Review of Physiology*, *76*, 333–363.

Kang, J., Huguenard, J. R., & Prince, D. A. (1996). Development of BK channels in neocortical pyramidal neurons. *Journal of Neurophysiology*, *76*(1), 188–198.

Kathiresan, T., Harvey, M., Orchard, S., Sakai, Y., & Sokolowski, B. (2009). A protein interaction network for the large conductance Ca(2+)-activated K(+) channel in the mouse cochlea. *Molecular & Cellular Proteomics*, *8*(8), 1972–1987.

Katsuki, H., Shinohara, A., Fujimoto, S., Kume, T., & Akaike, A. (2005). Tetraethylammonium exacerbates ischemic neuronal injury in rat cerebrocortical slice cultures. *European Journal of Pharmacology*, *508*(1–3), 85–91.

Kaufmann, W. A., Ferraguti, F., Fukazawa, Y., Kasugai, Y., Shigemoto, R., Laake, P., … Ottersen, O. P. (2009). Large-conductance calcium-activated potassium channels in Purkinje cell plasma membranes are clustered at sites of hypolemmal microdomains. *The Journal of Comparative Neurology*, *515*(2), 215–230.

Kaufmann, W. A., Kasugai, Y., Ferraguti, F., & Storm, J. F. (2010). Two distinct pools of large-conductance calcium-activated potassium channels in the somatic plasma membrane of central principal neurons. *Neuroscience*, *169*(3), 974–986.

Kendler, K. S., Kalsi, G., Holmans, P. A., Sanders, A. R., Aggen, S. H., Dick, D. M., … Gejman, P. V. (2011). Genomewide association analysis of symptoms of alcohol dependence in the molecular genetics of schizophrenia (MGS2) control sample. *Alcoholism, Clinical and Experimental Research*, *35*(5), 963–975.

Kim, H., Jo, S., Song, H. J., Park, Z. Y., & Park, C. S. (2007). Myelin basic protein as a binding partner and calmodulin adaptor for the BKCa channel. *Proteomics*, 7(15), 2591–2602.

Kim, E. Y., Ridgway, L. D., Zou, S., Chiu, Y. H., & Dryer, S. E. (2007). Alternatively spliced C-terminal domains regulate the surface expression of large conductance calcium-activated potassium channels. *Neuroscience*, *146*(4), 1652–1661.

Kim, E. Y., Zou, S., Ridgway, L. D., & Dryer, S. E. (2007). Beta1-subunits increase surface expression of a large-conductance Ca^{2+}-activated K^+ channel isoform. *Journal of Neurophysiology*, *97*(5), 3508–3516.

Knaus, H. G., McManus, O. B., Lee, S. H., Schmalhofer, W. A., Garcia-Calvo, M., Helms, L. M., … Garcia, M. L. (1994). Tremorgenic indole alkaloids potently inhibit smooth muscle high-conductance calcium-activated potassium channels. *Biochemistry*, *33*(19), 5819–5828.

Knaus, H. G., Schwarzer, C., Koch, R. O., Eberhart, A., Kaczorowski, G. J., Glossmann, H., … Sperk, G. (1996). Distribution of high-conductance Ca(2+)-activated K^+ channels in rat brain: Targeting to axons and nerve terminals. *The Journal of Neuroscience*, *16*(3), 955–963.

Koide, M., Bonev, A. D., Nelson, M. T., & Wellman, G. C. (2012). Inversion of neurovascular coupling by subarachnoid blood depends on large-conductance Ca^{2+}-activated K^+ (BK) channels. *Proceedings of the National Academy of Sciences of the United States of America*, *109*(21), E1387–E1395.

Kreifeldt, M., Cates-Gatto, C., Roberts, A. J., & Contet, C. (2015). BK channel beta1 subunit contributes to behavioral adaptations elicited by chronic intermittent ethanol exposure. *Alcoholism, Clinical and Experimental Research*, *39*(12), 2394–2402.

Kreifeldt, M., Le, D., Treistman, S. N., Koob, G. F., & Contet, C. (2013). BK channel beta1 and beta4 auxiliary subunits exert opposite influences on escalated ethanol drinking in dependent mice. *Frontiers in Integrative Neuroscience*, 7, 105.

Kros, C. J., Ruppersberg, J. P., & Rusch, A. (1998). Expression of a potassium current in inner hair cells during development of hearing in mice. *Nature, 394*(6690), 281–284.
Ksiazek, A., Ladno, W., Szulczyk, B., Grzelka, K., & Szulczyk, P. (2013). Properties of BK-type Ca(+) (+)-dependent K(+) channel currents in medial prefrontal cortex pyramidal neurons in rats of different ages. *Frontiers in Cellular Neuroscience*, 7, 185.
Kudo, T., Loh, D. H., Kuljis, D., Constance, C., & Colwell, C. S. (2011). Fast delayed rectifier potassium current: Critical for input and output of the circadian system. *The Journal of Neuroscience, 31*(8), 2746–2755.
Kuebler, D., Zhang, H., Ren, X., & Tanouye, M. A. (2001). Genetic suppression of seizure susceptibility in Drosophila. *Journal of Neurophysiology, 86*(3), 1211–1225.
Kuhlman, S. J., & McMahon, D. G. (2004). Rhythmic regulation of membrane potential and potassium current persists in SCN neurons in the absence of environmental input. *The European Journal of Neuroscience, 20*(4), 1113–1117.
Kulik, A., Nakadate, K., Hagiwara, A., Fukazawa, Y., Lujan, R., Saito, H., … Shigemoto, R. (2004). Immunocytochemical localization of the alpha 1A subunit of the P/Q-type calcium channel in the rat cerebellum. *The European Journal of Neuroscience, 19*(8), 2169–2178.
Kyle, B. D., Ahrendt, E., Braun, A. P., & Braun, J. E. (2013). The large conductance, calcium-activated K+ (BK) channel is regulated by cysteine string protein. *Scientific Reports, 3*, 2447.
Lancaster, B., & Nicoll, R. A. (1987). Properties of two calcium-activated hyperpolarizations in rat hippocampal neurones. *The Journal of Physiology, 389*, 187–203.
Langer, P., Grunder, S., & Rusch, A. (2003). Expression of Ca^{2+}-activated BK channel mRNA and its splice variants in the rat cochlea. *The Journal of Comparative Neurology, 455*(2), 198–209.
Langlet, F., Mullier, A., Bouret, S. G., Prevot, V., & Dehouck, B. (2013). Tanycyte-like cells form a blood-cerebrospinal fluid barrier in the circumventricular organs of the mouse brain. *The Journal of Comparative Neurology, 521*(15), 3389–3405.
Laumonnier, F., Roger, S., Guerin, P., Molinari, F., M'Rad, R., Cahard, D., … Briault, S. (2006). Association of a functional deficit of the BKCa channel, a synaptic regulator of neuronal excitability, with autism and mental retardation. *The American Journal of Psychiatry, 163*(9), 1622–1629.
LeDoux, M. S., & Lorden, J. F. (2002). Abnormal spontaneous and harmaline-stimulated Purkinje cell activity in the awake genetically dystonic rat. *Experimental Brain Research, 145*(4), 457–467.
Ledoux, J., Werner, M. E., Brayden, J. E., & Nelson, M. T. (2006). Calcium-activated potassium channels and the regulation of vascular tone. *Physiology (Bethesda), 21*, 69–78.
Lee, U. S., & Cui, J. (2009). {beta} Subunit-specific modulations of BK channel function by a mutation associated with epilepsy and dyskinesia. *The Journal of Physiology, 587*(Pt. 7), 1481–1498.
Lee, J., Ueda, A., & Wu, C. F. (2008). Pre- and post-synaptic mechanisms of synaptic strength homeostasis revealed by slowpoke and shaker K^+ channel mutations in Drosophila. *Neuroscience, 154*(4), 1283–1296.
Lein, E. S., Hawrylycz, M. J., Ao, N., Ayres, M., Bensinger, A., Bernard, A., … Jones, A. R. (2007). Genome-wide atlas of gene expression in the adult mouse brain. *Nature, 445*(7124), 168–176.
Li, M., Chang, S., Yang, L., Shi, J., McFarland, K., Yang, X., … Cui, J. (2014). Conopeptide Vt3.1 preferentially inhibits BK potassium channels containing beta4 subunits via electrostatic interactions. *The Journal of Biological Chemistry, 289*(8), 4735–4742.
Li, W., Gao, S. B., Lv, C. X., Wu, Y., Guo, Z. H., Ding, J. P., & Xu, T. (2007). Characterization of voltage-and Ca^{2+}-activated K^+ channels in rat dorsal root ganglion neurons. *Journal of Cellular Physiology, 212*(2), 348–357.

Li, Q., Madison, R., & Moore, S. D. (2014). Presynaptic BK channels modulate ethanol-induced enhancement of GABAergic transmission in the rat central amygdala nucleus. *The Journal of Neuroscience*, *34*(41), 13714–13724.

Li, H. J., Zhang, Y. J., Zhou, L., Han, F., Wang, M. Y., Xue, M. Q., & Qi, Z. (2014). Chlorpromazine confers neuroprotection against brain ischemia by activating BKCa channel. *European Journal of Pharmacology*, *735*, 38–43.

Liao, Y., Kristiansen, A. M., Oksvold, C. P., Tuvnes, F. A., Gu, N., Runden-Pran, E., … Storm, J. F. (2010). Neuronal Ca^{2+}-activated K^{+} channels limit brain infarction and promote survival. *PloS One*, *5*(12), e15601.

Liu, J., Ye, J., Zou, X., Xu, Z., Feng, Y., Zou, X., … Cang, Y. (2014). CRL4A(CRBN) E3 ubiquitin ligase restricts BK channel activity and prevents epileptogenesis. *Nature Communications*, *5*, 3924.

Liu, J. M., Yi, Z., Liu, S. Z., Chang, J. H., Dang, X. B., Li, Q. Y., & Zhang, Y. L. (2015). The mitochondrial division inhibitor mdivi-1 attenuates spinal cord ischemia-reperfusion injury both in vitro and in vivo: Involvement of BK channels. *Brain Research*, *1619*, 155–165.

Loane, D. J., Lima, P. A., & Marrion, N. V. (2007). Co-assembly of N-type Ca^{2+} and BK channels underlies functional coupling in rat brain. *Journal of Cell Science*, *120*(Pt. 6), 985–995.

Lorenz, S., Heils, A., Kasper, J. M., & Sander, T. (2007). Allelic association of a truncation mutation of the KCNMB3 gene with idiopathic generalized epilepsy. *American Journal of Medical Genetics. Part B, Neuropsychiatric Genetics*, *144*(1), 10–13.

Lu, R., Lukowski, R., Sausbier, M., Zhang, D. D., Sisignano, M., Schuh, C. D., … Schmidtko, A. (2014). BKCa channels expressed in sensory neurons modulate inflammatory pain in mice. *Pain*, *155*(3), 556–565.

Ly, C., Melman, T., Barth, A. L., & Ermentrout, G. B. (2011). Phase-resetting curve determines how BK currents affect neuronal firing. *Journal of Computational Neuroscience*, *30*(2), 211–223.

Ma, D., Nakata, T., Zhang, G., Hoshi, T., Li, M., & Shikano, S. (2007). Differential trafficking of carboxyl isoforms of Ca^{2+}-gated (Slo1) potassium channels. *FEBS Letters*, *581*(5), 1000–1008.

MacDonald, S. H., Ruth, P., Knaus, H. G., & Shipston, M. J. (2006). Increased large conductance calcium-activated potassium (BK) channel expression accompanied by STREX variant downregulation in the developing mouse CNS. *BMC Developmental Biology*, *6*, 37.

Magee, J. C., & Johnston, D. (1997). A synaptically controlled, associative signal for Hebbian plasticity in hippocampal neurons. *Science*, *275*(5297), 209–213.

Marcotti, W., Johnson, S. L., Holley, M. C., & Kros, C. J. (2003). Developmental changes in the expression of potassium currents of embryonic, neonatal and mature mouse inner hair cells. *The Journal of Physiology*, *548*(Pt. 2), 383–400.

Marrero, H. G., Treistman, S. N., & Lemos, J. R. (2015). Ethanol effect on BK channels is modulated by magnesium. *Alcoholism, Clinical and Experimental Research*, *39*(9), 1671–1679.

Marrion, N. V., & Tavalin, S. J. (1998). Selective activation of Ca^{2+}-activated K^{+} channels by co-localized Ca^{2+} channels in hippocampal neurons. *Nature*, *395*(6705), 900–905.

Martin, G. E., Hendrickson, L. M., Penta, K. L., Friesen, R. M., Pietrzykowski, A. Z., Tapper, A. R., & Treistman, S. N. (2008). Identification of a BK channel auxiliary protein controlling molecular and behavioral tolerance to alcohol. *Proceedings of the National Academy of Sciences of the United States of America*, *105*(45), 17543–17548.

Martin, G., Puig, S., Pietrzykowski, A., Zadek, P., Emery, P., & Treistman, S. (2004). Somatic localization of a specific large-conductance calcium-activated potassium channel

subtype controls compartmentalized ethanol sensitivity in the nucleus accumbens. *The Journal of Neuroscience*, *24*(29), 6563–6572.
Martinez-Espinosa, P. L., Yang, C., Gonzalez-Perez, V., Xia, X. M., & Lingle, C. J. (2014). Knockout of the BK beta2 subunit abolishes inactivation of BK currents in mouse adrenal chromaffin cells and results in slow-wave burst activity. *The Journal of General Physiology*, *144*(4), 275–295.
Martire, M., Barrese, V., D'Amico, M., Iannotti, F. A., Pizzarelli, R., Samengo, I., … Taglialatela, M. (2010). Pre-synaptic BK channels selectively control glutamate versus GABA release from cortical and hippocampal nerve terminals. *Journal of Neurochemistry*, *115*(2), 411–422.
Matthews, E. A., Weible, A. P., Shah, S., & Disterhoft, J. F. (2008). The BK-mediated fAHP is modulated by learning a hippocampus-dependent task. *Proceedings of the National Academy of Sciences of the United States of America*, *105*(39), 15154–15159.
McCartney, C. E., McClafferty, H., Huibant, J. M., Rowan, E. G., Shipston, M. J., & Rowe, I. C. (2005). A cysteine-rich motif confers hypoxia sensitivity to mammalian large conductance voltage- and Ca-activated K (BK) channel alpha-subunits. *Proceedings of the National Academy of Sciences of the United States of America*, *102*(49), 17870–17876.
McLarnon, J. G. (1995). Inactivation of a high conductance calcium dependent potassium current in rat hippocampal neurons. *Neuroscience Letters*, *193*(1), 5–8.
Meera, P., Wallner, M., & Toro, L. (2000). A neuronal beta subunit (KCNMB4) makes the large conductance, voltage- and Ca^{2+}-activated K^{+} channel resistant to charybdotoxin and iberiotoxin. *Proceedings of the National Academy of Sciences of the United States of America*, *97*(10), 5562–5567.
Mehranfard, N., Gholamipour-Badie, H., Motamedi, F., Janahmadi, M., & Naderi, N. (2014). The effect of paxilline on early alterations of electrophysiological properties of dentate gyrus granule cells in pilocarpine-treated rats. *Iranian Journal of Pharmaceutical Research*, *13*(Suppl.), 125–132.
Meredith, A. L., Wiler, S. W., Miller, B. H., Takahashi, J. S., Fodor, A. A., Ruby, N. F., & Aldrich, R. W. (2006). BK calcium-activated potassium channels regulate circadian behavioral rhythms and pacemaker output. *Nature Neuroscience*, *9*(8), 1041–1049.
Misonou, H., Menegola, M., Buchwalder, L., Park, E. W., Meredith, A., Rhodes, K. J., … Trimmer, J. S. (2006). Immunolocalization of the Ca^{2+}-activated K^{+} channel Slo1 in axons and nerve terminals of mammalian brain and cultured neurons. *The Journal of Comparative Neurology*, *496*(3), 289–302.
Mitra, P., & Slaughter, M. M. (2002a). Calcium-induced transitions between the spontaneous miniature outward and the transient outward currents in retinal amacrine cells. *The Journal of General Physiology*, *119*(4), 373–388.
Mitra, P., & Slaughter, M. M. (2002b). Mechanism of generation of spontaneous miniature outward currents (SMOCs) in retinal amacrine cells. *The Journal of General Physiology*, *119*(4), 355–372.
Monk, T. H. (2005). Aging human circadian rhythms: Conventional wisdom may not always be right. *Journal of Biological Rhythms*, *20*(4), 366–374.
Montgomery, J. R., & Meredith, A. L. (2012). Genetic activation of BK currents in vivo generates bidirectional effects on neuronal excitability. *Proceedings of the National Academy of Sciences of the United States of America*, *109*(46), 18997–19002.
Montgomery, J. R., Whitt, J. P., Wright, B. N., Lai, M. H., & Meredith, A. L. (2013). Mis-expression of the BK K(+) channel disrupts suprachiasmatic nucleus circuit rhythmicity and alters clock-controlled behavior. *American Journal of Physiology. Cell Physiology*, *304*(4), C299–C311.

Moore, R. Y. (2013). The suprachiasmatic nucleus and the circadian timing system. *Progress in Molecular Biology and Translational Science, 119*, 1–28.

Mori, A., Suzuki, S., Sakamoto, K., Nakahara, T., & Ishii, K. (2011). BMS-191011, an opener of large-conductance Ca^{2+}-activated potassium channels, dilates rat retinal arterioles in vivo. *Biological & Pharmaceutical Bulletin, 34*(1), 150–152.

Muller, A., Kukley, M., Uebachs, M., Beck, H., & Dietrich, D. (2007). Nanodomains of single Ca2+ channels contribute to action potential repolarization in cortical neurons. *The Journal of Neuroscience, 27*(3), 483–495.

Muller, C., Mas Gomez, N., Ruth, P., & Strauss, O. (2014). CaV1.3 L-type channels, maxiK Ca(2+)-dependent K(+) channels and bestrophin-1 regulate rhythmic photoreceptor outer segment phagocytosis by retinal pigment epithelial cells. *Cellular Signalling, 26*(5), 968–978.

Myrick, L. K., Deng, P. Y., Hashimoto, H., Oh, Y. M., Cho, Y., Poidevin, M. J., … Klyachko, V. A. (2015). Independent role for presynaptic FMRP revealed by an FMR1 missense mutation associated with intellectual disability and seizures. *Proceedings of the National Academy of Sciences of the United States of America, 112*(4), 949–956.

Neher, E. (1998). Vesicle pools and Ca^{2+} microdomains: New tools for understanding their roles in neurotransmitter release. *Neuron, 20*(3), 389–399.

Nelson, A. B., Gittis, A. H., & du Lac, S. (2005). Decreases in CaMKII activity trigger persistent potentiation of intrinsic excitability in spontaneously firing vestibular nucleus neurons. *Neuron, 46*(4), 623–631.

Nelson, A. B., Krispel, C. M., Sekirnjak, C., & du Lac, S. (2003). Long-lasting increases in intrinsic excitability triggered by inhibition. *Neuron, 40*(3), 609–620.

N'Gouemo, P., & Morad, M. (2014). Alcohol withdrawal is associated with a down-regulation of large-conductance Ca(2)(+)-activated K(+) channels in rat inferior colliculus neurons. *Psychopharmacology, 231*(9), 2009–2018.

Noskova, L., Stranecky, V., Hartmannova, H., Pristoupilova, A., Baresova, V., Ivanek, R., … Kmoch, S. (2011). Mutations in DNAJC5, encoding cysteine-string protein alpha, cause autosomal-dominant adult-onset neuronal ceroid lipofuscinosis. *American Journal of Human Genetics, 89*(2), 241–252.

Novak, P., Gorelik, J., Vivekananda, U., Shevchuk, A. I., Ermolyuk, Y. S., Bailey, R. J., … Korchev, Y. E. (2013). Nanoscale-targeted patch-clamp recordings of functional presynaptic ion channels. *Neuron, 79*(6), 1067–1077.

Obaid, A. L., Flores, R., & Salzberg, B. M. (1989). Calcium channels that are required for secretion from intact nerve terminals of vertebrates are sensitive to omega-conotoxin and relatively insensitive to dihydropyridines. Optical studies with and without voltage-sensitive dyes. *The Journal of General Physiology, 93*(4), 715–729.

Obermair, G. J., Szabo, Z., Bourinet, E., & Flucher, B. E. (2004). Differential targeting of the L-type Ca^{2+} channel alpha 1C (CaV1.2) to synaptic and extrasynaptic compartments in hippocampal neurons. *The European Journal of Neuroscience, 19*(8), 2109–2122.

Oddo, S., Caccamo, A., Shepherd, J. D., Murphy, M. P., Golde, T. E., Kayed, R., … LaFerla, F. M. (2003). Triple-transgenic model of Alzheimer's disease with plaques and tangles: Intracellular Abeta and synaptic dysfunction. *Neuron, 39*(3), 409–421.

Oh, K. H., Abraham, L. S., Gegg, C., Silvestri, C., Huang, Y. C., Alkema, M. J., … Kim, H. (2015). Presynaptic BK channel localization is dependent on the hierarchical organization of alpha-catulin and dystrobrevin and fine-tuned by CaV2 calcium channels. *BMC Neuroscience, 16*, 26.

O'Malley, D., & Harvey, J. (2007). MAPK-dependent actin cytoskeletal reorganization underlies BK channel activation by insulin. *The European Journal of Neuroscience, 25*(3), 673–682.

O'Malley, D., Irving, A. J., & Harvey, J. (2005). Leptin-induced dynamic alterations in the actin cytoskeleton mediate the activation and synaptic clustering of BK channels. *FASEB Journal, 19*(13), 1917–1919.

O'Neill, J. S., Maywood, E. S., Chesham, J. E., Takahashi, J. S., & Hastings, M. H. (2008). cAMP-dependent signaling as a core component of the mammalian circadian pacemaker. *Science*, *320*(5878), 949–953.

Pacheco Otalora, L. F., Hernandez, E. F., Arshadmansab, M. F., Francisco, S., Willis, M., Ermolinsky, B., … Garrido-Sanabria, E. R. (2008). Down-regulation of BK channel expression in the pilocarpine model of temporal lobe epilepsy. *Brain Research*, *1200*, 116–131.

Paisansathan, C., Xu, H., Vetri, F., Hernandez, M., & Pelligrino, D. A. (2010). Interactions between adenosine and K+ channel-related pathways in the coupling of somatosensory activation and pial arteriolar dilation. *American Journal of Physiology. Heart and Circulatory Physiology*, *299*(6), H2009–H2017.

Palacio, S., Velazquez-Marrero, C., Marrero, H. G., Seale, G. E., Yudowski, G. A., & Treistman, S. N. (2015). Time-dependent effects of ethanol on BK channel expression and trafficking in hippocampal neurons. *Alcoholism, Clinical and Experimental Research*, *39*(9), 1619–1631.

Panda, S., Antoch, M. P., Miller, B. H., Su, A. I., Schook, A. B., Straume, M., … Hogenesch, J. B. (2002). Coordinated transcription of key pathways in the mouse by the circadian clock. *Cell*, *109*(3), 307–320.

Pattillo, J. M., Yazejian, B., DiGregorio, D. A., Vergara, J. L., Grinnell, A. D., & Meriney, S. D. (2001). Contribution of presynaptic calcium-activated potassium currents to transmitter release regulation in cultured Xenopus nerve-muscle synapses. *Neuroscience*, *102*(1), 229–240.

Pelkey, K. A., Topolnik, L., Lacaille, J. C., & McBain, C. J. (2006). Compartmentalized Ca(2+) channel regulation at divergent mossy-fiber release sites underlies target cell-dependent plasticity. *Neuron*, *52*(3), 497–510.

Pelucchi, B., Grimaldi, A., & Moriondo, A. (2008). Vertebrate rod photoreceptors express both BK and IK calcium-activated potassium channels, but only BK channels are involved in receptor potential regulation. *Journal of Neuroscience Research*, *86*(1), 194–201.

Petersen, O. H., Tepikin, A., & Park, M. K. (2001). The endoplasmic reticulum: One continuous or several separate Ca(2+) stores? *Trends in Neurosciences*, *24*(5), 271–276.

Petrik, D., & Brenner, R. (2007). Regulation of STREX exon large conductance, calcium-activated potassium channels by the beta4 accessory subunit. *Neuroscience*, *149*(4), 789–803.

Petrik, D., Wang, B., & Brenner, R. (2011). Modulation by the BK accessory beta4 subunit of phosphorylation-dependent changes in excitability of dentate gyrus granule neurons. *The European Journal of Neuroscience*, *34*(5), 695–704.

Pietrzykowski, A. Z., Friesen, R. M., Martin, G. E., Puig, S. I., Nowak, C. L., Wynne, P. M., … Treistman, S. N. (2008). Posttranscriptional regulation of BK channel splice variant stability by miR-9 underlies neuroadaptation to alcohol. *Neuron*, *59*(2), 274–287.

Pietrzykowski, A. Z., Martin, G. E., Puig, S. I., Knott, T. K., Lemos, J. R., & Treistman, S. N. (2004). Alcohol tolerance in large-conductance, calcium-activated potassium channels of CNS terminals is intrinsic and includes two components: Decreased ethanol potentiation and decreased channel density. *The Journal of Neuroscience*, *24*(38), 8322–8332.

Pitts, G. R., Ohta, H., & McMahon, D. G. (2006). Daily rhythmicity of large-conductance Ca^{2+}-activated K^+ currents in suprachiasmatic nucleus neurons. *Brain Research*, *1071*(1), 54–62.

Piwonska, M., Wilczek, E., Szewczyk, A., & Wilczynski, G. M. (2008). Differential distribution of Ca^{2+}-activated potassium channel beta4 subunit in rat brain: Immunolocalization in neuronal mitochondria. *Neuroscience*, *153*(2), 446–460.

Poolos, N. P., & Johnston, D. (1999). Calcium-activated potassium conductances contribute to action potential repolarization at the soma but not the dendrites of hippocampal CA1 pyramidal neurons. *The Journal of Neuroscience*, *19*(13), 5205–5212.

Poulsen, A. N., Wulf, H., Hay-Schmidt, A., Jansen-Olesen, I., Olesen, J., & Klaerke, D. A. (2009). Differential expression of BK channel isoforms and beta-subunits in rat neurovascular tissues. *Biochimica et Biophysica Acta*, *1788*(2), 380–389.

Prakriya, M., & Lingle, C. J. (1999). BK channel activation by brief depolarizations requires Ca^{2+} influx through L- and Q-type Ca^{2+} channels in rat chromaffin cells. *Journal of Neurophysiology*, *81*(5), 2267–2278.

Pratt, C. P., He, J., Wang, Y., Barth, A. L., & Bruchez, M. P. (2015). Fluorogenic green-inside red-outside (GIRO) labeling approach reveals adenylyl cyclase-dependent control of BKalpha surface expression. *Bioconjugate Chemistry*, *26*(9), 1963–1971.

Price, D. L., Ludwig, J. W., Mi, H., Schwarz, T. L., & Ellisman, M. H. (2002). Distribution of rSlo Ca^{2+}-activated K^+ channels in rat astrocyte perivascular endfeet. *Brain Research*, *956*(2), 183–193.

Quignard, J. F., Harley, E. A., Duhault, J., Vanhoutte, P. M., & Feletou, M. (2003). K^+ channels in cultured bovine retinal pericytes: Effects of beta-adrenergic stimulation. *Journal of Cardiovascular Pharmacology*, *42*(3), 379–388.

Raffaelli, G., Saviane, C., Mohajerani, M. H., Pedarzani, P., & Cherubini, E. (2004). BK potassium channels control transmitter release at CA3-CA3 synapses in the rat hippocampus. *The Journal of Physiology*, *557*(Pt. 1), 147–157.

Ramirez-Latorre, J. A. (2012). Functional upregulation of Ca(2+)-activated K(+) channels in the development of substantia nigra dopamine neurons. *PloS One*, 7(12), e51610.

Rancz, E. A., & Hausser, M. (2006). Dendritic calcium spikes are tunable triggers of cannabinoid release and short-term synaptic plasticity in cerebellar Purkinje neurons. *The Journal of Neuroscience*, *26*(20), 5428–5437.

Rehak, R., Bartoletti, T. M., Engbers, J. D., Berecki, G., Turner, R. W., & Zamponi, G. W. (2013). Low voltage activation of KCa1.1 current by Cav3-KCa1.1 complexes. *PloS One*, *8*(4), e61844.

Reinhart, P. H., Chung, S., & Levitan, I. B. (1989). A family of calcium-dependent potassium channels from rat brain. *Neuron*, *2*(1), 1031–1041.

Riazi, M. A., Brinkman-Mills, P., Johnson, A., Naylor, S. L., Minoshima, S., Shimizu, N., … McDermid, H. E. (1999). Identification of a putative regulatory subunit of a calcium-activated potassium channel in the dup(3q) syndrome region and a related sequence on 22q11.2. *Genomics*, *62*(1), 90–94.

Robitaille, R., & Charlton, M. P. (1992). Presynaptic calcium signals and transmitter release are modulated by calcium-activated potassium channels. *Journal of Neuroscience*, *12*(1), 297–305.

Robitaille, R., Garcia, M. L., Kaczorowski, G. J., & Charlton, M. P. (1993). Functional colocalization of calcium and calcium-gated potassium channels in control of transmitter release. *Neuron*, *11*(4), 645–655.

Romani, A. M. (2008). Magnesium homeostasis and alcohol consumption. *Magnesium Research*, *21*(4), 197–204.

Runden-Pran, E., Haug, F. M., Storm, J. F., & Ottersen, O. P. (2002). BK channel activity determines the extent of cell degeneration after oxygen and glucose deprivation: A study in organotypical hippocampal slice cultures. *Neuroscience*, *112*(2), 277–288.

Sailer, C. A., Kaufmann, W. A., Kogler, M., Chen, L., Sausbier, U., Ottersen, O. P., … Knaus, H. G. (2006). Immunolocalization of BK channels in hippocampal pyramidal neurons. *The European Journal of Neuroscience*, *24*(2), 442–454.

Sakaba, T., Ishikane, H., & Tachibana, M. (1997). Ca^{2+}-activated K^+ current at presynaptic terminals of goldfish retinal bipolar cells. *Neuroscience Research*, *27*(3), 219–228.

Sakagami, Y., Yamamoto, K., Sugiura, S., Inokuchi, K., Hayashi, T., & Kato, N. (2005). Essential roles of Homer-1a in homeostatic regulation of pyramidal cell excitability: A possible link to clinical benefits of electroconvulsive shock. *The European Journal of Neuroscience*, *21*(12), 3229–3239.

Sakai, Y., & Sokolowski, B. (2015). The large conductance calcium-activated potassium channel affects extrinsic and intrinsic mechanisms of apoptosis. *Journal of Neuroscience Research*, *93*(5), 745–754.

Salzmann, M., Seidel, K. N., Bernard, R., Pruss, H., Veh, R. W., & Derst, C. (2010). BKbeta1 subunits contribute to BK channel diversity in rat hypothalamic neurons. *Cellular and Molecular Neurobiology*, *30*(6), 967–976.

Samaranayake, H., Saunders, J. C., Greene, M. I., & Navaratnam, D. S. (2004). Ca(2+) and K(+) (BK) channels in chick hair cells are clustered and colocalized with apical-basal and tonotopic gradients. *The Journal of Physiology*, *560*(Pt. 1), 13–20.

Samengo, I., Curro, D., Barrese, V., Taglialatela, M., & Martire, M. (2014). Large conductance calcium-activated potassium channels: Their expression and modulation of glutamate release from nerve terminals isolated from rat trigeminal caudal nucleus and cerebral cortex. *Neurochemical Research*, *39*(5), 901–910.

Santos, A. R., Trentin, A. P., Ferreira, J., Yunes, R. A., & Calixto, J. B. (2003). Mechanisms involved in the antinociception caused by compound MV8612 isolated from Mandevilla velutina in mice. *Brain Research*, *961*(2), 269–276.

Sarantopoulos, C. D., McCallum, J. B., Rigaud, M., Fuchs, A., Kwok, W. M., & Hogan, Q. H. (2007). Opposing effects of spinal nerve ligation on calcium-activated potassium currents in axotomized and adjacent mammalian primary afferent neurons. *Brain Research*, *1132*(1), 84–99.

Sausbier, M., Hu, H., Arntz, C., Feil, S., Kamm, S., Adelsberger, H., … Ruth, P. (2004). Cerebellar ataxia and Purkinje cell dysfunction caused by Ca^{2+}-activated K^{+} channel deficiency. *Proceedings of the National Academy of Sciences of the United States of America*, *101*(25), 9474–9478.

Sausbier, U., Sausbier, M., Sailer, C. A., Arntz, C., Knaus, H. G., Neuhuber, W., & Ruth, P. (2006). Ca^{2+}-activated K^{+} channels of the BK-type in the mouse brain. *Histochemistry and Cell Biology*, *125*(6), 725–741.

Savinaa, T. A., Levina, S. G., Poletaevab, I. I., Fedotovab, I. B., & Shchipakinaa, T. G. (2014). Audiogenic kindling changes the subunit composition of BK channels in dentate gyrus of Krushinskii–Molodkina rats. *Biochemistry (Moscow) Supplement Series A: Membrane and Cell Biology*, *8*(1), 111–115.

Schilling, T., & Eder, C. (2007). Ion channel expression in resting and activated microglia of hippocampal slices from juvenile mice. *Brain Research*, *1186*, 21–28.

Schilling, T., & Eder, C. (2015). Microglial K(+) channel expression in young adult and aged mice. *Glia*, *63*(4), 664–672.

Scholz, A., Gruss, M., & Vogel, W. (1998). Properties and functions of calcium-activated K^{+} channels in small neurones of rat dorsal root ganglion studied in a thin slice preparation. *The Journal of Physiology*, *513*(Pt. 1), 55–69.

Schuckit, M. A. (1994). Low level of response to alcohol as a predictor of future alcoholism. *The American Journal of Psychiatry*, *151*(2), 184–189.

Schuckit, M. A., Smith, T. L., Anderson, K. G., & Brown, S. A. (2004). Testing the level of response to alcohol: Social information processing model of alcoholism risk—A 20-year prospective study. *Alcoholism, Clinical and Experimental Research*, *28*(12), 1881–1889.

Schuckit, M. A., Wilhelmsen, K., Smith, T. L., Feiler, H. S., Lind, P., Lange, L. A., & Kalmijn, J. (2005). Autosomal linkage analysis for the level of response to alcohol. *Alcoholism, Clinical and Experimental Research*, *29*(11), 1976–1982.

Seidel, K. N., Derst, C., Salzmann, M., Holtje, M., Priller, J., Markgraf, R., … Pruss, H. (2011). Expression of the voltage- and Ca^{2+}-dependent BK potassium channel subunits BKbeta1 and BKbeta4 in rodent astrocytes. *Glia, 59*(6), 893–902.

Selemon, L. D. (2013). A role for synaptic plasticity in the adolescent development of executive function. *Translational Psychiatry, 3*, e238.

Servais, L., Bearzatto, B., Schwaller, B., Dumont, M., De Saedeleer, C., Dan, B., … Cheron, G. (2005). Mono- and dual-frequency fast cerebellar oscillation in mice lacking parvalbumin and/or calbindin D-28k. *The European Journal of Neuroscience, 22*(4), 861–870.

Servais, L., Hourez, R., Bearzatto, B., Gall, D., Schiffmann, S. N., & Cheron, G. (2007). Purkinje cell dysfunction and alteration of long-term synaptic plasticity in fetal alcohol syndrome. *Proceedings of the National Academy of Sciences of the United States of America, 104*(23), 9858–9863.

Shao, L. R., Halvorsrud, R., Borg-Graham, L., & Storm, J. F. (1999). The role of BK-type Ca^{2+}-dependent K^+ channels in spike broadening during repetitive firing in rat hippocampal pyramidal cells. *The Journal of Physiology, 521*(Pt. 1), 135–146.

Sheehan, J. J., Benedetti, B. L., & Barth, A. L. (2009). Anticonvulsant effects of the BK-channel antagonist paxilline. *Epilepsia, 50*(4), 711–720.

Shi, J., He, H. Q., Zhao, R., Duan, Y. H., Chen, J., Chen, Y., … Ji, Y. H. (2008). Inhibition of martentoxin on neuronal BK channel subtype (alpha + beta4): Implications for a novel interaction model. *Biophysical Journal, 94*(9), 3706–3713.

Shieh, C. C., Turner, S. C., Zhang, X. F., Milicic, I., Parihar, A., Jinkerson, T., … Gopalakrishnan, M. (2007). A-272651, a nonpeptidic blocker of large-conductance Ca^{2+}-activated K^+ channels, modulates bladder smooth muscle contractility and neuronal action potentials. *British Journal of Pharmacology, 151*(6), 798–806.

Shirazi-zand, Z., Ahmad-Molaei, L., Motamedi, F., & Naderi, N. (2013). The role of potassium BK channels in anticonvulsant effect of cannabidiol in pentylenetetrazole and maximal electroshock models of seizure in mice. *Epilepsy & Behavior, 28*(1), 1–7.

Shoudai, K., Nonaka, K., Maeda, M., Wang, Z. M., Jeong, H. J., Higashi, H., … Akaike, N. (2007). Effects of various K^+ channel blockers on spontaneous glycine release at rat spinal neurons. *Brain Research, 1157*, 11–22.

Shruti, S., Clem, R. L., & Barth, A. L. (2008). A seizure-induced gain-of-function in BK channels is associated with elevated firing activity in neocortical pyramidal neurons. *Neurobiology of Disease, 30*(3), 323–330.

Shruti, S., Urban-Ciecko, J., Fitzpatrick, J. A., Brenner, R., Bruchez, M. P., & Barth, A. L. (2012). The brain-specific Beta4 subunit downregulates BK channel cell surface expression. *PloS One*, 7(3), e33429.

Singh, H., Li, M., Hall, L., Chen, S., Sukur, S., Lu, R., … Toro, L. (2016). MaxiK channel interactome reveals its interaction with GABA transporter 3 and heat shock protein 60 in the mammalian brain. *Neuroscience, 317*, 76–107.

Singh, H., Lu, R., Bopassa, J. C., Meredith, A. L., Stefani, E., & Toro, L. (2013). MitoBK (Ca) is encoded by the Kcnma1 gene, and a splicing sequence defines its mitochondrial location. *Proceedings of the National Academy of Sciences of the United States of America, 110*(26), 10836–10841.

Skafidas, E., Testa, R., Zantomio, D., Chana, G., Everall, I. P., & Pantelis, C. (2014). Predicting the diagnosis of autism spectrum disorder using gene pathway analysis. *Molecular Psychiatry, 19*(4), 504–510.

Smith, M. R., Nelson, A. B., & Du Lac, S. (2002). Regulation of firing response gain by calcium-dependent mechanisms in vestibular nucleus neurons. *Journal of Neurophysiology, 87*(4), 2031–2042.

Sokolowski, B., Orchard, S., Harvey, M., Sridhar, S., & Sakai, Y. (2011). Conserved BK channel-protein interactions reveal signals relevant to cell death and survival. *PloS One, 6*(12), e28532.

Solaro, C. R., & Lingle, C. J. (1992). Trypsin-sensitive, rapid inactivation of a calcium-activated potassium channel. *Science, 257*(5077), 1694–1698.

Song, B., & Marvizon, J. C. (2005). N-methyl-D-aspartate receptors and large conductance calcium-sensitive potassium channels inhibit the release of opioid peptides that induce mu-opioid receptor internalization in the rat spinal cord. *Neuroscience, 136*(2), 549–562.

Soto, D., Comes, N., Ferrer, E., Morales, M., Escalada, A., Pales, J., ... Gasull, X. (2004). Modulation of aqueous humor outflow by ionic mechanisms involved in trabecular meshwork cell volume regulation. *Investigative Ophthalmology & Visual Science, 45*(10), 3650–3661.

Storm, J. F. (1987a). Action potential repolarization and a fast after-hyperpolarization in rat hippocampal pyramidal cells. *The Journal of Physiology, 385*, 733–759.

Storm, J. F. (1987b). Intracellular injection of a Ca^{2+} chelator inhibits spike repolarization in hippocampal neurons. *Brain Research, 435*(1–2), 387–392.

Strauss, O. (2005). The retinal pigment epithelium in visual function. *Physiological Reviews, 85*(3), 845–881.

Stumpff, F., Strauss, O., Boxberger, M., & Wiederholt, M. (1997). Characterization of maxi-K-channels in bovine trabecular meshwork and their activation by cyclic guanosine monophosphate. *Investigative Ophthalmology & Visual Science, 38*(9), 1883–1892.

Sun, X., Gu, X. Q., & Haddad, G. G. (2003). Calcium influx via L- and N-type calcium channels activates a transient large-conductance Ca^{2+}-activated K^{+} current in mouse neocortical pyramidal neurons. *The Journal of Neuroscience, 23*(9), 3639–3648.

Swensen, A. M., & Bean, B. P. (2003). Ionic mechanisms of burst firing in dissociated Purkinje neurons. *The Journal of Neuroscience, 23*(29), 9650–9663.

Tanimoto, N., Sothilingam, V., Euler, T., Ruth, P., Seeliger, M. W., & Schubert, T. (2012). BK channels mediate pathway-specific modulation of visual signals in the in vivo mouse retina. *The Journal of Neuroscience, 32*(14), 4861–4866.

Tao, Q., & Kelly, M. E. (1996). Calcium-activated potassium current in cultured rabbit retinal pigment epithelial cells. *Current Eye Research, 15*(3), 237–246.

Thieme, H., Stumpff, F., Ottlecz, A., Percicot, C. L., Lambrou, G. N., & Wiederholt, M. (2001). Mechanisms of action of unoprostone on trabecular meshwork contractility. *Investigative Ophthalmology & Visual Science, 42*(13), 3193–3201.

Tian, L., Jeffries, O., McClafferty, H., Molyvdas, A., Rowe, I. C., Saleem, F., ... Shipston, M. J. (2008). Palmitoylation gates phosphorylation-dependent regulation of BK potassium channels. *Proceedings of the National Academy of Sciences of the United States of America, 105*(52), 21006–21011.

Toro, B., Cox, N., Wilson, R. J., Garrido-Sanabria, E., Stefani, E., Toro, L., & Zarei, M. M. (2006). KCNMB1 regulates surface expression of a voltage and Ca^{2+}-activated K^{+} channel via endocytic trafficking signals. *Neuroscience, 142*(3), 661–669.

Triller, A., & Choquet, D. (2005). Surface trafficking of receptors between synaptic and extrasynaptic membranes: And yet they do move!. *Trends in Neurosciences, 28*(3), 133–139.

Tseng-Crank, J., Foster, C. D., Krause, J. D., Mertz, R., Godinot, N., DiChiara, T. J., & Reinhart, P. H. (1994). Cloning, expression, and distribution of functionally distinct Ca(2+)-activated K^{+} channel isoforms from human brain. *Neuron, 13*(6), 1315–1330.

Typlt, M., Mirkowski, M., Azzopardi, E., Ruettiger, L., Ruth, P., & Schmid, S. (2013). Mice with deficient BK channel function show impaired prepulse inhibition and spatial learning, but normal working and spatial reference memory. *PloS One, 8*(11), e81270.

Uebele, V. N., Lagrutta, A., Wade, T., Figueroa, D. J., Liu, Y., McKenna, E., ... Swanson, R. (2000). Cloning and functional expression of two families of beta-subunits of the large conductance calcium-activated K+ channel. *The Journal of Biological Chemistry, 275*(30), 23211–23218.

van Welie, I., & du Lac, S. (2011). Bidirectional control of BK channel open probability by CAMKII and PKC in medial vestibular nucleus neurons. *Journal of Neurophysiology, 105*(4), 1651–1659.

Velasquez-Martinez, M. C., Vazquez-Torres, R., Rojas, L. V., Sanabria, P., & Jimenez-Rivera, C. A. (2015). Alpha-1 adrenoreceptors modulate GABA release onto ventral tegmental area dopamine neurons. *Neuropharmacology, 88*, 110–121.

Velazquez-Marrero, C., Seale, G. E., Treistman, S. N., & Martin, G. E. (2014). Large conductance voltage- and Ca^{2+}-gated potassium (BK) channel beta4 subunit influences sensitivity and tolerance to alcohol by altering its response to kinases. *The Journal of Biological Chemistry, 289*(42), 29261–29272.

Velazquez-Marrero, C., Wynne, P., Bernardo, A., Palacio, S., Martin, G., & Treistman, S. N. (2011). The relationship between duration of initial alcohol exposure and persistence of molecular tolerance is markedly nonlinear. *The Journal of Neuroscience, 31*(7), 2436–2446.

Velinov, M., Dolzhanskaya, N., Gonzalez, M., Powell, E., Konidari, I., Hulme, W., … Zuchner, S. (2012). Mutations in the gene DNAJC5 cause autosomal dominant Kufs disease in a proportion of cases: Study of the Parry family and 8 other families. *PloS One*, 7(1), e29729.

Wallner, M., Meera, P., & Toro, L. (1999). Molecular basis of fast inactivation in voltage and Ca^{2+}-activated K^+ channels: A transmembrane beta-subunit homolog. *Proceedings of the National Academy of Sciences of the United States of America, 96*(7), 4137–4142.

Wang, Z. W. (2008). Regulation of synaptic transmission by presynaptic CaMKII and BK channels. *Molecular Neurobiology, 38*(2), 153–166.

Wang, B., Jaffe, D. B., & Brenner, R. (2014). Current understanding of iberiotoxin-resistant BK channels in the nervous system. *Frontiers in Physiology, 5*, 382.

Wang, L., Kang, H., Li, Y., Shui, Y., Yamamoto, R., Sugai, T., & Kato, N. (2015). Cognitive recovery by chronic activation of the large-conductance calcium-activated potassium channel in a mouse model of Alzheimer's disease. *Neuropharmacology, 92*, 8–15.

Wang, G. Y., Robinson, D. W., & Chalupa, L. M. (1998). Calcium-activated potassium conductances in retinal ganglion cells of the ferret. *Journal of Neurophysiology, 79*(1), 151–158.

Wang, B., Rothberg, B. S., & Brenner, R. (2009). Mechanism of increased BK channel activation from a channel mutation that causes epilepsy. *The Journal of General Physiology, 133*(3), 283–294.

Wang, F., Zhang, Y., Wang, L., Sun, P., Luo, X., Ishigaki, Y., … Kato, N. (2015). Improvement of spatial learning by facilitating large-conductance calcium-activated potassium channel with transcranial magnetic stimulation in Alzheimer's disease model mice. *Neuropharmacology, 97*, 210–219.

Wanner, S. G., Koch, R. O., Koschak, A., Trieb, M., Garcia, M. L., Kaczorowski, G. J., & Knaus, H. G. (1999). High-conductance calcium-activated potassium channels in rat brain: Pharmacology, distribution, and subunit composition. *Biochemistry, 38*(17), 5392–5400.

Weiger, T. M., Holmqvist, M. H., Levitan, I. B., Clark, F. T., Sprague, S., Huang, W. J., … Curtis, R. (2000). A novel nervous system beta subunit that downregulates human large conductance calcium-dependent potassium channels. *The Journal of Neuroscience, 20*(10), 3563–3570.

Welsh, D. K., Logothetis, D. E., Meister, M., & Reppert, S. M. (1995). Individual neurons dissociated from rat suprachiasmatic nucleus express independently phased circadian firing rhythms. *Neuron, 14*(4), 697–706.

Wilcock, D. M., Vitek, M. P., & Colton, C. A. (2009). Vascular amyloid alters astrocytic water and potassium channels in mouse models and humans with Alzheimer's disease. *Neuroscience, 159*(3), 1055–1069.

Williams, S. E., Wootton, P., Mason, H. S., Bould, J., Iles, D. E., Riccardi, D., ... Kemp, P. J. (2004). Hemoxygenase-2 is an oxygen sensor for a calcium-sensitive potassium channel. *Science*, *306*(5704), 2093–2097.

Wimmers, S., Halsband, C., Seyler, S., Milenkovic, V., & Strauss, O. (2008). Voltage-dependent Ca^{2+} channels, not ryanodine receptors, activate Ca^{2+}-dependent BK potassium channels in human retinal pigment epithelial cells. *Molecular Vision*, *14*, 2340–2348.

Womack, M. D., & Khodakhah, K. (2002). Characterization of large conductance Ca^{2+}-activated K^+ channels in cerebellar Purkinje neurons. *The European Journal of Neuroscience*, *16*(7), 1214–1222.

Wu, L. G., Westenbroek, R. E., Borst, J. G., Catterall, W. A., & Sakmann, B. (1999). Calcium channel types with distinct presynaptic localization couple differentially to transmitter release in single calyx-type synapses. *The Journal of Neuroscience*, *19*(2), 726–736.

Wulf, H., Hay-Schmidt, A., Poulsen, A. N., Klaerke, D. A., Olesen, J., & Jansen-Olesen, I. (2008). Molecular studies of BKCa channels in intracranial arteries: Presence and localization. *Cell and Tissue Research*, *334*(3), 359–369.

Wynne, P. M., Puig, S. I., Martin, G. E., & Treistman, S. N. (2009). Compartmentalized beta subunit distribution determines characteristics and ethanol sensitivity of somatic, dendritic, and terminal large-conductance calcium-activated potassium channels in the rat central nervous system. *The Journal of Pharmacology and Experimental Therapeutics*, *329*(3), 978–986.

Xia, X. M., Ding, J. P., & Lingle, C. J. (1999). Molecular basis for the inactivation of Ca^{2+}- and voltage-dependent BK channels in adrenal chromaffin cells and rat insulinoma tumor cells. *The Journal of Neuroscience*, *19*(13), 5255–5264.

Xie, J., & Black, D. L. (2001). A CaMK IV responsive RNA element mediates depolarization-induced alternative splicing of ion channels. *Nature*, *410*(6831), 936–939.

Xie, J., Jan, C., Stoilov, P., Park, J., & Black, D. L. (2005). A consensus CaMK IV-responsive RNA sequence mediates regulation of alternative exons in neurons. *RNA*, *11*(12), 1825–1834.

Xie, J., & McCobb, D. P. (1998). Control of alternative splicing of potassium channels by stress hormones. *Science*, *280*(5362), 443–446.

Xu, W., Liu, Y., Wang, S., McDonald, T., Van Eyk, J. E., Sidor, A., & O'Rourke, B. (2002). Cytoprotective role of Ca^{2+}-activated K^+ channels in the cardiac inner mitochondrial membrane. *Science*, *298*(5595), 1029–1033.

Xu, J. W., & Slaughter, M. M. (2005). Large-conductance calcium-activated potassium channels facilitate transmitter release in salamander rod synapse. *The Journal of Neuroscience*, *25*(33), 7660–7668.

Yamamoto, K., Ueta, Y., Wang, L., Yamamoto, R., Inoue, N., Inokuchi, K., ... Kato, N. (2011). Suppression of a neocortical potassium channel activity by intracellular amyloid-beta and its rescue with Homer1a. *The Journal of Neuroscience*, *31*(31), 11100–11109.

Yan, J., & Aldrich, R. W. (2012). BK potassium channel modulation by leucine-rich repeat-containing proteins. *Proceedings of the National Academy of Sciences of the United States of America*, *109*(20), 7917–7922.

Yang, J., Krishnamoorthy, G., Saxena, A., Zhang, G., Shi, J., Yang, H., ... Cui, J. (2010). An epilepsy/dyskinesia-associated mutation enhances BK channel activation by potentiating Ca^{2+} sensing. *Neuron*, *66*(6), 871–883.

Yang, C. T., Zeng, X. H., Xia, X. M., & Lingle, C. J. (2009). Interactions between beta subunits of the KCNMB family and Slo3: Beta4 selectively modulates Slo3 expression and function. *PLoS One*, *4*(7), e6135.

Ye, H., Buttigieg, J., Wan, Y., Wang, J., Figley, S., & Fehlings, M. G. (2012). Expression and functional role of BK channels in chronically injured spinal cord white matter. *Neurobiology of Disease*, *47*(2), 225–236.

Ye, H., Jalini, S., Mylvaganam, S., & Carlen, P. (2010). Activation of large-conductance Ca(2+)-activated K(+) channels depresses basal synaptic transmission in the hippocampal CA1 area in APP (swe/ind) TgCRND8 mice. *Neurobiology of Aging*, *31*(4), 591–604.

Yu, L., Eaton, A. F., Yue, Q., Bao, H. F., Ma, H. P., Cuppoletti, J., & Eaton, D. C. (2015). Unoprostone activation of BK (KCa1.1) channel splice variants. *Biochimica et Biophysica Acta*, *1848*(11 Pt. A), 2859–2867.

Zarei, M. M., Eghbali, M., Alioua, A., Song, M., Knaus, H. G., Stefani, E., & Toro, L. (2004). An endoplasmic reticulum trafficking signal prevents surface expression of a voltage- and Ca^{2+}-activated K^+ channel splice variant. *Proceedings of the National Academy of Sciences of the United States of America*, *101*(27), 10072–10077.

Zhang, Y., Bonnan, A., Bony, G., Ferezou, I., Pietropaolo, S., Ginger, M., … Frick, A. (2014). Dendritic channelopathies contribute to neocortical and sensory hyperexcitability in Fmr1(-/y) mice. *Nature Neuroscience*, *17*(12), 1701–1709.

Zhang, X. F., Gopalakrishnan, M., & Shieh, C. C. (2003). Modulation of action potential firing by iberiotoxin and NS1619 in rat dorsal root ganglion neurons. *Neuroscience*, *122*(4), 1003–1011.

Zhang, X. L., Mok, L. P., Katz, E. J., & Gold, M. S. (2010). BKCa currents are enriched in a subpopulation of adult rat cutaneous nociceptive dorsal root ganglion neurons. *The European Journal of Neuroscience*, *31*(3), 450–462.

Zhang, H., Schools, G. P., Lei, T., Wang, W., Kimelberg, H. K., & Zhou, M. (2008). Resveratrol attenuates early pyramidal neuron excitability impairment and death in acute rat hippocampal slices caused by oxygen-glucose deprivation. *Experimental Neurology*, *212*(1), 44–52.

Zhang, Z. B., Tian, M. Q., Gao, K., Jiang, Y. W., & Wu, Y. (2015). De novo KCNMA1 mutations in children with early-onset paroxysmal dyskinesia and developmental delay. *Movement Disorders*, *30*(9), 1290–1292.

Zhang, H., Xie, M., Schools, G. P., Feustel, P. F., Wang, W., Lei, T., … Zhou, M. (2009). Tamoxifen mediated estrogen receptor activation protects against early impairment of hippocampal neuron excitability in an oxygen/glucose deprivation brain slice ischemia model. *Brain Research*, *1247*, 196–211.

Zou, S., Jha, S., Kim, E. Y., & Dryer, S. E. (2008a). A novel actin-binding domain on Slo1 calcium-activated potassium channels is necessary for their expression in the plasma membrane. *Molecular Pharmacology*, *73*(2), 359–368.

Zou, S., Jha, S., Kim, E. Y., & Dryer, S. E. (2008b). The beta 1 subunit of L-type voltage-gated Ca^{2+} channels independently binds to and inhibits the gating of large-conductance Ca^{2+}-activated K^+ channels. *Molecular Pharmacology*, *73*(2), 369–378.

CHAPTER NINE

BK Channels and the Control of the Pituitary

P.J. Duncan[1], M.J. Shipston
Centre for Integrative Physiology, College of Medicine & Veterinary Medicine, University of Edinburgh, Edinburgh, United Kingdom
[1]Corresponding author: e-mail address: pduncan2@staffmail.ed.ac.uk

Contents

Abstract

The pituitary gland provides the important link between the nervous system and the endocrine system and regulates a diverse range of physiological functions. The pituitary is connected to the hypothalamus by the pituitary stalk and is comprised primarily of two lobes. The anterior lobe consists of five hormone-secreting cell types which are electrically excitable and display single-spike action potentials as well as complex bursting patterns. Bursting is of particular interest as it raises intracellular calcium to a greater extent than spiking and is believed to underlie secretagogue-induced hormone secretion. BK channels have been identified as a key regulator of bursting in anterior pituitary cells. Experimental data and mathematical modeling have demonstrated that

International Review of Neurobiology, Volume 128
ISSN 0074-7742
http://dx.doi.org/10.1016/bs.irn.2016.03.004

BK activation during the upstroke of an action potential results in a prolonged depolarization and an increase in intracellular calcium. In contrast, the posterior lobe is primarily composed of axonal projections of magnocellular neurosecretory cells which extend from the supraoptic and paraventricular nuclei of the hypothalamus. In these neuroendocrine cells, BK channel activation results in a decrease in excitability and hormone secretion. The opposite effect of BK channels in the anterior and posterior pituitary highlights the diverse role of BK channels in regulating the activity of excitable cells. Further studies of pituitary cell excitability and the specific role of BK channels would lead to a greater understanding of how pituitary cell excitability is regulated by both hypothalamic secretagogues and negative feedback loops, and could ultimately lead to novel treatments to pituitary-related disorders.

1. OVERVIEW AND ANATOMY OF THE PITUITARY GLAND

1.1 Introduction

The pituitary gland provides an important link between the nervous system and the endocrine system. The pituitary is responsible for regulating many aspects of physiology including growth, stress, reproduction, and water balance, while dysfunction can have many consequences on health. The most common causes of pituitary-related diseases are benign tumors, commonly referred to as pituitary adenomas, which often results in excessive hormone production but can also inhibit hormone production. Cushing's disease is one example where a pituitary adenoma secretes excess adrenocorticotrophic hormone (ACTH) resulting in hypercortisolism and includes symptoms such as hypertension, insulin resistance, and visceral weight gain (Newell-Price, 2003).

Cellular electrical excitability is a key regulator of hormone secretion, and while many pituitary cell types are electrically excitable, patterns of activity range from single-spike action potentials to complex bursting patterns. How the different patterns of electrical excitability are linked to control of hormone secretion in specific pituitary cell types, and the ionic mechanisms controlling these distinct patterns of electrical activity, are gradually beginning to emerge (Stojilkovic, Tabak, & Bertram, 2010; Tagliavini, Tabak, Bertam, & Pedersen, 2016). Increasing evidence supports important, but distinct, roles for large conductance calcium- and voltage-activated potassium (BK) channels (Toro, Wallner, Meera, & Tanaka, 1998) in regulating pituitary cell excitability and hormone secretion. In several anterior pituitary cell types, BK channels have been shown to play a paradoxical role where they promote bursting behavior that is hypothesized to promote hormone secretion. In contrast, BK channels in the posterior pituitary inhibit the activity of hypothalamic neuroendocrine cells leading to a suppression of hormone release.

In this chapter, we explore how advances in our ability to analyze, model, and interrogate distinct pituitary cell types in real time is providing new insights into the role of BK channels in regulating pituitary cell excitability and hormone secretion.

1.2 Pituitary Anatomy and Systems

The pituitary, approximately the size of a pea in humans, is located at the base of the brain below the hypothalamus (Singh & Panwar, 2006). It is comprised primarily of two anatomically and functionally distinct lobes (Fig. 1): the anterior lobe (adenohypophysis) and the posterior lobe (neurohypophysis) that have different embryological origins (Kelberman, Rizzoti, Lovell-Badge, Robinson, & Dattani, 2009). The different genetic regulation of cells in each lobe likely specifies the different functional roles of BK channels in the control of electrical excitability and hormone secretion in distinct pituitary cell types.

The anterior pituitary, derived from oral ectoderm, contains five hormone-secreting cell types that are regulated by hormones released by hypothalamic neurons into the pituitary portal circulation at the median eminence (Fig. 1). These secretory cells include somatotrophs (up to 50% of cells) and corticotrophs (2–5% of cells) in which BK channels have been demonstrated to play an important role in controlling cellular excitability. Somatotrophs secrete growth hormone (GH) which promotes growth and cell proliferation, whereas corticotrophs secrete ACTH which regulates glucocorticoid output from the adrenal glands in response to stress.

In contrast, the posterior lobe, derived from evagination of ectoderm ventrally from the developing brain, consists primarily of axonal projections of magnocellular neurosecretory cells which extend from the supraoptic (SON) and paraventricular (PVN) nuclei of the hypothalamus and release vasopressin (AVP) and oxytocin into the peripheral blood stream (Ooi, Tawadros, & Escalona, 2004). The peripheral effects of AVP, also known as antidiuretic hormone, include maintaining water balance through regulating water permeability of kidney collecting ducts and regulation of blood pressure by increasing vascular tone (Koshimizu et al., 2012). Oxytocin is important for lactation and uterine contraction during child birth (Gimpl & Fahrenholz, 2001). Both AVP and oxytocin also have central effects that control complex social behaviors.

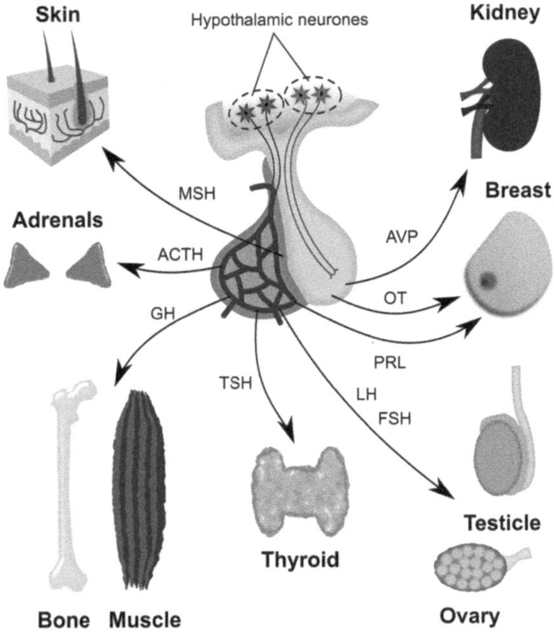

Fig. 1 The pituitary gland regulates a wide range of physiological functions. The anterior pituitary (*purple*; *gray* in the print version) consists of five hormone-secreting cell types that are regulated by hormones released by hypothalamic neurons into the pituitary portal circulation at the median eminence. The secretory cells include somatotrophs (growth hormone—GH, representing ~50% of cells), lactotrophs (prolactin—PRL, 10–25%), thyrotrophs (thyrotropin-releasing hormone—TSH, ~15%), gonadotrophs (luteinizing hormone—LH; follicle-stimulating hormone—FSH, 10–15%), and corticotrophs (adrenocorticotrophic hormone—ACTH, 2–5%). In addition, the anterior gland contains 5–10% nonendocrine folliculostellate cells and the adjacent intermediate lobe, which is ill-defined in humans, contains melanotrophs (melanotrophin-stimulating hormone—MSH). The posterior lobe consists of axonal projections from the supraoptic (SON) and paraventricular (PVN) nuclei of the hypothalamus which secrete arginine vasopressin (AVP) and oxytocin (OT).

1.3 Identification of Pituitary Cell Types

The identification of each of the five anterior pituitary cell types for real-time electrophysiology and calcium imaging assays has proven to be a challenge. Recent developments in genetic techniques now allow fluorescent proteins to be expressed in specific pituitary cell types, using specific promoters in transgenic animals (Budry et al., 2011; Duncan et al., 2015) or using viral transduction systems (Liang et al., 2011), thus allowing specific pituitary cell types to be visualized in real time for experimentation. These approaches now provide a priori identification of primary pituitary cells in contrast to previous post hoc identification, or manipulations that may significantly affect the physiological properties of cells, such as biotinylated peptides (Childs, Marchetti, & Brown, 1987), hemolytic plaque assays (Lee & Tse, 1997), and post hoc staining of fixed cells (Brunton et al., 2007). In addition, early studies of anterior pituitary cell excitability largely relied on immortalized cell lines which retain some, but not all, of the electrical and secretory properties of primary cells. For example, there are a large number of GH cell lines which have been extensively used to study somatotrophs, while clonal variants of AtT20 cells represent the only corticotroph cell line currently available (Ooi et al., 2004). Data from such cell lines have revealed that BK channels are regulated by phosphorylation cascades activated by several hypothalamic neuropeptides, including stimulatory peptides such as corticotropin-releasing hormone (CRH) in AtT20 cells (Shipston, Kelly, & Antoni, 1996) as well as inhibitory peptides such as somatostatin in the GH4C1 somatotroph cell line (White, Schonbrunn, & Armstrong, 1991).

2. ELECTRICAL EXCITABILITY OF ANTERIOR PITUITARY CELLS

2.1 Pituitary Cell Excitability Is Different Compared to Neurons

The ability of anterior pituitary cells to fire action potentials was an important discovery in the field of endocrinology (Kidokoro, 1975). It has been well established that the electrical properties of anterior pituitary cells differ from those of neurons. At synapses, neurotransmitter release can be triggered by a single action potential or a train of single action potentials, whereas prolonged depolarization is required for exocytosis in endocrine cells (Rettig & Neher, 2002). Many anterior pituitary cell types are electrically excitable, firing single-spike action potentials as well as complex bursting patterns (Stojilkovic et al., 2010). Bursting, often referred to as "pseudo-plateau bursting," is characterized by small-amplitude membrane oscillations during a prolonged membrane depolarization. Bursting is an important property of pituitary cells as it has been proposed to increase intracellular calcium concentration to a greater extent than spiking, thus acting to drive hormone secretion (Stojilkovic, Zemkova, & Van Goor, 2005; Tagliavini et al., 2016; Van Goor, Zivadinovic, Martinez-Fuentes, & Stojilkovic, 2001). Anterior pituitary cells which have high basal secretion (somatotrophs and lactotrophs) exhibit spontaneous bursting activity, whereas cells with low basal secretion (gonadotrophs) display predominantly single-spike action potentials at rest (Stojilkovic et al., 2005). Stimulation of anterior pituitary cells with the appropriate secretagogue is associated with a rise in intracellular Ca^{2+} levels (Bonnefont, Fiekers, Creff, & Mollard, 2000).

In neurons, TTX-sensitive voltage-gated Na^{+} channels are required for the upstroke of an action potential. In contrast, although TTX-sensitive Na^{+} channels are expressed in many anterior pituitary cells and can determine the amplitude of single sharp spikes, TTX has little effect on spontaneous activity in pituitary cells (Kuryshev, Haak, Childs, & Ritchie, 1997; Liang et al., 2011), as action potentials in pituitary cells are largely driven by Ca^{2+}. Indeed, spontaneous activity can be abolished by blocking L-type Ca^{2+} channels with nifedipine (Kuryshev, Childs, & Ritchie, 1996). Additionally, most previous studies in vitro have shown no difference in basal secretion of somatotrophs, lactotrophs, or gonadotrophs following TTX treatment (Van Goor, Zivadinovic, Martinez-Fuentes, et al., 2001). It has been proposed that TTX does not have a significant effect in regulating pituitary cell excitability because the majority of voltage-gated Na^{+} channels (>90%) will be inactive at the resting membrane potential typically seen in these cells (Van Goor, Zivadinovic, & Stojilkovic, 2001). An important caveat of these studies is that we do not fully understand the electrical behavior of anterior pituitary cells in vivo. For example, if the resting membrane potential of anterior pituitary cells is more negative than that observed in vitro, the TTX-sensitive Na^{+} channels could play a more prominent role.

Resting membrane potential of many anterior pituitary cell types in vitro is relatively depolarized, approximately −55 mV, compared with neurons, around −70 mV (Kuryshev et al., 1997; Lee & Tse, 1997). This has been attributed to a background Na^{+} conductance. Replacement of external Na^{+} with $NMDG^{+}$, a large organic cation, results in a robust and reversible hyperpolarization which brings membrane potential closer to that seen in neurons (Kucka, Bjelobaba, Tomić, & Stojilkovic, 2013; Liang et al., 2011; Tomić et al., 2011). Current evidence suggests that a member of the transient receptor potential channel family is a possible candidate (Tomić et al., 2011) although due to lack of high specificity pharmacological tools, the molecular identity of this conductance has yet to be determined.

3. PARADOXICAL ROLE OF BK CHANNELS AND BURSTING

3.1 BK Channels in the Anterior Pituitary

An important feature of several anterior pituitary cells, including somatotrophs and corticotrophs, is the ability to display pseudo-plateau bursting behavior, which results in high intracellular calcium concentrations and is proposed to underlie secretagogue-induced hormone section (Tagliavini et al., 2016; Van Goor, Zivadinovic, Martinez-Fuentes, et al., 2001). Initially, it was assumed that activation of a hyperpolarizing BK current would result in an inhibition of pituitary cell excitability and decrease in hormone secretion. It is now known that BK channels play an important role in the generation of bursting activity and that there is a positive correlation between BK expression, bursting, and hormone secretion in anterior pituitary cells. Rat somatotrophs have a high BK expression and display a high degree of spontaneous bursting, whereas gonadotrophs express very little BK in comparison and display predominantly spiking under basal conditions (Van Goor, Zivadinovic, Martinez-Fuentes, et al., 2001; Van Goor, Zivadinovic, & Stojilkovic, 2001). It is hypothesized that fast activation of BK channels during the upstroke of an action potential limits initial spike amplitude and prevents the activation of delayed rectifier K^+ channels. Membrane potential then oscillates around a depolarized level before being terminated by a slow K^+ conductance, resulting in a burst. Interestingly, genetic deletion of the BK channel pore-forming α-subunit in mice ($BK^{-/-}$) results in a ~40% reduction in plasma ACTH and corticosterone compared to wild-type litter mates following 30 min restraint stress (Brunton et al., 2007). However, an important caveat is these mice are global BK knockouts, and it is therefore unknown whether this hyporesponsive phenotype is due to changes at the level of the hypothalamus, pituitary, adrenals, or most likely, a combination of all three.

3.2 Experimental Investigation of BK Channels and Bursting in Corticotrophs

In contrast to the role of BK channels in controlling *spontaneous* bursting in somatotrophs, recent evidence has also revealed an important role for BK channels in *secretagogue-induced* bursting, as observed in corticotroph cells. Our recent studies demonstrate that murine corticotroph cells in vitro are spontaneously active and display predominantly single-spike action potentials under basal conditions, although a small proportion of cells (<6%) display spontaneous bursting events (Duncan et al., 2015; Liang et al., 2011). A 3-min stimulation of corticotrophs with a combination of the two hypothalamic ACTH secretagogues CRH and AVP, at physiological levels (0.2 and 2 nM, respectively), which approximate those released into portal circulation in response to stress (Gibbs & Vale, 1982; Sheward & Fink, 1991), results in a robust membrane depolarization (Fig. 2A). This depolarization is accompanied by both an increase in firing frequency and a

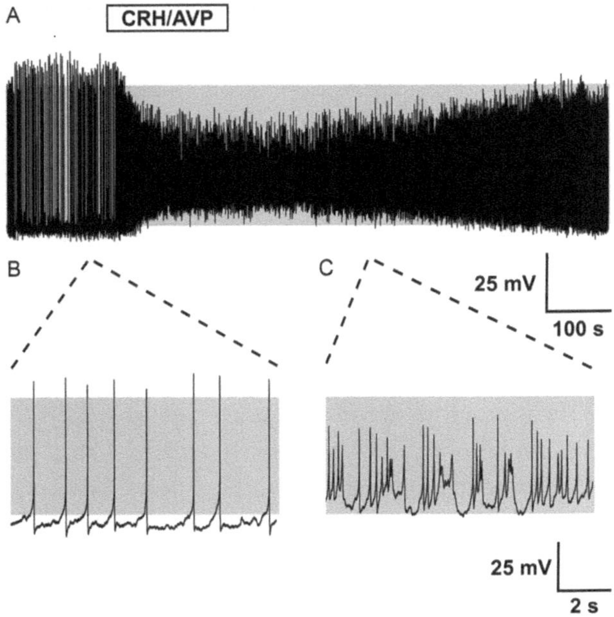

Fig. 2 Corticotroph excitability is regulated by hypothalamic neuropeptides CRH and AVP. Representative current clamp recording from a murine corticotroph cell exposed to 0.2 nM corticotropin-releasing hormone (CRH) and 2 nM arginine vasopressin (AVP) for 3 min (A). Under basal conditions, corticotrophs display predominantly single-spike action potentials (B). Stimulation with physiological concentrations of CRH/AVP results in a robust depolarization accompanied by an increase in firing frequency and a transition from spiking to complex bursting patterns (C). Following a washout period, firing frequency typically begins to decrease but remains elevated compared to baseline. *Gray* shading indicates membrane potential between −50 and +10 mV. *Figure adapted from data in Duncan, P. J., Şengül, S., Tabak, J., Ruth, P., Bertram, R., & Shipston, M. J. (2015). Large conductance Ca(2+)-activated K(+) (BK) channels promote secretagogue-induced transition from spiking to bursting in murine anterior pituitary corticotrophs.* The Journal of Physiology, 593, *1197–1211.*

transition from single-spike action potentials (Fig. 2B) to complex bursting patterns (Fig. 2C). Following washout of CRH/AVP, firing frequency and bursting typically begins to decrease but remains elevated compared to baseline for several minutes before returning to prestimulation levels. The transition to bursting is dependent upon functional BK channels as both pharmacological inhibition of BK channels with paxilline and genetic deletion of the BK channel pore-forming α-subunit largely prevented bursting even though the increase in spike frequency was retained. However, in both cases bursting was not completely abolished, suggesting that although BK channel is important for promoting bursting, they may not be the only component (Duncan et al., 2015). Indeed, previous studies have shown that inhibition of intermediate-conductance calcium-activated potassium (IK) channels with TRAM-34 resulted in an increase in bursting activity in female corticotrophs, while IK-knockout ($IK^{-/-}$) mice are hyperresponsive following restraint stress (Liang et al., 2011), suggesting that BK and IK channels have opposite effect in the regulation of bursting activity in corticotrophs. An intriguing observation is that TRAM-34 had no effect in male corticotrophs (Duncan et al., 2015), suggesting potential sexual dimorphisms in the control of bursting behavior.

Interestingly, secretagogue-induced bursting is driven primarily by CRH through BK-dependent mechanisms, while AVP promotes an increase in single-spike frequency through BK-independent mechanisms (Duncan et al., 2015). This observation provides clues into the mechanisms of bursting, and which pathways may act to regulate BK channels and drive bursting, as CRH and AVP activate distinct intracellular signaling cascades. In corticotrophs, CRH acts exclusively through CRH receptor type 1 (CRHR1), which couples to the $G\alpha_s$ pathway leading to the activation of adenylate cyclase, increasing cyclic AMP (cAMP) concentrations and protein kinase A (PKA) activity (Goncharova, 2013; Papadimitriou & Priftis, 2009). In contrast, AVP acts through AVP receptor 1B subtype (V1B) in corticotrophs, which couples to $G\alpha_q$ and leads to the activation of phospholipase C (PLC), which cleaves phosphatidylinositol 4,5-biphosphate to produce inositol 1,4,5-triphosphate and diacylglycerol, leading to the activation of protein kinase C (Murat et al., 2012). Somatotrophs are also primarily regulated by the cAMP/PKA pathway and show a high degree of bursting, while gonadotrophs, which show low bursting and low secretion, are regulated by the $G\alpha_q$ pathway (Stojilkovic et al., 2010). Taken together, this suggests that the cAMP/PKA pathway is likely the major driver to promote bursting in multiple anterior pituitary cell types.

The differential control of bursting and spike frequency by the distinct intracellular signaling pathways activated by CRH and AVP is likely to have important physiological consequences. For example, CRH has been shown to be the major secretagogue in humans, rodents, and other species, whereas AVP is to more dominant in sheep (Goncharova, 2013) and distinct stressors can differentially control CRH and/or AVP release (Ulrich-Lai & Herman, 2009). Taken together, the ionic mechanisms engaged to control ACTH secretion may be different between species and/or in response to different homeostatic challenges that may control the amplitude, or pattern, of ACTH secretion.

The above discussion is based on electrophysiological analysis of metabolically intact murine corticotrophs in vitro using physiological concentrations of CRH and AVP. It is important to note that we do not know the electrical behavior of corticotrophs in vivo. Furthermore, some studies have reported that corticotrophs in vitro are not electrically excitable, or proposed mechanisms of regulation using CRH and AVP concentrations in orders of magnitude above physiological levels (Lee, Smart, Rubinstein, Low, & Tse, 2011; Lee & Tse, 1997; Yamashita et al., 2009). In most of these studies, differences in experimental technique likely lead to conflicting data. For example, it is now widely appreciated that conventional whole-cell recordings of anterior pituitary cells, including somatotrophs and corticotrophs, result in the rapid cessation of spontaneous activity (Liang et al., 2011; Tsaneva-Atanasova, Sherman, van Goor, & Stojilkovic, 2007). In contrast, using perforated-patch clamp approaches, that largely retains the normal metabolic state of the cell by preventing cellular dialysis (Rae, Cooper, Gates, & Watsky, 1991), reveals spontaneous excitability and robust regulation by neuropeptide signaling pathways. It is important to note that due to their small size, pituitary cells have a high input resistance and as a consequence small changes in background current would be sufficient to produce a large effect on membrane potential. For example, a cell with an input resistance of 3 GΩ could be depolarized by as much as 15 mV by a reduction of the background K^+ current by 5 pA (Lee & Tse, 1997). Therefore, even a small amount of BK current could have a large impact on electrical excitability. Clearly, understanding the electrical properties of corticotrophs in vivo, while technically challenging, is important for our understanding of the role of BK channels at the system level.

3.3 Modeling the Role of BK Channels on Pituitary Cell Excitability

Mathematical modeling is a powerful tool that can be used to determine the role of ion channels in regulating excitability and to make predictions, which can be tested experimentally. Recent modeling of the role of BK channel in both spontaneous bursting in somatotrophs and CRH-induced transition to bursting in corticotrophs has provided significant insight into the role of BK channels in bursting.

The first pituitary cell model was based on the classic Hodgkin–Huxley model (Hodgkin & Huxley, 1952) and incorporated L- and T-type voltage-gated Ca^{2+} currents, a delayed rectifier K^+ current, Ca^{2+}-activated K^+ currents, and a leak current (LeBeau, Robson, McKinnon, Donald, & Sneyd,

1997). This corticotroph model showed that CRH increased firing frequency and broadened action potential but did not show bursting behavior. Early mathematical models of corticotrophs had a resting membrane potential of around −70 mV (LeBeau et al., 1997; Shorten, Robson, McKinnon, & Wall, 2000), which is hyperpolarized compared to that seen experimentally. Modification of these models to incorporate a background Na^+ conductance and an inwardly rectifying K^+ current resulted in a resting membrane potential comparable to native cells in vitro (Liang et al., 2011).

BK channels were demonstrated to generate bursting in somatotrophs, which increased intracellular Ca^{2+} to a greater extent than spiking (Van Goor, Li, & Stojilkovic, 2001). Thus, a newer model of somatotrophs predicted the existence of two populations of BK channel important in the generation of bursting activity; BK channels in close proximity to L-type Ca^{2+} channels (BK-near), which activate rapidly due to high Ca^{2+} concentrations in a microdomain, followed by BK channels located distantly from Ca^{2+} channels (BK-far), which activate as global Ca^{2+} concentration increases and are responsible for the termination of a burst (Tsaneva-Atanasova et al., 2007). Ca^{2+} concentrations in microdomains reach equilibrium within microseconds (Simon & Llinás, 1985) and therefore gating of BK-near channels is essentially driven by membrane potential.

To test this model, a dynamic clamp study on the somatotroph cell line (GH4C1) showed spontaneously that bursting cells could be switched to spiking using the BK channel blocker iberiotoxin which then could be reversed by adding back an artificial fast-activating BK conductance (Tabak, Tomaiuolo, Gonzalez-Iglesias, Milescu, & Bertram, 2011). In order to promote bursting, BK-near channel activation has to be sufficiently fast (<10 ms); otherwise the BK current inhibits bursting (Vo, Tabak, Bertram, & Wechselberger, 2014). This model considers BK channel activation to be a purely voltage-dependent process for the BK channels located adjacent to L-type Ca^{2+} channels (Van Goor, Li, et al., 2001). As Ca^{2+} concentrations in microdomains reach equilibrium within microseconds (Simon & Llinás, 1985), the gating of BK-near is essentially driven by membrane potential (Tabak et al., 2011). This highlights one potential problem with mathematical modeling. Although mathematical modeling and dynamic clamp can be a useful tool when deciphering the roles of individual channels in regulating electrical activity, assumptions of channel properties or localization could lead to an oversimplification of complex biological systems, resulting in discrepancies between the model and experimental data. For example, the critical evidence for spatial distribution of different BK channel populations with L-type Ca^{2+} channels in anterior pituitary cells is currently lacking and represents a significant experimental challenge to address.

A modification of the two BK channel population hypothesis was also incorporated into a corticotroph model which has been used to predict the mechanism by which the two secretagogues, CRH and AVP, differentially regulate excitability (Duncan et al., 2015). The corticotroph model consists of six ionic currents: an L-type Ca^{2+} current, a delayed rectifier K^+ current, an inward rectifier K^+ current, a nonselective Na^+ current, a fast-activating BK-near current, and a slow-activating BK-far current (Fig. 3A). In this model, the upstroke of an action potential is largely driven by Ca^{2+} influx through L-type Ca^{2+} channels. Under basal conditions, at the peak of the action potential, delayed rectifier K^+ channels rapidly activate causing a fast repolarization, which results in a spike (Fig. 3B). However, in the presence of CRH, activation of a fast BK current (BK-near) during the upstroke of an action potential is predicted to limit spike amplitude and prevent the activation of delayed rectifier K^+ channels. Membrane potential then oscillates around a depolarized state before a slow BK current (BK-far) activates due to a rise in global Ca^{2+} and results in burst termination (Fig. 3C). In the model, CRH-induced bursting is simulated by increasing L-type Ca^{2+} current while significantly right-shifting the voltage activation curve of BK-near compared to prestimulation levels. As in the somatotroph model, a critical determinant of bursting is that, in the presence of CRH, the activation time constant of BK-near is sufficiently fast (ie, <10 ms; Duncan et al., 2015) so that spike amplitude is truncated. This condition of fast activation would be predicted to be satisfied for a BK channel exposed to Ca^{2+} concentrations close to an open L-type Ca^{2+} channel. In contrast, AVP alone has no effect on BK-near, rather it increases the nonselective Na^+ current, which results in a slight depolarization and an increase in single-spike frequency but not bursting. This is consistent with experimental data where depolarizing current injection alone only stimulates an increase in firing frequency without a transition to bursting (Duncan et al., 2015). Interestingly, this also raises the question whether the AVP-dependent increase in spike frequency is as efficient at stimulating ACTH secretion compared to CRH-dependent bursting activity. Recent mathematical modeling and analysis of secretion in GH4C1 cells support the hypothesis that bursting can be more efficient at stimulating secretion than an increase in single-spike frequency (Tagliavini et al., 2016). Whether the same is true in corticotrophs remains to be determined; however, CRH is more potent than AVP in

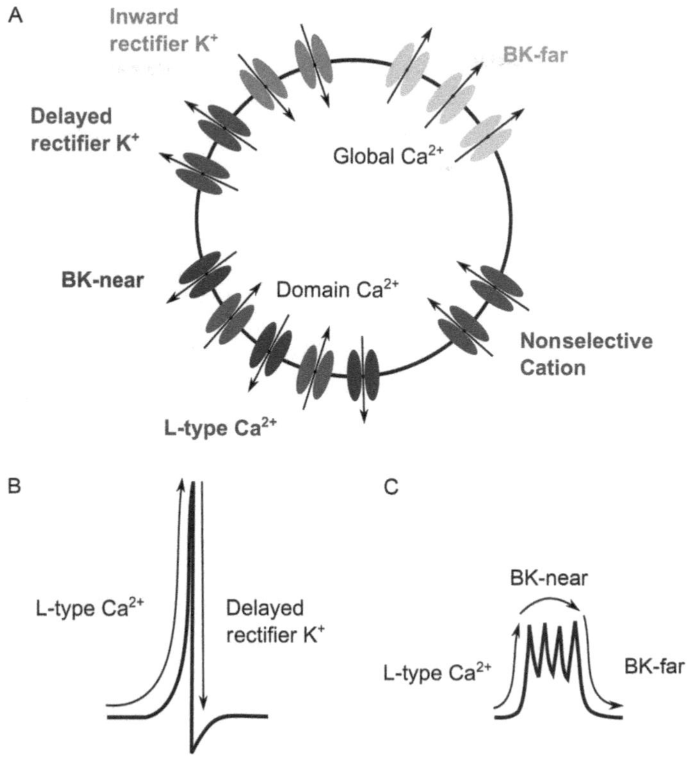

Fig. 3 Modeling the role of BK channels in pituitary cell excitability. Schematic diagram of the ionic currents in the pituitary corticotroph model (A). BK channels in close proximity to L-type Ca^{2+} channels (BK-near) rapidly activate due to high concentrations of domain Ca^{2+}, whereas BK channels located distantly from L-type Ca^{2+} channels (BK-far) slowly activate as global Ca^{2+} rises. In the absence of a fast BK current, L-type Ca^{2+} current is responsible for the upstroke of an action potential and at the peak delayed rectifier K^+ channels activate, which results in repolarization (B). Activation of a fast BK current (BK-near) during the upstroke of an action potential limits spike amplitude and the activation of delayed rectifier K^+ channels, membrane potential, oscillates around a depolarized level resulting in a burst which is terminated as global Ca^{2+} rises and activates BK-far channels (C).

stimulating secretion in population-based assays from murine corticotrophs (Brunton et al., 2007). Moreover, while it is well established that CRH and AVP synergize to promote ACTH secretion, in part as AVP promotes CRH-stimulated cAMP accumulation (Antoni, 1993), whether the effect of increasing spike frequency and bursting upon coapplication of CRH and AVP also contributes to the synergistic effect of CRH/AVP on ACTH secretion remains to be determined.

3.4 Properties and Molecular Identification of BK Channels Controlling Bursting

The two BK channel population model has led to an elegant working hypothesis of how BK channels may control spontaneous or secretagogue-evoked bursting. However, the molecular identity or spatial distribution of the proposed distinct BK channel populations in somatotrophs or corticotrophs is currently not known. As somatotrophs display spontaneous bursting, the two populations of BK channels in the somatotroph model have the same intrinsic properties but only differ in that BK-near sees local Ca^{2+} concentration (domain Ca^{2+}), while BK-far responds to the cytosolic Ca^{2+} concentration (global Ca^{2+}; Tsaneva-Atanasova et al., 2007). However, as corticotrophs transition to bursting following CRH stimulation, which is driven by the cAMP/PKA pathway, the corticotroph model suggests that CRH modifies the properties, or local environment, of the BK-near population close to Ca^{2+} channels. In the model, CRH stimulation of bursting is dependent upon activation of L-type Ca^{2+} channels together with a significant right shift in the activation curve for BK-near, while BK-near activates fast enough (<10 ms) to limit spike amplitude.

The murine anterior pituitary gland has been shown, at least at the mRNA level, to express different splice variants of the BK channel pore-forming α-subunit (ZERO and STREX) as well as the accessory β2- and β4-subunits (Brenner et al., 2005; Chatterjee et al., 2009). Furthermore, mRNA expression of ZERO, β2, β4, but not STREX mRNA, is significantly upregulated in a "social-instability" chronic stress paradigm in mice. As discussed in chapters "Modulation of BK Channel Function by Auxiliary Beta and Gamma Subunits" by Yan and Li and "Posttranscriptional and Posttranslational Regulation of BK Channels" by Shipston and Tian of the present volume, alternative splicing and assembly with regulatory β-subunits can dramatically modify BK channel properties and regulation. For example, the STREX (*st*ress axis-*re*gulated *ex*on) variant has increased Ca^{2+} sensitivity and dramatically slower deactivation kinetics compared to ZERO channels, which lack the STREX insert (Saito, Nelson, Salkoff, & Lingle, 1997; Xie & McCobb, 1998). The STREX and ZERO splice variants are also differentially regulated by PKA phosphorylation. PKA-dependent phosphorylation of the STREX variant results in inhibition by shifting the activation curve to the right, whereas the ZERO isoform is activated (Tian, Duncan, et al., 2001).

Although the splice variant and accessory subunit expression in distinct anterior pituitary cell types are largely unknown, corticotrophs and somatotrophs both express ZERO and STREX splice variant mRNA (Shipston, Duncan, Clark, Antoni, & Tian, 1999; Zhou et al., 2012). Theoretically, a wide variety of BK channel splice variant and/or β-subunit combinations could have voltage activation curves and fast activation kinetics, if located next to open L-type Ca^{2+} channels, to contribute to bursting in somatotrophs. In the corticotroph model, under unstimulated conditions, BK-near is modeled with a significantly left-shifted voltage activation curve with the CRH-evoked transition to bursting dependent upon a significant right shift of the BK-near activation curve while displaying an activation <10 ms next to an open L-type Ca^{2+} channel. Intriguingly, the STREX splice variant has a significantly left-shifted voltage activation curve compared to other variants (Chen et al., 2005; Shipston et al., 1999). Moreover, it is the only known variant in which cAMP/PKA-dependent phosphorylation elicits a significant right shift in the activation curve and cAMP/PKA-phosphorylated STREX channels would have fast activation kinetics (<10 ms) at concentrations of Ca^{2+} predicted close to an open L-type Ca^{2+} channel (Chen et al., 2005; Tian, Hammond, Florance, Antoni, & Shipston, 2001). While these data suggest that an important molecular component of BK-near in corticotrophs is likely to include the STREX variant, the molecular identity remains elusive. For example, cAMP/PKA phosphorylation of the channel complex may differentially control coupling with accessory β-subunits or modify the functional coupling between L-type Ca^{2+} channels and BK-near. Molecular identification of BK-near represents a very significant challenge although if pharmacological tools or antibodies could be developed to reliably distinguish between different BK channel splice variants, it would allow us to interrogate the role and spatial distributions of BK channel variants in the control of bursting.

Importantly, as BK channels control bursting in corticotrophs the molecular composition of BK-near may also provide a target for glucocorticoid feedback by preventing the CRH-induced transition to bursting. The HPA axis is kept in balance through an elegant negative feedback mechanism where elevation of plasma glucocorticoids can feedback directly on the pituitary (or indirectly through the hypothalamus) over different timescales to inhibit ACTH release. If negative feedback loops are compromised, it can lead to hormone excess and ultimately disease. For example, patients suffering from depression often show impaired glucocorticoid negative feedback and effective antidepressant therapy has been revealed to reestablish effective negative feedback and restore normal glucocorticoid levels (Jameison & Dinan, 2001).

The regulation of BK channels by glucocorticoids is well documented through multiple mechanisms in different systems. For example, glucocorticoids can exert a very rapid (nongenomic) regulation of cellular excitability via direct effects on BK channel activity dependent on accessory β-subunits (King et al., 2006; Lovell, King, & McCobb, 2004). On the timescale of minutes to hours, glucocorticoids may prevent cAMP/PKA-dependent regulation of BK channel STREX variant phosphorylation through control of protein phosphatase activity associated with the channel (Tian, Duncan, et al., 2001; Tian, Hammond, et al., 2001; Tian, Knaus, & Shipston, 1998). At a longer timescale, glucocorticoids can regulate pre-mRNA alternative splicing of the BK channel including changing the ratio of STREX and ZERO splice variants expressed in the murine pituitary (Lai & McCobb, 2002, 2006). Clearly, establishing whether glucocorticoids control CRH-dependent bursting, including via control of BK-near, will be important to define molecular mechanisms of glucocorticoid feedback as well as potentially providing additional insights into the likely molecular composition of BK channels in corticotrophs.

4. BK CHANNELS IN HYPOTHALAMIC NEURONS REGULATING THE PITUITARY

4.1 Magnocellular and Parvocellular Neurosecretory Cells

The chapter "BK Channels in Neurons" by Barth and Contet of this volume discussed the role of BK channels in neurons where they are primarily responsible for controlling the repolarization and hyperpolarization phases of the action potential and are important in regulating refractory period, which is a major determinant of firing frequency (Wang, Jaffe, & Brenner, 2014). Unlike the anterior pituitary, which consists of excitable endocrine cells, the posterior pituitary is primarily composed of axonal projections from the SON and PVN of the hypothalamus.

Magnocellular secretory cells (MNCs) are neuroendocrine cells, which synthesize and secrete AVP and OT into systemic circulation at the posterior lobe. Hormone secretion during physiological stimulation of MNCs is dependent on the firing patterns of these neurons and is enhanced during a train of action potentials (Poulain & Wakerley, 1982). BK channels are an important component in regulating frequency and firing patterns of neurons. Recordings from AVP-expressing cells from the SON revealed that firing frequency can be enhanced by BK channel antagonists iberiotoxin (IbTX) and charybdotoxin (ChTX), and that activity can be completely abolished using BK channel activator NS1619 (Ohbuchi et al., 2010). Inhibition of BK channels increases firing frequency through attenuating the afterhyperpolarization (AHP) magnitude and thus shortening the refractory period (Greffrath et al., 2004). In AVP-expressing neurons, nitric oxide (NO) is an important signaling molecule and acts to suppress neuronal activity in times of normal hydration through activation of BK channels (Kadekaro et al., 2006).

There is evidence that BK channels display different electrophysiological and pharmacological properties depending on their location within MNCs. BK channels located at nerve terminals have a low Ca^{2+} sensitivity and are activated by ethanol, while cell body BK channels have a higher Ca^{2+} sensitivity and are insensitive to ethanol (Dopico, Widmer, Wang, Lemos, & Treistman, 1999). A follow-up study revealed that the differential properties of somatic and terminal BK channels are mediated by the association of β-subunits; β1 is highly expressed in dendrites and the soma, while β4 is primarily expressed in nerve terminals and the soma to a lesser extent (Wynne, Puig, Martin, & Treistman, 2009). BK channels in the presynaptic terminal are likely to provide local negative feedback of voltage-dependent Ca^{2+} influx during depolarization–secretion coupling, thus limiting secretion, while cell body BK channels may be important in regulating patterns of cell excitability (Dopico et al., 1999). BK channel potentiation by ethanol is decreased by chronic exposure and is thought to be an important factor in alcohol tolerance (Knott, Dopico, Dayanithi, Lemos, & Treistman, 2002). BK channel sensitivity to ethanol is splice variant dependent with BK channels expressing the ALCOREX insert (*alco*hol-*re*gulated *ex*on) showing potentiation by ethanol (Pietrzykowski et al., 2008). The same study revealed that chronic alcohol exposure upregulates microRNA miR-9 which mediates the selective downregulation of BK channel α-subunits containing the ALCOREX exon, leading to alcohol tolerance.

The PVN also consists of a large population of parvocellular neurosecretory cells which project to the median eminence at the brain and release hypothalamic secretagogues, such as CRH, TRH, and AVP, into portal circulation to stimulate cells of the anterior pituitary (Kiss, 1988). In response to stress, CRH is released from the PVN, a small population of neuroendocrine cells comprising of approximately 4000 neurons in rats (Herman et al., 2003). Excitability of these cells is largely driven by glutamate

(Evanson & Herman, 2015) and currently little is known about the role of BK channels in regulating excitability. However, there is evidence for high levels of the BK accessory β1-subunit in parvocellular cells of the PVN, as well as magnocellular cells of the PVN and SON, which is surprising as the β1-subunit is not normally found in neurons (Salzmann et al., 2010), but instead has a high expression in vascular smooth muscle cells (Brenner et al., 2000). Association with β1-subunits results in an increased voltage and Ca^{2+} sensitivity of BK channels (Plüger et al., 2000), which is likely to play an important role in regulating hypothalamic neuroendocrine cell excitability. A potential functional role for BK channels in controlling PVN excitability has also been revealed through analysis of the response to acute stress in mice. A short restraint stress paradigm in mice significantly increases PVN neuronal activation, as measured by an increase in early-gene mRNA expression, even though BK channel mRNA and protein levels are relatively low in the murine PVN (Brunton et al., 2007). In contrast, stress-induced activation of PVN neurons is significantly reduced in mice lacking the BK channel pore-forming α-subunit ($BK^{-/-}$ mice), suggesting that BK channels may play an important role in PVN neurons through the control of intrinsic properties and/or their neural inputs (Brunton et al., 2007).

5. AIMS AND CHALLENGES FOR UNDERSTANDING THE ROLE OF BK CHANNELS IN THE PITUITARY

The pituitary gland regulates a wide variety of physiological processes and acts as a "hub" integrating inputs from the brain while regulating outputs from target endocrine glands. The anterior pituitary contains hormone-secreting cells, which are electrically excitable, and electrophysiological studies have demonstrated that patterns of excitability are an important mechanism in regulating hormone secretion. BK channels play a prominent role in the generation of bursting activity, which is important in secretagogue-induced hormone release. The posterior pituitary is composed of axonal projections from hypothalamic neuroendocrine cells where BK channel activation results in a hyperpolarizing current which reduces activity and suppresses hormone secretion. The opposite effect of BK channels in anterior and posterior cell types clearly demonstrates the diverse roles of BK channels in regulating excitable cells and highlights the need to investigate the role of BK channels in a context-specific manner. Indeed, these studies further support data from other endocrine and neural systems that the cellular context in which BK channels are expressed is an important determinant of function. For example, in rat adrenal chromaffin cells electrophysiological and modeling data support a model in which BK channels (including the STREX variant) in fact promote increased action potential frequency by allowing fast repolarization and hyperpolarization to recover voltage-gated Na^+ and Ca^{2+} channels from inactivation (Lovell et al., 2004; Lovell & McCobb, 2001). Moreover, BK channels can also have distinct functional roles in different neurons. In hippocampal neurons, BK channels assembled with β4-subunits limit repetitive firing by slowing action potential repolarization, allowing increased Ca^{2+} influx to activate small-conductance, voltage-independent potassium (SK) channels (Brenner et al., 2005). In contrast, BK channels support high-frequency action potentials in cerebellar Purkinje neurons again by promoting fast repolarization and hyperpolarization to relieve voltage-gated Na^+ channels from inactivation (Sausbier et al., 2004).

5.1 Anterior Pituitary Cell Excitability in Networks

The majority of previous electrophysiological studies of pituitary cells have been performed on acutely dispersed cells or pituitary cell lines. We know that pituitary cell excitability is not an in vitro artifact, as action potentials have also been observed in situ in rat pituitary slices (Bonnefont & Mollard, 2003). The five anterior pituitary cell types were previously believed to be randomly distributed throughout the gland. However, with the advent of genetic labeling and advanced imaging techniques, it is now known that anterior pituitary cells form extensive structural and functional networks, which are dynamic in nature (Le Tissier et al., 2012). Three-dimensional mapping of the pituitary gland has revealed that corticotrophs form a network located on the lateral tips and also extend across the ventral surface (Budry et al., 2011). It is important to note that anterior pituitary cells form both homotypic and heterotypic cell–cell connections via gap junctions (Hodson et al., 2012), which implies that neighboring cells could have a large influence on excitability. Spontaneous activity of corticotrophs in vitro is highly variable and although more than 90% of cells displayed spontaneous activity with predominately single-spike activity, firing frequency varied widely (0.01–1 Hz) under unstimulated conditions (Duncan et al., 2015). It is currently unknown whether this variability is an artifact of the dissociation process, which disrupts the corticotroph network. It is possible that heterogeneity could still exist in the intact pituitary and there may exist more than one subpopulation of corticotrophs, which may be differentially located within the network. Alternatively, it is possible that there is only a single population of corticotrophs and that spontaneous activity is a purely stochastic event controlled by background noise. It is therefore important for future studies to focus on excitability of intact pituitary networks and to determine whether BK-dependent behaviors are retained at the network level.

5.2 Pharmacological Tools for Investigating Pituitary Excitability

When investigating the mechanisms of pituitary cell excitability, deciphering the role of a specific ion channel remains a significant challenge. Currently, pharmacological and genetic tools are the main means of investigating the role of ion channels in pituitary cell excitability. Pharmacologically, there are compounds available to block BK channels (paxilline and IbTX) with high specificity. Early BK activators such as NS1619 were shown to have nonspecific effects, but newer compounds such as NS11021 are being developed with greater specificity (Bentzen et al., 2007). Currently, BK channel blockers and activators target BK channels largely indiscriminately. For example, while BK channels that include the regulatory β4-subunit are resistant to inhibition by IbTX, but not paxilline, these inhibitors cannot readily discriminate between different splice variants such as STREX and ZERO. Thus, if the two BK population model is correct, the sequential activation of BK-near followed by BK-far is important, and interpreting functional roles of each population by simply activating or inhibiting both populations simultaneously will remain difficult. The spatial localization of BK channels in pituitary cells is currently unknown and we do not currently have adequate tools to rigorously interrogate the potentially distinct molecular composition of such native BK channel populations experimentally.

5.3 Dynamic Clamp Integrates Modeling and Experimentation

Mathematical modeling has provided important insight into the role of BK channels in the regulation of pituitary cell excitability. Models have been constructed based on experimental observations, which can in turn make predictions that can be tested experimentally. The results can then feedback and further refine the model which can be used in dynamic clamp approaches to either add or subtract a specific current artificially and measure the effect on excitability. Dynamic clamp has already been used to investigate the role of BK channels in somatotrophs and has shown that adding a fast BK current to spiking cells can switch to bursting, while a bursting cell can be switched to spiking by subtracting BK current (Tabak et al., 2011). Clearly, application of dynamic clamp to other pituitary models has the potential to provide significant insight into BK channel function.

Although dynamic clamp can be useful to interrogate the role of specific currents in cell excitability, it is not without limitations. For example, while dynamic clamp can be used to determine the types of current (ie, slow/fast potassium), it cannot confirm a molecular identity for a specific current. In addition, care must be taken when subtracting a current using dynamic clamp. If a negative current subtracted was greater than the amount in the cell, it would result in the application of a positive current.

In a modeling study of 512 simulations where L-type Ca^{2+}, delayed rectifier K^+, small-conductance K^+ (SK), and leak conductances were randomly chosen ($\pm$50% of their default values), addition of a BK conductance switched spiking cells to bursting cells in 70% of cases (Tabak et al., 2011). Experimental data showed that pharmacological blockade or genetic deletion of BK channels resulted in a significant reduction in secretagogue-induced bursting activity; however, in both cases bursting was not completely abolished (Duncan et al., 2015). Taken together, we can conclude that although BK channels greatly facilitate bursting behavior, they are unlikely to be the only current involved.

5.4 Conclusions

It is well established that BK channels play a diverse array of physiological roles depending on tissue expression. In the pituitary gland, BK channels generally have an excitatory effect in anterior pituitary cells where they play an important role in promoting bursting behavior, while in the posterior pituitary BK channels decrease excitability of neuroendocrine cells resulting in inhibition of hormone secretion. Electrophysiological studies combined with mathematical modeling and dynamic clamp have been used to interrogate the role of BK channels in regulating anterior pituitary cell excitability. Future studies analyzing the specific role of BK channels in controlling pituitary cell excitability would lead to a greater understanding of how electrical excitability is coupled to regulation of hormone secretion by hypothalamic secretagogues, as well as negative feedback loops, and could ultimately lead to novel treatments for pituitary-related disorders.

ACKNOWLEDGMENTS

Work in the authors' laboratory is generously supported by the Medical Research Council, British Heart Foundation, and Diabetes UK.

REFERENCES

Antoni, F. A. (1993). Vasopressinergic control of pituitary adrenocorticotropin secretion comes of age. *Frontiers in Neuroendocrinology*, *14*, 76–122.

Bentzen, B. H., Nardi, A., Calloe, K., Madsen, L. S., Olesen, S., & Grunnet, M. (2007). The small molecule NS11021 is a potent and specific activator of Ca2+-activated big-conductance K+ channels. *Molecular Pharmacology*, *72*, 1033–1044.

Bonnefont, X., Fiekers, J., Creff, A., & Mollard, P. (2000). Rhythmic bursts of calcium transients in acute anterior pituitary slices. *Endocrinology*, *141*, 868–875.

Bonnefont, X., & Mollard, P. (2003). Electrical activity in endocrine pituitary cells in situ: A support for a multiple-function coding. *FEBS Letters*, *548*, 49–52.

Brenner, R., Chen, Q. H., Vilaythong, A., Toney, G. M., Noebels, J. L., & Aldrich, R. W. (2005). BK channel beta4 subunit reduces dentate gyrus excitability and protects against temporal lobe seizures. *Nature Neuroscience*, *8*, 1752–1759.

Brenner, R., Perez, G. J., Bonev, A. D., Eckman, D. M., Kosek, J. C., Wiler, S. W., ... Aldrich, R. W. (2000). Vasoregulation by the beta1 subunit of the calcium-activated potassium channel. *Nature*, *407*, 870–876.

Brunton, P. J., Sausbier, M., Wietzorrek, G., Sausbier, U., Knaus, H.-G., Russell, J. A., ... Shipston, M. J. (2007). Hypothalamic-pituitary-adrenal axis hyporesponsiveness to restraint stress in mice deficient for large-conductance calcium- and voltage-activated potassium (BK) channels. *Endocrinology*, *148*, 5496–5506.

Budry, L., Lafont, C., El Yandouzi, T., Chauvet, N., Conéjero, G., Drouin, J., & Mollard, P. (2011). Related pituitary cell lineages develop into interdigitated 3D cell networks. *Proceedings of the National Academy of Sciences of the United States of America*, *108*, 12515–12520.

Chatterjee, O., Taylor, L. A., Ahmed, S., Nagaraj, S., Hall, J. J., Finckbeiner, S. M., ... McCobb, D. P. (2009). Social stress alters expression of large conductance calcium-activated potassium channel subunits in mouse adrenal medulla and pituitary glands. *Journal of Neuroendocrinology*, *21*, 167–176.

Chen, L., Tian, L., MacDonald, S. H.-F., McClafferty, H., Hammond, M. S. L., Huibant, J.-M., ... Shipston, M. J. (2005). Functionally diverse complement of large conductance calcium- and voltage-activated potassium channel (BK) alpha-subunits generated from a single site of splicing. *The Journal of Biological Chemistry*, *280*, 33599–33609.

Childs, G. V., Marchetti, C., & Brown, A. M. (1987). Involvement of sodium channels and two types of calcium channels in the regulation of adrenocorticotropin release. *Endocrinology*, *120*, 2059–2069.

Dopico, A. M., Widmer, H., Wang, G., Lemos, J. R., & Treistman, S. N. (1999). Rat supraoptic magnocellular neurones show distinct large conductance, Ca2+-activated K+ channel subtypes in cell bodies versus nerve endings. *The Journal of Physiology*, *519*, 101–114.

Duncan, P. J., Şengül, S., Tabak, J., Ruth, P., Bertram, R., & Shipston, M. J. (2015). Large conductance Ca(2+) -activated K(+) (BK) channels promote secretagogue-induced transition from spiking to bursting in murine anterior pituitary corticotrophs. *The Journal of Physiology*, *593*, 1197–1211.

Evanson, N. K., & Herman, J. P. (2015). Role of paraventricular nucleus glutamate signaling in regulation of HPA axis stress responses. *Interdisciplinary Information Sciences*, *21*, 253–260.

Gibbs, D. M., & Vale, W. (1982). Presence of corticotrophin releasing factor-like immunoreactivity in hypophysial portal blood. *Endocrinology*, *111*, 1418–1420.

Gimpl, G., & Fahrenholz, F. (2001). The oxytocin receptor system: Structure, function, and regulation. *Physiological Reviews*, *81*, 629–683.

Goncharova, N. D. (2013). Stress responsiveness of the hypothalamic-pituitary-adrenal axis: Age-related features of the vasopressinergic regulation. *Frontiers in Endocrinology*, *4*, 1–15.

Greffrath, W., Magerl, W., Disque-Kaiser, U., Martin, E., Reuss, S., & Boehmer, G. (2004). Contribution of Ca2+-activated K+ channels to hyperpolarizing after-potentials and discharge pattern in rat supraoptic neurones. *Journal of Neuroendocrinology, 16*, 577–588.

Herman, J. P., Figueiredo, H., Mueller, N. K., Ulrich-Lai, Y., Ostrander, M. M., Choi, D. C., & Cullinan, W. E. (2003). Central mechanisms of stress integration: Hierarchical circuitry controlling hypothalamo–pituitary–adrenocortical responsiveness. *Frontiers in Neuroendocrinology, 24*, 151–180.

Hodgkin, A. L., & Huxley, A. F. (1952). A quantitative description of membrane current and its application to conduction and excitation in nerve. *Journal of Physiology, 117*, 500–544.

Hodson, D. J., Schaeffer, M., Romanò, N., Fontanaud, P., Lafont, C., Birkenstock, J., ... Mollard, P. (2012). Existence of long-lasting experience-dependent plasticity in endocrine cell networks. *Nature Communications, 3*, 605.

Jameison, K., & Dinan, T. G. (2001). Glucocorticoids and cognitive function: From physiology to pathophysiology. *Human Psychopharmacology, 16*, 293–302.

Kadekaro, M., Su, G., Chu, R., Lei, Y., Li, J., & Fang, L. (2006). Nitric oxide up-regulates the expression of calcium-dependent potassium channels in the supraoptic nuclei and neural lobe of dehydrated rats. *Neuroscience Letters, 411*, 117–122.

Kelberman, D., Rizzoti, K., Lovell-Badge, R., Robinson, I. C. A. F., & Dattani, M. T. (2009). Genetic regulation of pituitary gland development in human and mouse. *Endocrine Reviews, 30*, 790–829.

Kidokoro, Y. (1975). Spontaneous calcium action potentials in a clonal pituitary cell line and their relationship to prolactin secretion. *Nature, 258*, 741–742.

King, J. T., Lovell, P. V., Rishniw, M., Kotlikoff, M. I., Zeeman, M. L., & Mccobb, D. P. (2006). β2 and β4 subunits of BK channels confer differential sensitivity to acute modulation by steroid hormones. *Journal of Neurophysiology, 95*, 2878–2888.

Kiss, J. Z. (1988). Dynamism of chemoarchitecture in the hypothalamic paraventricular nucleus. *Brain Research Bulletin, 20*, 699–708.

Knott, T. K., Dopico, A. M., Dayanithi, G., Lemos, J., & Treistman, S. N. (2002). Integrated channel plasticity contributes to alcohol tolerance in neurohypophysial terminals. *Molecular Pharmacology, 62*, 135–142.

Koshimizu, T.-A., Nakamura, K., Egashira, N., Hiroyama, M., Nonoguchi, H., & Tanoue, A. (2012). Vasopressin V1a and V1b receptors: From molecules to physiological systems. *Physiological Reviews, 92*, 1813–1864.

Kucka, M., Bjelobaba, I., Tomić, M., & Stojilkovic, S. S. (2013). The role of cyclic nucleotides in pituitary lactotroph functions. *Frontiers in Endocrinology, 4*, 1–7.

Kuryshev, Y. A., Childs, G. V., & Ritchie, A. K. (1996). Corticotropin-releasing hormone stimulates Ca2+ entry through L- and P-type Ca2+ channels in rat corticotropes. *Endocrinology, 137*, 2269–2277.

Kuryshev, Y. A., Haak, L., Childs, G. V., & Ritchie, A. K. (1997). Corticotropin releasing hormone inhibits an inwardly rectifying potassium current in rat corticotropes. *The Journal of Physiology, 502*, 265–279.

Lai, G.-J., & McCobb, D. P. (2002). Opposing actions of adrenal androgens and glucocorticoids on alternative splicing of Slo potassium channels in bovine chromaffin cells. *Proceedings of the National Academy of Sciences of the United States of America, 99*, 7722–7727.

Lai, G.-J., & McCobb, D. P. (2006). Regulation of alternative splicing of Slo K+ channels in adrenal and pituitary during the stress-hyporesponsive period of rat development. *Endocrinology, 147*, 3961–3967.

LeBeau, A. P., Robson, A. B., McKinnon, A. E., Donald, R. A., & Sneyd, J. (1997). Generation of action potentials in a mathematical model of corticotrophs. *Biophysical Journal, 73*, 1263–1275.

Lee, A. K., Smart, J. L., Rubinstein, M., Low, M. J., & Tse, A. (2011). Reciprocal regulation of TREK-1 channels by arachidonic acid and CRH in mouse corticotropes. *Endocrinology, 152*, 1901–1910.

Lee, A. K., & Tse, A. (1997). Mechanism underlying corticotropin-releasing hormone (CRH) triggered cytosolic Ca2+ rise in identified rat corticotrophs. *Journal of Physiology, 504*, 367–378.

Le Tissier, P. R., Hodson, D. J., Lafont, C., Fontanaud, P., Schaeffer, M., & Mollard, P. (2012). Anterior pituitary cell networks. *Frontiers in Neuroendocrinology, 33*, 252–266.

Liang, Z., Chen, L., McClafferty, H., Lukowski, R., MacGregor, D., King, J. T., … Shipston, M. J. (2011). Control of hypothalamic-pituitary-adrenal stress axis activity by the intermediate conductance calcium-activated potassium channel, SK4. *The Journal of Physiology, 589*, 5965–5986.

Lovell, P. V., King, J. T., & McCobb, D. P. (2004). Acute modulation of adrenal chromaffin cell BK channel gating and cell excitability by glucocorticoids. *Journal of Neurophysiology, 91*, 561–570.

Lovell, P. V., & McCobb, D. P. (2001). Pituitary control of BK potassium channel function and intrinsic firing properties of adrenal chromaffin cells. *The Journal of Neuroscience, 21*, 3429–3442.

Murat, B., Devost, D., Andrés, M., Mion, J., Boulay, V., Corbani, M., … Guillon, G. (2012). V1b and CRHR1 receptor heterodimerization mediates synergistic biological actions of vasopressin and CRH. *Molecular Endocrinology, 26*, 502–520.

Newell-Price, J. (2003). Proopiomelanocortin gene expression and DNA methylation: Implications for Cushing's syndrome and beyond. *Journal of Endocrinology, 177*, 365–372.

Ohbuchi, T., Yokoyama, T., Saito, T., Suzuki, H., Fujihara, H., Katoh, A., … Ueta, Y. (2010). Modulators of BK and SK channels alter electrical activity in vitro in single vasopressin neurons isolated from the rat supraoptic nucleus. *Neuroscience Letters, 484*, 26–29.

Ooi, G. T., Tawadros, N., & Escalona, R. M. (2004). Pituitary cell lines and their endocrine applications. *Molecular and Cellular Endocrinology, 228*, 1–21.

Papadimitriou, A., & Priftis, K. N. (2009). Regulation of the hypothalamic-pituitary-adrenal axis. *Neuroimmunomodulation, 16*, 265–271.

Pietrzykowski, A. Z., Friesen, R. M., Martin, G. E., Puig, S. I., Nowak, C. L., Wynne, P. M., … Treistman, S. N. (2008). Posttranscriptional regulation of BK channel splice variant stability by miR-9 underlies neuroadaptation to alcohol. *Neuron, 59*, 274–287.

Plüger, S., Faulhaber, J., Fürstenau, M., Löhn, M., Waldschütz, R., Gollasch, M., … Pongs, O. (2000). Mice with disrupted BK channel beta1 subunit gene feature abnormal Ca2+ spark/STOC coupling and elevated blood pressure. *Circulation Research, 87*, E53–E60.

Poulain, D. A., & Wakerley, J. B. (1982). Electrophysiology of hypothalamic magnocellular neurones secreting oxytocin and vasopressin. *Neuroscience, 7*, 773–808.

Rae, J., Cooper, K., Gates, P., & Watsky, M. (1991). Low access resistance perforated patch recordings using amphotericin B. *Journal of Neuroscience Methods, 37*, 15–26.

Rettig, J., & Neher, E. (2002). Emerging roles of presynaptic proteins in Ca++-triggered exocytosis. *Science, 298*, 781–785.

Saito, M., Nelson, C., Salkoff, L., & Lingle, C. J. (1997). A cysteine-rich domain defined by a novel exon in a Slo variant in rat adrenal chromaffin cells and PC12 cells. *Journal of Biological Chemistry, 272*, 11710–11717.

Salzmann, M., Seidel, K. N., Bernard, R., Prüss, H., Veh, R. W., & Derst, C. (2010). BKbeta1 subunits contribute to BK channel diversity in rat hypothalamic neurons. *Cellular and Molecular Neurobiology, 30*, 967–976.

Sausbier, M., Hu, H., Arntz, C., Feil, S., Kamm, S., Adelsberger, H., … Ruth, P. (2004). Cerebellar ataxia and Purkinje cell dysfunction caused by Ca2+-activated K+ channel

deficiency. *Proceedings of the National Academy of Sciences of the United States of America, 101*, 9474–9478.
Sheward, J., & Fink, G. (1991). Effects of cortiocsterone on the secretion of corticotrophin-releasing factor, arginine vasopressin and oxytocin into hypophysial portal blood in long-term hypophysectomized rats. *Journal of Endocrinology, 129*, 91–98.
Shipston, M. J., Duncan, R. R., Clark, A. G., Antoni, F. A., & Tian, L. (1999). Molecular components of large conductance calcium-activated potassium (BK) channels in mouse pituitary corticotropes. *Molecular Endocrinology, 13*, 1728–1737.
Shipston, M. J., Kelly, J. S., & Antoni, F. A. (1996). Glucocorticoids block protein kinase A inhibition of calcium-activated potassium channels. *The Journal of Biological Chemistry, 271*, 9197–9200.
Simon, S. M., & Llinás, R. R. (1985). Compartmentalization of the submembrane calcium activity during calcium influx and its significance in transmitter release. *Biophysical Journal, 48*, 485–498.
Singh, N. P., & Panwar, V. K. (2006). Case report of a pituitary macroadenoma treated with artemether. *Integrative Cancer Therapies, 5*, 391–394.
Shorten, P. R., Robson, A. B., McKinnon, A. E., & Wall, D. J. (2000). CRH-induced electrical activity and calcium signalling in pituitary corticotrophs. *Journal of Theoretical Biology, 206*, 395–405.
Stojilkovic, S. S., Tabak, J., & Bertram, R. (2010). Ion channels and signaling in the pituitary gland. *Endocrine Reviews, 31*, 845–915.
Stojilkovic, S. S., Zemkova, H., & Van Goor, F. (2005). Biophysical basis of pituitary cell type-specific Ca2+ signaling-secretion coupling. *Trends in Endocrinology and Metabolism, 16*, 152–159.
Tabak, J., Tomaiuolo, M., Gonzalez-Iglesias, A. E., Milescu, L. S., & Bertram, R. (2011). Fast-activating voltage- and calcium-dependent potassium (BK) conductance promotes bursting in pituitary cells: A dynamic clamp study. *The Journal of Neuroscience, 31*, 16855–16863.
Tagliavini, A., Tabak, J., Bertam, R., & Pedersen, M. G. (2016). Is bursting more effective than spiking in evoking pituitary hormone secretion? A spatiotemporal simulation study of calcium and granule dynamics. *American Journal of Physiology. Endocrinology and Metabolism*. http://dx.doi.org/10.1152/ajpendo.00500.2015. (Epub ahead of print).
Tian, L., Duncan, R. R., Hammond, M. S. L., Coghill, L. S., Wen, H., Rusinova, R., … Shipston, M. J. (2001). Alternative splicing switches potassium channel sensitivity to protein phosphorylation. *Journal of Biological Chemistry, 276*, 7717–7720.
Tian, L., Hammond, M. S., Florance, H., Antoni, F. A., & Shipston, M. J. (2001). Alternative splicing determines sensitivity of murine calcium-activated potassium channels to glucocorticoids. *Journal of Physiology, 537*, 57–68.
Tian, L., Knaus, H.-G., & Shipston, M. J. (1998). Glucocorticoid regulation of calcium-activated potassium channels mediated by serine/threonine protein phosphatase. *The Journal of Biological Chemistry, 273*, 13531–13536.
Tomić, M., Kucka, M., Kretschmannova, K., Li, S., Nesterova, M., Stratakis, C. A., & Stojilkovic, S. S. (2011). Role of nonselective cation channels in spontaneous and protein kinase A-stimulated calcium signaling in pituitary cells. *American Journal of Physiology. Endocrinology and Metabolism, 301*, E370–E379.
Toro, L., Wallner, M., Meera, P., & Tanaka, Y. (1998). Maxi-KCa, a unique member of the voltage-gated K channel superfamily. *News in Physiological Sciences, 13*, 112–117.
Tsaneva-Atanasova, K., Sherman, A., van Goor, F., & Stojilkovic, S. S. (2007). Mechanism of spontaneous and receptor-controlled electrical activity in pituitary somatotrophs: Experiments and theory. *Journal of Neurophysiology, 98*, 131–144.
Ulrich-Lai, Y. M., & Herman, J. P. (2009). Neural regulation of endocrine and autonomic stress responses. *Nature Reviews. Neuroscience, 10*, 397–409.

Van Goor, F., Li, Y.-X., & Stojilkovic, S. S. (2001). Paradoxical role of large-conductance calcium-activated K+ (BK) channels in controlling action potential-driven Ca2+ entry in anterior pituitary Cells. *The Journal of Neuroscience*, *21*, 5902–5915.

Van Goor, F., Zivadinovic, D., Martinez-Fuentes, A. J., & Stojilkovic, S. S. (2001). Dependence of pituitary hormone secretion on the pattern of spontaneous voltage-gated calcium influx. Cell type-specific action potential secretion coupling. *The Journal of Biological Chemistry*, *276*, 33840–33846.

Van Goor, F., Zivadinovic, D., & Stojilkovic, S. S. (2001). Differential expression of ionic pituitary cells. *Molecular Endocrinology*, *15*, 1222–1236.

Vo, T., Tabak, J., Bertram, R., & Wechselberger, M. (2014). A geometric understanding of how fast activating potassium channels promote bursting in pituitary cells. *Journal of Computational Neuroscience*, *36*, 259–278.

Wang, B., Jaffe, D. B., & Brenner, R. (2014). Current understanding of iberiotoxin-resistant BK channels in the nervous system. *Frontiers in Physiology*, *5*, 1–11.

White, R. E., Schonbrunn, A., & Armstrong, D. L. (1991). Somatostatin stimulates Ca2+-activated K+ channels through protein dephosphorylation. *Nature*, *351*, 570–573.

Wynne, P. M., Puig, S. I., Martin, G. E., & Treistman, S. N. (2009). Compartmentalized beta subunit distribution determines characteristics and ethanol sensitivity of somatic, dendritic, and terminal large-conductance calcium-activated potassium channels in the rat central nervous system. *The Journal of Pharmacology and Experimental Therapeutics*, *329*, 978–986.

Xie, J., & McCobb, D. P. (1998). Control of alternative splicing of potassium channels by stress hormones. *Science*, *280*, 443–446.

Yamashita, M., Oki, Y., Iino, K., Hayashi, C., Yogo, K., Matsushita, F., … Nakamura, H. (2009). The role of store-operated Ca2+ channels in adrenocorticotropin release by rat pituitary cells. *Regulatory Peptides*, *156*, 57–64.

Zhou, X., Wulfsen, I., Korth, M., McClafferty, H., Lukowski, R., Shipston, M. J., … Wieland, T. (2012). Palmitoylation and membrane association of the stress axis regulated insert (STREX) controls BK channel regulation by protein kinase C. *The Journal of Biological Chemistry*, *287*, 32161–32171.

CHAPTER TEN

BK Channels in the Vertebrate Inner Ear

S.J. Pyott*[,1], R.K. Duncan[†]
*University Medical Center Groningen, Groningen, The Netherlands
[†]Kresge Hearing Research Institute, University of Michigan, Ann Arbor, MI, United States
[1]Corresponding author: e-mail address: s.pyott@umcg.nl

Contents

Abstract

The perception of complex acoustic stimuli begins with the deconstruction of sound into its frequency components. This spectral processing occurs first and foremost in the inner ear. In vertebrates, two very different strategies of frequency analysis have evolved. In nonmammalian vertebrates, the sensory hair cells of the inner ear are intrinsically electrically tuned to a narrow band of acoustic frequencies. This electrical tuning relies on the interplay between BK channels and voltage-gated calcium channels. Systematic variations in BK channel density and kinetics establish a gradient in electrical resonance that enables the coding of a broad range of acoustic frequencies. In contrast, mammalian hair cells are extrinsically tuned by mechanical properties of the cochlear duct. Even so, mammalian hair cells also express BK channels. These BK channels play critical roles in various aspects of mammalian auditory signaling, from developmental maturation to protection against acoustic trauma. This review summarizes the anatomical localization, biophysical properties, and functional contributions of BK channels in vertebrate inner ears. Areas of future research, based on an updated understanding of

International Review of Neurobiology, Volume 128
ISSN 0074-7742
http://dx.doi.org/10.1016/bs.irn.2016.03.016

the biology of both BK channels and the inner ear, are also highlighted. Investigation of BK channels in the inner ear continues to provide fertile research grounds for examining both BK channel biophysics and the molecular mechanisms underlying signal processing in the auditory periphery.

1. INTRODUCTION TO THE AUDITORY PERIPHERY

All around us at any given moment is an ever-changing, complex auditory scene. In order to follow a friend's voice through a crowded room or segregate a bubbling brook from buzzing cicadas, the auditory system must separate rich acoustic signals into individual frequency components and transmit this information to the brain. This spectral analysis begins in the auditory periphery, primarily within the auditory sensory epithelium of the inner ear. The auditory epithelium is comprised of sensory hair cells, support cells, and the neurons that connect the hair cells to the central nervous system. A place code exists along this epithelium. Specifically, any one hair cell is tuned to a narrow band of acoustic frequencies in the audible spectrum. The location of the hair cells then creates a "tonotopic" or frequency–place map along the length of the epithelium. In all vertebrate species, the sensory hair cells and their neural connections are exquisitely organized in this tonotopic pattern, but the mechanisms involved in achieving frequency tuning of the individual hair cells are distinct between mammals and nonmammals.

In nonmammalian vertebrates, an intrinsic, electrical tuning mechanism is primarily responsible for resolving frequency information. Electrical tuning arises from electrical resonance in the hair cell's membrane potential. Current stimulation results in a damped oscillation of the membrane voltage response, and the frequency of this oscillation is systematically graded along the auditory epithelium. In this way, hair cells at the apex of the cochlea resonate at low frequencies and those at the base are tuned to high frequencies. The interplay between BK channels and voltage-gated calcium channels (VGCCs) are the primary determinants of this phenomenon, underscoring the importance of BK channels in auditory processing in these species. The structure and function of BK channels in electrical tuning is considered further in Section 2.

In contrast, mammalian hair cells are extrinsically or mechanically tuned by accessory structures of the inner ear. Although mammalian hair cells are not electrically tuned as they are in other vertebrates, BK channels nevertheless carry a significant component of the total outward potassium conductances in these hair cells. Moreover, recent research has shown that the expression of BK channels is not restricted to the hair cells. In sum, a variety of evidence indicates that BK channels play critical roles in mammalian auditory signaling. BK channels in the mammalian inner ear are considered further in Section 3.

As in other systems, BK channels in the auditory system have been (sometimes confusingly) referred to by many names, including MaxiK, BK, BK_{Ca}, Slo1, $K_{Ca}1.1$ channels, and (in inner hair cells) $I_{K,f}$. Despite differences in nomenclature, these channels are encoded by the KCNMA1 gene and generally characterized by a large conductance to potassium and dual activation by depolarization and intracellular calcium. More general reviews of BK channels can be found elsewhere in this book (as well as Toro et al., 2014). As this chapter illustrates, the investigation of BK channels in the vertebrate inner ear has provided reciprocal insight into the fundamental molecular mechanisms underlying BK channel structure and function and systems-level auditory processing.

2. BK CHANNEL STRUCTURE AND FUNCTION IN NONMAMMALIAN AUDITORY ORGANS

2.1 Electrical Tuning of Nonmammalian Hair Cells

Auditory physiology in the 1970s was marked by a raging debate about the sources of tuning in the cochlea. Recordings from single auditory neurons revealed sharp, narrow-band tuning to sound stimulation (Kiang, 1965). Likewise, intracellular recordings from presynaptic hair cells were sharply tuned (Russell & Sellick, 1978). Yet, mechanical vibrations in the cochlear partition, that is, displacements of the basilar membrane, appeared poorly sensitive and broadly tuned (Evans & Wilson, 1975). The discrepancy suggested the actions of another amplifier somewhere within the sensory epithelium, and so the hunt for a "second filter" ensued. In many ways, this search provided the context for major discoveries about electrical and mechanical tuning mechanisms, including insight into the diversity and function of BK channels in the inner ear.

Over the next two decades, the search for a second filter revealed a startling evolutionary divergence in coding strategies between mammals and nonmammals. Advances in recording techniques and preparation health revealed sharply tuned mechanics in mammalian cochleae (Khanna & Leonard, 1982), discarding the need for a second filter in those species. However, comparable measurements in nonmammalian auditory organs continued to show broadly tuned basilar membranes (turtle, O'Neill & Bearden, 1995). Moreover, in some lower vertebrates, a basilar membrane is entirely absent even though a place code exists along the epithelium (frog; Pitchford & Ashmore, 1987).

In a series of now landmark papers, Andrew Crawford and Robert Fettiplace described an intrinsic electrical tuning in turtle hair cells that served as the primary means of frequency tuning in the turtle cochlea (Crawford & Fettiplace, 1980, 1981; Fettiplace & Crawford, 1980). Current injection into these hair cells caused a damped oscillation in the membrane voltage response that matched the tonotopic position of the cells and the narrow-band tuning of auditory nerve fibers at those characteristic frequencies (Fettiplace & Crawford, 1980). Feedback from voltage-gated potassium conductances was suggested as a possible mechanism, since application of the potassium channel blocker tetraethylammonium (TEA) reduced or eliminated membrane potential oscillations upon current injection (Art, Crawford, Fettiplace, & Fuchs, 1985). Electrical tuning was ultimately described in other auditory organs of nonmammalian vertebrates, including frog (Pitchford & Ashmore, 1987), alligator (Fuchs & Evans, 1988), and chicken (Fuchs, Nagai, & Evans, 1988), but was absent in hair cells from the mammalian cochlea (Kros & Crawford, 1990).

The advent of whole-cell voltage-clamp recording techniques solidified the role of voltage-gated ion channels in hair cell electrical tuning. The major outward potassium conductance in these cells was consistent with the features of BK channels, namely sensitivity to extracellular TEA and intracellular cesium (Art & Fettiplace, 1987), a large single-channel conductance (Fuchs & Evans, 1988; Hudspeth & Lewis, 1988b), and activation by calcium influx through VGCCs (Hudspeth & Lewis, 1988a). The interplay between VGCCs and BK channels, along with a leak conductance and membrane capacitance, is necessary and sufficient to achieve membrane resonance and electrical tuning (Hudspeth & Lewis, 1988a). In this model, damped oscillations are initiated by depolarizing current steps and rapid activation of VGCCs. The resulting calcium influx further depolarizes the hair cell membrane and, with some delay, activates BK channels. The delayed but fast activation of large-conductance BK channels opposes a regenerative calcium action potential and drives the voltage response downward. As the membrane potential becomes more negative, VGCCs close and BK channels begin to deactivate. The persisting current injection initiates the next phase of depolarization. The combined delays in activation and deactivation of these channels cause neither to reach their resting values during each oscillation, resulting in a damped resonance that eventually reaches a steady-state level. The two inescapable conclusions are that (1) the channels must be in close proximity to one another to provide rapid feedback and (2) some parameter must systematically change along the auditory epithelium to enable a smooth tonotopic gradient in electrical frequency tuning.

2.2 Interplay of BK Channels and VGCCs

Several functional studies support the close proximity of VGCCs and BK channels in nonmammalian hair cells. In these hair cells, and in contrast to that in mammalian analogs (see Section 3.2.1), BK channel activation requires calcium influx through VGCCs (Art & Fettiplace, 1987). Moreover, the high degree of calcium buffering in hair cells limits the degree of calcium diffusion and thereby restricts the size of BK channel VGCC domains to within 250 nm (Roberts, 1993). Additionally, BK channel activation is not eliminated by high intracellular concentrations of the fast calcium chelator BAPTA (Roberts, Jacobs, & Hudspeth, 1990), suggesting proximity at the single-channel level, possibly within tens of nanometers (Fakler & Adelman, 2008). Consequently, clusters of VGCC and BK channels are found in loose patch recordings from the basolateral membrane of frog saccular hair cells (Roberts et al., 1990). The average number of clusters per hair cell matched the average number of synaptic active zones, suggesting a tight correlation between these two critical calcium signaling domains. Similarly, cell-attached patch recordings at fluorescent calcium hotspots typically contained a small number of BK channels (ie, about 60% of the patches contained both channel types; Issa & Hudspeth, 1994). Importantly, membrane patches from regions outside calcium hotspots contained no BK currents. However, the number of recordings was small, so the basolateral membrane surface was not thoroughly scanned.

Immunocytochemistry studies have provided further evidence of channel clustering and colocalization (Li et al., 2009; Rodriguez-Contreras & Yamoah, 2001; Samaranayake, Saunders, Greene, & Navaratnam, 2004), but differences in epitope targeting, particularly in the C-terminus of the highly alternatively spliced BK channels, dramatically affect clustering and subcellular localization (Kim et al., 2010). Interestingly, although over 100 interaction partners have been identified for cochlear BK channels (Kathiresan, Harvey, Orchard, Sakai, & Sokolowski, 2009; Sokolowski, Orchard, Harvey, Sridhar, & Sakai, 2011) and physical coassembly of BK channels and VGCCs has been described in the rat brain (Berkefeld et al., 2006; Grunnet & Kaufmann, 2004; Loane, Lima, & Marrion, 2007), biochemical evidence supporting such direct interaction of BK channels and VGCCs in hair cells remains elusive (Lesage, Hibino, & Hudspeth, 2004; Sokolowski et al., 2011). Consequently, the fidelity of colocalization and the mechanisms behind targeting to the ribbon synapse remains an active area of research.

2.3 Tonotopic Distribution of BK Channel Properties

BK currents are the primary determinants of tuning frequency, since they gate more slowly than VGCCs. Therefore, BK currents are the rate-limiting step in electrical resonance (Art & Fettiplace, 1987). Accordingly, it is the kinetics of this current that is smoothly graded along the auditory epithelium, with higher resonant frequencies arising from faster BK currents (Art & Fettiplace, 1987; Fuchs & Evans, 1988; Fuchs et al., 1988; Smotherman & Narins, 1999). The startling conclusion from these data is that some molecular feature contributing to BK channel conductance is systematically graded along the cochlea. Modeling schemes suggest that gradients in a small number of variables can achieve resonant frequencies spanning 40–4000 Hz, a range sufficient to encode relevant acoustic signals in chicks and lower vertebrates (Wu & Fettiplace, 1996). The speed and size of the BK channel potassium conductance could be influenced by the calcium signal (eg, influx, diffusion, and extrusion) or intrinsic differences in BK channel gating. Although the magnitude of the calcium current increases with increasing resonant frequency, reduction of extracellular calcium by two orders of magnitude had little impact on BK channel relaxation time constants in turtle hair cells (Art, Fettiplace, & Wu, 1993). Likewise, tuning frequency is largely insensitive to intracellular BAPTA (Ricci, Gray-Keller, & Fettiplace, 2000), indicating that resonance is set by activity within BAPTA-resistant microdomains independent of calcium diffusion and extrusion.

Single-channel recordings of BK channel activity were first obtained from inside-out patches from frog saccular hair cells (Hudspeth & Lewis, 1988b), confirming a high unitary conductance ($\approx$200 pS) in symmetrical potassium and prototypical sensitivity to both membrane voltage and intracellular calcium concentration. Subsequently, extensive recordings in auditory epithelia demonstrated that variations in intrinsic BK channel kinetics correlated with estimates of resonant frequency in isolated turtle hair cells (Art, Wu, & Fettiplace, 1995) and direct measures of tonotopic position in chick epithelia (Duncan & Fuchs, 2003). Deactivation rates were strongly associated with resonant frequency in both species, but other gating parameters were independent of tuning or only weakly correlated (Duncan & Fuchs, 2003). Fig. 1 illustrates the variations in BK channel density and kinetics with hair cell tonotopic location that serves electrical tuning in non-mammalian hair cells.

Early predictions from findings in turtle cochlea suggested that BK channels from high-frequency hair cells would exhibit lower calcium sensitivity

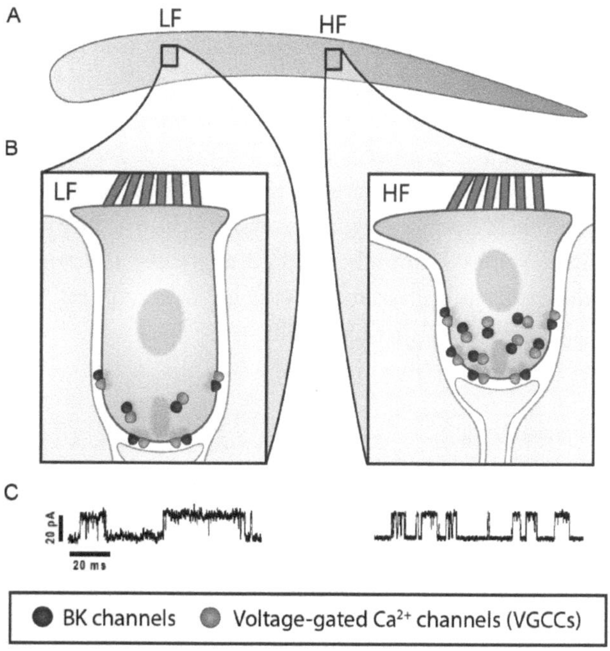

Fig. 1 Variations in BK channel density and kinetics with tonotopic location. (A) A schematic of the chick auditory epithelium with *boxes* indicating the positions of hair cells at two tonotopic locations, a low-frequency (LF) hair cell tuned to 250 Hz and a high-frequency (HF) hair cell tuned to 825 Hz. (B) Hair cells from these two locations are illustrated to show relative differences in cell morphology and ion channel density. BK and VGCC channels are clustered so that calcium influx from VGCCs activates neighboring BK channels (*arrows*). The density of these clusters increases with increasing tuning frequency. (C) Single-channel recordings from hair cells at these locations (R. K. Duncan, unpublished data) demonstrate kinetic differences that underlie electrical tuning. The mean open duration from these recordings was 9.8 and 1.9 ms for the LF and HF hair cells, respectively. Correspondingly, deactivation rates from whole-cell BK recordings increase systematically with tuning frequency. Recordings were made at −60 mV in the presence of 5 μM $[Ca^{2+}]_i$ using inside-out patches in symmetric potassium solutions.

(Art et al., 1993). However, single-channel data from these hair cells showed no tonotopic gradient in calcium affinity (Art et al., 1995). In chick, a weak correlation between calcium sensitivity and tuning was found, possibly due to the expanded range in resonant frequencies encoded in that animal. Interestingly, the average $K_D(0)$—the $[Ca^{2+}]_i$ required for half-activation at 0 mV—was greater in patches from the high-frequency end of the cochlea, but the variance (diversity) in calcium sensitivity increased with frequency as well. High-affinity channels were distributed throughout the cochlea, but the distribution of $K_D(0)$ broadened to include BK channels with reduced calcium affinity as location moved toward high frequencies. These data underscore the diversity in channel properties at any one location and stand in contrast to early models of tuning that mixed a small number of BK channel variants expressed at discrete locations to achieve smooth gradients in electrical tuning (Wu & Fettiplace, 1996). Indeed, in the chick, excised patches with greater numbers of channels showed better correlation between the predicted resonant frequency for that cell and its tonotopic location; patches with single or few channels showed a greater discrepancy between predicted and actual tuning frequency. The data suggest significant BK channel diversity even within single cells.

2.4 Splice Variation

What then is the source of BK channel diversity in nonmammalian auditory epithelia? BK channels are comprised of four pore-forming α subunits encoded by a single gene, KCNMA1, as well as several potential auxiliary subunits. Initial cloning experiments in the turtle identified two primary splice sites in the BK α subunit mRNA (Jones, Gray-Keller, Art, & Fettiplace, 1999). Alternative exons at these two sites reside in the cytoplasmic tail between the two calcium regulatory domains RCK1 and RCK2 (see Chapter "Posttranscriptional and Posttranslational Regulation of BK Channels" by Shipston and Tian). Alternative splicing in the turtle cochlea results in six naturally occurring variants with a range of calcium affinity and gating kinetics when heterologously expressed in oocytes (Jones et al., 1999). Interestingly, kinetics varied systematically with insertion/deletion (indel) length. Hypothetical regional distribution of these variants, some combined with accessory $\beta 1$ subunits, could account for tuning in the turtle. However, it remains unclear whether the variants are expressed in these distributions in the epithelium. Moreover, evidence of $\beta 1$ expression in the turtle cochlea is lacking.

Electrical tuning in chick spans a larger range of frequencies than in turtle, so a larger number of unique isoforms may be required. Indeed, there appears to be a larger number of splice sites utilized in the chick (ie, seven alternative splice sites within the coding region plus alternative start sites; see for review, Fettiplace & Fuchs, 1999). Sixteen alternative exons have been identified (Miranda-Rottmann, Kozlov, & Hudspeth, 2010). If each were incorporated independently and combined with others in a nonobligatory manner, then a staggering array of unique variants could be formed, particularly if unique isoforms formed heteromultimers with intermediate properties.

The potential diversity is large, but the question remains whether alternative splicing is distributed in such a way to govern or contribute to tonotopic gradients in tuning. Central to the hypothesis is the expectation that variants are expressed in restricted regions in the auditory epithelium. Navaratnam and colleagues divided the chick epithelium into quadrants and examined exon expression by PCR (Navaratnam, Bell, Tu, Cohen, & Oberholtzer, 1997). Several alternative exons were differentially distributed, and individual hair cells expressed multiple variants at the same splice site. Unfortunately, these data gave no indication about naturally occurring combinations of alternative exons or the relative abundance of these variants among the total pool of BK channel transcripts.

In a recent study, more than 3000 full-length variants of the BK channel α subunit were cloned and sequenced from the chick cochlea (Miranda-Rottmann et al., 2010). Transcripts were examined based on C-terminal splicing at site 7, because variation at this site has a profound influence over surface expression and clustering (Kim et al., 2010). Notably, transcripts ending in the C-terminal isoform QEDRL, which may be the predominant cell surface isoform, showed little alternative splicing. Over 90% of the QEDRL transcripts included no alternative exons at upstream splice sites. The authors found no selective inclusion of alternative exons at specific regions along the auditory epithelium and the limited degree of long, complex splice forms argues against indel length as a major contributor to tuning.

2.5 Regulatory β Subunits

In every model of electrical tuning, BK channel β subunits are employed to increase the range of calcium affinities and gating kinetics to allow for tuning in low-frequency hair cells. Four BK β subunits have been described (KCNMB1–4; see, for example, Contreras, Neely, Alvarez, Gonzalez, &

Latorre, 2012). β1 and β4 have been cloned from the chick cochlea, and gene expression studies suggest higher expression of both subunits in apical than basal hair cells (Bai, Surguchev, & Navaratnam, 2011; Ramanathan, Michael, Jiang, Hiel, & Fuchs, 1999). When coexpressed with BK channel α subunit variants in HEK293 cells or oocytes, β subunits increase calcium affinity and slow deactivation kinetics while maintaining the relative differences between alternative α splice forms (Bai et al., 2011; Ramanathan, Michael, & Fuchs, 2000). These features are consistent with enriched β expression in the low-frequency end of the cochlea. However, relaxation time constants in native BK channels from chick cochlea are about one order of magnitude faster than heterologously expressed BK channels coassembled with β1 or β4 (Bai et al., 2011; Duncan & Fuchs, 2003; Ramanathan et al., 2000).

Although siRNA knockdown of β1 and β4 in cultured chick cochlea suggests a role in BK channel surface expression (Bai et al., 2011), functional evidence is lacking. Electrophysiological data from the siRNA experiments will be important for establishing the necessity for either subunit in electrical tuning. Pharmacological experiments could also shed light on the functional impact of β subunit expression. Coassembly with β subunits modulates BK channel sensitivity to steroidal hormones and toxins. For example, BK channel activation by estradiol or the xenoestrogen tamoxifen requires coexpression with either β1 (Dick, Rossow, Smirnov, Horowitz, & Sanders, 2001; Dick & Sanders, 2001) or β4 (Behrens et al., 2000). However, when tamoxifen was applied to chick hair cell BK channels, only 1 of about 20 patches showed an increase in open probability (Tong & Duncan, 2009). The data did not support a major role for β subunits, except possibly in the most apical end of the cochlea. Since β4 confers resistance to the BK channel blocker iberiotoxin (Wang, Jaffe, & Brenner, 2014), it would be interesting to examine toxin sensitivity along the auditory epithelium in chick.

2.6 Other Possibilities

Other accessory subunits or interaction partners could play major roles in tuning. While recently identified BK γ subunits have not been identified in nonmammalian genomes (Dolan et al., 2007), the possibility remains that orthologues of these subunits or other protein partners regulate BK channel function in nonmammalian hair cells. Using tandem mass spectrometry, over 100 putative BK channel interaction partners have been identified in the chick cochlea (Sokolowski et al., 2011). Informatics analysis indicated that these interactions may be involved in cell death/survival pathways, trafficking, calcium binding, and kinase activity. In particular, protein kinases may have a strong impact on electrical tuning, since BK channels exhibit numerous phosphorylation sites that confer significant changes to channel gating (Yan et al., 2008). These possibilities are just now beginning to be explored. For example, the cyclin-dependent kinase CKD5 is differentially expressed along the chick auditory epithelium, increasing in expression with increasing tuning frequency (Bai, Surguchev, Joshi, Gross, & Navaratnam, 2012). Interaction with CDK5 results in faster deactivation kinetics, consistent with tuning in high-frequency hair cells.

Gene expression arrays from apical and basal segments of the chick auditory epithelium have been screened by two separate groups to determine candidate signaling pathways that might govern the myriad number of tonotopically distributed parameters in the inner ear, including BK channels and VGCCs (Frucht, Uduman, Kleinstein, Santos-Sacchi, & Navaratnam, 2011; Kowalik & Hudspeth, 2011). Three major pathways were indicated, including Notch, Wnt, and FGF. MicroRNAs and protein kinases were also identified as potential regulators of tonotopic distributions. Future experiments will need to combine screens for interaction partners with screens for apical-to-basal molecular gradients to identify new candidate molecules involved in BK channel diversity and electrical tuning.

3. BK CHANNELS IN THE MAMMALIAN COCHLEA

3.1 Anatomy of the Mammalian Auditory Sensory Epithelium

In mammals, the auditory sensory epithelium, called the organ of Corti, contains two distinct classes of hair cells: a single row of inner hair cells and three to four rows of outer hair cells. Both inner and outer hair cells transduce vibrations produced by sound waves into graded changes in membrane potential. The inner hair cells, however, are the primary sensory cells. When depolarized, the inner hair cells release glutamate and activate the primary auditory neurons, called the spiral ganglion cells, which relay (afferent) information to the brain. In contrast, the outer hair cells relay little afferent information. Instead, by virtue of their electromotility, depolarization generates mechanical forces necessary for the amplification of sound-induced vibrations of the basilar membrane. They are, therefore, essential for amplifying the mechanical signals transduced by the inner hair cells. The outer hair cells receive abundant inhibitory (efferent) feedback from the brainstem that modulates cochlear amplification. The anatomy of the mammalian cochlea and organ of Corti is more fully reviewed elsewhere (for example, Raphael & Altschuler, 2003).

In contrast to hair cells from other vertebrates, mammalian inner hair cells are extrinsically or mechanically tuned by accessory structures of the inner ear, most notably the basilar membrane. As first elucidated by von Békésy (1960), changes in mechanical properties of the basilar membrane, especially in the width and thickness, are responsible for separating sound frequencies spatially along the length of the overlying auditory sensory epithelium that contains the hair cells. Active properties of the outer hair cells refine this spatial frequency tuning. In this way, specific frequencies of sound maximally excite a subset of spatially localized inner hair cells, creating a tonotopic or frequency–place map. The distinct sensory and mechanical roles of inner and outer hair cells are nicely reviewed by Fettiplace and Hackney (2006). Although mammalian hair cells are not electrically tuned, BK channels nevertheless carry a significant component of the total outward potassium conductances in both inner and outer hair cells. Moreover, BK channel expression is not restricted to the hair cells. Work, largely over the last two decades, has shown that BK channels contribute in various and often unexpected ways to mammalian hearing.

3.2 BK Channel Expression in the Mammalian Auditory Sensory Epithelium

3.2.1 Inner Hair Cells

BK currents, initially referred to as $I_{K,f}$ to distinguish their fast rate of activation from other more slowly activating potassium currents, were first observed by Kros and Crawford (1990) using whole-cell patch-clamp recordings from isolated guinea pig inner hair cells. This conductance was found to be pharmacologically characteristic of BK channels: sensitive to TEA (Kros & Crawford, 1990), resistant to 4-AP (Kros & Crawford, 1990), and blocked by the highly specific BK channel blockers, iberiotoxin (Hafidi, Beurg, & Dulon, 2005; Kros, Ruppersberg, & Rusch, 1998; Marcotti, Johnson, & Kros, 2004a; Pyott, Glowatzki, Trimmer, & Aldrich, 2004), and charybdotoxin (Dulon, Sugasawa, Blanchet, & Erostegui, 1995; Kros et al., 1998).

Additional molecular evidence that $I_{K,f}$ is indeed carried by BK channels came from rtPCR and in situ hybridization. These techniques identified BK channel (α subunit) transcripts in the rat organ of Corti (Brändle et al., 2001) as well as various BK channel splice variants in rat inner hair cells (Langer, Grunder, & Rusch, 2003). The diversity and tonotopic patterns of BK channel splice variant expression in inner and outer mouse inner hair cells have also been carefully documented (Beisel et al., 2007). Immunohistochemistry verified expression of the BK channel protein in guinea pig (Skinner et al., 2003) and mouse inner hair cells (Hafidi et al., 2005; Pyott et al., 2004). Finally, definitive evidence that the BK channel is responsible for $I_{K,f}$ in inner hair cells came from BK α knockout mice. Importantly, $I_{K,f}$ was absent in whole-cell patch-clamp recordings of inner hair cells from these mice (Pyott et al., 2004; Rüttiger et al., 2004).

Immunofluorescent localization of BK channels revealed an unexpected subcellular localization in inner hair cells. In nonmammalian hair cells BK channels are colocalized with VGCCs at transmitter release sites (see Section 2.2). This colocalization of BK and VGCCs is essential for electrical tuning. In contrast, BK channels are predominantly and possibly exclusively localized to the necks of mammalian inner hair cells (Hafidi et al., 2005; Pyott et al., 2004) away from the afferent ribbons and VGCCs (Fig. 2). The functional importance of this subcellular localization is not clear although it undoubtedly reflects the fundamentally different role BK channels play in nonmammalian and mammalian hair cells. Importantly, expression in other subcellular locations of the inner hair cells that remains undetected by immunofluorescence cannot be excluded. Moreover, BK channels in both inner (see Wersinger et al., 2010) and outer (Engel et al., 2006; Maison et al., 2013; Wersinger et al., 2010) hair cells appear to show tonotopic gradients of increasing expression from apical (low-frequency) to basal (high-frequency) regions. The stronger expression in high-frequency regions suggests a contribution of BK channels to high-frequency hearing in mammals (see Section 3.3.4).

The biophysical properties of BK channels in inner hair cells are curiously different from those generally attributed to BK channels observed in other cells, including nonmammalian hair cells. Importantly, the dual activation of BK channels by both membrane depolarization and intracellular calcium is key to their regulation of membrane excitability in various cells (reviewed in Toro et al., 2014) including the electrical tuning of nonmammalian hair cells. Whereas BK currents in the hair cells of nonmammalian species depend on calcium entry into the cell (see Section 2.2), BK currents in mammalian (guinea pig and mouse) inner hair cells are curiously unaffected by the removal of extracellular calcium (Kros & Crawford, 1990; Marcotti et al., 2004a).

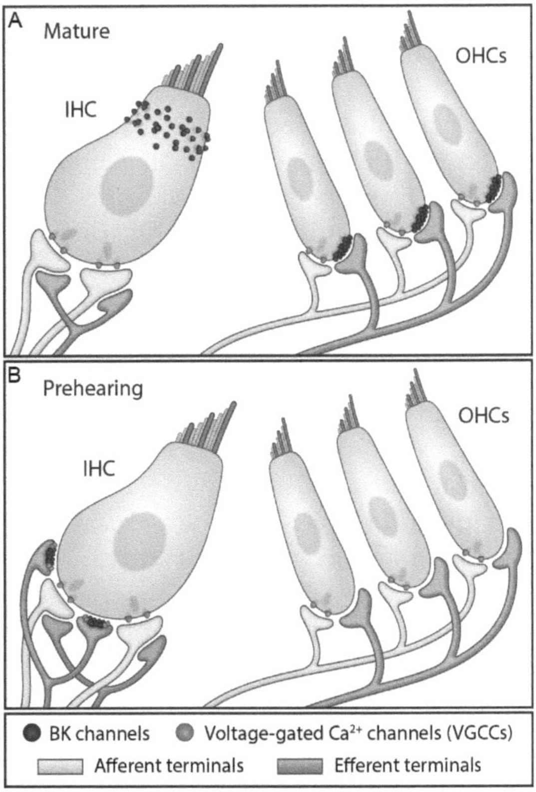

Fig. 2 Localization of BK channels in the mammalian organ of Corti. (A) In the mature organ of Corti (after the onset of hearing), BK channels are found in clusters in the necks of the inner hair cells and conspicuously distant from ribbon active zones (Pyott et al., 2004). In the higher frequency outer hair cells (>4 kHz), BK channels are found in single large clusters directly postsynaptic to (medial olivocochlear) efferent terminals (Wersinger, McLean, Fuchs, & Pyott, 2010). These clusters peak in size between 16 and 32 kHz (Maison, Pyott, Meredith, & Liberman, 2013). (B) Before the onset of hearing, BK channels are not present in either the inner or outer hair cells. However, during the period of transient efferent innervation of the inner hair cells (which peaks around postnatal days 7–9 in rat; Katz et al., 2004), BK channels are found in single smaller clusters in the presynaptic efferent terminals contacting the inner hair cells. These BK channels are activated by calcium influx from L-type VGCCs (Zorrilla de San Martin, Pyott, Ballestero, & Katz, 2010). The additional presence of BK channels in efferent terminals contacting the (type I) afferent terminals, which are themselves contacting the inner hair cells, or in nascent (prehearing) efferent terminals contacting the outer hair cells (especially higher frequency outer hair cells) has not been fully investigated (see Section 3.2.4).

Rigorous characterization of the voltage and calcium dependence of BK channel activation using whole-cell and excised patch-clamp recordings of rat inner hair cells showed that BK channel activation does not rely on VGCCs and is remarkably insensitive to local intracellular calcium concentration (Thurm, Fakler, & Oliver, 2005). The activation of BK currents at negative membrane potentials even in the absence of intracellular calcium led these investigators to speculate that BK channels in mammalian inner hair cells may function as purely voltage-gated potassium channels. In fact, the lack of activation by calcium may be a requirement of their subcellular localization away from VGCCs. This conclusion is, however, challenged by findings that intracellular photorelease of calcium activated a charybdotoxin-sensitive current in dissociated inner hair cells isolated from guinea pig (Dulon et al., 1995) and that BK currents in inner hair cells in the intact organ of Corti are modulated by voltage-dependent release from intracellular calcium stores (Marcotti et al., 2004a). Differences between the preparations may affect BK currents' properties. Nevertheless, these inconsistent findings warrant more investigation into the biophysics of BK channels in mammalian inner hair cells.

The biophysical properties of BK channels are known to be altered by alternative splicing and association with regulatory subunits (recently reviewed in Toro et al., 2014; see Chapter "Posttranscriptional and Post-translational Regulation of BK Channels" by Shipston and Tian). As in non-mammalian hair cells (reviewed in Fettiplace & Fuchs, 1999), various alternatively spliced BK channel α (pore-forming) subunit transcripts (Langer et al., 2003; Sakai, Harvey, & Sokolowski, 2011) have been observed in the mammalian inner ear and hair cells. The contribution of alternative splicing of the BK channel in the cochlea is unclear. Based on differences in detection by monoclonal antibodies that recognize distinct epitopes of the BK channel (S. J. Pyott, unpublished findings), distinct BK channel splice variants may be present in the inner and outer hair cells (see more in Section 3.2.2). Perhaps alternatively spliced variants enable the appropriate trafficking of BK channels to the neck of the inner hair cells and the synaptic poles of the outer hair cells.

While various β (regulatory) subunit (β1, β2, and β4) transcripts have been observed in the mammalian cochlea inner ear and hair cells (Langer et al., 2003), their contribution to auditory processing is also unclear. BK channel currents in inner hair cells from BK β1/β4 double-knockout mice are comparable to those from wild-type mice (Pyott et al., 2007). Moreover, inner and outer hair cell function as assessed by auditory brainstem responses and distortion product otoacoustic emissions are also comparable between BK β1/β4 double-knockout and wild-type mice (Pyott et al., 2007). These findings suggest that the BK β1 and β4 subunits do not regulate BK channel expression in the hair cells and are not required for normal function of the cochlea. Possible contributions of the BK β2 and β3 subunits have not been investigated yet. Importantly, the BK β2 subunit (Uebele et al., 2000; Wallner, Meera, & Toro, 1999; Xia, Ding, & Lingle, 1999) as well as particular splice variants of the BK β3 subunit (Lingle, Zeng, Ding, & Xia, 2001; Uebele et al., 2000; Xia, Ding, Zeng, Duan, & Lingle, 2000) impart inactivation to BK channels in other systems and may contribute to the inactivation of these currents in frog hair cells (Armstrong & Roberts, 2001), as well as in mouse inner hair cells, as reported by some (Pyott et al., 2004) but not others (Marcotti et al., 2004a). The presence of other recently identified regulatory subunits, BK γ1–γ4 (LRRC26, LRRC52, LRRC55, LRRC58, respectively; Zhang & Yan, 2014) has also not been investigated. Interestingly, BK γ subunits can produce BK channels with increased voltage sensitivity even in the absence of intracellular calcium (Yan & Aldrich, 2010, 2012). The expression of BK γ subunits may explain the curious biophysics of BK channels in mammalian inner hair cells.

3.2.2 Outer Hair Cells

Identification of BK channels using electrophysiology in mammalian outer hair cells has been much less straightforward compared to inner hair cells. In retrospect, these discrepancies are likely due to developmental and tonotopic gradients of BK channel expression in the mammalian organ of Corti that complicated experimental identification. The effects of enzymatic treatments used to dissociate the hair cells may also affect BK channel properties (see Armstrong & Roberts, 2001; Spreadbury, Kros, & Meech, 2004) and confound identification.

The first examination of currents in isolated guinea pig outer hair cells using patch-clamp electrophysiology was undertaken by Ashmore and Meech (1986). Based on measured single-channel conductances, they reported expression of a large-conductance calcium-activated potassium, presumably BK, current. The presence of high-conductance potassium channels was subsequently confirmed by Gitter, Fromter, and Zenner (1992) and van den Abbeele, Teulon, and Huy (1999) using cell-attached and excised inside-out patches recordings. Housley and Ashmore (1992) used patch-clamp electrophysiology to examine whole-cell currents in outer hair cells isolated from the guinea pig organ of Corti. In outer hair cells isolated from each turn, they observed a TEA- and nifedipine-sensitive potassium current. These results indicated the presence of a potassium channel dependent on calcium influx via L-type calcium channels in outer hair cells along the length of the organ of Corti. Finally, Nenov, Norris, and Bobbin (1997) identified a charybdotoxin-sensitive current in isolated long (lower frequency) outer hair cells isolated from guinea pig. In contrast, patch-clamp recordings from undissociated outer hair cells from adult guinea pig by Mammano and Ashmore (1996) identified distinct potassium channels in outer hair cells from apical turns and basal turns and argue against the expression of BK currents in outer hair cells from either region.

Identification of BK α transcripts using in situ hybridization in the rat organ of Corti revealed labeling of outer hair cells that appeared later than the more robust expression observed in the inner hair cells just before the onset of hearing (Langer et al., 2003). BK channel transcript and also protein expression was confirmed in outer hair cells in the guinea pig organ of Corti by in situ hybridization and immunocytochemistry (Skinner et al., 2003). The first examination to utilize BK α knockout mice identified BK channel immunoreactivity in the inner hair cells and also associated with efferent terminals contacting the outer hair cells (Rüttiger et al., 2004). In this study, BK α knockout mice displayed outer hair cell degeneration in the high-frequency cochlear turns. This finding motivated Engel et al. (2006) to examine the developmental and tonotopic gradient of BK channel immunoreactivity in outer hair cells. These authors observed BK channel immunoreactivity in mouse outer hair cells that increased from low- to high-frequency turns.

The tonotopic expression and functional contribution of BK channels in mammalian outer hair cells was further elucidated by the work of Pyott and colleagues (Fig. 2). Systematic examination of BK channel expression by whole-cell patch-clamp recordings and immunofluorescence in the rat organ of Corti indicated expression of BK channels after the onset of hearing and only in higher frequency outer hair cells (Wersinger et al., 2010). Importantly, this work also showed that acetylcholine application exclusively activated BK currents in high-frequency outer hair cells as opposed to SK currents in the lower frequency outer hair cells. This finding was unexpected since all previous studies in vertebrate hair cells had shown that acetylcholine, the major efferent neurotransmitter, mediated hair cell inhibition via the combined action of nicotinic acetylcholine receptors and small-conductance calcium-activated potassium, or SK, channels. [See, for example, work in chick (Yuhas & Fuchs, 1999), turtle (Holt, Lysakowski, & Goldberg, 2006), and low-frequency outer hair cells from the guinea pig (Nenov, Norris, & Bobbin, 1996) and rat (Oliver et al., 2000).] Subsequent work verified that BK channels indeed shape efferent inhibitory synaptic responses in the high-frequency outer hair cells (Rohmann, Wersinger, Braude, Pyott, & Fuchs, 2015). Finally, this gradient of BK channel expression was verified in mouse outer hair cells and the absence of the BK channel (in BK α knockout mice) results in an in vivo loss of outer hair cell efferent suppression and increased susceptibility to noise-induced temporary threshold shifts (Maison et al., 2013). Initial work by Pyott et al. (2004, 2007) did not observe BK channel immunoreactivity in mouse outer hair cells but later attributed this discrepancy to experimental conditions (overfixation) that abolished immunoreactivity specifically in the outer hair cells.

As in inner hair cells, the biophysical properties of BK channels in outer hair cells appear unusual. Specifically, voltage-independent activation of BK channels appears to occur. Evoked inhibitory postsynaptic currents in outer hair cells isolated from high-frequency turns of both mouse and rat organ of Corti are shaped by activation of BK channels (Rohmann et al., 2015). In these conditions, calcium influx through closely positioned acetylcholine receptors appears to raise intracellular calcium to levels sufficient to activate both SK and BK channels even in the absence of depolarization. The molecular mechanisms underlying this voltage-independent gating of BK channels in outer hair cells have not been explored but may be conferred by alternative splicing and/or the association of regulatory subunits (see Section 3.2.1).

3.2.3 Spiral Ganglion Neurons

BK channel expression has been mostly directly investigated in the sensory hair cells. However, various reports indicate the expression of BK channels in the spiral ganglion neurons. The spiral ganglion neurons are the first action potential generating neurons in the mature auditory pathway. Therefore, expression of BK channels, which regulate action potential latency and firing frequency in other neurons (N'Gouemo, 2014; see Chapter "BK Channels in Neurons" by Barth and Contet), is not necessarily surprising in the spiral ganglion neurons. Both in situ hybridization (Skinner et al., 2003) and immunocytochemistry (Hafidi et al., 2005; Skinner et al., 2003) demonstrated BK channel expression in the cell bodies of the spiral ganglion neurons (in addition to sensory hair cells) in the intact auditory sensory epithelium isolated from guinea pig (adult; Skinner et al., 2003) and mouse (postnatal day 12; Hafidi et al., 2005). BK channel immunoreactivity was detected in dissociated spiral ganglion neurons isolated from postnatal day 3 to 8 mice (Adamson, Reid, Mo, Bowne-English, & Davis, 2002). In this work, immunoreactivity increased in intensity from apex to base. Moreover, blockade of BK channels with charybdotoxin transformed spiral ganglion neuron firing patterns from rapidly adapting (firing one or two successive action potentials) to slowly adapting (firing multiple successive action potentials). In light of this work, BK channel expression in the spiral ganglion cells may appear before the onset of hearing and, therefore, before expression in the sensory hair cells. Importantly, these findings collectively indicate that BK channels are positioned to modulate afferent transmission (discussed later in Section 3.3.2).

3.2.4 Efferent Terminals

In addition to relaying afferent sensory information to the brain, hair cells of the vertebrate cochlea also receive efferent innervation from the brain (Guinan, 1996). In the mammalian auditory system, cholinergic neurons originating in the medial superior olive in the brainstem supply inhibitory efferent innervation to the outer hair cells (Liberman, Dodds, & Pierce, 1990). As discussed in Section 3.2.2, BK channels, present in the postsynaptic high-frequency outer hair cells, carry this inhibitory efferent current (Maison et al., 2013; Rohmann et al., 2015; Wersinger et al., 2010). Just before the onset of hearing efferent terminals originating from the medial superior olive transiently contact the inner hair cells (Simmons, 2002). This efferent inhibition modulates inner hair cell spiking (Glowatzki & Fuchs, 2002; Marcotti, Johnson, & Kros, 2004b) and has, therefore, been proposed to shape establishment of the afferent auditory pathway (Clause et al., 2014). Work by Zorrilla de San Martin and colleagues (2010) used a combination of whole-cell patch-clamp recordings, quantal analysis, and immunofluorescence to show that calcium entering the presynaptic efferent terminals via L-type calcium channels activates BK channels. Activation of BK channels serves to inhibit the presynaptic terminals and negatively regulate acetylcholine release from the terminal. This work suggests that BK α knockout mice should show increased efferent inhibition of the inner hair cells before the onset of hearing. This prediction and its effect on afferent development have not been investigated yet. This work also suggests that BK channels may be present in the medial efferent terminals contacting the outer hair cells before the onset of hearing. Although electrophysiological examination of apical (low-frequency) cochlear turns from prehearing mice suggests the absence of BK channels from outer hair cell efferent terminals (Ballestero et al., 2011), these experiments do not exclude the possibility of BK channel expression in efferent terminals contacting the higher frequency outer hair cells. Additionally, the role of BK channels in a second efferent pathway, arising from neurons originating in the lateral superior olive in the brainstem (Warr, 1992), has not been investigated yet. Fig. 2 illustrates the localization of BK channels in efferent terminals contacting the inner hair cells before the onset of hearing.

3.3 Function of BK Channels in the Mammalian Auditory System

3.3.1 Maturation of the Auditory System

BK channels appear to be involved in the maturation of mammalian hearing. Similar to early observations made in chick hair cells (Fuchs & Sokolowski, 1990; Li et al., 2009), BK channels are first observed in mammalian inner (Hafidi et al., 2005; Kros et al., 1998; Marcotti, Johnson, Holley, & Kros, 2003; Pyott et al., 2004) and outer (Rohmann et al., 2015) hair cells at the onset of hearing. The appearance of the BK channel and downregulation of VGCCs have been hypothesized to convert immature inner hair cells, which fire calcium-driven action potentials, into mature inner hair cells, which instead display graded receptor potential changes (Kros et al., 1998). Therefore, the expression of BK channels appears to be a defining biophysical indicator of the mature inner hair cell.

Consistent with this conclusion, in mice, mutations that interfere with the normal physiology of the inner hair cells and lead to deafness, including genetic deletion of the L-type calcium channel $Ca_V1.3$ (Brandt, Striessnig, & Moser, 2003; Nemzou, Bulankina, Khimich, Giese, & Moser, 2006) and hypothyroidism (Dettling et al., 2014; Dror et al., 2014; Winter et al., 2007), result in delayed or failed developmental appearance of BK currents or channels in the inner hair cells. Moreover, genetic deletion of the BK channel results in subtle morphological changes of the efferent terminals contacting the outer hair cells (Maison et al., 2013; Rohmann et al., 2015). Specifically, quantitative electron and light microscopic image analysis revealed that postsynaptic cisterns of BK α knockout outer hair cells are smaller than those of wild-type mice and that efferent presynaptic terminals of BK α knockout mice retain their smaller (prehearing) sizes compared to wild-type mice (Rohmann et al., 2015). Thus, BK channels appear to contribute to the final structural maturation of the descending efferent innervation of the outer hair cells.

Finally, although gross aspects of afferent maturation appear to be intact in BK α knockout mice, a careful characterization of the contribution of BK channels to maturation and organization of the ascending auditory pathways has not been carried out. Specifically, prehearing expression of BK channels in efferent terminals that contact the immature inner hair cells (see Section 3.2.4; Zorrilla de San Martin et al., 2010) may regulate efferent modulation of inner hair cell activity and shape tonotopic map refinement as observed in mice that lack a particular subunit (α9) of the nicotinic acetylcholine receptors (Clause et al., 2014).

3.3.2 Afferent Synaptic Transmission

The expression of BK channels in the inner hair cells (Section 3.2.1) and spiral ganglion neurons (Section 3.2.3) indicates that BK channels are poised to regulate afferent transmission in the cochlea. Indeed, intracochlear perfusion of the BK channel toxins charybdotoxin or iberiotoxin reversibly reduces the compound action potential recorded from the auditory nerve, suggesting that BK channels regulate transmission at the synapses between the inner hair cells and spiral ganglion neurons (Skinner et al., 2003).

Careful dissection of the possible pre- and postsynaptic roles of BK channels was revealed from analysis of single-fiber recordings made from the auditory nerve of BK α knockout and wild-type mice (Oliver et al., 2006). These experiments revealed a number of properties that were unchanged in BK α knockout mice, indicating that BK channels do not contribute to spontaneous rates, dynamic range, or the shapes of tuning curves of individual auditory nerve fibers. The preserved tuning curves in BK α knockout mice is further evidence that BK channels do not contribute to the tuning of mammalian hair cells as they do in hair cells from other nonmammalian vertebrates.

In contrast, individual auditory nerve fibers from BK α knockout mice show reduced temporal precision of spike timing and reduced sound-evoked discharge rates (Oliver et al., 2006). Presynaptic BK channels (present in the inner hair cells) likely contribute to the precision of spike timing, whereas postsynaptic BK channels (present in the spiral ganglion cells) likely determine evoked discharge rates. Increased variance of the first spike latency (an indicator of reduced temporal precision of spike timing) at the tone onset is consistent with the slowed voltage responses observed in the inner hair cells of BK α knockout compared to wild-type mice. These voltage responses control transmitter release from the inner hair cells and are slower in the inner hair cells from BK α knockout mice due to the loss of voltage-dependent conductance, which in turn increases the membrane time constant. Single auditory neurons from BK α knockout mice show reduced evoked firing rates compared to auditory neurons from wild-type mice. Because auditory neurons from knockout mice show longer absolute refractory periods, these postsynaptic BK channels presumably contribute to action potential recovery, likely by repolarizing the neurons and relieving sodium channel inactivation.

Ultimately, BK channels are important determinants of temporal precision in the mammalian auditory periphery (Oliver et al., 2006; Shaheen, Valero, & Liberman, 2015). This precision, importantly, also appears necessary to contribute to stimulus coding in the central auditory system (Kurt et al., 2012).

3.3.3 Protection from Acoustic Overexposure

BK channels are also involved in protection from acoustic overexposure. Although initial investigation showed that BK α knockout mice were more resistant to noise-induced hearing loss (Pyott et al., 2007), direct investigation reported greater noise-induced temporary threshold shifts in knockout compared to wild-type mice (Maison et al., 2013). Because activation of the outer hair cell cholinergic efferent system minimizes damage from acoustic trauma (Maison, Luebke, Liberman, & Zuo, 2002), weakened efferent inhibition would be predicted to result in increased vulnerability to acoustic injury. Thus, the observed increased sensitivity to noise exposure in the absence of BK channels is consistent with the involvement of BK channels in mediating efferent inhibition of the outer hair cells (Section 3.2.4).

Curiously, however, BK α knockout mice show comparable susceptibility to permanent thresholds shifts following noise exposure (Maison et al., 2013). Given that different patterns of structural damage distinguish noise-induced temporary vs permanent threshold shifts (Wang, Hirose, & Liberman, 2002), BK channels are likely involved in the reversible (but not irreversible) processes underlying noise-induced hearing loss. Further investigation of BK α knockout mice may, therefore, provide insight into the mechanisms contributing to the distinct pathologies of noise-induced hearing loss.

3.3.4 High-Frequency Hearing

The tonotopic increase in BK channel expression in high-frequency hair cells (Sections 3.2.1 and 3.2.2) suggests that BK channels are a necessary specialization of the expanded auditory frequency range typical of mammals. In both inner and outer hair cells, large conductance, rapidly activating BK currents result in lower input resistances and shorter membrane time constants, providing better voltage control during rapidly changing receptor potentials. In inner hair cells, the developmental appearance of BK channels decreases the membrane time constant and likely contributes to the prevention of action potentials (Section 3.2.1). Thus, the BK channel, while not necessary for frequency tuning per se, is nevertheless required to transform hair cells into signal transducers with fast signal following responses. In outer hair cells, BK channels, compared to SK channels, additionally provide a larger and faster conductance change in response to acetylcholine (Rohmann et al., 2015), increasing the gain and speed of efferent inhibition. Thus, BK channels, which serve a different function in nonmammalian hair cells, are preadapted to serve demands required of high-frequency hearing in mammalian inner and outer hair cells.

3.4 BK Channels in Vestibular Hair Cells

The mammalian vestibular structures use sensory hair cells to detect head rotations. Hair cell BK channels were first identified in the frog sacculus (Hudspeth & Lewis, 1988a, 1988b), an auditory as well as vestibular structure. Thus, it follows that BK channels may also be expressed in mammalian vestibular hair cells. The mammalian vestibular sensory epithelia contain two types of hair cells differentiated by their shape, distribution within the epithelia, and synaptic contacts. Type I hair cells are surrounded by a calyx-shaped afferent ending, whereas type II hair cells are contacted by bouton-shaped afferent as well as efferent endings (reviewed in Lysakowski & Goldberg, 2004). Based on findings in the auditory sensory epithelium, a variety of expression patterns might be predicted for the BK channel, including expression in (1) one or both classes of vestibular hair cells, (2) afferent calyx and/or bouton endings, and (3) efferent terminals. Moreover, expression patterns may change developmentally and/or regionally.

Whole-cell patch-clamp recordings of type II vestibular hair cells from mature guinea pigs identified an acetylcholine-induced BK current (Kong et al., 2007, 2005). This current is elicited by local, extracellular application of acetylcholine and blocked by the BK channel blockers charybdotoxin and iberiotoxin but not the SK channel blocker apamin. Moreover, the BK current is blocked by the L-type calcium channel blocker nifedipine. This observation, together with subsequent work (Guo et al., 2012; Yu, Guo, Wang, Zhou, & Kong, 2014), suggests that acetylcholine acts via muscarinic acetylcholine receptors to open L-type calcium channels. The influx of calcium via L-type calcium channels, in turn, activates BK channels. Thus, BK channels appear to mediate efferent inhibition of the type II vestibular hair cells but utilize muscarinic rather than nicotinic (α9/10-containing) acetylcholine receptors as in the auditory system (reviewed in Katz & Elgoyhen, 2014).

In contrast to this work, immunofluorescence to physically localize BK channels in the intact vestibular sensory epithelia (Schweizer, Savin, Luu, Sultemeier, & Hoffman, 2009) indicates expression only in a small subset of type I vestibular hair cells. Curiously, this expression peaks in juvenile rats and declines into adulthood. Unfortunately, BK α knockout mice have not been utilized yet to investigate the expression and/or functional contributions of the BK channel to the vestibular periphery. Global knockout of the BK channel causes gate abnormalities that may indicate dysfunction of the vestibular periphery (Meredith, Thorneloe, Werner, Nelson, & Aldrich, 2004). Inner ear-specific knockout of BK α via local Cre-mediated recombination of a floxed BK α allele in mice (Zemen et al., 2015) would be especially useful to identify peripheral contributions of the BK channel to balance.

4. CONCLUSION

Because of their dual activation by membrane voltage and intracellular calcium, BK channels are important and widely expressed molecular integrators of extracellular and intracellular signals of cell activity. Alternative splicing and association with regulatory subunits further refines the biophysical properties of BK channels and tailors their activity in cell type-specific ways. This coupled activation by membrane voltage and intracellular calcium and systematic variation in biophysical properties is exquisitely illustrated in the electrical tuning of nonmammalian hair cells. In these sensory cells, BK channels are essential for the frequency-specific responses necessary to decompose sound into its individual frequency components. Although the mammalian inner ear employs a different strategy of frequency decomposition, the sensory hair cells as well as synapses that regulate hair cell activity also express BK channels. In these structures, the subcellular localization of BK channels is regulated both tonotopically and developmentally. In the mammalian inner ear, BK channels appear to be preadapted to serve a variety of essential yet unexpected roles, including maturation of the sensory epithelium, high-frequency hearing, and protection from acoustic trauma.

Further investigation of BK channels in the inner ear will provide important insights into the molecular mechanisms underlying auditory processing in the periphery as well as BK channel biophysical properties and subcellular localization.

REFERENCES

Adamson, C. L., Reid, M. A., Mo, Z. L., Bowne-English, J., & Davis, R. L. (2002). Firing features and potassium channel content of murine spiral ganglion neurons vary with cochlear location. *The Journal of Comparative Neurology*, *447*, 331–350.

Armstrong, C. E., & Roberts, W. M. (2001). Rapidly inactivating and non-inactivating calcium-activated potassium currents in frog saccular hair cells. *The Journal of Physiology*, *536*, 49–65.

Art, J. J., Crawford, A. C., Fettiplace, R., & Fuchs, P. A. (1985). Efferent modulation of hair cell tuning in the cochlea of the turtle. *The Journal of Physiology*, *360*, 397–421.

Art, J. J., & Fettiplace, R. (1987). Variation of membrane properties in hair cells isolated from the turtle cochlea. *The Journal of Physiology*, *385*, 207–242.

Art, J. J., Fettiplace, R., & Wu, Y. C. (1993). The effects of low calcium on the voltage-dependent conductances involved in tuning of turtle hair cells. *The Journal of Physiology*, *470*, 109–126.

Art, J. J., Wu, Y. C., & Fettiplace, R. (1995). The calcium-activated potassium channels of turtle hair cells. *The Journal of General Physiology*, *105*, 49–72.

Ashmore, J. F., & Meech, R. W. (1986). Ionic basis of membrane potential in outer hair cells of guinea pig cochlea. *Nature*, *322*, 368–371.

Bai, J. P., Surguchev, A., Joshi, P., Gross, L., & Navaratnam, D. (2012). CDK5 interacts with Slo and affects its surface expression and kinetics through direct phosphorylation. *American Journal of Physiology. Cell Physiology*, *302*, C766–C780.

Bai, J. P., Surguchev, A., & Navaratnam, D. (2011). Beta4-subunit increases Slo responsiveness to physiological Ca2+ concentrations and together with beta1 reduces surface expression of Slo in hair cells. *American Journal of Physiology. Cell Physiology*, *300*, C435–C446.

Ballestero, J., Zorrilla de San Martin, J., Goutman, J., Elgoyhen, A. B., Fuchs, P. A., & Katz, E. (2011). Short-term synaptic plasticity regulates the level of olivocochlear inhibition to auditory hair cells. *The Journal of Neuroscience*, *31*, 14763–14774.

Behrens, R., Nolting, A., Reimann, F., Schwarz, M., Waldschutz, R., & Pongs, O. (2000). hKCNMB3 and hKCNMB4, cloning and characterization of two members of the large-conductance calcium-activated potassium channel beta subunit family. *FEBS Letters*, *474*, 99–106.

Beisel, K. W., Rocha-Sanchez, S. M., Ziegenbein, S. J., Morris, K. A., Kai, C., Kawai, J., et al. (2007). Diversity of Ca^{2+}-activated K^{+} channel transcripts in inner ear hair cells. *Gene*, *386*, 11–23.

Berkefeld, H., Sailer, C. A., Bildl, W., Rohde, V., Thumfart, J. O., Eble, S., et al. (2006). BKCa-Cav channel complexes mediate rapid and localized Ca^{2+}-activated K^{+} signaling. *Science*, *314*, 615–620.

Brändle, U., Frohnmayer, S., Krieger, T., Zenner, H. P., Ruppersberg, J. P., & Maassen, M. M. (2001). Expression of Ca(2+)-activated K(+) channel subunits and splice variants in the rat cochlea. *Hearing Research*, *161*, 23–28.

Brandt, A., Striessnig, J., & Moser, T. (2003). CaV1.3 channels are essential for development and presynaptic activity of cochlear inner hair cells. *The Journal of Neuroscience*, *23*, 10832–10840.

Clause, A., Kim, G., Sonntag, M., Weisz, C. J., Vetter, D. E., Rubsamen, R., et al. (2014). The precise temporal pattern of prehearing spontaneous activity is necessary for tonotopic map refinement. *Neuron, 82*, 822–835.

Contreras, G. F., Neely, A., Alvarez, O., Gonzalez, C., & Latorre, R. (2012). Modulation of BK channel voltage gating by different auxiliary beta subunits. *Proceedings of the National Academy of Sciences of the United States of America, 109*, 18991–18996.

Crawford, A. C., & Fettiplace, R. (1980). The frequency selectivity of auditory nerve fibres and hair cells in the cochlea of the turtle. *The Journal of Physiology, 306*, 79–125.

Crawford, A. C., & Fettiplace, R. (1981). An electrical tuning mechanism in turtle cochlear hair cells. *The Journal of Physiology, 312*, 377–412.

Dettling, J., Franz, C., Zimmermann, U., Lee, S. C., Bress, A., Brandt, N., et al. (2014). Autonomous functions of murine thyroid hormone receptor TRalpha and TRbeta in cochlear hair cells. *Molecular and Cellular Endocrinology, 382*, 26–37.

Dick, G. M., Rossow, C. F., Smirnov, S., Horowitz, B., & Sanders, K. M. (2001). Tamoxifen activates smooth muscle BK channels through the regulatory beta 1 subunit. *The Journal of Biological Chemistry, 276*, 34594–34599.

Dick, G. M., & Sanders, K. M. (2001). (Xeno)estrogen sensitivity of smooth muscle BK channels conferred by the regulatory beta1 subunit: A study of beta1 knockout mice. *The Journal of Biological Chemistry, 276*, 44835–44840.

Dolan, J., Walshe, K., Alsbury, S., Hokamp, K., O'Keeffe, S., Okafuji, T., et al. (2007). The extracellular leucine-rich repeat superfamily; a comparative survey and analysis of evolutionary relationships and expression patterns. *BMC Genomics, 8*, 320.

Dror, A. A., Lenz, D. R., Shivatzki, S., Cohen, K., Ashur-Fabian, O., & Avraham, K. B. (2014). Atrophic thyroid follicles and inner ear defects reminiscent of cochlear hypothyroidism in Slc26a4-related deafness. *Mammalian Genome, 25*, 304–316.

Dulon, D., Sugasawa, M., Blanchet, C., & Erostegui, C. (1995). Direct measurements of Ca(2+)-activated K+ currents in inner hair cells of the guinea-pig cochlea using photolabile Ca^{2+} chelators. *Pflügers Archiv, 430*, 365–373.

Duncan, R. K., & Fuchs, P. A. (2003). Variation in large-conductance, calcium-activated potassium channels from hair cells along the chicken basilar papilla. *The Journal of Physiology, 547*, 357–371.

Engel, J., Braig, C., Ruttiger, L., Kuhn, S., Zimmermann, U., Blin, N., et al. (2006). Two classes of outer hair cells along the tonotopic axis of the cochlea. *Neuroscience, 143*, 837–849.

Evans, E. F., & Wilson, J. P. (1975). Cochlear tuning properties: Concurrent basilar membrane and single nerve fiber measurements. *Science, 190*, 1218–1221.

Fakler, B., & Adelman, J. P. (2008). Control of K(Ca) channels by calcium nano/microdomains. *Neuron, 59*, 873–881.

Fettiplace, R., & Crawford, A. C. (1980). The origin of tuning in turtle cochlear hair cells. *Hearing Research, 2*, 447–454.

Fettiplace, R., & Fuchs, P. A. (1999). Mechanisms of hair cell tuning. *Annual Review of Physiology, 61*, 809–834.

Fettiplace, R., & Hackney, C. M. (2006). The sensory and motor roles of auditory hair cells. *Nature Reviews. Neuroscience, 7*, 19–29.

Frucht, C. S., Uduman, M., Kleinstein, S. H., Santos-Sacchi, J., & Navaratnam, D. S. (2011). Gene expression gradients along the tonotopic axis of the chicken auditory epithelium. *Journal of the Association for Research in Otolaryngology, 12*, 423–435.

Fuchs, P. A., & Evans, M. G. (1988). Voltage oscillations and ionic conductances in hair cells isolated from the alligator cochlea. *Journal of Comparative Physiology. A, Sensory, Neural, and Behavioral Physiology, 164*, 151–163.

Fuchs, P. A., Nagai, T., & Evans, M. G. (1988). Electrical tuning in hair cells isolated from the chick cochlea. *The Journal of Neuroscience, 8*, 2460–2467.

Fuchs, P. A., & Sokolowski, B. H. (1990). The acquisition during development of Ca-activated potassium currents by cochlear hair cells of the chick. *Proceedings of the Biological Sciences*, *241*, 122–126.

Gitter, A. H., Fromter, E., & Zenner, H. P. (1992). C-type potassium channels in the lateral cell membrane of guinea-pig outer hair cells. *Hearing Research*, *60*, 13–19.

Glowatzki, E., & Fuchs, P. A. (2002). Transmitter release at the hair cell ribbon synapse. *Nature Neuroscience*, *5*, 147–154.

Grunnet, M., & Kaufmann, W. A. (2004). Coassembly of big conductance Ca^{2+}-activated K^+ channels and L-type voltage-gated Ca^{2+} channels in rat brain. *The Journal of Biological Chemistry*, *279*, 36445–36453.

Guinan, J. J., Jr. (1996). The physiology of olivocochlear efferents. In P. Dallos, A. N. Popper, & R. R. Fay (Eds.), *The cochlea* (pp. 435–502).New York: Springer-Verlag. http://link.springer.com/chapter/10.1007%2F978-1-4612-0757-3_8#page-1.

Guo, C. K., Wang, Y., Zhou, T., Yu, H., Zhang, W. J., & Kong, W. J. (2012). M2 muscarinic ACh receptors sensitive BK channels mediate cholinergic inhibition of type II vestibular hair cells. *Hearing Research*, *285*, 13–19.

Hafidi, A., Beurg, M., & Dulon, D. (2005). Localization and developmental expression of BK channels in mammalian cochlear hair cells. *Neuroscience*, *130*, 475–484.

Holt, J. C., Lysakowski, A., & Goldberg, J. M. (2006). Mechanisms of efferent-mediated responses in the turtle posterior crista. *The Journal of Neuroscience*, *26*, 13180–13193.

Housley, G. D., & Ashmore, J. F. (1992). Ionic currents of outer hair cells isolated from the guinea-pig cochlea. *The Journal of Physiology*, *448*, 73–98.

Hudspeth, A. J., & Lewis, R. S. (1988a). A model for electrical resonance and frequency tuning in saccular hair cells of the bull-frog, Rana catesbeiana. *The Journal of Physiology*, *400*, 275–297.

Hudspeth, A. J., & Lewis, R. S. (1988b). Kinetic analysis of voltage- and ion-dependent conductances in saccular hair cells of the bull-frog, Rana catesbeiana. *The Journal of Physiology*, *400*, 237–274.

Issa, N. P., & Hudspeth, A. J. (1994). Clustering of Ca^{2+} channels and Ca(2+)-activated K^+ channels at fluorescently labeled presynaptic active zones of hair cells. *Proceedings of the National Academy of Sciences of the United States of America*, *91*, 7578–7582.

Jones, E. M., Gray-Keller, M., Art, J. J., & Fettiplace, R. (1999). The functional role of alternative splicing of Ca(2+)-activated K+ channels in auditory hair cells. *Annals of the New York Academy of Sciences*, *868*, 379–385.

Kathiresan, T., Harvey, M., Orchard, S., Sakai, Y., & Sokolowski, B. (2009). A protein interaction network for the large conductance Ca(2+)-activated K(+) channel in the mouse cochlea. *Molecular & Cellular Proteomics*, *8*, 1972–1987.

Katz, E., & Elgoyhen, A. B. (2014). Short-term plasticity and modulation of synaptic transmission at mammalian inhibitory cholinergic olivocochlear synapses. *Frontiers in Systems Neuroscience*, *8*, 224.

Katz, E., Elgoyhen, A. B., Gomez-Casati, M. E., Knipper, M., Vetter, D. E., Fuchs, P. A., et al. (2004). Developmental regulation of nicotinic synapses on cochlear inner hair cells. *The Journal of Neuroscience*, *24*, 7814–7820.

Khanna, S. M., & Leonard, D. G. (1982). Basilar membrane tuning in the cat cochlea. *Science*, *215*, 305–306.

Kiang, N. Y.-S. (1965). Stimulus coding in the auditory nerve and cochlear nucleus. *Acta Oto-Laryngologica*, *59*, 186–200.

Kim, J. M., Beyer, R., Morales, M., Chen, S., Liu, L. Q., & Duncan, R. K. (2010). Expression of BK-type calcium-activated potassium channel splice variants during chick cochlear development. *The Journal of Comparative Neurology*, *518*, 2554–2569.

Kong, W. J., Guo, C. K., Zhang, X. W., Chen, X., Zhang, S., Li, G. Q., et al. (2007). The coupling of acetylcholine-induced BK channel and calcium channel in guinea pig saccular type II vestibular hair cells. *Brain Research*, *1129*, 110–115.

Kong, W. J., Guo, C. K., Zhang, S., Hao, J., Wang, Y. J., & Li, Z. W. (2005). The properties of ACh-induced BK currents in guinea pig type II vestibular hair cells. *Hearing Research*, *209*, 1–9.

Kowalik, L., & Hudspeth, A. J. (2011). A search for factors specifying tonotopy implicates DNER in hair-cell development in the chick's cochlea. *Developmental Biology*, *354*, 221–231.

Kros, C. J., & Crawford, A. C. (1990). Potassium currents in inner hair cells isolated from the guinea-pig cochlea. *The Journal of Physiology*, *421*, 263–291.

Kros, C. J., Ruppersberg, J. P., & Rusch, A. (1998). Expression of a potassium current in inner hair cells during development of hearing in mice. *Nature*, *394*, 281–284.

Kurt, S., Sausbier, M., Ruttiger, L., Brandt, N., Moeller, C. K., Kindler, J., et al. (2012). Critical role for cochlear hair cell BK channels for coding the temporal structure and dynamic range of auditory information for central auditory processing. *The FASEB Journal*, *26*, 3834–3843.

Langer, P., Grunder, S., & Rusch, A. (2003). Expression of Ca2+-activated BK channel mRNA and its splice variants in the rat cochlea. *The Journal of Comparative Neurology*, *455*, 198–209.

Lesage, F., Hibino, H., & Hudspeth, A. J. (2004). Association of beta-catenin with the alpha-subunit of neuronal large-conductance Ca2+-activated K+ channels. *Proceedings of the National Academy of Sciences of the United States of America*, *101*, 671–675.

Li, Y., Atkin, G. M., Morales, M. M., Liu, L. Q., Tong, M., & Duncan, R. K. (2009). Developmental expression of BK channels in chick cochlear hair cells. *BMC Developmental Biology*, *9*, 67.

Liberman, M. C., Dodds, L. W., & Pierce, S. (1990). Afferent and efferent innervation of the cat cochlea: Quantitative analysis with light and electron microscopy. *The Journal of Comparative Neurology*, *301*, 443–460.

Lingle, C. J., Zeng, X. H., Ding, J. P., & Xia, X. M. (2001). Inactivation of BK channels mediated by the NH(2) terminus of the beta3b auxiliary subunit involves a two-step mechanism: Possible separation of binding and blockade. *The Journal of General Physiology*, *117*, 583–606.

Loane, D. J., Lima, P. A., & Marrion, N. V. (2007). Co-assembly of N-type Ca^{2+} and BK channels underlies functional coupling in rat brain. *Journal of Cell Science*, *120*, 985–995.

Lysakowski, A., & Goldberg, J. M. (2004). Morphophysiology of the vestibular sensory periphery. In S. M. Highstein, R. R. Fay, & A. N. Popper (Eds.), *The vestibular system* (pp. 57–152). New York: Springer-Verlag.

Maison, S. F., Luebke, A. E., Liberman, M. C., & Zuo, J. (2002). Efferent protection from acoustic injury is mediated via alpha9 nicotinic acetylcholine receptors on outer hair cells. *The Journal of Neuroscience*, *22*, 10838–10846.

Maison, S. F., Pyott, S. J., Meredith, A. L., & Liberman, M. C. (2013). Olivocochlear suppression of outer hair cells in vivo: Evidence for combined action of BK and SK2 channels throughout the cochlea. *Journal of Neurophysiology*, *109*, 1525–1534.

Mammano, F., & Ashmore, J. F. (1996). Differential expression of outer hair cell potassium currents in the isolated cochlea of the guinea-pig. *The Journal of Physiology*, *496*(Pt. 3), 639–646.

Marcotti, W., Johnson, S. L., Holley, M. C., & Kros, C. J. (2003). Developmental changes in the expression of potassium currents of embryonic, neonatal and mature mouse inner hair cells. *The Journal of Physiology*, *548*, 383–400.

Marcotti, W., Johnson, S. L., & Kros, C. J. (2004a). Effects of intracellular stores and extracellular Ca(2+) on Ca(2+)-activated K(+) currents in mature mouse inner hair cells. *The Journal of Physiology*, *557*, 613–633.

Marcotti, W., Johnson, S. L., & Kros, C. J. (2004b). A transiently expressed SK current sustains and modulates action potential activity in immature mouse inner hair cells. *The Journal of Physiology, 560*, 691–708.

Meredith, A. L., Thorneloe, K. S., Werner, M. E., Nelson, M. T., & Aldrich, R. W. (2004). Overactive bladder and incontinence in the absence of the BK large conductance Ca2+-activated K+ channel. *The Journal of Biological Chemistry, 279*, 36746–36752.

Miranda-Rottmann, S., Kozlov, A. S., & Hudspeth, A. J. (2010). Highly specific alternative splicing of transcripts encoding BK channels in the chicken's cochlea is a minor determinant of the tonotopic gradient. *Molecular and Cellular Biology, 30*, 3646–3660.

Navaratnam, D. S., Bell, T. J., Tu, T. D., Cohen, E. L., & Oberholtzer, J. C. (1997). Differential distribution of Ca2+-activated K+ channel splice variants among hair cells along the tonotopic axis of the chick cochlea. *Neuron, 19*, 1077–1085.

Nemzou, N. R., Bulankina, A. V., Khimich, D., Giese, A., & Moser, T. (2006). Synaptic organization in cochlear inner hair cells deficient for the CaV1.3 (alpha1D) subunit of L-type Ca^{2+} channels. *Neuroscience, 141*, 1849–1860.

Nenov, A. P., Norris, C., & Bobbin, R. P. (1996). Acetylcholine response in guinea pig outer hair cells. II. Activation of a small conductance Ca(2+)-activated K+ channel. *Hearing Research, 101*, 149–172.

Nenov, A. P., Norris, C., & Bobbin, R. P. (1997). Outwardly rectifying currents in guinea pig outer hair cells. *Hearing Research, 105*, 146–158.

N'Gouemo, P. (2014). BK_{Ca} channel dysfunction in neurological diseases. *Frontiers in Physiology, 5*, 373. http://dx.doi.org/10.3389/fphys.2014.00373.

Oliver, D., Klocker, N., Schuck, J., Baukrowitz, T., Ruppersberg, J. P., & Fakler, B. (2000). Gating of Ca2+-activated K+ channels controls fast inhibitory synaptic transmission at auditory outer hair cells. *Neuron, 26*, 595–601.

Oliver, D., Taberner, A. M., Thurm, H., Sausbier, M., Arntz, C., Ruth, P., et al. (2006). The role of BKCa channels in electrical signal encoding in the mammalian auditory periphery. *The Journal of Neuroscience, 26*, 6181–6189.

O'Neill, M. P., & Bearden, A. (1995). Laser-feedback measurements of turtle basilar membrane motion using direct reflection. *Hearing Research, 84*, 125–138.

Pitchford, S., & Ashmore, J. F. (1987). An electrical resonance in hair cells of the amphibian papilla of the frog Rana temporaria. *Hearing Research, 27*, 75–83.

Pyott, S. J., Glowatzki, E., Trimmer, J. S., & Aldrich, R. W. (2004). Extrasynaptic localization of inactivating calcium-activated potassium channels in mouse inner hair cells. *The Journal of Neuroscience, 24*, 9469–9474.

Pyott, S. J., Meredith, A. L., Fodor, A. A., Vazquez, A. E., Yamoah, E. N., & Aldrich, R. W. (2007). Cochlear function in mice lacking the BK channel alpha, beta1, or beta4 subunits. *The Journal of Biological Chemistry, 282*, 3312–3324.

Ramanathan, K., Michael, T. H., & Fuchs, P. A. (2000). Beta subunits modulate alternatively spliced, large conductance, calcium-activated potassium channels of avian hair cells. *The Journal of Neuroscience, 20*, 1675–1684.

Ramanathan, K., Michael, T. H., Jiang, G. J., Hiel, H., & Fuchs, P. A. (1999). A molecular mechanism for electrical tuning of cochlear hair cells. *Science, 283*, 215–217.

Raphael, Y., & Altschuler, R. A. (2003). Structure and innervation of the cochlea. *Brain Research Bulletin, 60*, 397–422.

Ricci, A. J., Gray-Keller, M., & Fettiplace, R. (2000). Tonotopic variations of calcium signalling in turtle auditory hair cells. *The Journal of Physiology, 524*(Pt. 2), 423–436.

Roberts, W. M. (1993). Spatial calcium buffering in saccular hair cells. *Nature, 363*, 74–76.

Roberts, W. M., Jacobs, R. A., & Hudspeth, A. J. (1990). Colocalization of ion channels involved in frequency selectivity and synaptic transmission at presynaptic active zones of hair cells. *The Journal of Neuroscience, 10*, 3664–3684.

Rodriguez-Contreras, A., & Yamoah, E. N. (2001). Direct measurement of single-channel Ca(2+) currents in bullfrog hair cells reveals two distinct channel subtypes. *The Journal of Physiology, 534*, 669–689.

Rohmann, K. N., Wersinger, E., Braude, J. P., Pyott, S. J., & Fuchs, P. A. (2015). Activation of BK and SK channels by efferent synapses on outer hair cells in high-frequency regions of the rodent cochlea. *The Journal of Neuroscience, 35*, 1821–1830.

Russell, I. J., & Sellick, P. M. (1978). Intracellular studies of hair cells in the mammalian cochlea. *The Journal of Physiology, 284*, 261–290.

Rüttiger, L., Sausbier, M., Zimmermann, U., Winter, H., Braig, C., Engel, J., et al. (2004). Deletion of the Ca2+-activated potassium (BK) alpha-subunit but not the BKbeta1-subunit leads to progressive hearing loss. *Proceedings of the National Academy of Sciences of the United States of America, 101*, 12922–12927.

Sakai, Y., Harvey, M., & Sokolowski, B. (2011). Identification and quantification of full-length BK channel variants in the developing mouse cochlea. *Journal of Neuroscience Research, 89*, 1747–1760.

Samaranayake, H., Saunders, J. C., Greene, M. I., & Navaratnam, D. S. (2004). Ca(2+) and K(+) (BK) channels in chick hair cells are clustered and colocalized with apical-basal and tonotopic gradients. *The Journal of Physiology, 560*, 13–20.

Schweizer, F. E., Savin, D., Luu, C., Sultemeier, D. R., & Hoffman, L. F. (2009). Distribution of high-conductance calcium-activated potassium channels in rat vestibular epithelia. *The Journal of Comparative Neurology, 517*, 134–145.

Shaheen, L. A., Valero, M. D., & Liberman, M. C. (2015). Towards a diagnosis of cochlear neuropathy with envelope following responses. *Journal of the Association for Research in Otolaryngology, 16*, 727–745.

Simmons, D. D. (2002). Development of the inner ear efferent system across vertebrate species. *Journal of Neurobiology, 53*, 228–250.

Skinner, L. J., Enee, V., Beurg, M., Jung, H. H., Ryan, A. F., Hafidi, A., et al. (2003). Contribution of BK Ca2+-activated K+ channels to auditory neurotransmission in the Guinea pig cochlea. *Journal of Neurophysiology, 90*, 320–332.

Smotherman, M. S., & Narins, P. M. (1999). The electrical properties of auditory hair cells in the frog amphibian papilla. *The Journal of Neuroscience, 19*, 5275–5292.

Sokolowski, B., Orchard, S., Harvey, M., Sridhar, S., & Sakai, Y. (2011). Conserved BK channel-protein interactions reveal signals relevant to cell death and survival. *PloS One, 6*, e28532.

Spreadbury, I. C., Kros, C. J., & Meech, R. W. (2004). Effects of trypsin on large-conductance Ca(2+)-activated K(+) channels of guinea-pig outer hair cells. *Hearing Research, 190*, 115–127.

Thurm, H., Fakler, B., & Oliver, D. (2005). Ca2+-independent activation of BKCa channels at negative potentials in mammalian inner hair cells. *The Journal of Physiology, 569*, 137–151.

Tong, M., & Duncan, R. K. (2009). Tamoxifen inhibits BK channels in chick cochlea without alterations in voltage-dependent activation. *American Journal of Physiology. Cell Physiology, 297*, C75–C85.

Toro, L., Li, M., Zhang, Z., Singh, H., Wu, Y., & Stefani, E. (2014). MaxiK channel and cell signalling. *Pflügers Archiv, 466*, 875–886.

Uebele, V. N., Lagrutta, A., Wade, T., Figueroa, D. J., Liu, Y., McKenna, E., et al. (2000). Cloning and functional expression of two families of beta-subunits of the large conductance calcium-activated K+ channel. *The Journal of Biological Chemistry, 275*, 23211–23218.

van den Abbeele, T., Teulon, J., & Huy, P. T. (1999). Two types of voltage-dependent potassium channels in outer hair cells from the guinea pig cochlea. *The American Journal of Physiology, 277*, C913–C925.

von Békésy, G. (1960). *Experiments in hearing.* New York: McGraw-Hill.
Wallner, M., Meera, P., & Toro, L. (1999). Molecular basis of fast inactivation in voltage and Ca2+-activated K+ channels: A transmembrane beta-subunit homolog. *Proceedings of the National Academy of Sciences of the United States of America, 96*, 4137–4142.
Wang, Y., Hirose, K., & Liberman, M. C. (2002). Dynamics of noise-induced cellular injury and repair in the mouse cochlea. *Journal of the Association for Research in Otolaryngology, 3*, 248–268.
Wang, B., Jaffe, D. B., & Brenner, R. (2014). Current understanding of iberiotoxin-resistant BK channels in the nervous system. *Frontiers in Physiology, 5*, 382.
Warr, W. B. (1992). Organization of olivocochlear efferent systems in mammals. In D. Webster, A. Popper, & R. Fay (Eds.), *Mammalian auditory pathways: Neuroanatomy* (pp. 410–448). New York: Springer.
Wersinger, E., McLean, W. J., Fuchs, P. A., & Pyott, S. J. (2010). BK channels mediate cholinergic inhibition of high frequency cochlear hair cells. *PloS One, 5*, e13836.
Winter, H., Braig, C., Zimmermann, U., Engel, J., Rohbock, K., & Knipper, M. (2007). Thyroid hormone receptor alpha1 is a critical regulator for the expression of ion channels during final differentiation of outer hair cells. *Histochemistry and Cell Biology, 128*, 65–75.
Wu, Y. C., & Fettiplace, R. (1996). A developmental model for generating frequency maps in the reptilian and avian cochleas. *Biophysical Journal, 70*, 2557–2570.
Xia, X. M., Ding, J. P., & Lingle, C. J. (1999). Molecular basis for the inactivation of Ca2+- and voltage-dependent BK channels in adrenal chromaffin cells and rat insulinoma tumor cells. *The Journal of Neuroscience, 19*, 5255–5264.
Xia, X. M., Ding, J. P., Zeng, X. H., Duan, K. L., & Lingle, C. J. (2000). Rectification and rapid activation at low Ca2+ of Ca2+-activated, voltage-dependent BK currents: Consequences of rapid inactivation by a novel beta subunit. *The Journal of Neuroscience, 20*, 4890–4903.
Yan, J., & Aldrich, R. W. (2010). LRRC26 auxiliary protein allows BK channel activation at resting voltage without calcium. *Nature, 466*, 513–516.
Yan, J., & Aldrich, R. W. (2012). BK potassium channel modulation by leucine-rich repeat-containing proteins. *Proceedings of the National Academy of Sciences of the United States of America, 109*, 7917–7922.
Yan, J., Olsen, J. V., Park, K. S., Li, W., Bildl, W., Schulte, U., et al. (2008). Profiling the phospho-status of the BKCa channel alpha subunit in rat brain reveals unexpected patterns and complexity. *Molecular & Cellular Proteomics, 7*, 2188–2198.
Yu, H., Guo, C. K., Wang, Y., Zhou, T., & Kong, W. J. (2014). Gentamicin blocks the ACh-induced BK current in guinea pig type II vestibular hair cells by competing with Ca(2)(+) at the L-type calcium channel. *International Journal of Molecular Sciences, 15*, 6757–6771.
Yuhas, W. A., & Fuchs, P. A. (1999). Apamin-sensitive, small-conductance, calcium-activated potassium channels mediate cholinergic inhibition of chick auditory hair cells. *Journal of Comparative Physiology. A, Sensory, Neural, and Behavioral Physiology, 185*, 455–462.
Zemen, B. G., Lai, M. H., Whitt, J. P., Khan, Z., Zhao, G., & Meredith, A. L. (2015). Generation of Kcnma1fl-tdTomato, a conditional deletion of the BK channel alpha subunit in mouse. *Physiological Reports, 3*, e12612. http://dx.doi.org/10.14814/phy2.12612.
Zhang, J., & Yan, J. (2014). Regulation of BK channels by auxiliary γ subunits. *Frontiers in Physiology, 5*, 401.
Zorrilla de San Martin, J., Pyott, S., Ballestero, J., & Katz, E. (2010). Ca(2+) and Ca(2+)-activated K(+) channels that support and modulate transmitter release at the olivocochlear efferent-inner hair cell synapse. *The Journal of Neuroscience, 30*, 12157–12167.

CHAPTER ELEVEN

BK Channels in the Vascular System

G. Krishnamoorthy-Natarajan*,[1], M. Koide†
*Nutriflour International Inc. LLC, Norwalk, CT, United States
†University of Vermont, Burlington, VT, United States
[1]Corresponding author: e-mail address: nutriflourint2015@gmail.com

Contents

Abstract

Autoregulation of blood flow is essential for the preservation of organ function to ensure continuous supply of oxygen and essential nutrients and removal of metabolic waste. This is achieved by controlling the diameter of muscular arteries and arterioles that exhibit a myogenic response to changes in arterial blood pressure, nerve activity and tissue metabolism. Large-conductance voltage and Ca^{2+}-dependent K^+ channels

International Review of Neurobiology, Volume 128
ISSN 0074-7742
http://dx.doi.org/10.1016/bs.irn.2016.03.017

(BK channels), expressed exclusively in smooth muscle cells (SMCs) in the vascular wall of healthy arteries, play a critical role in regulating the myogenic response. Activation of BK channels by intracellular, local, and transient ryanodine receptor-mediated "Ca^{2+} sparks," provides a hyperpolarizing influence on the SMC membrane potential thereby decreasing the activity of voltage-dependent Ca^{2+} channels and limiting Ca^{2+} influx to promote SMC relaxation and vasodilation. The BK channel α subunit, a large tetrameric protein with each monomer consisting of seven-transmembrane domains, a long intracellular C-terminal tail and an extracellular N-terminus, associates with the β1 and γ subunits in vascular SMCs. The BK channel is regulated by factors originating within the SMC or from the endothelium, perivascular nerves and circulating blood, that significantly alter channel gating properties, Ca^{2+} sensitivity and expression of the α and/or β1 subunit. The BK channel thus serves as a central receiving dock that relays the effects of the changes in several such concomitant autocrine and paracrine factors and influences cardiovascular health. This chapter describes the primary mechanism of regulation of myogenic response by BK channels and the alterations to this mechanism wrought by different vasoactive mediators.

1. INTRODUCTION

1.1 Peripheral Vascular Resistance and Autoregulation of Blood Flow

The average volume of blood in a typical human, a healthy 30-year-old male weighing about 150–160 lbs and 5′7″ tall, is about 5 L, which is supplied to an estimated 37.2×10^{12} (37.2 billion) cells that compose the various tissues and organs in the human body. The primary purpose of blood flow is to ensure the continuous supply of oxygen and nutrients and removal of carbon dioxide and residual metabolic waste to ensure uninterrupted functioning of the underlying tissue/organ. It also facilitates the transport of hormones and drugs to and between organs as well as sustains homeostasis by maintaining bodily heat and pH. Blood flow to a tissue at any given moment in time is determined by the perfusion pressure (defined as the difference between mean arterial pressure (MAP) and the venous pressure), the metabolic need of the tissue and the peripheral vascular resistance or PVR (regulated by the network of precapillary arterioles and postcapillary venules in the microcirculation). Indeed, it can be stated that PVR is the single most important determinant in regulating organ function due to its influence on the MAP, cardiac output (CO), distribution of CO between and within organs and capillary hydrostatic pressure. PVR depends on the radius and length of blood vessels and blood viscosity according to the Hagen–Poiseuille equation, which describes laminar flow inside a hollow cylindrical vessel along a pressure gradient as:

$$Q = \Delta P \cdot \pi \cdot r^4 / 8 \cdot L \cdot \eta,$$

leading to a formula for

$$\mathrm{PVR} = 8 \cdot L \cdot \eta / \left(\pi \cdot r^4\right),$$

where Q = rate of blood flow, ΔP = transmural pressure gradient, PVR = resistance to blood flow, L = length of the blood vessel, η = viscosity of blood, and r = radius of blood vessel. Compared to vessel length or blood viscosity, the radius of a blood vessel or "tone" is a more dynamic entity since it is subject to regulation as either an increase in radius (vasodilation) or a decrease in radius (vasoconstriction) by local, systemic and hormonal stimuli which fluctuate on a relatively rapid time-scale in response to changes in tissue metabolism, cardiac or nervous activity and hemodynamic factors such as arterial pressure (Bianconi et al., 2013; Delp & Laughlin, 1998; Duling & Klitzman, 1980; Folkow, 1955; Guyenet, 2006; Haddy, Overbeck, & Daugherty, 1968; Iadecola, 2004; Olsson, 1981; Schubert & Mulvany, 1999). Thus, small changes in the radius of the blood vessel have a profound impact on PVR and blood flow. The ability of blood vessels to adjust *tone* and vascular resistance to regulate blood flow and maintain constant tissue perfusion in response to changes in perfusion pressure is termed as *autoregulation*.

1.2 Vascular Wall Structure: Form Determines Function

The blood vessel is a multilayered cylindrical entity whose structure embodies the complex physiology involved in the regulation of *tone*. In general, blood vessels are composed of three heterogeneous layers—an inner layer *tunica intima* composed of longitudinally oriented elastic elements and a single layer of squamous endothelial cells (ECs), a middle layer *tunica media* composed of circumferentially oriented elastic elements and smooth muscle cells (SMCs), and an outer layer *tunica adventitia* composed of fibrous connective tissue, the *vaso vasorum* (a network of blood vessels supplying the outer layers of the vessel wall), lymphatic vessels, unmyelinated perivascular sympathetic nerves, fibroblasts, progenitor and immune cells (Stenmark et al., 2013; Tortora & Derrickson, 2009). In medium arteries, the internal elastic lamina and the external elastic lamina clearly separate the *intima* from the *media* and the *media* from the *adventitia,* respectively, and are composed of more elastic fibers. Blood vessels along the circulatory network differ in the admixture of these layers depending on the size and category of the blood vessels due to a progressive thinning of vessel wall and lumen size down the pressure gradient as blood vessels branch from large (aorta, pulmonary trunk, and subclavian artery) to medium (most blood vessels in the body) to small precapillary arterioles and postcapillary venules to culminate in a network of capillaries on or inside organ tissues (Burton, 1954; Hendry, Farley, & McLafferty, 2012).

Differences in the admixture of these layers in the vessel wall confer distinct contractile properties to the vessel wall. For instance, the abundance of SMCs in the *tunica media* of medium-sized arteries and precapillary arterioles relative to the number of elastic fibers in the *tunica media* and *adventitia* in large arteries leads to elasticity in large arteries and muscularity in medium-sized arteries and precapillary arterioles. Thus, the task of regulating PVR and blood flow to match CO to tissue metabolic needs falls to muscular arteries and arterioles, which regulate the contractile state of the SMCs in the *tunica media* and regulate vascular *tone*.

1.3 Myogenic Response for the Autoregulation of Blood Flow

The adjustment of *tone* in response to hemodynamic factors was termed as the *myogenic response* by Bayliss, who first described the intrinsic behavior of muscular arteries and arterioles to constrict in response to local increases and dilate in response to local decreases in transmural pressure (Bayliss, 1902). Myogenic reactivity is a prominent feature of the cerebral circulation but is also observed in the circulatory networks supplying the heart, mesentery and kidney. This inherent property of SMCs is responsible for the establishment of a basal *myogenic tone*, which is a partially restricted state of arteries at a constant physiological blood pressure in vivo, as well as *myogenic reactivity*, defined as further vasoconstriction or vasodilation as may be elicited from the basal myogenic tone for the autoregulation of blood flow and capillary hydrostatic pressure (Davis & Hill, 1999; Osol, Brekke, McElroy-Yaggy, & Gokina, 2002). The response, however, may be tempered or enhanced by vasoactive factors released locally or by the endothelium or by perivascular nerve activity. Large-conductance Ca^{2+}-activated K^{+} ion channels (BK channels) play an important role in regulating the vascular *tone* in general and the myogenic response in particular by modulating smooth muscle contractility (Paterno, Heistad, & Faraci, 2000).

1.4 The SMC as the Determinant of the Myogenic State

SMC contractility is driven by interactions between contractile proteins, myosin and actin, arranged in thick and thin filaments, respectively, which are adhered and supported by intermediate filaments. Myosin thick filaments are composed of two myosin heavy chains (MW of 200 and 204 kDa) and two sets of myosin light chain subunits—the regulatory (20 kDa) LC_{20} and the alkali (17 kDa) LC_{17}. Thin filaments comprising actin, tropomyosin and other proteins are attached to cell membranes via membrane plaques containing several proteins such as vinculin, metavinculin, α-actinin and talin and are supported by dense bodies, held together by intermediate filaments comprising vimentin and/or desmin. Heads of myosin molecules in thick filaments bind to actin in thin filaments to form crossbridges and generate or release tension depending on the crossbridge angle. The cycling of crossbridges and initiation of contraction is activated by the phosphorylation of Ser-19 on LC_{20} by myosin light chain kinase (MLCK), a specialized Ca^{2+}-calmodulin-dependent protein kinase, which in turn is activated by the binding of the Ca^{2+}-calmodulin complex following an increase in intracellular Ca^{2+} (Stull, Lin, Krueger, Trewhella, & Zhi, 1998; Webb, 2003). The phosphorylation of the myosin molecule by MLCK is balanced by the action of myosin light chain phosphatase (MLCP) on the myosin molecule as well as the modulation of MLCK activity by protein kinases such as cAMP, cGMP, and protein kinase C (PKC). The activation and regulation of MLCK by Ca^{2+} is thus the principal event that determines the contractile state of the SMC and the myogenic response (Davis & Hill, 1999; Hathaway, March, Lash, Adam, & Wilensky, 1991).

2. BK CHANNELS IN VASCULAR SMOOTH MUSCLE

2.1 Ca^{2+} Addresses the "State of the Smooth Muscle"

It is apparent that Ca^{2+} handling in arteriolar SMCs underlies its regulation of vascular *myogenic tone* and is achieved via the interplay of several cytosolic and transmembrane molecular entities expressed in the SMC (Berridge & Dupont, 1994; Clapham, 1995; Karaki et al., 1997). Global increases in Ca^{2+} caused by Ca^{2+} entry via membrane ion channels (mainly through membrane depolarization and activation of voltage-dependent Ca^{2+} channels, VDCCs) or Ca^{2+} release from intracellular Ca^{2+} stores (sarcoplasmic reticulum, SR) as inositol trisphosphate (IP_3) receptor-mediated Ca^{2+} waves lead ultimately to the activation of LC_{20} by MLCK and vasoconstriction.

Vasorelaxation is mediated by cellular processes that extrude Ca^{2+} from the cytosol (through exchangers and transporters), inhibit Ca^{2+} entry (mainly through activation of K^+ channels and membrane hyperpolarization) or promote uptake of cytosolic Ca^{2+} into intracellular stores (via the SR Ca^{2+}-ATPase, SERCA) and drive dephosphorylation of LC_{20} via the action of MLCP (Mizuno et al., 2008; Somlyo & Somlyo, 2003).

Changes in arterial pressure and blood flow produce shear and tensile forces in the vascular wall, which are sensed by SMC and cytoskeletal proteins that transduce wall mechanical signals into contractile events in SMCs via mechanisms that are as yet little understood (Sun, Li, & Meininger, 2012; Ye, Nesmith, & Parker, 2014). Nevertheless, pressure changes are associated with altered global Ca^{2+}, membrane potential and vascular tone. The resting membrane potential of a SMC from an artery with basal myogenic tone is between –50 and –40 mV, which is sufficient for the activation of VDCCs and influx of Ca^{2+} to raise global Ca^{2+} from ~100 to ~200 nM and induce a tonic vasoconstriction (Knot & Nelson, 1998). Change in arterial pressure shifts membrane potential toward either a hyperpolarized state of ~60 mV and a reduced global Ca^{2+} of ~100 nM or a depolarized state of ~30 mV and a rise in global Ca^{2+} to ~300–400 nM, which changes are found to be sufficient, respectively, for maximal relaxation or contraction of the SMC and subsequent vasodilation or vasoconstriction (Hill-Eubanks, Werner, Heppner, & Nelson, 2011). Thus, changes in global Ca^{2+} in a SMC lie at the heart of SMC contractility and vascular tone regulation.

2.2 The RyR/BK Complex: Think Local, Act Global

Even as global changes in Ca^{2+} drives the contractile state of the SMC, Ca^{2+} concentration is far from uniform throughout the cytosol as the term "global" implies. Measurements using fluorescent markers at very high temporal resolutions have uncovered dynamic Ca^{2+} signals in the SMC in the form of Ca^{2+} sparklets, Ca^{2+} flashes, Ca^{2+} puffs, Ca^{2+} sparks and Ca^{2+} waves, that can transiently increase local Ca^{2+} concentrations to μM levels (Hill-Eubanks et al., 2011) and regulate local signals that mediate either contraction or relaxation of the SMC (Berridge & Dupont, 1994; Clapham, 1995; Karaki et al., 1997).

Among the chief mechanisms critical for maintaining intracellular Ca^{2+} homeostasis and SMC relaxation, is the role played by Ca^{2+} release from the SR via ryanodine receptors (RyRs) in the form of Ca^{2+} sparks. RyRs are large tetrameric complexes expressed in the SR membrane and are formed by the assembly of four subunits of ~560 kDa each, with four membrane spanning domains, a large cytosolic amino terminal and a SR-luminal carboxy terminal (Lanner, Georgiou, Joshi, & Hamilton, 2010). Three isoforms of RyRs—RyR1, RyR2, and RyR3—are expressed in vascular SMCs (Neylon, Richards, Larsen, Agrotis, & Bobik, 1995; Yang et al., 2005). Spontaneous release of Ca^{2+} from the SR via clusters of four to six RyRs (Cheng & Lederer, 2008), composed mainly of RyR1 and RyR2 in vascular SMCs, contribute to the formation of a "Ca^{2+} spark", a fundamental feature of vascular SMCs. First identified in cardiac muscle (Cheng, Lederer, & Cannell, 1993), followed by skeletal muscle (Klein et al., 1996), Ca^{2+} sparks in vascular SMCs were first characterized by Nelson et al. (1995) in rat cerebral arteries and are distinguished from other transient Ca^{2+} signals by their distinctive dynamic properties with a rise-time of ~20–40 ms, a duration (decay to half-time) of ~50–60 ms and a spatial spread of ~12.6 μm corresponding to ~1% of the cell surface area (Nelson et al., 1995).

In vascular SMCs, RyRs are closely juxtaposed with BK channels that are expressed in the plasma membrane. BK channels are formed as homotetramers, comprising four pore-forming α subunits. Each α subunit has seven-transmembrane domains with a voltage-sensing domain in S4, a long extracellular N-terminus, and a long cytosolic C-terminal that contains binding sites for Ca^{2+}, Mg^{2+}, and several regulatory molecules and domains. Auxiliary β subunits or the recently discovered γ subunits may associate with the α subunits in a tissue-specific manner and regulate the Ca^{2+} and voltage sensitivity of gating (Hoshi, Pantazis, & Olcese, 2013; Zhang & Yan, 2014). Of the four known auxiliary β subunit families (β_{1-4}), the SMCs exclusively express the β_1 subunit, whose interaction with the α subunit imparts greater Ca^{2+} sensitivity and slower kinetics of activation and deactivation, allowing the channel to open at lower Ca^{2+} concentrations at physiological membrane potentials and for longer durations compared to the native BK channel composed of only α subunits (Contreras et al., 2013). Four γ subunits, which are leucine-rich repeat containing proteins, have been identified to date, namely LRRC26 (γ1), LRRC52 (γ2), LRRC55 (γ3) and LRRC38 (γ4). The γ1 subunit mRNA has been found in low copy numbers in aortic SMCs (Yan & Aldrich, 2010) and protein expression has been identified in rat cerebral artery myocytes (Evanson, Bannister, Leo, & Jaggar, 2014). All four γ subunits increase the voltage sensitivity of the BKα subunit in the absence of Ca^{2+}, with the γ1 subunit alone shifting the activation voltage of the α subunit to lower potentials by ~130 mV, suggesting high channel conductance in cells at physiological resting potentials when intracellular Ca^{2+} is extremely low (Yan & Aldrich, 2010).

At physiological membrane potentials, the K_D of the BK channel for Ca^{2+} is ~20 μM. A single Ca^{2+} spark from a single RyR cluster, may increase local Ca^{2+} concentrations up to ~30 μM and activate up to 30 BK channels to produce an outward K^+ current in the form of spontaneous transient outward K^+ currents (STOCs) resulting from a 100-fold increase in open probability (Jaggar, Porter, Lederer, & Nelson, 2000). This one-to-one coupling between a Ca^{2+} spark and activation of the underlying population of BK channels, along with the considerably large single-channel conductance of the BK channel (~150–250 pS) results in a large membrane hyperpolarizing K^+ current that significantly reduces VDCC activity, leading to a decrease in Ca^{2+} influx and vasorelaxation (Knot, Standen, & Nelson, 1998). The local transient action of Ca^{2+} sparks and activation of BK channels drives a global change in membrane potential and global Ca^{2+}, which ultimately relaxes smooth muscle and increases arterial diameter (Fig. 1). Ca^{2+} sparks have since been characterized in SMCs of nearly every organ blood vessel, including coronary arteries (Porter et al., 1998), mesenteric arteries (Krishnamoorthy, Sonkusare, Heppner, & Nelson, 2014; Miriel, Mauban, Blaustein, & Gil Wier, 1999), rat portal vein (Arnaudeau, Macrez-Leprêtre, & Mironneau, 1996), unpressurized cerebral arteries (Jaggar, Stevenson, & Nelson, 1998), and unpressurized spiral modiolar artery (Krishnamoorthy, Regehr, Berge, Scherer, & Wangemann, 2011), making it, along with BK channels STOCs

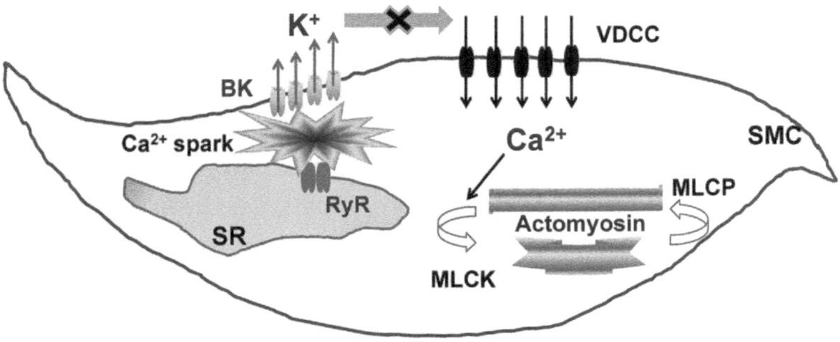

Fig. 1 BK channel regulates SMC intracellular Ca^{2+} homeostasis. Simplified schematic showing the mechanism by which, the activation of BK channels by RyR-mediated Ca^{2+} sparks potentiates a large outward K^+ current that hyperpolarizes SMC membrane potential, decreases VDCC activity, limits Ca^{2+} influx, relieves SMC constriction and modulates the myogenic response. *RyR*, ryanodine receptor; *VDCC*, voltage-dependent Ca^{2+} channel; *MLCK*, myosin light chain kinase; *MLCP*, myosin light chain phosphatase; *SR*, sarcoplasmic reticulum; *SMC*, smooth muscle cell; *actomyosin*, contractile fibers.

(Jaggar et al., 2000), a universal regulator of vascular myogenic tone. The importance of such a negative feedback mechanism for the regulation of myogenic tone in muscular arteries and arterioles is evident from observations that pharmacological inhibition of BK channels by low concentrations of TEA, paxilline or iberiotoxin, or genetic manipulation of BK channel α, β, or γ subunit expression leads to the loss of this potent hyperpolarizing signal, leading to membrane depolarization, increase in VDCC activation, increase in Ca^{2+} influx, vasoconstriction and increase in myogenic tone (Brenner et al., 2000; Evanson et al., 2014; Knot et al., 1998; Krishnamoorthy et al., 2014; Nelson et al., 1995).

2.3 IP_3-Mediated Activation

In a much less characterized mechanism, recent evidence has alluded to the activation of BK channels by IP_3 in SMCs of rat and mouse cerebral and porcine coronary arteries. In these studies, direct application of IP_3 in excised cell patches increased BK channel activation and inhibition of the IP_3 receptor (IP_3R) abolished this effect indicating that IP_3-mediated Ca^{2+} release from the SR is not involved in this mechanism. Rather, it is more likely a direct interaction of IP_3R with the BK channel protein itself, whereby it modulates the Ca^{2+} sensitivity of activation (Kim, Weir, Macdonald, & Zhang, 1998; Yang et al., 2013; Zhao et al., 2010). However, it is not yet evident if this is a universal mechanism for the regulation of vascular tone.

2.4 Endothelial BK Channels

In blood vessels, BK channels are almost exclusively expressed in the SMC membrane, where they regulate membrane potential. In general, BK channels are absent in ECs. Few studies that have reported the presence of BK channels in the endothelium, find poor expression of channels that elicit iberiotoxin-sensitive currents only at highly positive potentials (Kohler et al., 2000; Ledoux, Bonev, & Nelson, 2008; Marchenko & Sage, 1996). mRNA expression analysis of the porcine aorta indicated the presence of the BKα subunit in both SMCs and ECs, whereas the expression of the β1 subunit in only SMCs. These observations were further supported by the activation of BK currents in ECs by a BKα subunit-specific opener NS1619, but not a BKβ1 subunit-specific opener DHS-1 (Papassotiriou et al., 2000). The β1 subunit confers greater Ca^{2+} sensitivity to channel activation such that at a given Ca^{2+} concentration, the voltage dependence of activation of BKα + β channel is shifted left compared to that of BKα alone

(Orio, Rojas, Ferreira, & Latorre, 2002; Papassotiriou et al., 2000), implying that lower cytosolic Ca^{2+} concentrations are required to achieve maximal activation at physiological membrane potentials. Thus, the far-right shifted voltage dependence of activation of endothelial BK channels may be attributed to the absence of regulatory BKβ subunits (Papassotiriou et al., 2000). It is as yet unknown whether ECs express any of the BKγ subunits. However, ECs may express functional BK channels under special conditions, such as following chronic hypoxia, wherein the apparent Ca^{2+} sensitivity and voltage dependence of activation of this EC BK channel is greater to that observed otherwise in EC BK channels, despite the absence of β subunits (Riddle, Hughes, & Walker, 2011). Various reasons have been proposed to explain such observations, including the removal of an inhibitory effect by the scaffolding protein, caveolin-1 following chronic hypoxia or an effect of endogenous H_2S on the BK channel moiety (Jackson-Weaver et al., 2013). Regardless of the underlying mechanism, activation of EC BK channels following chronic hypoxia has been shown to aid in reversing/inhibiting vasoconstrictor effects by reducing myogenic reactivity (Hughes, Riddle, Paffett, Gonzalez Bosc, & Walker, 2010). Another study posits that when expressed, endothelial BK channels modulate myogenic reactivity by regulating NO production (Brakemeier et al., 2003). Further studies are required to better elucidate and understand the expression and function of BK channels in ECs.

3. VASCULAR REGULATION BY BK CHANNELS

Regulation of vascular tone is mediated by vasoactive signals intrinsic to the SMC as well as of extrinsic origin from the endothelium, perivascular nerves and circulating hormones, all of which regulate the contractile state of the SMC. Several of these vasoactive agents are known modulators of BK channel activity or Ca^{2+} sparks or both and in turn affect the regulation of myogenic tone by BK channels.

3.1 Phosphorylation

A major mechanism of regulation of BK channel activity is via phosphorylation of the BK channel protein subunit by several kinases, including protein kinase A (PKA), protein kinase G-1 (PKG-1) and PKC, leading to apparent changes in the channel's voltage and Ca^{2+} sensitivity (Kume, Takai, Tokuno, & Tomita, 1989; Schubert & Nelson, 2001) (see also Chapter "Posttranscriptional and Posttranslational Regulation of BK Channels" by

Shipston and Tian of the present volume). In SMCs, the action of PKA and PKG-1 increases channel open probability without affecting its voltage sensitivity (Minami, Fukuzawa, Nakaya, Zeng, & Inoue, 1993; Robertson, Schubert, Hescheler, & Nelson, 1993). PKA and PKG-1 may also indirectly increase BK channel activation via phosphorylation of phospholamban, which disinhibits the SR Ca^{2+} ATPase and increases the SR Ca^{2+} load leading to an increase in the amplitude and frequency of Ca^{2+} sparks (Porter et al., 1998; Wellman, Santana, Bonev, & Nelson, 2001). Phosphorylation by PKC, on the other hand, inhibits BK channel activation in SMCs (Schubert, Noack, & Serebryakov, 1999; Taguchi, Kaneko, & Kubo, 2000) and seems to predispose the BK channel to be phosphorylated by either PKA or PKG-1 (Barman, Zhu, & White, 2004). It would appear that the actions of PKA and PKG-1 promote membrane hyperpolarization and vasorelaxation, whereas the actions of PKC seemingly promote membrane depolarization and vasoconstriction. Thus, phosphorylation of BK channels by protein kinases is a common mechanism mediating the effects of agonists, vasoconstrictors and vasodilators. Vasodilations produced by 14,15-epoxyeicosatrienoic acid (14,15-EET) in normotensive mesenteric arteries (Yang et al., 2014), by somatosensory activation in cerebral pial arteries (Paisansathan, Xu, Vetri, Hernandez, & Pelligrino, 2010) and by perinatal hypoxia in the adult murine pulmonary circulation (Marino, Beny, Peyter, Diaceri, & Tolsa, 2011) have been shown to be mediated by PKA-dependent activation of BK channels and are diminished or inhibited by PKA inhibitors. The best-known example of PKG-dependent activation of BK channels is the vasodilation of penile arteries by phosphodiesterase inhibitors acting via the NO/cGMP/PKG pathway, which are abolished/diminished by PKG inhibitors (Prieto et al., 2006; Sánchez et al., 2008). Several pathological conditions are linked to a PKC-dependent loss of the hyperpolarizing influence of BK channels and are reversed by prior inhibition of PKC such as inhibition of BK currents in aortic SMCs following gamma irradiation (Kizub, Pavlova, Ivanova, & Soloviev, 2010), impairment of neurovascular coupling in type-1 diabetes mellitus rats caused by a suppression of BK channels (Vetri, Xu, Paisansathan, & Pelligrino, 2012), suppression of the BKβ1 subunit expression in diabetic mouse arteries and hyperglycemic human coronary artery SMCs (Lu et al., 2012), significantly higher phenylephrine-induced contractions in mesenteric arteries of adult offspring rats exposed to prenatal hypertension (Bo et al., 2015) and angiotensin-II induced vasoconstriction caused by PKC-dependent internalization and degradation of surface BKα subunits in cerebral arteries (Leo et al., 2015).

3.2 Nitric Oxide

Nitric oxide (NO) produced and released from the vascular endothelium in every blood vessel is a potent and the most-studied vasodilator and modulates several downstream targets leading to SMC relaxation, earning it the name "endothelium-derived relaxing factor". NO is synthesized by the enzyme NO synthase (NOS), which converts L-arginine to NO and L-citrulline in an oxygen-dependent reaction (Moncada, Palmer, & Higgs, 1991). Three isoforms of NOS exist—endothelial NOS (eNOS) and neuronal NOS (nNOS) constitutively expressed in ECs and neuronal cells, respectively, with relatively greater expression of eNOS than nNOS in blood vessels, except in the brain, where the expression levels of eNOS and nNOS are comparable. The expression of the third isoform, inducible NOS (iNOS), as the name suggests, is induced under pathological conditions. In myogenically reactive blood vessels, a basal level of NO, constitutively produced predominantly by eNOS, generates a relaxing signal that attenuates basal myogenic tone (Geary, Krause, & Duckles, 1998). The activation of eNOS being Ca^{2+} dependent, factors that cause EC intracellular Ca^{2+} to increase, activate NOS and increase production of NO. After diffusing into SMC, NO promotes SMC relaxation in part by greatly increasing the activity of the NO-sensitive, heme-based sensor protein, soluble guanylate cyclase (sGC), which leads to a rapid synthesis of cGMP (Arnold, Mittal, Katsuki, & Murad, 1977; Pfeifer et al., 1999). cGMP has numerous downstream targets in the vascular SMC, chief among them being PKG-1, which as mentioned earlier, serves to increase BK channel activation either directly via phosphorylation or indirectly via augmentation of Ca^{2+} sparks (Mandalà, Heppner, Bonev, & Nelson, 2007). Thus, inhibition of the NOS/cGMP/PKG pathway leads to decrease in BK channel activity, vasoconstriction and increase in myogenic tone, as stated in previous sections. Recent work has shown that NO may also increase BK channel activation by stimulating rapid anterograde trafficking of BKβ1 subunit-containing endosomes, in a PKG/PKA-dependent fashion, to cause a three-fold increase in the surface expression of β1 subunits and significant increase in BK channel activation (Leo et al., 2014).

3.3 Estrogen

Naturally occurring estrogens, such as 17β-estradiol, are major vasorelaxants, especially in the coronary and uterine microvasculature (Darkow, Lu, & White, 1997; Han et al., 2006; Rosenfeld, Cornfield, &

Roy, 2001; Rosenfeld, White, Roy, & Cox, 2000; White, Darkow, & Lang, 1995; White et al., 2002). Acute application of 17β-estradiol has been shown to directly activate BK channels, but requires the presence of the α subunit and atleast two β1 subunits of the BK channel for estrogen to be effective (De Wet et al., 2006; Valverde et al., 1999; White et al., 2002). The activation likely results from multiple sites of interaction between the BK channel complex (α+β1) and 17β-estradiol (Liu et al., 2010; Morrow et al., 2006). However, estrogen may also indirectly activate BK channels via activation of estrogen receptor ERα in SMCs (Han et al., 2006), presumably leading to enhanced bioavailability of NO (Broughton, Miller, & Sobey, 2010) and formation of cGMP (Rosenfeld et al., 2000) to induce vasodilation (Geary et al., 1998). Further, there is evidence to suggest that prolonged exposure to estrogen enhances BK channel function, estrogen responsiveness and basal vascular tone in uterine and mammary arteries via upregulation of the BKβ1 subunit mRNA leading to altered BKα:β1 stoichiometry (Nagar, Liu, & Rosenfeld, 2005).

3.4 Lipid Metabolism

Eicosanoids, which encompasses 20-carbon chain fatty acids and their metabolites, are the most biologically active lipid mediators of vascular tone regulation. The term "eicosanoid" includes prostaglandins, thromboxanes, leukotrienes, epoxyeicosatrienoic acids (EETs) and other oxygenated derivatives, all of which are produced primarily by three classes of enzymes—cyclooxygenases (COX-1 and COX-2), lipoxygenases (LOX) and cytochrome P450 monooxygenases (CYP450), from three basic polyunsaturated fatty acids (PUFAs)—8*c*,11*c*,14*c*-eicosatrienoic (dihomo-γ-linolenic or 20:3($n-6$)), 5*c*,8*c*,11*c*,14*c*-eicosatetraenoic (arachidonic or 20:4($n-6$) or AA) and 5*c*,8*c*,11*c*,14*c*,17*c*-eicosapentaenoic (20:5($n-3$) or EPA).

Of these, AA, by far the most studied, is an essential fatty acid and a major component of phospholipids that is necessary for the integrity of cell membranes. Free AA is metabolized from membrane phospholipids upon the stimulation of phospholipase A_2 (PLA_2) by various physiological and pathological factors, including hormones and cytokines. Further metabolism of AA in ECs and SMCs by one or more of the three aforementioned enzymes generates several vasoactive products that are either vasoconstricting or vasorelaxing in nature (Fig. 2), several of which act via BK channels (reviewed in Dopico & Bukiya, 2014).

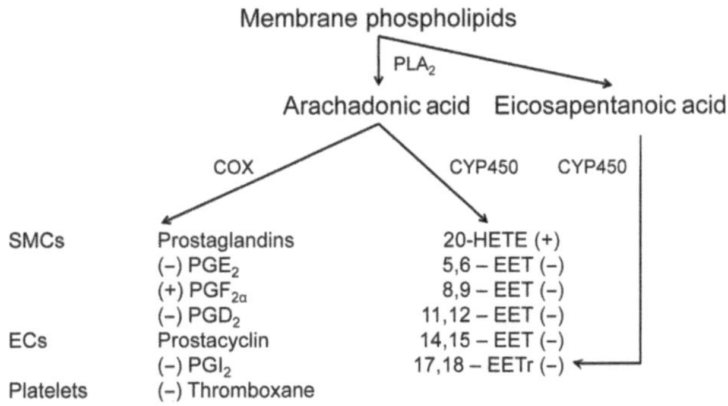

Fig. 2 Lipid metabolism in the vascular wall. Simplified signaling schematic depicting part of the metabolic cascade involved in the conversion of membrane phospholipids into vasoactive factors that affect BK channel activation. *PLA_2*, phospholipase A_2; *COX*, cyclooxygenase; *CYP450*, cytochrome P450 monooxygenase; *PG*, prostaglandin; *EET*, epoxyeicosatrienoic acid; *HETE*, hydroxyeicosatetraenoic acid; *EETr*, epoxyeicosatetraenoic acid; *SMC*, smooth muscle cell; *EC*, endothelial cell; (+) denotes vasoconstrictor, (–) denotes vasodilator.

3.4.1 Fatty Acids

To begin with, in patch-clamp experiments, fatty acids can directly activate the BK channel protein complex in the absence of cytosolic components in excised inside-out patches of vascular SMCs from pulmonary (Clarke, Petrou, Walsh, & Singer, 2002, 2003; Kirber, Ordway, Clapp, Walsh, & Singer, 1992), coronary (Ahn, Kim, Lee, Kang, & Kang, 1994), mesenteric (Dopico, Kirber, Singer, & Walsh, 1994) and human umbilical (Martín et al., 2014) arteries in an as-yet unidentified mechanism. AA has also been shown to cause vasodilation of pulmonary arteries, but whether it involves a direct activation of BK channel subunit is unclear (Guerard et al., 2004). However, a direct activation of BK channel protein underlies the vasorelaxation and lowering of blood pressure produced by omega-3 fatty acids (Hoshi, Wissuwa, et al., 2013).

3.4.2 Metabolites of AA by the COX Pathway

Metabolism of AA by the COX enzymes involves the catalysis of enzymes COX-1 and COX-2 in the synthesis of vasoconstricting and vasodilating prostanoids which include prostaglandins, prostacyclin and thromboxane. It is suggested that the function of COX-1 is to maintain homeostatic conditions by monitoring and rapidly responding to changes in circulating hormones. It is a constitutive enzyme that produces prostaglandins in cells and is highly expressed in those tissues where prostaglandins have a specialized role, such as in the vascular endothelium. It is suggested that COX-1 may function only at high concentrations of AA, such as that may occur during platelet aggregation, cell injury or inflammation. In contrast, COX-2 is constitutively expressed only in the brain and kidney and is inducible in other tissues, where it is expressed under the control of the proinflammatory transcription factor NF-κB and produces prostanoids in cells that are involved in inflammation and which are primarily pathophysiological in nature. Induction of COX-2 expression is also regulated by sphingosine-1-phosphate (S1P) and inhibited by nonsteroidal antiinflammatory drugs (NSAIDs) such as ibuprofen and acetaminophen (Wong, Wong, Tian, Lau, & Huang, 2010).

Products of the COX pathway may be either vasodilating (prostaglandin I_2 (PGI_2), prostaglandin E_2 (PGE_2) and prostaglandin D_2 (PGD_2)) or vasoconstricting (prostaglandin $F_{2\alpha}$ ($PGF_{2\alpha}$) and thromboxane A_2 (TXA_2)). PGE_2 and $PGF_{2\alpha}$ are produced in SMCs of blood vessels, PGI_2 is produced in the endothelium and TXA_2 is produced by the platelets, which is enhanced following platelet activation during inflammation and tissue injury. There is evidence for the regulation of BK channel function by prostanoids in vascular SMCs. PGI_2 and PGE_2 have been shown to increase BK channel currents in retinal pericytes (Burnette & White, 2006) and coronary artery SMCs (Zhu, Han, & White, 2002). However, there is no evidence of a direct interaction between these prostanoids and the BK channel protein. Instead, PGI_2 invokes vasorelaxation in a cAMP-independent, G_s protein-dependent mechanism (Tanaka, Yamaki, Koike, & Toro, 2004), whereas PGE_2-induced BK channel activation may involve cAMP-stimulated cross-activation of PKG but not PKA (Zhu et al., 2002). On the other hand, the prostanoid TXA_2 may inhibit BK channel activity by a direct interaction between channel-forming BKβ1 subunits and TXA_2 receptors (Liu et al., 2012).

3.4.3 *Metabolites of AA by the Cytochrome P450 Epoxygenase Pathway*

Metabolites of AA generated by the cytochrome P450 (CYP) pathway include EETs, dihydroxyeicosatetraenoic acids (DiHETE) and 19- and 20-hydroxyeicosatetraenoic acids (19- and 20-HETE, respectively). In the cardiovascular system, the major metabolites EETs and 20-HETE are produced in the ECs and SMCs, respectively, and operate as paracrine factors and second messengers to play prominent roles in the regulation of the contractile state in renal, cerebral and pulmonary vasculature.

Small blood vessels throughout the body contain the CYP mRNA and protein and when incubated with AA, produce 20-HETE in the SMCs (Harder et al., 1994). 20-HETE causes robust vasoconstriction in pressurized renal (Imig et al., 1996; McGiff & Quilley, 1999), cerebral (Harder et al., 1994), coronary, mesenteric and skeletal muscle arteries (Miyata & Roman, 2005) via inhibition of BK channels (Zou et al., 1996), thereby promoting membrane depolarization and enhanced Ca^{2+} entry via VDCCs (Gebremedhin et al., 1998). The mechanism of BK channel inhibition may occur via multiple pathways in different arteries. For instance, in cerebral arteries, 20-HETE BK channel inhibition may be a result of PKC activation (Lange, Gebremedhin, Narayanan, & Harder, 1997). In renal artery SMCs, 20-HETE may activate mitogen-activated protein kinase (MAPK) in addition to PKC (Sun, Falck, Harder, & Roman, 1999). However, the reversal of the effect of 20-HETE on BK channel activity and vascular tone in the renal arteries is not achieved by inhibitors of PKC and MAPK, but achieved by inhibitors of tyrosine kinases (Sun et al., 1999). Thus, 20-HETE may employ several mechanisms to act as a counteractive measure against BK channel activation and membrane hyperpolarization to regulate the myogenic response in these arteries. In vivo, CYP450 inhibitors may reduce the vasoconstrictor responses to application of vasopressin, angiotensin-II, endothelin and norepinephrine (Imig & Deichmann, 1997; McGiff & Quilley, 1999; Oyekan, Youseff, Fulton, Quilley, & McGiff, 1999). This is possibly achieved by relieving the 20-HETE-induced inhibition of BK channels. Recent studies have shown a cross-talk between the 20-HETE-mediated vasoconstriction and NO-mediated vasodilation in renal and cerebral arteries (Alonso-Galicia, Hudetz, Shen, Harder, & Roman, 1999; Sun et al., 1998). In addition to a cGMP-dependent BK channel activation pathway, NO-dependent vasorelaxation involves a cGMP-independent mechanism in these arteries. It is suggested that this cGMP-independent mechanism may involve the inhibition of 20-HETE-mediated BK channel inhibition by NO.

Four regioisomers of EETs—5,6-EET, 8,9-EET, 11,12-EET, and 14,15-EET are synthesized in the endothelium by CYP450 with AA as the substrate. A fifth, 17,18-EETr, is synthesized by the CYP450 from EPA as substrate. EETs, DiHETE and 17,18-EETr are all potent vasodilators of renal, cerebral, coronary and mesenteric arteries and arterioles and play a role in local blood flow regulation as well as systemic blood pressure. Given the nature of their synthesis and function, EETs, more specifically 11,12-EET, are termed as the endothelium-derived hyperpolarizing factor

(EDHF). They are also produced in brain astrocytes, cardiac tissue, airway and parenchymal tissues in the lung and renal tubular cells, where they act as paracrine factors that regulate local vascular tone. The primary mechanism of EETs-dependent vasorelaxation is the activation of BK channels in vascular SMCs as observed in coronary (Zhang et al., 2001), cerebral (Hercule et al., 2007), mammary (Archer et al., 2003), mesenteric (Dimitropoulou et al., 2007; Hercule et al., 2007) and pulmonary (Loot et al., 2012; Morin, Sirois, Echave, Rizcallah, & Rousseau, 2009) arteries. The activation occurs without increasing channel unitary conductance (Fukao, Mason, Kenyon, Horowitz, & Keef, 2001) but by lengthening opening times and shortening closing times (Lu et al., 2001). Interestingly, activation of BK channels by 11,12-EET in excised patches from coronary artery SMCs does not require β1 subunits (Fukao et al., 2001) and similar action by 17,18-EETr was preserved in cerebral and mesenteric vascular SMCs from BKβ1$^{-/-}$ mice (Hercule et al., 2007). In other studies, Imig and Deichmann (1997) report the involvement of PP2A, a protein phosphatase, in the activation of BK channels by 11,12-EET to relax renal arterioles, while Earley, Heppner, Nelson, and Brayden (2005) suggest the activation of TRPV4 (transient receptor potential cation channel, subfamily 5, member 4) by 11,12-EET in cerebral artery SMCs leading to augmentation of Ca^{2+} sparks and BK channel activation, and Fan, Muroya, and Roman (2015) indicate binding of EETs to an EET receptor leading to the activation of adenylyl cyclase, augmented production of cAMP, activation of PKA and enhanced activation of BK channel in renal afferent arterioles. All of these highlight the primacy of the BK channel in preserving the EDHF functionality. More detailed summaries of the action of EETs in the regulation of vascular tone and the involvement of BK channels in this function may be obtained from Dopico and Bukiya (2014), Spector (2009), and Roman, Maier, Sun, Harder, and Alonso-Galicia (2000).

3.5 Reactive Oxygen Species

Reactive oxygen species (ROS), includes peroxides, superoxide, hydroxyl radical and singlet oxygen, which are produced through multiple mechanisms as a natural byproduct of cellular metabolism. A major source of ROS formation is the underlying tissue or any layer of the wall structure, when NADPH oxidase (NOX) complexes in mitochondria produce the superoxide radical as an end byproduct of oxidative phosphorylation by means of the electron transport chain. Superoxide is converted by the enzyme superoxide dismutase to hydrogen peroxide (H_2O_2), which is then converted by catalase into water. The deleterious effects of ROS manifests when excessive amounts are produced leading to oxidative stress. Depending on the cellular redox status, the presence of the endothelium, experimental conditions or the concentrations studies, ROS, superoxide and H_2O_2, may either constrict or dilate blood vessels by acting, in part through BK channels. ROS have been shown to constrict SMCs via part inhibition of BK channels (Brzezinska, Gebremedhin, Chilian, Kalyanaraman, & Elliott, 2000; Liu et al., 2002). Superoxide itself dilates cerebral arteries in a BK-dependent fashion either by direct activation or indirectly by conversion to H_2O_2 (Wei, Kontos, & Beckman, 1996). One of the many pathways by which H_2O_2 induces SMC hyperpolarization and relaxation is by either directly activating BK channels or stimulating sGC (Hayabuchi, Nakaya, Matsuoka, & Kuroda, 1998; Thengchaisri & Kuo, 2003). However, H_2O_2 may also potently inhibit BK channel by modifying cysteines on the BKα subunit (Tang, Garcia, Heinemann, & Hoshi, 2004).

BK channel expression and activation is also regulated by acute or chronic hypoxia and subsequently plays unique roles in the uterine vascular adaptation during pregnancy (Zhu, Xiao, & Zhang, 2013) and in the pulmonary microcirculation (Bonnet & Archer, 2007). As a protective mechanism to prevent deleterious effects of chronic hypoxia-induced maternal cardiac dysfunction on fetal development, an upregulation of BK channel expression and a reduced myogenic tone is reported in uterine arteries during pregnancy (Hu et al., 2011). In the pulmonary vasculature, chronic hypoxia-induced pulmonary hypertension is associated with depolarization of pulmonary artery SMCs and a reduction in the mRNA of expressed K^+ channels, including BK channels (Yuan, Wang, Juhaszova, Gaine, & Rubin, 1998). Paradoxically, there are also reports of an increase in BKα and BKβ1 subunit expression in these cells, suggesting a compensatory mechanism to counter hypoxia-induced vasoconstriction (Ahn et al., 2012; Resnik, Herron, Fu, Ivy, & Cornfield, 2006).

4. REGULATION OF NEUROGENIC TONE BY BK CHANNELS

4.1 Neurotransmitters

Blood vessels are innervated by sympathetic and parasympathetic nerve endings, which, upon stimulation, release neurotransmitters into the perivascular space and regulate vascular tone. The diameters of distal small arterioles in the microcirculation are regulated by myogenic activity as well as sympathetic nerve activity, whereas in more proximal arteries, nervous activity plays a significant role in regulating PVR, blood pressure and vascular tone. The importance of neurogenic regulation of vascular tone is emphasized by the finding that sympathetic nervous system hyperactivity is a characteristic feature of all manifestations of human hypertension, obesity and heart failure (Haddock & Hill, 2011; Mancia, Di Rienzo, Parati, & Grassi, 1997) and is the primary physiological response to stress in the form of the "flight or fight" reflex. In mesenteric arteries, in which neurogenic regulation of vascular tone has been much studied, sympathetic nerves release the purinergic neurotransmitter, ATP, and the adrenergic neurotransmitter, norepinephrine (NE) to constrict arteries (Garcha & Hughes, 1997; Muramatsu, 1987), while parasympathetic nerves release acetylcholine (ACh), calcitonin gene-related peptide (CGRP) and substance P to dilate arteries via endothelium-dependent pathways (Climent et al., 2012; Mori, Suzuki, Sakamoto, Nakahara, & Ishii, 2011), although CGRP has also been shown to activate BK channels via a PKA-dependent mechanism (Bol, Leybaert, & Vanheel, 2012). Sympathetic neurotransmitters, ATP and NE, bind to respective postjunctional purinergic P_2X receptors and α adrenergic receptors in SMCs (Muramatsu, Ohmura, & Oshita, 1989). Activation of the purinergic P_2X_1 receptor channels by ATP causes an influx of Na^+ and Ca^{2+} ions to generate an excitatory junction potential, which are manifested in SMCs as spontaneous local Ca^{2+} increases called junctional Ca^{2+} transients (jCaTs) (Lamont & Wier, 2002). jCaTs depolarize the SMC membrane and activate VDCCs leading to rapid Ca^{2+} influx and rapid constriction (Lamont, Vainorius, & Wier, 2003; Lamont, Vial, Evans, & Wier, 2006). Thus, ATP contributes to the early rapid and transient phase of nerve-evoked vasoconstriction. Activation of α_1-adrenergic receptors by NE promotes constriction through mechanisms that involve Ca^{2+} entry via VDCCs, Ca^{2+} release from intracellular stores as well as Ca^{2+} sensitization of the contractile proteins (Villalba, Stankevicius, Garcia-Sacristan, Simonsen, & Prieto, 2007). Thus, NE contributes to the late sustained phase of nerve-evoked vasoconstriction (Muramatsu, 1987; Muramatsu et al., 1989). Recently, BK channels have been shown to regulate neurogenic tone by opposing increases in intracellular Ca^{2+} through membrane hyperpolarization during both the rapid and the slow phase of nerve-evoked vasoconstriction (Krishnamoorthy et al., 2014). Moreover, based on observations of enhanced myogenic contractility in arteries from $BK\beta1^{-/-}$ mice, it may be speculated that loss of the β1 subunit would lead to a loss of this hyperpolarizing signal and enhance nerve-evoked constrictions in these arteries.

K^+ ions in the extracellular space, behaving as a paracrine agent, have a significant role to play in the regulation of vascular tone in the context of the neurovascular unit in the brain as explained next.

4.2 Neurovascular Coupling

Neurovascular coupling, also referred to as functional hyperemia, is a mechanism of local cerebral blood flow (CBF) regulation that enables the adequate delivery of oxygen and nutrients to localized areas of active neurons in brain. Since 1890, when Roy and Sherrington first discovered this intrinsic regulation of local CBF within the brain (Roy & Sherrington, 1890), the molecular basis of functional hyperemia has been extensively studied using both in vivo and ex vivo approaches. It is now known that this fine-tuning of local CBF to match regional brain function involves the coordinated activity of the neurovascular unit; neurons, astrocytes, and parenchymal arterioles (Gordon, Mulligan, & MacVicar, 2007; Iadecola & Nedergaard, 2007; Koehler, Roman, & Harder, 2009; Koide, Sukhotinsky, Ayata, & Wellman, 2013). The following steps are involved in the neurovascular coupling signaling cascade: (1) neuronal activation causes release of neurotransmitters (eg, glutamate). (2) Activation of neurotransmitter receptors (eg, metabotropic glutamate receptor) on astrocytes triggers propagation of intracellular Ca^{2+} waves that lead to increased cytosolic Ca^{2+} in astrocyte processes (endfeet) that encase nearby parenchymal arterioles. (3) Ca^{2+}-dependent vasodilatory signals from astrocyte endfeet promote parenchymal arteriolar dilation and increased CBF.

One of the major vasodilatory signals involved in neurovascular coupling are K^+ ions released into the perivascular space by efflux through BK channels on astrocyte endfeet (Filosa et al., 2006). An immunohistological study, performed by double staining of glial fibrillary acidic protein (astrocyte marker) and rSlo (ie, the pore-forming subunit of the BK channel), demonstrated that BK channels are highly expressed in astrocyte endfeet adjacent to small vessels in rat hippocampus slices (Price, Ludwig, Mi, Schwarz, & Ellisman, 2002). These BK channels located on astrocyte endfeet are activated (~160-fold increase in open-state probability) by increased cytosolic Ca^{2+} in response to evoked neuronal activation in freshly prepared mouse cortical brain slices (Filosa et al., 2006). As activation of endfeet BK channels causes K^+ efflux, it is conceivable that this increase in endfoot BK channel activity will cause substantial elevation in K^+ concentration in the restricted perivascular space (ie, between astrocytic endfeet and arteriolar smooth muscle, ~20 nm; Nagelhus et al., 1999). In experimental studies using isolated cerebral arteries, modest increases in extracellular K^+ concentration (<20 mM) caused near maximum dilation via activation of inward rectifier potassium (K_{ir}) channels on SMCs (Girouard et al., 2010; Knot, Zimmermann, & Nelson, 1996). Due to the electrophysiological properties of BK channels (ie, large conductance or high K^+ permeability) and the small volume of perivascular space, relatively modest increases in endfeet BK channel activity are capable of increasing perivascular K^+ concentration to this vasodilatory range (5–20 mM). As stated by Filosa et al. "the opening of a single-BK channel for 0.2 s should be sufficient to provide enough K^+ ions (about 10^6) in the restricted space between astrocytic endfeet and SMCs to elevate the local K^+ concentration to 10 mM" (Filosa et al., 2006). Consistent with this concept, the BK channel blocker paxilline greatly diminished neuronally evoked arteriolar dilation (neurovascular coupling) in brain slices (Fig. 3) as well as whisker stimulation-induced CBF increases in somatosensory cortex in vivo (functional hyperemia) (Girouard et al., 2010). Thus, neurally evoked Ca^{2+} increases in astrocyte endfeet and the subsequent activation of endfeet BK channels is thought to play a key role in the regulation of arteriolar diameter in neurovascular coupling.

Recent studies have also demonstrated that astrocyte endfeet BK channels contribute to the pathological neurovascular coupling after subarachnoid hemorrhage (SAH), a common form of hemorrhagic stroke (Koide, Bonev, Nelson, & Wellman, 2012; Koide et al., 2013; Pappas, Koide, & Wellman, 2015). In freshly prepared brain slices from SAH model animals, local neuronal activation triggered by electrical field stimulation (EFS) led to arterial constriction, rather than vasodilation normally observed in brain slices from healthy animals (Koide et al., 2012). This inversion of neurovascular coupling (ie, EFS-induced vasoconstriction) results from the additional activation of endfoot BK channels due to the emergence of high-amplitude spontaneous Ca^{2+} signals (eHACSs) in astrocyte endfeet (Pappas et al., 2015). Based on the observation that Ca^{2+} elevations during eHACSs reach greater than twice the concentration (eg, 738 ± 72 nM) compared to "control-like" events (eg, 302 ± 10 nM) (Pappas et al., 2015), it is likely that eHACSs increased basal perivascular K^+ concentration due to increased endfeet BK channel activity. This higher basal perivascular K^+ concentration combined with neuronally triggered endfeet BK channel activation is thought to increase perivascular K^+ in excess of

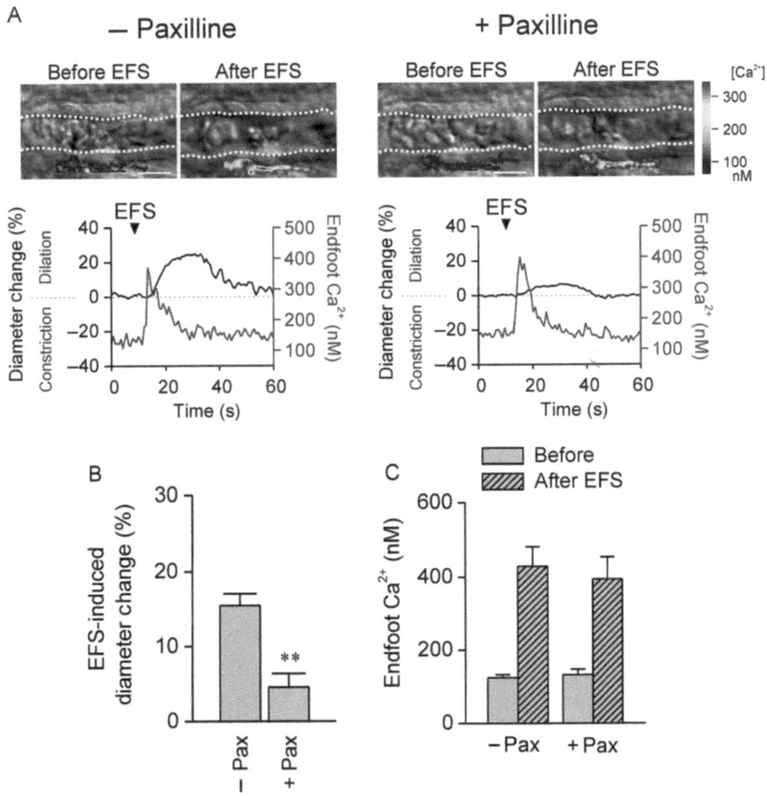

Fig. 3 Role of BK channels in neurovascular coupling. (A) Paxilline (Pax), a BK channel blocker, abolished arteriolar dilation following focal neuronal activation in freshly prepared rat cortical brain slices. Arteriolar diameter and astrocytic endfeet Ca^{2+} concentration were simultaneously measured using the fluorescent Ca^{2+} indicator, fluo-4, and two-photon imaging. Focal neuronal activation was triggered by electrical field stimulation (EFS). *Upper*: Infrared-differential interference contrast (IR-DIC) images overlapped pseudocolor-mapped Ca^{2+} levels in astrocyte endfeet in rat cortical brain slices. *Red* (*gray* in the print version) *dashes outline* the intraluminal diameter of parenchymal arterioles. Scale bar: 10 μm. *Lower*: EFS-induced changes in arteriolar diameter (*black lines*) and endfoot Ca^{2+} concentration (*red* lines in the color version or *gray* lines in the print version) obtained from brain slices depicted in upper images. (B) and (C) Summary of EFS-induced changes in arteriolar diameter (B) and astrocytic endfoot Ca^{2+} (C) before and after a 10-min incubation of brain slices with paxilline (1 μM). $^{**}P < 0.01$ vs before paxilline treatment (–Pax). *Modified from Koide, M., Bonev, A. D., Nelson, M. T., & Wellman, G. C. (2012). Inversion of neurovascular coupling by subarachnoid blood depends on large-conductance Ca²⁺-activated K⁺ (BK) channels.* Proceedings of the National Academy of Sciences of the United States of America, 109*(21), E1387–E1395.*

the threshold (20 mM K^+) leading to vasoconstriction, rather than vasodilation. Indeed, studies have shown that treatment of brain slices with paxilline to block BK channels or abolishment of spontaneous endfoot Ca^{2+} signals (treatment with the sarco/endoplasmic reticulum Ca^{2+}-ATPase inhibitor, cyclopiazonic acid) prevented SAH-induced inversion of neurovascular coupling (Koide et al., 2012; Pappas et al., 2015). These studies suggested that endfeet BK channel activation associated with abnormal astrocyte Ca^{2+} signaling is a major cause of inversion of neurovascular coupling after SAH, and targeting endfeet BK channels has potential as a therapeutic target to alleviate brain ischemia after SAH.

In summary, adequate CBF regulation is crucial for maintaining proper brain function, and neurovascular coupling is one of the mechanisms to enable temporally and spatially matched local CBF delivery with neuronal metabolic demands. Astrocyte endfeet BK channels play a key role in both the physiological neurovascular coupling response and the pathological inversion of neurovascular coupling from vasodilation to vasoconstriction after SAH. Future studies are likely to reveal pathological roles of BK channels in other CNS diseases (see also Chapter "BK Channels in Neurons" by Barth and Contet) and may also lead to new therapeutic strategies that possibly combine targeting endfeet and vascular smooth muscle BK channels to promote increased CBF and the maintenance of neurovascular coupling.

5. VASCULAR PATHOLOGY CAUSED BY BK CHANNEL DYSREGULATION

Given the ubiquitous presence of BK channels in vascular SMCs, their primary role in the regulation of myogenic response and the various factors that regulate BK channel function, it is evident that any alteration, reduction or inhibition of the channel will lead to pathology and disease. BK channel dysfunction may introduce a pathological state or an existing underlying pathology may lead to BK channel dysfunction as either a compensatory or aggravating mechanism. Recent technological advances and better pharmacological and genetic tools have enabled us to gain insight into the relative importance and primacy of BK channel in maintaining the optimum vascular myogenic tone and regulate blood flow and organ function.

5.1 Hypertension

BK channel dysfunction has been linked with hypertension, underscoring the importance of this molecular entity in the regulation of PVR. In some models of hypertension, BK channel currents are higher in SMCs of arteries from hypertensive models compared to those from normotensive controls, indicating an upregulation in the BK channel expression (Cox, 2002; Eichhorn & Dobrev, 2007). An overexpression of the BKβ1 subunit in an aging rat model (Chang, Wu, & Wang, 2006) and of the BKα subunit in spontaneously hypertensive rats (SHR) (Liu, Hudetz, Knaus, & Rusch, 1998) has also been reported. The BK expression in cerebral arteries from SHR had no change, however (Nishimaru, Eghbali, Stefani, & Toro, 2004). These reports are consistent with the idea of BK channels overcompensating to counter the increased depolarization and VDCC activity in the SMCs due to the rise in blood pressure. However, other hypertensive models have reported a decrease in BK channel function, which may be associated with downregulation of the expression of the BKβ1 subunit in SHR arteries (Amberg, Bonev, Rossow, Nelson, & Santana, 2003) and in angiotensin-II-dependent hypertensive rats (Amberg et al., 2003) or a decrease in the expression of the BKα subunit in L-nitro-arginine-hypertensive rats (Bratz, Dick, Partridge, & Kanagy, 2005) or a decrease in expression of both BKα and BKβ1 subunits in aging rats (Marijic et al., 2001; Nishimaru et al., 2004).

These observations are consistent with the findings from the $BK\alpha^{-/-}$ mouse model, which showed a moderate increase in blood pressure (Sausbier et al., 2005) and the $BK\beta1^{-/-}$ mouse model, which reported significantly higher blood pressure and a greater contractile response in isolated pressurized arteries and in isolated aortic rings (Brenner et al., 2000; Pluger et al., 2000) compared to wild-type controls. The increased contractile responses in the $BK\beta1^{-/-}$ mice may be attributed to the uncoupling of Ca^{2+} sparks to STOCs in SMCs of the arteries, in which the frequency of STOCs is significantly reduced in comparison to the normal amplitude and frequency of Ca^{2+} sparks (Brenner et al., 2000; Pluger et al., 2000). The loss of this significant hyperpolarizing influence would lead to SMC membrane depolarization and increase in intracellular Ca^{2+}. In a feed-forward mechanism, the increased SMC depolarization in $BK\beta1^{-/-}$ mice is found to increase the production of superoxide by NOX and endothelial dysfunction due to the diminished PKG-1 activity in these mice (Oelze et al., 2006), which would further depolarize SMC and enhance vasoconstriction. The higher blood pressure in the $BK\beta1^{-/-}$ mice is also linked with deficient K^+ handling in the cortical collecting duct, leading to increased plasma K^+ and mild aldosteronism and hypertension (Pluznick, Wei, Carmines, & Sansom, 2003). The slight increase in blood pressure in the

$BK\alpha^{-/-}$ mice is attributed to the complete abolition of STOCs, a reduced efficacy of the cGMP/PKG pathway as well as, similar to the $BK\beta1^{-/-}$ mice, to primary aldosteronism (Sausbier et al., 2005). These mice also experience erectile dysfunction due to the reduced efficacy of the cGMP/PKG pathway (Werner, Meredith, Aldrich, & Nelson, 2008; Werner, Zvara, Meredith, Aldrich, & Nelson, 2005). The seemingly opposite effects on BK channel protein expression in response to the same underlying pathology of hypertension may be attributed to several factors including differences in the genesis of hypertension in each model, differences in the vascular beds which were used in experimentation, influence of parallel mechanisms that are affected in a specific hypertensive model on the ultimate fate of BK channel function.

As evidence of a genetic predisposition toward innate preventive mechanism of BK channel function, a genetic epidemiological study identified a single-nucleotide polymorphism in the BKβ1 subunit gene, G352A, corresponding to an E65K mutation in the human BK channel protein. E65K resulted in a gain of function of the BK channel, exhibiting greater Ca^{2+} sensitivity and providing a more efficient negative feedback effect on SMC contractility (Fernandez-Fernandez et al., 2004). This augmentation in BK channel function is associated with a decrease in diastolic hypertension in aging women, thus providing protection against infarction and stroke (Senti et al., 2005).

5.2 Diabetes

BK channel dysfunction is observed and implicated in diabetes-related vascular disease. Lu, He, Katusic, and Lee (2006) found in human coronary SMCs that, in response to high glucose, BK channel function is impaired with reduced current density and slower channel activation and deactivation kinetics, which was also mimicked by H_2O_2, strongly suggesting that high glucose increases the production of superoxide and the formation of H_2O_2, which is known for its inhibition of BK channel. A mutation on the hSlo gene, C911A, however, may provide protection from the high glucose-induced effects of H_2O_2 (Lu et al., 2006).

Reduced expression of the BKα and/or BKβ1 subunit may also underlie reduced BK channel activity as seen in streptozotocin-treated rats and mice (Dong et al., 2008; McGahon, Zhang, Scholfield, Curtis, & McGeown, 2007). However, reduced protein expression was not the cause for reduced activity of BK channels in a Zucker diabetic rat model (Burnham,

Johnson, & Weston, 2006). In this case, it could be argued that BK channels, being less sensitive or "refractory" to superoxide, may provide a compensatory mechanism to counter the cardiovascular effects of diabetes (Liu & Gutterman, 2002).

A common pathology in people suffering from chronic diabetes is the onset and progression of diabetic retinopathy, where retinal vasoconstriction precedes structural changes. Thus, the effect of diabetes on retinal blood flow assumes significance. Similar to other vascular beds, the expression of BKβ1 subunit is reduced in retinal arterioles from insulin resistant rats (McGahon, Dash, et al., 2007; McGahon, Zhang, et al., 2007). Mori et al. (2011) found that ACh-induced vasodilation was repressed and was insensitive to BK channel inhibition in retinal arterioles from diabetic rats, suggesting that BK channel dysfunction may contribute to abnormal retinal hemodynamics in diabetes and the pathogenesis of diabetic retinopathy by enhancing retinal vasoconstriction and reduced retinal blood flow.

5.3 Therapeutic Potential for BK Channels

Despite overwhelming evidence suggesting a crucial role for BK channel in several cardiovascular pathologies, there is a dearth of promising clinical solutions targeting BK channels in humans, with the exception of a potential treatment for bronchial asthma and chronic obstructive pulmonary disease (Nardi & Olesen, 2008). The reasons for this may be manifold, not the least being the lack of specificity and potency to target BK channels in humans, made more difficult by the extensive splicing of the BKα subunit, to which is added another level of specificity in the form of association with none, one or more of the four BKβ1 or BKγ subunits. A second reason may be the difficulty posed by ubiquity of BK channel expression in every cell or tissue, which opens up the potential for any number of interplays between inhibitory or compensatory mechanisms within and between cells, tissues and organ systems, so as to render any specific therapy redundant or ineffective.

Regardless, existing small molecule therapies for other smooth muscle or vascular pathologies may already target BK channels, such as NO donors, PDE inhibitors, cilostazol, PGI_2 analogs, hydralazine, acetazolamide, and xenoestrogens (reviewed in Félétou, 2009). Drugs that may suppress BK channel function include verapamil or simvastatin. These therapies continue to provide opportunities to refine our understanding of how BK channel contributes to physiology and pathology. The first promising therapy targeting the human BK channel has been in the reversal of erectile dysfunction in humans by direct gene transfer (Melman, Bar-Chama, McCullough, Davies, & Christ, 2006). However, considering that vascular dysfunction is associated with marked changes in the expression of the BK channel protein, identifying and targeting specific pathways involved in the transcription and/or translation of the BK channel protein may yield better results. Such efforts may lead to the identification of common and universal pathways targeting several molecular entities involved in the progression of pathology.

REFERENCES

Ahn, Y. T., Kim, Y. M., Adams, E., Lyu, S. C., Alvira, C. M., & Cornfield, D. N. (2012). Hypoxia-inducible factor-1alpha regulates KCNMB1 expression in human pulmonary artery smooth muscle cells. *American Journal of Physiology. Lung Cellular and Molecular Physiology, 302*(3), L352–L359.

Ahn, D. S., Kim, Y. B., Lee, Y. H., Kang, B., & Kang, D. (1994). Fatty acids directly increase the activity of Ca^{2+}-activated K^+ channels in rabbit coronary smooth muscle cells. *Yonsei Medical Journal, 35*(1), 10–24.

Alonso-Galicia, M., Hudetz, A. G., Shen, H., Harder, D. R., & Roman, R. J. (1999). Contribution of 20-HETE to vasodilator actions of nitric oxide in the cerebral microcirculation. *Stroke. A Journal of Cerebral Circulation, 30*(12), 2727–2734. discussion 2734.

Amberg, G. C., Bonev, A. D., Rossow, C. F., Nelson, M. T., & Santana, L. F. (2003). Modulation of the molecular composition of large conductance, Ca^{2+} activated K^+ channels in vascular smooth muscle during hypertension. *Journal of Clinical Investigation, 112*(5), 717–724.

Archer, S. L., Gragasin, F. S., Wu, X., Wang, S., McMurtry, S., Kim, D. H., et al. (2003). Endothelium-derived hyperpolarizing factor in human internal mammary artery is 11,12-epoxyeicosatrienoic acid and causes relaxation by activating smooth muscle BK_{Ca} channels. *Circulation, 107*(5), 769–776.

Arnaudeau, S., Macrez-Leprêtre, N., & Mironneau, J. (1996). Activation of calcium sparks by angiotensin II in vascular myocytes. *Biochemical and Biophysical Research Communications, 222*(3), 809–815.

Arnold, W. P., Mittal, C. K., Katsuki, S., & Murad, F. (1977). Nitric oxide activates guanylate cyclase and increases guanosine 3′:5′-cyclic monophosphate levels in various tissue preparations. *Proceedings of the National Academy of Sciences of the United States of America, 74*(8), 3203–3207.

Barman, S. A., Zhu, S., & White, R. E. (2004). PKC activates BK_{Ca} channels in rat pulmonary arterial smooth muscle via cGMP-dependent protein kinase. *American Journal of Physiology. Lung Cellular and Molecular Physiology, 286*(6), L1275–L1281.

Bayliss, W. M. (1902). On the local reactions of the arterial wall to changes of internal pressure. *Journal of Physiology, 28*(3), 220–231.

Berridge, M. J., & Dupont, G. (1994). Spatial and temporal signalling by calcium. *Current Opinion in Cell Biology, 6*(2), 267–274.

Bianconi, E., Piovesan, A., Facchin, F., Beraudi, A., Casadei, R., Frabetti, F., et al. (2013). An estimation of the number of cells in the human body. *Annals of Human Biology, 40*(6), 463–471.

Bo, L., Jiang, L., Zhou, A., Wu, C., Li, J., Gao, Q., et al. (2015). Maternal high-salt diets affected pressor responses and microvasoconstriction via PKC/BK channel signaling pathways in rat offspring. *Molecular Nutrition & Food Research, 59*(6), 1190–1199.

Bol, M., Leybaert, L., & Vanheel, B. (2012). Influence of methanandamide and CGRP on potassium currents in smooth muscle cells of small mesenteric arteries. *Pflügers Archiv-European Journal of Physiology, 463*(5), 669–677.

Bonnet, S., & Archer, S. L. (2007). Potassium channel diversity in the pulmonary arteries and pulmonary veins: Implications for regulation of the pulmonary vasculature in health and during pulmonary hypertension. *Pharmacology and Therapeutics, 115*(1), 56–69.

Brakemeier, S., Eichler, I., Knorr, A., Fassheber, T., Köhler, R., & Hoyer, J. (2003). Modulation of Ca^{2+}-activated K^+ channel in renal artery endothelium in situ by nitric oxide and reactive oxygen species. *Kidney International, 64*(1), 199–207.

Bratz, I. N., Dick, G. M., Partridge, L. D., & Kanagy, N. L. (2005). Reduced molecular expression of K^+ channel proteins in vascular smooth muscle from rats made hypertensive with N{omega}-nitro-L-arginine. *American Journal of Physiology. Heart and Circulatory Physiology, 289*(3), H1277–H1283.

Brenner, R., Peréz, G. J., Bonev, A. D., Eckman, D. M., Kosek, J. C., Wiler, S. W., et al. (2000). Vasoregulation by the β1 subunit of the calcium-activated potassium channel. *Nature, 407*(6806), 870–876.

Broughton, B. R., Miller, A. A., & Sobey, C. G. (2010). Endothelium-dependent relaxation by G protein-coupled receptor 30 agonists in rat carotid arteries. *American Journal of Physiology. Heart and Circulatory Physiology, 298*(3), H1055–H1061.

Brzezinska, A. K., Gebremedhin, D., Chilian, W. M., Kalyanaraman, B., & Elliott, S. J. (2000). Peroxynitrite reversibly inhibits Ca^{2+}-activated K^+ channels in rat cerebral artery smooth muscle cells. *American Journal of Physiology. Heart and Circulatory Physiology, 278*(6), H1883–H1890.

Burnette, J. O., & White, R. E. (2006). PGI 2 opens potassium channels in retinal pericytes by cyclic AMP-stimulated, cross-activation of PKG. *Experimental Eye Research, 83*(6), 1359–1365.

Burnham, M. P., Johnson, I. T., & Weston, A. H. (2006). Reduced Ca^{2+}-dependent activation of large-conductance Ca^{2+}-activated K^+ channels from arteries of type 2 diabetic zucker diabetic fatty rats. *American Journal of Physiology. Heart and Circulatory Physiology, 290*(4), H1520–H1527.

Burton, A. C. (1954). Relation of structure to function of the tissues of the wall of blood vessels. *Physiological Reviews, 34*(6).

Chang, T., Wu, L., & Wang, R. (2006). Altered expression of BK channel beta1 subunit in vascular tissues from spontaneously hypertensive rats. *American Journal of Hypertension, 19*(7), 678–685.

Cheng, H., & Lederer, W. J. (2008). Calcium sparks. *Physiological Reviews, 88*(4), 1491–1545.

Cheng, H., Lederer, W. J., & Cannell, M. B. (1993). Calcium sparks: Elementary events underlying excitation-contraction coupling in heart muscle. *Science (New York, N.Y.), 262*(5134), 740–744.

Clapham, D. E. (1995). Calcium signaling. *Cell, 80*(2), 259–268.

Clarke, A. L., Petrou, S., Walsh, J. V., Jr., & Singer, J. J. (2002). Modulation of BK_{Ca} channel activity by fatty acids: Structural requirements and mechanism of action. *American Journal of Physiology. Cell Physiology, 283*(5), C1441–C1453.

Clarke, A. L., Petrou, S., Walsh, J. V., Jr., & Singer, J. J. (2003). Site of action of fatty acids and other charged lipids on BK_{Ca} channels from arterial smooth muscle cells. *American Journal of Physiology Cell Physiology, 284*(3), C607–C619.

Climent, B., Schubert, R., Stankevicius, E., García-Sacristán, A., Simonsen, U., & Rivera, L. (2012). Large conductance Ca^{2+}-activated K^+ channels modulate endothelial cell outward currents and nitric oxide release in the intact rat superior mesenteric artery. *Biochemical and Biophysical Research Communications, 417*(3), 1007–1013.

Contreras, G. F., Castillo, K., Enrique, N., Carrasquel-Ursulaez, W., Castillo, J. P., Milesi, V., et al. (2013). A BK (Slo1) channel journey from molecule to physiology. *Channels*, 7(6), 442–458.

Cox, R. H. (2002). Changes in the expression and function of arterial potassium channels during hypertension. *Vascular Pharmacology, 38*(1), 13–23.
Darkow, D. J., Lu, L., & White, R. E. (1997). Estrogen relaxation of coronary artery smooth muscle is mediated by nitric oxide and cGMP. *American Journal of Physics, 272*(6 Pt. 2), H2765–H2773.
Davis, M. J., & Hill, M. A. (1999). Signaling mechanisms underlying the vascular myogenic response. *Physiological Reviews, 79*(2), 387–423.
De Wet, H., Allen, M., Holmes, C., Stobbart, M., Lippiat, J. D., De Wet, H., et al. (2006). Modulation of the BK channel by estrogens: Examination at single channel level. *Molecular and Membrane Biology, 23*(5), 420–429.
Delp, M., & Laughlin, M. (1998). Regulation of skeletal muscle perfusion during exercise. *Acta Physiologica Scandinavica, 162*(3), 411–419.
Dimitropoulou, C., West, L., Field, M. B., White, R. E., Reddy, L. M., Falck, J. R., et al. (2007). Protein phosphatase 2A and Ca^{2+}-activated K^{+} channels contribute to 11, 12-epoxyeicosatrienoic acid analog mediated mesenteric arterial relaxation. *Prostaglandins & Other Lipid Mediators, 83*(1), 50–61.
Dong, L., Zheng, Y., Van Riper, D., Rathore, R., Liu, Q., Singer, H. A., et al. (2008). Functional and molecular evidence for impairment of calcium-activated potassium channels in type-1 diabetic cerebral artery smooth muscle cells. *Journal of Cerebral Blood Flow and Metabolism, 28*(2), 377–386.
Dopico, A. M., & Bukiya, A. N. (2014). Lipid regulation of BK channel function. *Frontiers in Physiology, 5*, 312.
Dopico, A. M., Kirber, M. T., Singer, J. J., & Walsh, J. V., Jr. (1994). Membrane stretch directly activates large conductance Ca^{2+}-activated K^{+} channels in mesenteric artery smooth muscle cells. *American Journal of Hypertension*, 7(1), 82–89.
Duling, B. R., & Klitzman, B. (1980). Local control of microvascular function: Role in tissue oxygen supply. *Annual Review of Physiology, 42*(1), 373–382.
Earley, S., Heppner, T. J., Nelson, M. T., & Brayden, J. E. (2005). TRPV4 forms a novel Ca^{2+} signaling complex with ryanodine receptors and BK_{Ca} channels. *Circulation Research, 97*(12), 1270–1279.
Eichhorn, B., & Dobrev, D. (2007). Vascular large conductance calcium-activated potassium channels: Functional role and therapeutic potential. *Naunyn-Schmiedeberg's Archives of Pharmacology, 376*(3), 145–155.
Evanson, K. W., Bannister, J. P., Leo, M. D., & Jaggar, J. H. (2014). LRRC26 is a functional BK channel auxiliary gamma subunit in arterial smooth muscle cells. *Circulation Research, 115*(4), 423–431.
Fan, F., Muroya, Y., & Roman, R. J. (2015). Cytochrome P450 eicosanoids in hypertension and renal disease. *Current Opinion in Nephrology and Hypertension, 24*(1), 37–46.
Félétou, M. (2009). Calcium-activated potassium channels and endothelial dysfunction: Therapeutic options? *British Journal of Pharmacology, 156*(4), 545–562.
Fernandez-Fernandez, J. M., Tomas, M., Vazquez, E., Orio, P., Latorre, R., Senti, M., et al. (2004). Gain-of-function mutation in the KCNMB1 potassium channel subunit is associated with low prevalence of diastolic hypertension. *Journal of Clinical Investigation, 113*(7), 1032–1039.
Filosa, J. A., Bonev, A. D., Straub, S. V., Meredith, A. L., Wilkerson, M. K., Aldrich, R. W., et al. (2006). Local potassium signaling couples neuronal activity to vasodilation in the brain. *Nature Neuroscience, 9*(11), 1397–1403.
Folkow, B. (1955). Nervous control of the blood vessels. *Physiological Reviews, 35*(3), 629.
Fukao, M., Mason, H. S., Kenyon, J. L., Horowitz, B., & Keef, K. D. (2001). Regulation of BK_{Ca} channels expressed in human embryonic kidney 293 cells by epoxyeicosatrienoic acid. *Molecular Pharmacology, 59*(1), 16–23.
Garcha, R. S., & Hughes, A. D. (1997). Action of ryanodine on neurogenic responses in rat isolated mesenteric small arteries. *British Journal of Pharmacology, 122*(1), 142–148.

Geary, G. G., Krause, D. N., & Duckles, S. P. (1998). Estrogen reduces myogenic tone through a nitric oxide-dependent mechanism in rat cerebral arteries. *American Journal of Physics*, *275*(1 Pt. 2), H292–H300.
Gebremedhin, D., Lange, A. R., Narayanan, J., Aebly, M. R., Jacobs, E. R., & Harder, D. R. (1998). Cat cerebral arterial smooth muscle cells express cytochrome P450 4A2 enzyme and produce the vasoconstrictor 20-HETE which enhances l-type Ca^{2+} current. *Journal of Physiology*, *507*(3), 771–781.
Girouard, H., Bonev, A. D., Hannah, R. M., Meredith, A., Aldrich, R. W., & Nelson, M. T. (2010). Astrocytic endfoot Ca^{2+} and BK channels determine both arteriolar dilation and constriction. *Proceedings of the National Academy of Sciences of the United States of America*, *107*(8), 3811–3816.
Gordon, G. R., Mulligan, S. J., & MacVicar, B. A. (2007). Astrocyte control of the cerebrovasculature. *Glia*, *55*(12), 1214–1221.
Guerard, P., Goirand, F., Fichet, N., Bernard, A., Rochette, L., Morcillo, E. J., et al. (2004). Arachidonic acid relaxes human pulmonary arteries through K^+ channels and nitric oxide pathways. *European Journal of Pharmacology*, *501*(1), 127–135.
Guyenet, P. G. (2006). The sympathetic control of blood pressure. *Nature Reviews Neuroscience*, 7(5), 335–346.
Haddock, R. E., & Hill, C. E. (2011). Sympathetic overdrive in obesity involves purinergic hyperactivity in the resistance vasculature. *The Journal of Physiology*, *589*(13), 3289–3307.
Haddy, F., Overbeck, H., & Daugherty, R. (1968). Peripheral vascular resistance. *Annual Review of Medicine*, *19*(1), 167–194.
Han, G., Yu, X., Lu, L., Li, S., Ma, H., Zhu, S., et al. (2006). Estrogen receptor alpha mediates acute potassium channel stimulation in human coronary artery smooth muscle cells. *The Journal of Pharmacology and Experimental Therapeutics*, *316*(3), 1025–1030.
Harder, D. R., Gebremedhin, D., Narayanan, J., Jefcoat, C., Falck, J. R., Campbell, W. B., et al. (1994). Formation and action of a P-450 4A metabolite of arachidonic acid in cat cerebral microvessels. *The American Journal of Physics*, *266*(5 Pt. 2), H2098–H2107.
Hathaway, D. R., March, K. L., Lash, J. A., Adam, L. P., & Wilensky, R. L. (1991). Vascular smooth muscle. A review of the molecular basis of contractility. *Circulation*, *83*(2), 382–390.
Hayabuchi, Y., Nakaya, Y., Matsuoka, S., & Kuroda, Y. (1998). Hydrogen peroxide-induced vascular relaxation in porcine coronary arteries is mediated by Ca^{2+}-activated K channels. *Heart and Vessels*, *13*(1), 9–17.
Hendry, C., Farley, A., & McLafferty, E. (2012). Blood vessels, circulation and blood pressure. *Nursing Standard (Royal College of Nursing (Great Britain): 1987)*, *27*(11), 35–40.
Hercule, H. C., Salanova, B., Essin, K., Honeck, H., Falck, J. R., Sausbier, M., et al. (2007). The vasodilator 17, 8-epoxyeicosatetraenoic acid targets the pore-forming BK α channel subunit in rodents. *Experimental Physiology*, *92*(6), 1067–1076.
Hill-Eubanks, D. C., Werner, M. E., Heppner, T. J., & Nelson, M. T. (2011). Calcium signaling in smooth muscle. *Cold Spring Harbor Perspectives in Biology*, *3*(9), a004549.
Hoshi, T., Pantazis, A., & Olcese, R. (2013). Transduction of voltage and Ca^{2+} signals by Slo1 BK channels. *Physiology (Bethesda, Md.)*, *28*(3), 172–189.
Hoshi, T., Wissuwa, B., Tian, Y., Tajima, N., Xu, R., Bauer, M., et al. (2013). Omega-3 fatty acids lower blood pressure by directly activating large-conductance ca^{2+}-dependent K^+ channels. *Proceedings of the National Academy of Sciences of the United States of America*, *110*(12), 4816–4821.
Hu, X. Q., Xiao, D., Zhu, R., Huang, X., Yang, S., Wilson, S., et al. (2011). Pregnancy upregulates large-conductance Ca^{2+}-activated K^+ channel activity and attenuates myogenic tone in uterine arteries. *Hypertension*, *58*(6), 1132–1139.
Hughes, J. M., Riddle, M. A., Paffett, M. L., Gonzalez Bosc, L. V., & Walker, B. R. (2010). Novel role of endothelial BK_{Ca} channels in altered vasoreactivity following hypoxia. *American Journal of Physiology. Heart and Circulatory Physiology*, *299*(5), H1439–H1450.

Iadecola, C. (2004). Neurovascular regulation in the normal brain and in Alzheimer's disease. *Nature Reviews Neuroscience*, *5*(5), 347–360.
Iadecola, C., & Nedergaard, M. (2007). Glial regulation of the cerebral microvasculature. *Nature Neuroscience*, *10*(11), 1369–1376.
Imig, J. D., & Deichmann, P. C. (1997). Afferent arteriolar responses to ANG II involve activation of PLA2 and modulation by lipoxygenase and P-450 pathways. *The American Journal of Physics*, *273*(2 Pt. 2), F274–F282.
Imig, J. D., Zou, A. P., Stec, D. E., Harder, D. R., Falck, J. R., & Roman, R. J. (1996). Formation and actions of 20-hydroxyeicosatetraenoic acid in rat renal arterioles. *The American Journal of Physics*, *270*(1 Pt. 2), R217–R227.
Jackson-Weaver, O., Osmond, J. M., Riddle, M. A., Naik, J. S., Gonzalez Bosc, L. V., Walker, B. R., et al. (2013). Hydrogen sulfide dilates rat mesenteric arteries by activating endothelial large-conductance Ca^{2+}-activated K^+ channels and smooth muscle Ca^{2+} sparks. *American Journal of Physiology. Heart and Circulatory Physiology*, *304*(11), H1446–H1454.
Jaggar, J. H., Porter, V. A., Lederer, W. J., & Nelson, M. T. (2000). Calcium sparks in smooth muscle. *American Journal of Physiology Cell Physiology*, *278*(2), C235–C256.
Jaggar, J. H., Stevenson, A. S., & Nelson, M. T. (1998). Voltage dependence of Ca^{2+} sparks in intact cerebral arteries. *The American Journal of Physics*, *274*(6 Pt. 1), C1755–C1761.
Karaki, H., Ozaki, H., Hori, M., Mitsui-Saito, M., Amano, K., Harada, K., et al. (1997). Calcium movements, distribution, and functions in smooth muscle. *Pharmacological Reviews*, *49*(2), 157–230.
Kim, C. J., Weir, B. K., Macdonald, R. L., & Zhang, H. (1998). Erythrocyte lysate releases Ca^{2+} from IP3-sensitive stores and activates Ca^{2+}-dependent K^+ channels in rat basilar smooth muscle cells. *Neurological Research*, *20*(1), 23–30.
Kirber, M. T., Ordway, R. W., Clapp, L. H., Walsh, J. V., & Singer, J. J. (1992). Both membrane stretch and fatty acids directly activate large conductance Ca^{2+}-activated K^+ channels in vascular smooth muscle cells. *FEBS Letters*, *297*(1), 24–28.
Kizub, I. V., Pavlova, O. O., Ivanova, I. V., & Soloviev, A. I. (2010). Protein kinase C-dependent inhibition of BK_{Ca} current in rat aorta smooth muscle cells following γ-irradiation. *International Journal of Radiation Biology*, *86*(4), 291–299.
Klein, M., Cheng, H., Santana, L., Jiang, Y., Lederer, W., & Schneider, M. (1996). Two mechanisms of quantized calcium release in skeletal muscle. *Nature*, *379*(6564), 455–458.
Knot, H. J., & Nelson, M. T. (1998). Regulation of arterial diameter and wall $[Ca^{2+}]$ in cerebral arteries of rat by membrane potential and intravascular pressure. *The Journal of Physiology*, *508*(1), 199–209.
Knot, H. J., Standen, N. B., & Nelson, M. T. (1998). Ryanodine receptors regulate arterial diameter and wall $[Ca^{2+}]$ in cerebral arteries of rat via Ca^{2+}-dependent K^+ channels. *The Journal of Physiology*, *508*(1), 211–221.
Knot, H. J., Zimmermann, P. A., & Nelson, M. T. (1996). Extracellular K^+-induced hyperpolarizations and dilatations of rat coronary and cerebral arteries involve inward rectifier K^+ channels. *The Journal of Physiology*, *492*(Pt. 2), 419–430.
Koehler, R. C., Roman, R. J., & Harder, D. R. (2009). Astrocytes and the regulation of cerebral blood flow. *Trends in Neurosciences*, *32*(3), 160–169.
Kohler, R., Degenhardt, C., Kühn, M., Runkel, N., Paul, M., & Hoyer, J. (2000). Expression and function of endothelial Ca^{2+}-activated K^+ channels in human mesenteric artery: A single-cell reverse transcriptase-polymerase chain reaction and electrophysiological study in situ. *Circulation Research*, *87*(6), 496–503.
Koide, M., Bonev, A. D., Nelson, M. T., & Wellman, G. C. (2012). Inversion of neurovascular coupling by subarachnoid blood depends on large-conductance Ca^{2+}-activated

K^+ (BK) channels. *Proceedings of the National Academy of Sciences of the United States of America, 109*(21), E1387–E1395.
Koide, M., Sukhotinsky, I., Ayata, C., & Wellman, G. C. (2013). Subarachnoid hemorrhage, spreading depolarizations and impaired neurovascular coupling. *Stroke Research and Treatment, 2013*, 819340.
Krishnamoorthy, G., Regehr, K., Berge, S., Scherer, E. Q., & Wangemann, P. (2011). Calcium sparks in the intact gerbil spiral modiolar artery. *BMC Physiology, 11*. 15-6793-11-15.
Krishnamoorthy, G., Sonkusare, S. K., Heppner, T. J., & Nelson, M. T. (2014). Opposing roles of smooth muscle BK channels and ryanodine receptors in the regulation of nerve-evoked constriction of mesenteric resistance arteries. *American Journal of Physiology. Heart and Circulatory Physiology, 306*(7), H981–H988.
Kume, H., Takai, A., Tokuno, H., & Tomita, T. (1989). Regulation of Ca^{2+}-dependent K^+-channel activity in tracheal myocytes by phosphorylation. *Nature, 341*(6238), 152–154.
Lamont, C., Vainorius, E., & Wier, W. G. (2003). Purinergic and adrenergic Ca^{2+} transients during neurogenic contractions of rat mesenteric small arteries. *The Journal of Physiology, 549*(3), 801–808.
Lamont, C., Vial, C., Evans, R. J., & Wier, W. G. (2006). P2X1 receptors mediate sympathetic postjunctional Ca^{2+} transients in mesenteric small arteries. *American Journal of Physiology. Heart and Circulatory Physiology, 291*(6), H3106–H3113.
Lamont, C., & Wier, W. G. (2002). Evoked and spontaneous purinergic junctional Ca^{2+} transients (jCaTs) in rat small arteries. *Circulation Research, 91*(6), 454–456.
Lange, A., Gebremedhin, D., Narayanan, J., & Harder, D. (1997). 20-Hydroxyeicosatetraenoic acid-induced vasoconstriction and inhibition of potassium current in cerebral vascular smooth muscle is dependent on activation of protein kinase C. *Journal of Biological Chemistry, 272*(43), 27345–27352.
Lanner, J. T., Georgiou, D. K., Joshi, A. D., & Hamilton, S. L. (2010). Ryanodine receptors: Structure, expression, molecular details, and function in calcium release. *Cold Spring Harbor Perspectives in Biology, 2*(11), a003996.
Ledoux, J., Bonev, A. D., & Nelson, M. T. (2008). Ca^{2+}-activated K^+ channels in murine endothelial cells: Block by intracellular calcium and magnesium. *The Journal of General Physiology, 131*(2), 125–135.
Leo, M. D., Bannister, J. P., Narayanan, D., Nair, A., Grubbs, J. E., Gabrick, K. S., et al. (2014). Dynamic regulation of beta1 subunit trafficking controls vascular contractility. *Proceedings of the National Academy of Sciences of the United States of America, 111*(6), 2361–2366.
Leo, M. D., Bulley, S., Bannister, J. P., Kuruvilla, K. P., Narayanan, D., & Jaggar, J. H. (2015). Angiotensin II stimulates internalization and degradation of arterial myocyte plasma membrane BK channels to induce vasoconstriction. *American Journal of Physiology. Cell Physiology, 309*(6), C392–C402.
Liu, Y., & Gutterman, D. D. (2002). The coronary circulation in diabetes: Influence of reactive oxygen species on K^+ channel-mediated vasodilation. *Vascular Pharmacology, 38*(1), 43–49.
Liu, M. Y., Hattori, Y., Sato, A., Ichikawa, R., Zhang, X. H., & Sakuma, I. (2002). Ovariectomy attenuates hyperpolarization and relaxation mediated by endothelium-derived hyperpolarizing factor in female rat mesenteric artery: A concomitant decrease in connexin-43 expression. *Journal of Cardiovascular Pharmacology, 40*(6), 938–948.
Liu, Y., Hudetz, A. G., Knaus, H. G., & Rusch, N. J. (1998). Increased expression of Ca^{2+}-sensitive K^+ channels in the cerebral microcirculation of genetically hypertensive rats: Evidence for their protection against cerebral vasospasm. *Circulation Research, 82*(6), 729–737.

Liu, B., Luo, W., Zhang, Y., Li, H., Zhu, N., Huang, D., et al. (2012). Role of cyclooxygenase-1-mediated prostacyclin synthesis in endothelium-dependent vasoconstrictor activity of porcine interlobular renal arteries. *American Journal of Physiology. Renal Physiology, 302*(9), F1133–F1140.

Liu, G., Niu, X., Wu, R. S., Chudasama, N., Yao, Y., Jin, X., et al. (2010). Location of modulatory beta subunits in BK potassium channels. *The Journal of General Physiology, 135*(5), 449–459.

Loot, A. E., Moneke, I., Keserü, B., Oelze, M., Syzonenko, T., Daiber, A., et al. (2012). 11,12-EET stimulates the association of BK channel α and β 1 subunits in mitochondria to induce pulmonary vasoconstriction. *PloS One*, 7(9), e46065.

Lu, T., Chai, Q., Yu, L., d'Uscio, L. V., Katusic, Z. S., He, T., et al. (2012). Reactive oxygen species signaling facilitates FOXO-3a/FBXO-dependent vascular BK channel beta1 subunit degradation in diabetic mice. *Diabetes, 61*(7), 1860–1868.

Lu, T., He, T., Katusic, Z. S., & Lee, H. C. (2006). Molecular mechanisms mediating inhibition of human large conductance Ca^{2+}-activated K^+ channels by high glucose. *Circulation Research, 99*(6), 607–616.

Lu, T., Katakam, P. V., VanRollins, M., Weintraub, N. L., Spector, A. A., & Lee, H. (2001). Dihydroxyeicosatrienoic acids are potent activators of Ca^{2+}-activated K^+ channels in isolated rat coronary arterial myocytes. *Journal of Physiology, 534*(3), 651–667.

Mancia, G., Di Rienzo, M., Parati, G., & Grassi, G. (1997). Sympathetic activity, blood pressure variability and end organ damage in hypertension. *Journal of Human Hypertension, 11*(Suppl. 1), S3–S8.

Mandalà, M., Heppner, T. J., Bonev, A. D., & Nelson, M. T. (2007). Effect of endogenous and exogenous nitric oxide on calcium sparks as targets for vasodilation in rat cerebral artery. *Nitric Oxide, 16*(1), 104–109.

Marchenko, S. M., & Sage, S. O. (1996). Calcium-activated potassium channels in the endothelium of intact rat aorta. *The Journal of Physiology, 492*(Pt. 1), 53–60.

Marijic, J., Li, Q., Song, M., Nishimaru, K., Stefani, E., & Toro, L. (2001). Decreased expression of voltage- and Ca^{2+}-activated K^+ channels in coronary smooth muscle during aging. *Circulation Research, 88*(2), 210–216.

Marino, M., Beny, J. L., Peyter, A. C., Diaceri, G., & Tolsa, J. F. (2011). Perinatal hypoxia enhances cyclic adenosine monophosphate-mediated BK_{Ca} channel activation in adult murine pulmonary artery. *Journal of Cardiovascular Pharmacology, 57*(2), 154–165.

Martín, P., Moncada, M., Enrique, N., Asuaje, A., Capuccino, J. M. V., Gonzalez, C., et al. (2014). Arachidonic acid activation of BK_{Ca} (Slo1) channels associated to the β1-subunit in human vascular smooth muscle cells. *Pflügers Archiv-European Journal of Physiology, 466*(9), 1779–1792.

McGahon, M. K., Dash, D. P., Arora, A., Wall, N., Dawicki, J., Simpson, D. A., et al. (2007). Diabetes downregulates large-conductance Ca^{2+}-activated potassium beta 1 channel subunit in retinal arteriolar smooth muscle. *Circulation Research, 100*(5), 703–711.

McGahon, M. K., Zhang, X., Scholfield, C. N., Curtis, T. M., & McGeown, J. G. (2007). Selective downregulation of the BKβ1 subunit in diabetic arteriolar myocytes. *Channels, 1*(3), 141–143.

McGiff, J. C., & Quilley, J. (1999). 20-HETE and the kidney: Resolution of old problems and new beginnings. *The American Journal of Physics, 277*(3 Pt. 2), R607–R623.

Melman, A., Bar-Chama, N., McCullough, A., Davies, K., & Christ, G. (2006). H maxi-K gene transfer in males with erectile dysfunction: Results of the first human trial. *Human Gene Therapy, 17*(12), 1165–1176.

Minami, K., Fukuzawa, K., Nakaya, Y., Zeng, X., & Inoue, I. (1993). Mechanism of activation of the Ca^{2+}-activated K^+ channel by cyclic AMP in cultured porcine coronary artery smooth muscle cells. *Life Sciences, 53*(14), 1129–1135.

Miriel, V. A., Mauban, J. R., Blaustein, M. P., & Gil Wier, W. (1999). Local and cellular Ca^{2+} transients in smooth muscle of pressurized rat resistance arteries during myogenic and agonist stimulation. *The Journal of Physiology, 518*(3), 815–824.

Miyata, N., & Roman, R. J. (2005). Role of 20-hydroxyeicosatetraenoic acid (20-HETE) in vascular system. *Journal of Smooth Muscle Research, 41*(4), 175–193.

Mizuno, Y., Isotani, E., Huang, J., Ding, H., Stull, J. T., & Kamm, K. E. (2008). Myosin light chain kinase activation and calcium sensitization in smooth muscle in vivo. *American Journal of Physiology. Cell Physiology, 295*(2), C358–C364.

Moncada, S., Palmer, R. M., & Higgs, E. A. (1991). Nitric oxide: Physiology, pathophysiology, and pharmacology. *Pharmacological Reviews, 43*(2), 109–142.

Mori, A., Suzuki, S., Sakamoto, K., Nakahara, T., & Ishii, K. (2011). Role of calcium-activated potassium channels in acetylcholine-induced vasodilation of rat retinal arterioles in vivo. *Naunyn-Schmiedeberg's Archives of Pharmacology, 383*(1), 27–34.

Morin, C., Sirois, M., Echave, V., Rizcallah, E., & Rousseau, E. (2009). Relaxing effects of 17(18)-EpETE on arterial and airway smooth muscles in human lung. *American Journal of Physiology. Lung Cellular and Molecular Physiology, 296*(1), L130–L139.

Morrow, J. P., Zakharov, S. I., Liu, G., Yang, L., Sok, A. J., & Marx, S. O. (2006). Defining the BK channel domains required for beta1-subunit modulation. *Proceedings of the National Academy of Sciences of the United States of America, 103*(13), 5096–5101.

Muramatsu, I. (1987). The effect of reserpine on sympathetic, purinergic neurotransmission in the isolated mesenteric artery of the dog: A pharmacological study. *British Journal of Pharmacology, 91*(3), 467–474.

Muramatsu, I., Ohmura, T., & Oshita, M. (1989). Comparison between sympathetic adrenergic and purinergic transmission in the dog mesenteric artery. *The Journal of Physiology, 411*, 227–243.

Nagar, D., Liu, X. T., & Rosenfeld, C. R. (2005). Estrogen regulates {beta}1-subunit expression in Ca^{2+}-activated K^+ channels in arteries from reproductive tissues. *American Journal of Physiology. Heart and Circulatory Physiology, 289*(4), H1417–H1427.

Nagelhus, E. A., Horio, Y., Inanobe, A., Fujita, A., Haug, F., Nielsen, S., et al. (1999). Immunogold evidence suggests that coupling of K siphoning and water transport in rat retinal Müller cells is mediated by a coenrichment of Kir4. 1 and AQP4 in specific membrane domains. *Glia, 26*(1), 47–54.

Nardi, A., & Olesen, S. (2008). BK channel modulators: A comprehensive overview. *Current Medicinal Chemistry, 15*(11), 1126–1146.

Nelson, M. T., Cheng, H., Rubart, M., Santana, L. F., Bonev, A. D., Knot, H. J., et al. (1995). Relaxation of arterial smooth muscle by calcium sparks. *Science (New York, N.Y.), 270*(5236), 633–637.

Neylon, C. B., Richards, S. M., Larsen, M. A., Agrotis, A., & Bobik, A. (1995). Multiple types of ryanodine receptor/Ca^{2+} release channels are expressed in vascular smooth muscle. *Biochemical and Biophysical Research Communications, 215*(3), 814–821.

Nishimaru, K., Eghbali, M., Stefani, E., & Toro, L. (2004). Function and clustered expression of MaxiK channels in cerebral myocytes remain intact with aging. *Experimental Gerontology, 39*(5), 831–839.

Oelze, M., Warnholtz, A., Faulhaber, J., Wenzel, P., Kleschyov, A. L., Coldewey, M., et al. (2006). NADPH oxidase accounts for enhanced superoxide production and impaired endothelium-dependent smooth muscle relaxation in BKbeta1$^{-/-}$ mice. *Arteriosclerosis, Thrombosis, and Vascular Biology, 26*(8), 1753–1759.

Olsson, R. (1981). Local factors regulating cardiac and skeletal muscle blood flow. *Annual Review of Physiology, 43*(1), 385–395.

Orio, P., Rojas, P., Ferreira, G., & Latorre, R. (2002). New disguises for an old channel: MaxiK channel beta-subunits. *News in Physiological Sciences: An International Journal of Physiology Produced Jointly by the International Union of Physiological Sciences and the American Physiological Society, 17*, 156–161.

Osol, G., Brekke, J. F., McElroy-Yaggy, K., & Gokina, N. I. (2002). Myogenic tone, reactivity, and forced dilatation: A three-phase model of in vitro arterial myogenic behavior. *American Journal of Physiology. Heart and Circulatory Physiology, 283*(6), H2260–H2267.

Oyekan, A., Youseff, T., Fulton, D., Quilley, J., & McGiff, J. (1999). Renal cytochrome P450 ω-hydroxylase and epoxygenase activity are differentially modified by nitric oxide and sodium chloride. *Journal of Clinical Investigation, 104*(8), 1131.

Paisansathan, C., Xu, H., Vetri, F., Hernandez, M., & Pelligrino, D. A. (2010). Interactions between adenosine and K^+ channel-related pathways in the coupling of somatosensory activation and pial arteriolar dilation. *American Journal of Physiology. Heart and Circulatory Physiology, 299*(6), H2009–H2017.

Papassotiriou, J., Kohler, R., Prenen, J., Krause, H., Akbar, M., Eggermont, J., et al. (2000). Endothelial K^+ channel lacks the Ca^{2+} sensitivity-regulating beta subunit. *FASEB Journal: Official Publication of the Federation of American Societies for Experimental Biology, 14*(7), 885–894.

Pappas, A. C., Koide, M., & Wellman, G. C. (2015). Astrocyte Ca^{2+} signaling drives inversion of neurovascular coupling after subarachnoid hemorrhage. *The Journal of Neuroscience: The Official Journal of the Society for Neuroscience, 35*(39), 13375–13384.

Paterno, R., Heistad, D. D., & Faraci, F. M. (2000). Potassium channels modulate cerebral autoregulation during acute hypertension. *American Journal of Physiology. Heart and Circulatory Physiology, 278*(6), H2003–H2007.

Pfeifer, A., Ruth, P., Dostmann, W., Sausbier, M., Klatt, P., & Hofmann, F. (1999). Structure and function of cGMP-dependent protein kinases. *Reviews of physiology, biochemistry and pharmacology: Vol. 135.* (pp. 105–149). New York and Berlin, Germany: Springer-Verlag.

Pluger, S., Faulhaber, J., Furstenau, M., Lohn, M., Waldschutz, R., Gollasch, M., et al. (2000). Mice with disrupted BK channel beta1 subunit gene feature abnormal Ca^{2+} spark/STOC coupling and elevated blood pressure. *Circulation Research, 87*(11), E53–E60.

Pluznick, J. L., Wei, P., Carmines, P. K., & Sansom, S. C. (2003). Renal fluid and electrolyte handling in BK_{Ca}-beta1$^{-/-}$ mice. *American Journal of Physiology. Renal Physiology, 284*(6), F1274–F1279.

Porter, V. A., Bonev, A. D., Knot, H. J., Heppner, T. J., Stevenson, A. S., Kleppisch, T., et al. (1998). Frequency modulation of Ca^{2+} sparks is involved in regulation of arterial diameter by cyclic nucleotides. *The American Journal of Physics, 274*(5 Pt. 1), C1346–C1355.

Price, D. L., Ludwig, J. W., Mi, H., Schwarz, T. L., & Ellisman, M. H. (2002). Distribution of rSlo Ca^{2+}-activated K^+ channels in rat astrocyte perivascular endfeet. *Brain Research, 956*(2), 183–193.

Prieto, D., Rivera, L., Benedito, S., Recio, P., Villalba, N., Hernández, M., et al. (2006). Ca^{2+}-activated K^+ (K_{Ca}) channels are involved in the relaxations elicited by sildenafil in penile resistance arteries. *European Journal of Pharmacology, 531*(1), 232–237.

Resnik, E., Herron, J., Fu, R., Ivy, D. D., & Cornfield, D. N. (2006). Oxygen tension modulates the expression of pulmonary vascular BK_{Ca} channel alpha- and beta-subunits. *American Journal of Physiology. Lung Cellular and Molecular Physiology, 290*(4), L761–L768.

Riddle, M. A., Hughes, J. M., & Walker, B. R. (2011). Role of caveolin-1 in endothelial BK_{Ca} channel regulation of vasoreactivity. *American Journal of Physiology. Cell Physiology, 301*(6), C1404–C1414.

Robertson, B. E., Schubert, R., Hescheler, J., & Nelson, M. T. (1993). cGMP-dependent protein kinase activates Ca^{2+}-activated K^+ channels in cerebral artery smooth muscle cells. *American Journal of Physics, 265*(1 Pt. 1), C299–C303.

Roman, R. J., Maier, K. G., Sun, C., Harder, D. R., & Alonso-Galicia, M. (2000). Renal and cardiovascular actions of 20-hydroxyeicosatetraenoic acid and epoxyeicosatrienoic acids. *Clinical and Experimental Pharmacology and Physiology, 27*(11), 855–865.

Rosenfeld, C. R., Cornfield, D. N., & Roy, T. (2001). Ca^{2+}-activated K^+ channels modulate basal and E_2beta-induced rises in uterine blood flow in ovine pregnancy. *American Journal of Physiology. Heart and Circulatory Physiology, 281*(1), H422–H431.
Rosenfeld, C. R., White, R. E., Roy, T., & Cox, B. E. (2000). Calcium-activated potassium channels and nitric oxide coregulate estrogen-induced vasodilation. *American Journal of Physiology. Heart and Circulatory Physiology, 279*(1), H319–H328.
Roy, C. S., & Sherrington, C. S. (1890). On the regulation of the blood-supply of the brain. *The Journal of Physiology, 11*(1-2), 85–158.
Sánchez, A., Villalba, N., Martínez, A. C., García-Sacristán, A., Hernández, M., & Prieto, D. (2008). Mechanisms of the relaxant effect of vardenafil in rat penile arteries. *European Journal of Pharmacology, 586*(1), 283–287.
Sausbier, M., Arntz, C., Bucurenciu, I., Zhao, H., Zhou, X. B., Sausbier, U., et al. (2005). Elevated blood pressure linked to primary hyperaldosteronism and impaired vasodilation in BK channel-deficient mice. *Circulation, 112*(1), 60–68.
Schubert, R., & Mulvany, M. J. (1999). The myogenic response: Established facts and attractive hypotheses. *Clinical Science, 96*, 313–326.
Schubert, R., & Nelson, M. T. (2001). Protein kinases: Tuners of the BK_{Ca} channel in smooth muscle. *Trends in Pharmacological Sciences, 22*(10), 505–512.
Schubert, R., Noack, T., & Serebryakov, V. N. (1999). Protein kinase C reduces the K_{Ca} current of rat tail artery smooth muscle cells. *American Journal of Physics, 276*(3 Pt. 1), C648–C658.
Senti, M., Fernandez-Fernandez, J. M., Tomas, M., Vazquez, E., Elosua, R., Marrugat, J., et al. (2005). Protective effect of the KCNMB1 E65K genetic polymorphism against diastolic hypertension in aging women and its relevance to cardiovascular risk. *Circulation Research, 97*(12), 1360–1365.
Somlyo, A. P., & Somlyo, A. V. (2003). Ca^{2+} sensitivity of smooth muscle and nonmuscle myosin II: Modulated by G proteins, kinases, and myosin phosphatase. *Physiological Reviews, 83*(4), 1325–1358.
Spector, A. A. (2009). Arachidonic acid cytochrome P450 epoxygenase pathway. *Journal of Lipid Research, 50*(Suppl.), S52–S56.
Stenmark, K. R., Yeager, M. E., El Kasmi, K. C., Nozik-Grayck, E., Gerasimovskaya, E. V., Li, M., et al. (2013). The adventitia: Essential regulator of vascular wall structure and function. *Annual Review of Physiology, 75*, 23–47.
Stull, J. T., Lin, P., Krueger, J. K., Trewhella, J., & Zhi, G. (1998). Myosin light chain kinase: Functional domains and structural motifs. *Acta Physiologica Scandinavica, 164*(4), 471–482.
Sun, C. W., Alonso-Galicia, M., Taheri, M. R., Falck, J. R., Harder, D. R., & Roman, R. J. (1998). Nitric oxide-20-hydroxyeicosatetraenoic acid interaction in the regulation of K^+ channel activity and vascular tone in renal arterioles. *Circulation Research, 83*(11), 1069–1079.
Sun, C. W., Falck, J. R., Harder, D. R., & Roman, R. J. (1999). Role of tyrosine kinase and PKC in the vasoconstrictor response to 20-HETE in renal arterioles. *Hypertension, 33*(1 Pt. 2), 414–418.
Sun, Z., Li, Z., & Meininger, G. A. (2012). Mechanotransduction through fibronectin-integrin focal adhesion in microvascular smooth muscle cells: Is calcium essential? *American Journal of Physiology. Heart and Circulatory Physiology, 302*(10), H1965–H1973.
Taguchi, K., Kaneko, K., & Kubo, T. (2000). Protein kinase C modulates Ca^{2+}-activated K^+ channels in cultured rat mesenteric artery smooth muscle cells. *Biological and Pharmaceutical Bulletin, 23*(12), 1450–1454.
Tanaka, Y., Yamaki, F., Koike, K., & Toro, L. (2004). New insights into the intracellular mechanisms by which PGI2 analogues elicit vascular relaxation: Cyclic AMP-independent, Gs-protein mediated-activation of MaxiK channel. *Current Medicinal Chemistry-Cardiovascular & Hematological Agents, 2*(3), 257–265.

Tang, X. D., Garcia, M. L., Heinemann, S. H., & Hoshi, T. (2004). Reactive oxygen species impair Slo1 BK channel function by altering cysteine-mediated calcium sensing. *Nature Structural and Molecular Biology*, *11*(2), 171–178.

Thengchaisri, N., & Kuo, L. (2003). Hydrogen peroxide induces endothelium-dependent and -independent coronary arteriolar dilation: Role of cyclooxygenase and potassium channels. *American Journal of Physiology. Heart and Circulatory Physiology*, *285*(6), H2255–H2263.

Tortora, G., & Derrickson, B. (2009). *Principles of anatomy and physiology: Organisation, support and movement and control systems of the human body*. Hoboken, NJ: John Wiley and Sons.

Valverde, M. A., Rojas, P., Amigo, J., Cosmelli, D., Orio, P., Bahamonde, M., et al. (1999). Acute activation of maxi-K channels (hSlo) by estradiol binding to the beta subunit. *Science (New York, NY)*, *285*(5435), 1929–1931.

Vetri, F., Xu, H., Paisansathan, C., & Pelligrino, D. A. (2012). Impairment of neurovascular coupling in type 1 diabetes mellitus in rats is linked to PKC modulation of BK_{Ca} and Kir channels. *American Journal of Physiology. Heart and Circulatory Physiology*, *302*(6), H1274–H1284.

Villalba, N., Stankevicius, E., Garcia-Sacristan, A., Simonsen, U., & Prieto, D. (2007). Contribution of both Ca^{2+} entry and Ca^{2+} sensitization to the alpha1-adrenergic vasoconstriction of rat penile small arteries. *American Journal of Physiology. Heart and Circulatory Physiology*, *292*(2), H1157–H1169.

Webb, R. C. (2003). Smooth muscle contraction and relaxation. *Advances in Physiology Education*, *27*(1-4), 201–206.

Wei, E. P., Kontos, H. A., & Beckman, J. S. (1996). Mechanisms of cerebral vasodilation by superoxide, hydrogen peroxide, and peroxynitrite. *The American Journal of Physics*, *271*(3 Pt. 2), H1262–H1266.

Wellman, G. C., Santana, L. F., Bonev, A. D., & Nelson, M. T. (2001). Role of phospholamban in the modulation of arterial Ca^{2+} sparks and Ca^{2+}-activated K^+ channels by cAMP. *American Journal of Physiology. Cell Physiology*, *281*(3), C1029–C1037.

Werner, M. E., Meredith, A. L., Aldrich, R. W., & Nelson, M. T. (2008). Hypercontractility and impaired sildenafil relaxations in the BK_{Ca} channel deletion model of erectile dysfunction. *American Journal of Physiology. Regulatory, Integrative and Comparative Physiology*, *295*(1), R181–R188.

Werner, M. E., Zvara, P., Meredith, A. L., Aldrich, R. W., & Nelson, M. T. (2005). Erectile dysfunction in mice lacking the large-conductance calcium-activated potassium (BK) channel. *The Journal of Physiology*, *567*(2), 545–556.

White, R. E., Darkow, D. J., & Lang, J. L. (1995). Estrogen relaxes coronary arteries by opening BK_{Ca} channels through a cGMP-dependent mechanism. *Circulation Research*, 77(5), 936–942.

White, R. E., Han, G., Maunz, M., Dimitropoulou, C., El-Mowafy, A. M., Barlow, R. S., et al. (2002). Endothelium-independent effect of estrogen on Ca^{2+}-activated K^+ channels in human coronary artery smooth muscle cells. *Cardiovascular Research*, *53*(3), 650–661.

Wong, S. L., Wong, W. T., Tian, X. Y., Lau, C. W., & Huang, Y. (2010). Prostaglandins in action: Indispensable roles of cyclooxygenase-1 and-2 in endothelium-dependent contractions. *Advances in Pharmacology*, *60*, 61.

Yan, J., & Aldrich, R. W. (2010). LRRC26 auxiliary protein allows BK channel activation at resting voltage without calcium. *Nature*, *466*(7305), 513–516.

Yang, Y., Li, P., Cheng, J., Cai, F., Lei, M., Tan, X., et al. (2013). IP3 decreases coronary artery tone via activating the BK_{Ca} channel of coronary artery smooth muscle cells in pigs. *Biochemical and Biophysical Research Communications*, *439*(3), 363–368.

Yang, X. R., Lin, M. J., Yip, K. P., Jeyakumar, L. H., Fleischer, S., Leung, G. P., et al. (2005). Multiple ryanodine receptor subtypes and heterogeneous ryanodine receptor-gated Ca^{2+} stores in pulmonary arterial smooth muscle cells. *American Journal of Physiology. Lung Cellular and Molecular Physiology, 289*(2), L338–L348.

Yang, C., Yang, J., Xu, X., Yan, S., Pan, S., Pan, X., et al. (2014). Vasodilatory effect of 14,15-epoxyeicosatrienoic acid on mesenteric arteries in hypertensive and aged rats. *Prostaglandins & Other Lipid Mediators, 112*, 1–8.

Ye, G. J., Nesmith, A. P., & Parker, K. K. (2014). The role of mechanotransduction on vascular smooth muscle myocytes cytoskeleton and contractile function. *The Anatomical Record, 297*(9), 1758–1769.

Yuan, X., Wang, J., Juhaszova, M., Gaine, S. P., & Rubin, L. J. (1998). Attenuated K^+ channel gene transcription in primary pulmonary hypertension. *Lancet, 351*(9104), 726–727.

Zhang, Y., Oltman, C. L., Lu, T., Lee, H. C., Dellsperger, K. C., & VanRollins, M. (2001). EET homologs potently dilate coronary microvessels and activate BK_{Ca} channels. *American Journal of Physiology. Heart and Circulatory Physiology, 280*(6), H2430–H2440.

Zhang, J., & Yan, J. (2014). Regulation of BK channels by auxiliary gamma subunits. *Frontiers in Physiology, 5*, 401. 10.3389.

Zhao, G., Neeb, Z. P., Leo, M. D., Pachuau, J., Adebiyi, A., Ouyang, K., et al. (2010). Type 1 IP3 receptors activate BK_{Ca} channels via local molecular coupling in arterial smooth muscle cells. *The Journal of General Physiology, 136*(3), 283–291.

Zhu, S., Han, G., & White, R. E. (2002). PGE2 action in human coronary artery smooth muscle: Role of potassium channels and signaling cross-talk. *Journal of Vascular Research, 39*(6), 477–488.

Zhu, R., Xiao, D., & Zhang, L. (2013). Potassium channels and uterine vascular adaptation to pregnancy and chronic hypoxia. *Current Vascular Pharmacology, 11*(5), 737–747.

Zou, A. P., Fleming, J. T., Falck, J. R., Jacobs, E. R., Gebremedhin, D., Harder, D. R., et al. (1996). 20-HETE is an endogenous inhibitor of the large-conductance Ca^{2+}-activated K^+ channel in renal arterioles. *The American Journal of Physics, 270*(1 Pt. 2), R228–R237.

CHAPTER TWELVE

Developing Molecular Pharmacology of BK Channels for Therapeutic Benefit

G.J. Kaczorowski[1], M.L. Garcia
Kanalis Consulting, L.L.C., Edison, NJ, United States
[1]Corresponding author: e-mail address: gkaczorowski@optonline.net

Contents

Abstract

High conductance, calcium-activated potassium (BK) channels (KCa1.1) are important in regulating physiologic responses in many types of tissues and, as such, present opportunities for development of new therapeutic agents. Both channel agonists and inhibitors could have therapeutic utility, depending on medical application under consideration. However, characterization of molecular pharmacology of BK channels is incomplete and has been difficult to accomplish because of paucity of chemical leads that are acceptable templates for Medicinal Chemistry investigation. Only through continued prosecution of new high-throughput screening campaigns can this situation be rectified. Examples are presented of BK channel agonist and inhibitor discovery paradigms which will be useful for progressing BK channel future drug discovery strategies.

1. POTENTIAL THERAPEUTIC BENEFITS OF TARGETING BK CHANNELS

BK channels are widely found in both electrically excitable and non-excitable tissues, and they play a critical role in coupling changes in cellular

International Review of Neurobiology, Volume 128
ISSN 0074-7742
http://dx.doi.org/10.1016/bs.irn.2016.02.013

membrane potential and/or intracellular Ca concentration to regulation of cell excitability, cellular signal transduction, and general cell homeostasis (Contreras et al., 2013). Additionally, it has been postulated that BK-like channels are expressed in intracellular organelles, such as mitochondria of cardiomyocytes (Singh et al., 2013), where they may have a protective function during cardiac ischemia. In many excitable cells, BK channels are localized in plasma membrane complexes with voltage-gated Ca channels; their activation via Ca entry causes cell hyperpolarization which prevents further Ca influx by closing proximal Ca channels, thereby providing a negative-feedback regulatory mechanism for Ca entry into the cell. In smooth muscle, proximity of BK channels to sarcoplasmic reticulum Ca release sites elicits the same type of negative-feedback regulatory loop. However, in nonexcitable cells, BK channels supply positive-feedback regulation for Ca entry by providing requisite counter ion (K) efflux, thereby stimulating Ca influx, or they can form a K flux pathway in epithelia that facilitates transmembrane flux of chloride along with water.

Because of their functional properties and extensive pattern of expression, BK channels regulate a wide range of processes in animals: they control smooth muscle contractility, neuronal excitability, neurotransmitter release, endocrine secretion, epithelial function, and activation/migration of nonexcitable cells like fibroblasts, among other putative physiologic roles (eg, in cell metastasis). Therefore, it is expected that direct pharmacological modulation of BK channels, either their activation or inhibition, would present a plethora of therapeutic opportunities (Bentzen, Olesen, Ronn, & Grunnet, 2014; Garcia, Shen, & Kaczorowski, 2007; Nardi & Olesen, 2008). However, given broad and complex influence which BK channels have on integrated system physiology, along with their wide tissue distribution, serious consideration must be given to health benefit vs potential safety liabilities, and subsequent therapeutic index, when developing a BK channel modulator for any potential medical indication. This can be much less of an issue in other situations of ion channel drug development, for example, when targeting a subtype member of a large super family of voltage-gated ion channels that has limited tissue distribution and a highly specialized, unique physiologic function (eg, Kir1.1 in the kidney; Garcia & Kaczorowski, 2014).

Development of BK channel agonists which target smooth muscle has historically been the most commonly explored drug discovery approach. Based on first principle considerations of smooth muscle physiology, stimulatory effect of BK channel inhibitors on smooth muscle contractility both ex vivo and in vivo, indirect activation of BK channels in some smooth muscle tissues by second messenger pathways (eg, coupling of beta-2 receptor stimulation to BK channel activation and subsequent relaxation of airway smooth muscle), phenotypes of mice in which various BK channel subunits have been genetically ablated, and human phenotypes where properties of components of native BK channel complexes are altered by mutation, it is predicted that activation of BK channels in this tissue type could result in therapeutic benefit(s) (for a comprehensive review, see Yang & Cui, 2015). BK channels directly influence contractile activity of several types of smooth muscle (eg, vascular, airway, uterine, and bladder), and given that activation of other vascular K channels (eg, by K-ATP channel agonists) is a proven pharmacological means to achieve smooth muscle relaxation, it is expected that a BK channel agonist would function similarly, but perhaps with a different and superior tissue selectivity profile. Therefore, indications for BK channel agonists exist in hypertension, erectile dysfunction (depending on the ability to target the appropriate vascular bed), asthma, dysmenorrhea, and overactive bladder.

In marked contrast, possible utility of activating BK channels in neuronal tissue as a therapeutic approach must be approached with caution (Brenner et al., 2005; Du et al., 2005). Human mutations that enhance BK activity can cause generalized epilepsy and paroxysmal dyskinesia, while BK channel gain-of-function mutations in rodents cause spontaneous seizures and temporal lobe epilepsy; manifestation of these potential side effects during use of a peripheral BK agonist might be avoided by limiting brain penetration/exposure, or by focusing on activation of specific BK channel molecular complexes with unique beta subunit compositions in target tissues of interest that are dissimilar from those of central nervous system (CNS) BK channels. Perhaps one beneficial utility of a neuronal BK agonist would be to activate channels in the peripheral nervous system (PNS) to reduce electrical activity of pain circuits and elicit analgesia; activation of neuronal K channels to block pain signaling has a pharmacological precedent with Kv7.2/Kv7.3 channel activators (eg, flupirtine, an analog of retigabine; Tsantoulas & McMahon, 2014) functioning in the PNS. Another potential application of a BK agonist is to activate channels in nephron segments of kidney to stimulate K secretion and thereby limit plasma hyperkalemia (Holtzclaw, Grimm, & Sansom, 2011). However, bringing any of these therapeutic approaches to fruition requires discovery of more robust pharmacological tools than currently exist, as outlined later.

Development of BK channel blockers could also be of therapeutic benefit in certain pathophysiologies, but a more cautious strategic approach is required than with a BK channel agonist, so as not to disrupt numerous normal physiologic functions of this channel (for a comprehensive review, see Yang & Cui, 2015). For example, systemic exposure of a BK channel inhibitor in plasma would cause smooth muscle hyperexcitability, potentially resulting in vasoconstriction (hypertension) and spontaneous urinary bladder activity (frequent and uncontrolled urination). Broad exposure in the CNS could result in impaired motor behavior and cerebellar ataxia (eg, staggers syndrome in animals is caused by ingestion of indole diterpene BK channel inhibitors; see later), disrupted circadian rhythm, generalized and temporal epilepsy (N'Gouemo, 2011), seizures, and potentially neuronal cell death in specific brain regions due to cellular Ca overload. Moreover, disruption of BK channel activity in kidney can result in diminution of flow-induced K secretion and hyperaldosteronism, also contributing to development of hypertension. Therefore, if BK channels are to be targeted with an inhibitor, care must be taken to prevent broad exposure of BK channel containing tissues to drug during dosing. To this end, that can be accomplished by application of an agent locally, directly onto the target, or synthesizing a small molecule modulator with metabolic liabilities to insure rapid drug metabolism and fast clearance, or using the drug acutely rather than chronically to achieve therapeutic benefit. Disease indications are described later where modulation of epithelial transport, treatment of a hypotensive crisis, or block of fibroblast activation/migration could be treated with a BK channel inhibitor, employing such outlined strategies.

2. ISSUES FACING THE BK CHANNEL AS A DRUG TARGET

Given the important role of BK channels in such a wide variety of tissues, it is rather remarkable that drug development efforts focused on this target have largely failed to date. There are a number of reasons for this. With respect to BK channel agonists, several issues account for this lack of progress. In the past, Bristol-Myers Squibb focused on developing BK channel activators as neuroprotective agents to treat stoke. The development compound identified, BMS-204,352 (MaxiPost), failed to demonstrate efficacy in a placebo controlled clinical trial with acute stroke patients, despite being active in animal models of focal ischemia-induced brain damage (Jensen, 2002). MaxiPost did not display adverse effects in human safety trials that would compromise therapeutic benefit, such as systemic blood pressure reduction or diminished cerebral blood flow, so it was concluded that the BK agonist mechanism was invalid for stroke therapy. We believe that there are other reasons for this negative clinical outcome. In our profiling of MaxiPost, we found it to be a weak BK channel agonist, requiring high concentrations (≥10 μM) to demonstrate significant activity. In addition, others have shown that MaxiPost displays significant off target activities, having interactions with other ion channel targets such as Kv7.X and GABA-A channels in the same concentration range as BK channel activation (Korsgaard et al., 2005). Thus, potency and selectivity of this drug candidate are issues. More importantly, focal ischemia animal models used to develop MaxiPost are notorious for their lack of translatability to human clinical stroke trials. There is a long history of compounds, representing many different mechanisms of action, being active in these animal models, but failing to demonstrate protection in treating acute stroke in patients; the acute stroke indication still represents a serious unmet medical need. Unfortunately, failure of this BK agonist program at that time decreased appetite of Pharma to invest in further BK agonist drug development efforts, but we believe that the BK channel activation hypothesis was never adequately tested in human subjects.

Another confounding factor in the agonist field is lack of specificity of many BK channel activators described in the literature (Bentzen et al., 2014). For example, the structural series exemplified by NS-1619 have significant Cav1.2 channel blocker activity. Since blocking L-type Ca channels also relaxes smooth muscle, it is difficult to differentiate cause of smooth muscle relaxation in vivo when both pharmacological activities are present in the same molecule. However, in vitro, a pure BK channel agonist will not relax smooth muscle contracted by high K depolarization, whereas a Ca entry blocker will; this simple experimental test can be used as a diagnostic of compound selectivity. The complete extent of polypharmacy which the NS series of BK channel agonists possesses is unclear. Similarly, other BK channel agonists have significant phosphodiesterase (PDE) inhibitory activity (see later); such enzyme inhibitors relax smooth muscle via a number of different mechanisms, besides K channel activation (Xin, Cheng, Soder, & Petkov, 2012), so these compounds cannot be used to obtain proof of concept (POC) for therapeutic utility of BK channel activation. PDE inhibitors also have pronounced effects on cardiac contractility. Therefore, identification of a prototypical, potent and selective, small molecule BK channel agonist is critically needed to support proof of principle studies for smooth muscle therapies of interest.

With respect to developing BK channel inhibitors, a number of different cardiovascular and CNS liabilities are likely to present when such compounds are dosed systemically in animals, presumptively diminishing Pharma's interest in this avenue of drug discovery. This is especially true with highly brain penetrant compounds that can cause ataxia, tremors, seizures, and potentially even mortality. Clearly, limiting systemic exposure is necessary for successful development of a BK channel blocker, and risk–benefit ratio must be rigorously assessed before initiating a project; based on precedent, industry is unlikely to be supportive of such efforts without a well-conceived strategy for execution.

Nonetheless, even if a decision were made to pursue a BK channel agonist or antagonist drug project, historically, the most significant limitation which prevents implementation is difficulty in identifying acceptable small molecule chemical leads to support the effort. The BK channel, at present, has a relatively poor molecular pharmacology. Except for previously identified peptide and natural product channel modulators (see later), there have been few structural templates representing BK channel small molecule activators or inhibitors that can provide an adequate substrate to sustain a Medicinal Chemistry synthetic program. Thus, construction of an appropriate HTS is the first, most critical element for any BK project, in order to deal with this problem. Once an acceptable, chemically tractable, small molecule lead is identified, one with good BK channel potency, limited ancillary pharmacology, and an appropriate mechanism of action, then Medicinal Chemistry can properly prosecute a drug development program. Contrary to a view often alluded to in literature, it is not feasible to rationally design ion channel drugs, or, for that matter, drugs directed against most other therapeutic targets. Lack of three-dimensional ion channel structural information, incomplete understanding of the dynamics of protein conformational changes that occur during channel gating, and having limited structural information on details of most drug–ion channel protein interactions make de novo drug design impossible at present. Therefore, finding an appropriate chemical lead is most critical; it is a product of hard work, patience, and most importantly, luck. By optimizing a mechanism-based HTS and interrogating large chemical libraries rigorously, probability of success is increased; this is a rational discovery approach (Kaczorowski, McManus, Priest, & Garcia, 2008). This review will focus on progress that has been made to date in identifying BK channel modulators using such a strategy.

To put material presented into proper context, it should be noted that rational- or molecular-based ion channel drug discovery in general is a very difficult task which has met with limited success since its inception 35 years ago, a time that coincides with development of the patch voltage clamp technique. The clinical failures of drug candidates targeting TRPV1 or Nav1.7 for pain, various cardiac ion channels for ventricular and atrial arrhythmias, and a suite of CNS ligand-gated ion channels for various neurology applications are examples of this reality. Even though ca. 15% of registered drugs function by targeting ion channels and membrane transporters, most of these agents were discovered/developed using in vivo animal pharmacology and phenotypic screening approaches (McManus, 2014). It is hoped that probability of success in future endeavors will increase by revisiting phenotypic assay formats, combined with new high-throughput functional ion channel screening and primary assay technologies, application of human genetics to guide better target selection, and nanopharmacology techniques to deliver ion channel drugs to their targets in a more efficient manner. Rationale supporting selection of ion channels, including BK, as therapeutic targets is very strong; it is execution of the effort where substantial improvement is needed.

3. NATURAL PRODUCT SCREENING AS A WAY TO DEVELOP MOLECULAR PHARMACOLOGY OF BK CHANNELS

Our group at Merck became interested in exploring BK channels as a therapeutic target in the mid-1980s. No potent or selective channel modulators were known at that time and only the modestly active blocker, tetraethylammonium ion (IC_{50} of 100 μM), was available for study. Newly described patch-clamp electrophysiology was a labor-intensive, arduous, and low capacity technique to measure ion channel activity; a maximum of only three to five compounds could be tested per week using this protocol. Thus, during this period, there was no high capacity, functional ion channel assay that could support a screening campaign. In order to develop the molecular pharmacology of BK channels, we first sought to identify a potent probe with which to establish a binding assay with purified plasma membrane vesicles containing a measurable amount of BK channel protein. In 1985, Miller and colleagues published the first paper identifying charybdotoxin (ChTX), described as a small 7 kDa protein, which they purified from venom of the Israeli scorpion, *Leiurus quinquestriatus* var. *hebraeus*, and which blocked skeletal muscle BK channels reconstituted in lipid bilayers in a potent (nM) and reversible fashion (Miller, Moczydlowski, Latorre, & Phillips, 1985). We independently purified ChTX to homogeneity from the same venom source using patch clamp of BK channels in GH3 pituitary cells as primary screening assay, characterized physical properties of the resulting 37 amino acid, 4.2 kDa peptide, sequenced ChTX to deduce its primary structure and made the peptide synthetically by solid state synthesis to confirm identity and biological activity of native peptide (Sugg, Garcia, Reuben, Patchett, & Kaczorowski, 1990). Fortunately, all these efforts were successfully accomplished because the purified ChTX preparation in the initial literature report was not homogeneous; it was a mixture of at least three peptides, all blocking different K channels, which was the basis for initial confusion in the field as to why two reported preparations behaved differently in terms of their bioactivity (Garcia, Garcia-Calvo, Hidalgo, Lee, & MacKinnon, 1994).

ChTX acts as a pore blocker to inhibit BK channels (MacKinnon & Miller, 1988). When ChTX was radiolabeled by iodination of Y_{36} with ^{125}I and shown to retain functional activity, a specific, high affinity, equilibrium binding assay (pM ligand affinity of [^{125}I]ChTX in low ionic strength buffer) was established with purified bovine aortic sarcolemmal plasma membrane vesicles, which contained a density of native BK channels that gave an acceptable signal to noise ratio in ligand binding experiments. This work thus identified the first potent probe with which to develop biochemistry and molecular pharmacology of vascular smooth muscle BK channels. Importantly, it provided an assay which allowed not only purification to homogeneity of functionally active BK channels from both bovine aorta and trachea smooth muscle (as demonstrated in bilayer recordings of reconstituted purified channels) but also discovery and molecular characterization of the first auxilliary subunit (beta-1) of the BK family of beta subunits (Garcia-Calvo et al., 1994). Purification of the ChTX receptor and demonstration of its functional BK channel activity validated the ChTX binding site characterized in vascular and airway smooth muscle sarcolemmal membranes as representing the BK channel.

Having established a reproducible and high capacity [^{125}I]ChTX binding protocol in aortic sarcolemmal membranes, this technique was used to screen for novel BK channel modulators. Initially, because ChTX is not a selective blocker of BK channels (it also inhibits KCa3.1, Kv1.2, Kv1.3, and at lower potency, Kv1.6 channels), ca. 100 venoms from different animal sources were used to test for other peptidyl blockers with potentially higher BK specificity. Only two crude venoms were found to inhibit the binding reaction to smooth muscle, but not binding of peptide to brain membranes which represents Kv1.X channels; activity from the Indian scorpion, *Buthus tamulus*, which was a minor component, was purified to homogeneity. Amino acid sequence of this 4.3 kDa peptide, termed iberiotoxin (IbTX), was determined, validated by solid state synthesis, and the peptide was shown to be a potent BK channel blocker (Galvez et al., 1990). IbTX is 68% homologous to ChTX. However, unlike ChTX, IbTX does not block any other ChTX-sensitive K channel and was subsequently shown to be a highly selective blocker of native BK channels containing beta-1 subunits. IbTX has been produced both synthetically and biosynthetically, labeled with ^{125}I for use as a high-affinity BK channel probe, and its pore blocking mechanism (similar to ChTX) has been well characterized. IbTX is currently widely used as a very selective BK channel peptidyl inhibitor by the field, but its blocking activity is modulated by beta subunits present in the BK channel complex, as for example, with beta-4 markedly diminishing BK channel blocking activity of IbTX (for a comprehensive review, see Torres, Granados, & Latorre, 2014).

Strategy of using [^{125}I]ChTX binding as a screen to identify modulators of BK channels relies on assumption that actives can affect the peptide–channel interaction competitively, allosterically, or noncompetitively, and that any functional activity of identified hits will need to be determined by patch clamp. The binding assay casts a broad net which in principle could identify channel inhibitors or activators, along with agents that nonspecifically interfere with the binding reaction. Since Merck historically had much commercial success in identifying drugs by screening natural products for chemical leads (eg, lovastatin (Mevacor®), ivermectin (Ivomec®), caspofungin (Cancidas®), and a number of antibiotics), [^{125}I]ChTX binding assay using aortic sarcolemmal membrane vesicles was prosecuted seriously and performed for an extended period (ca. 3 years) in both Rahway and at the high-throughput natural product screening laboratory in Madrid, Spain. A number of different sources were screened for leads: extracts of fermentation broths in which various types of microorganisms were grown; plant extracts; extracts of medical herbs obtained from traditional medicine sources in various countries; extracts of marine organisms grown in cultures. At the same time, 500 synthetic compounds from the proprietary, small molecule Merck sample collection were routinely screened per week (in all, ca. 100,000 small molecules were tested). The binding assay was robust and performed well, but the hit rate was extremely low, and despite intense effort, there were only a few interesting compounds discovered from natural product extracts, and none from the synthetic small molecule collection tested.

At beginning of the screening exercise, much excitement was generated when several fermentation broth extracts inhibited [^{125}I]ChTX binding, and also activated BK channels in single channel recordings from excised patches of GH3 cells, in a specific and reversible fashion. The enthusiasm was short lived, however, when structure of these purified activities was shown to be the fatty acid, linoleic acid. It is curious that systematic examination of related fatty acids demonstrated a defined structure–activity relationship (SAR) for agonist activity in terms of fatty acid alkyl chain length and position of unsaturation in the molecule; for example, linoleic acid functioned as a BK channel agonist at nM concentrations, while linolenic acid, a closely related structural analog with one more degree of unsaturation, was relatively inactive in these experiments. Interestingly, literature reports that application of linoleic acid will relax pressor-contracted aortic and coronary artery smooth muscle (Engler, 1992; Pomposiello, Alva, Wilde, & Carretero, 1998). Because ion channels are embedded in plasma membranes in a lipid annulus, modification of this structure by incorporation of a fatty acid could influence dynamics of protein conformational changes, thereby modifying gating behavior. Effects of lipids on BK channel function and the physiological role of such regulation remains an active area of investigation (Dopico & Bukiya, 2014). Although our discovery of activation of BK channels by fatty acids might have been one of the first such observations in the field, it has no obvious commercial application, so its mechanism was never pursued or formally reported. But this observation did serve as POC for utilizing [^{125}I]ChTX binding as a screen to detect BK channel agonists.

Several notable "basic" discoveries were made concerning BK channel activators and inhibitors from the natural product [^{125}I]ChTX binding campaign; two different structural classes of channel agonists and a family of potent indole alkaloid channel inhibitors were identified that represent major successes of this effort. BK channels play an important role in regulating airway smooth muscle tone and release of bronchoconstrictive agents from nerves in the lung; thus, BK channel agonists might be expected to relax airway smooth muscle and to treat neurogenic inflammation in asthma (Stretton, Miura, Belvisi, & Barnes, 1992; Valverde et al., 2011). During our screening of traditional medicines, we were conscious of searching for medicinal herb extracts that were used, for example, to treat asthma. Such a medicinal herb with this application in Ghana (and for treating dysmenorrhea, as well) is the plant, *Desmodium adscendens*, whose extracts were already known to inhibit electrically paced contraction of gut and airway smooth muscle. Based on this pharmacological profile, several compounds had previously been isolated from *D. adscendens* and shown to relax smooth muscle or possess antiinflammatory activity. Strikingly, when tested, this plant extract scored as a positive in the [^{125}I]ChTX binding assay, so inhibitory activity was fractionated, purified to homogeneity, and subjected to structural determination; the resulting active was a mixture of three known and structurally related triterpenoid glycosides (McManus et al., 1993), the most potent of which was dehydrosoyasaponin I (**I**; Fig. 1). Dehydrosoyasaponin I was a partial (maximal 60%) inhibitor of [^{125}I]ChTX binding, where it displayed a K_i of 120 nM; an analysis of equilibrium ligand binding inhibition revealed (**I**) affected both ligand K_d and B_{max} through an allosteric mechanism. More importantly, (**I**) increased open probability of single BK channels from bovine tracheal smooth muscle sarcolemmal membranes reconstituted in planar lipid bilayers; 500 nM (**I**) had no effect when added to the outside face of the channel, but 100 nM applied to the inner face increased channel open probability by 80-fold, in a

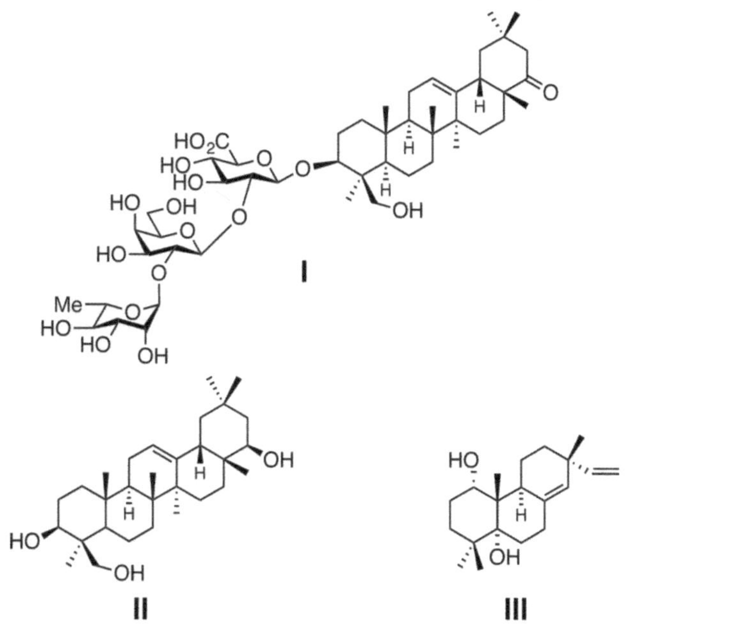

Fig. 1 Structures of natural product BK channel agonists.

concentration-dependent and reversible manner; threshold for the effect was ≤ 10 nM. Identical results regarding potency and sidedness of (**I**) action were obtained in single channel recordings from excised inside-out patches of bovine aortic smooth muscle cells. Mechanistic studies reveal that (**I**) does not substitute for Ca in activating BK channels; (**I**) requires presence of beta-1 to manifest agonist activity, as it has no effect on the alpha, pore-forming, subunit when tested on this entity, alone; three to four molecules of (**I**) bind to BK channels to elicit maximal effect (consistent with the 4:1 stoichiometry of beta-1 to alpha subunits in the smooth muscle channel complex); and (**I**) activates channels by shifting voltage and Ca dependence for BK channel opening so that channels open at more hyperpolarized potentials and lower Ca concentrations (Giangiacomo, Kamassah, Harris, & McManus, 1998; McManus et al., 1995). The selectivity of (**I**) was very good as it had no effect when tested at 10 μM on K-ATP, KCa3.1, Kv1.3, Cav1.2, Na–Ca exchange, or Na, K-ATPase activities.

Remarkably, SAR studies with this structural class demonstrate that both triterpene and carbohydrate moieties are critical for BK channel agonist activity, as aglycone triterpene molecules (eg, **II**; Fig. 1) are completely inactive at low μM concentrations (McManus et al., 1993). Carbohydrate chemistry is difficult to prosecute for drug development, so this series of glycotriterpenes was never considered as a program lead by medicinal chemists, despite their potency, selectivity, and mechanism of action. Moreover, these agents possessed no pharmacological activity in acute smooth muscle contractility experiments, as expected, because of their intracellular site of action and membrane impermeant nature of triterpene glycosides. However, it is known from its use in Ghana that this medicinal herb must be dosed chronically to elicit therapeutic benefit; for example, an alcoholic tea is brewed from leaves of the plant and drank for several days to treat uterine cramps. It is possible that if glycotriterpenes are active as BK agonists in vivo, they are transformed in liver into metabolites which are the active species that function in humans. This hypothesis was never pursued further and this class of glycotriterpenes remains only scientifically, but not therapeutically, interesting. However, this work is a representative example of the tenants of a rational drug discovery approach (Kaczorowski et al., 2008).

A structurally different, but perhaps related, BK channel modulator was identified from an extract of a fermentation broth in which an unidentified coelomycete fungus was cultured (Singh et al., 1994); this extract inhibited [^{125}I]ChTX binding incompletely, to a maximum of 90%, with a K_i of 1 μM. The bioassay-guided purification resulted in isolation and identification of a novel dihydroxyisoprimane, termed maxikdiol (**III**; Fig. 1). Scatchard analyses and demonstration that (**III**) enhances rate of [^{125}I]ChTX dissociation from channels indicate that this compound functions as an allosteric inhibitor of the binding reaction. A patch-clamp investigation of single channel activity in excised inside-out membrane patches from bovine aortic smooth muscle cells indicates that (**III**) activates BK channels in a reversible fashion; threshold for channel activation was ca. 1 μM, with dose-dependent stimulation of BK activity observed in the 3–10 μM concentration range of test compound. However, (**III**) does not require presence of beta-1 to function as a channel agonist, as it activates the BK channel pore-forming subunit, as well. No off target effects were observed for (**III**) against any of the other K channels on which it was tested, suggesting some specificity in compound's action. It is not known whether (**III**) shares a common binding site on the inner face of the BK channel with the glycotriterpene BK channel agonists previously identified (a hypothesis based on some moderate structural similarity between compound classes), but (**III**) is ca. 100-fold less active than (**I**) as a channel agonist (eg, similar to (**II**), a triterpene aglycone structure). Perhaps diminution of compound activity is due to lack of carbohydrate substituents in the structure of (**III**). Because of its weak agonist activity and chemically uninteresting structure, (**III**) was not investigated further.

When extracts of solid fermentation cultures of a *Phomopsis* species of endophytic fungi were examined in smooth muscle [^{125}I]ChTX binding, interestingly, a stimulation of ligand binding was observed. Purification and structural elucidation of the agent causing enhanced binding identified paxilline (**IV**; Fig. 2), which is a member of a family of well-known tremorgenic indole diterpene alkaloids (Knaus et al., 1994). This chemical series also includes aflatrem, paspalitrem A and C, penitrem A, verrucologen, paspalanine, and paspalicine (**V**; Fig. 2). These molecules are known to cause tremors and ataxia in cattle when they feed on grasses infested with fungal species that produce such alkaloids (Cheeke, 1995),

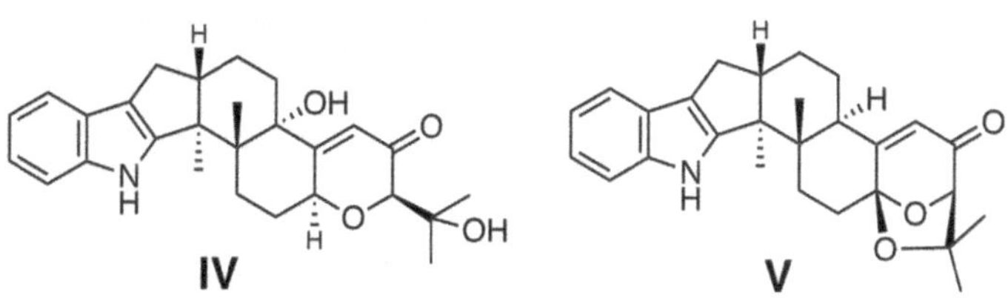

IV **V**

Fig. 2 Structures of natural product BK channel inhibitors.

but mechanistic basis of this toxicity was unclear at time of discovery. There had been speculation that CNS bioactivity was related to alkaloid induced increases in neurotransmitter release and activation of GABA-A modulated Cl channels (Norris et al., 1980; Yao, Peter, Baur, & Siegl, 1989). After identification of (**IV**), we profiled many indole diterpenes of this structural class in [^{125}I]ChTX binding and found them to have a complex, allosteric interaction with peptide binding to the pore of the BK channel; some alkaloids stimulated binding, some were partial or complete inhibitors of the reaction, and paspalinine had only modest inhibitory activity. However, in functional assays, where single BK channels were monitored in excised inside-out membrane patches from bovine aortic smooth muscle cells, all compounds were very potent channel inhibitors with K_i values of <10 nM; there was no obvious correlation between effects on binding (ie, extent of modulation or EC_{50}/IC_{50} value) and functional inhibitory potencies.

Indole diterpene BK channel inhibitors were active when added at either membrane surface and inhibition was rapid, but only slowly reversible because of hydrophobic nature of the molecules; based on single channel studies, mechanistic properties of this family are not homogeneous as there were distinct blocking patterns detected in recordings, although generally it appears that inhibitory activity is related to the state of the channel since compounds are less effective when BK channel open probability is very high; inhibition of BK alpha or BK alpha/beta-1 channels was shown to occur with equivalent potency (Sanchez & McManus, 1996). More recent mechanistic studies with paxilline indicate that the molecule inhibits BK channels by binding to either the open or closed state of the channel, but that it has a strong preference for binding to the closed state (Zhou & Lingle, 2014). Several indole diterpene representatives were tested for specificity against other ion channels in functional assays: at μM concentrations, there was no effect on KCa3.1, Kv1.2, Kv1.3, Cav1.2, or Nav1.2; (**IV**) blocked Kv2.1 with an IC_{50} of 2 μM. Also, limited Medicinal Chemistry SAR studies with (**IV**) as template indicate that most analogs prepared display a significant loss in potency when compared with parent compound (Knaus et al., 1994). This is an indication that a highly specific site exists on BK channels to which these indole alkaloids bind.

One compound from this class, paspalacine ((**V**), a deshydroxyl analog at C-19; Fig. 2) is reported to be nontremorgenic, but was still found to be a potent BK inhibitor (Knaus et al., 1994). Thus, this lack of tremorgenicity is probably more related to poor brain exposure of compound, than dissociation of tremor from BK channel inhibition. Tremorgenicity and ataxia observed with indole diterpenes are very likely mechanism based, as we have observed this adverse effect with distinct structural classes of BK channel blockers. Consistent with this idea, mice in which the BK channel has been genetically ablated are not affected by these indole alkaloids (Imlach et al., 2008). The tremorgenic effect in animals is usually transient, reversible, and not typically lethal; there are no lasting long-term effects when animals ingest these compounds. Discovery that this family of tremorgenic indole diterpenes affords a series of potent and selective BK channel inhibitors which is accessible to the field for use as a standard is made possible because many of these compounds are commercially available. This is probably the most significant discovery made during the Merck natural product screening effort (see later).

Strategy of using a binding assay to identify new ion channel drug leads is not novel. It was a successful approach employed broadly in the past to identify new Ca entry blockers that target Cav1.2 channels for treatment of hypertension when high-affinity, radiolabeled ligands for that channel became available. Indeed, contract research organizations (CRO's) offer Pharma a service of profiling putative development candidates for ancillary pharmacology by first performing a battery of ligand binding assays that include measurements directed against Ca, Na, K, or GABA channels, as a preliminary way to assess potential safety issues; test agents that score positively are followed up by these CRO's in ion channel functional assays for confirmation and further study. We had been successful in screening natural products to identify leads for other ion channel programs at Merck that garnered Medicinal Chemistry support using ligand binding approaches. For example, the bis-benzylisoquinoline alkaloid, tetrandrine, was identified from a Chinese medicinal herb used in traditional medicine to treat angina by its inhibition of [^{3}H]diltiazem binding; it was shown to inhibit Cav1.2 at a site where the therapeutically used, but structurally dissimilar antihypertensive drug, diltiazem, interacts, a novel finding given the undeveloped molecular pharmacology at the benzothiazepine binding site on Cav1.2 (King et al., 1988). Unfortunately, [^{125}I]ChTX binding to BK channels in smooth muscle had marginal success in identifying a lead suitable for Medicinal Chemistry, despite a significant and extended screening effort. Therefore, when new technologies became available which allowed possibility of using a functional HTS approach to screen for BK channel modulators, that strategy was adopted, and [^{125}I]ChTX binding was abandoned.

4. DISCOVERY OF SMALL MOLECULE BK CHANNEL AGONISTS

In principle, there are at least two ways to develop a functional BK channel assay to support screening for agonists or inhibitors of this channel; a cell based, biochemical approach in which either membrane potential or ion flux is measured using well-characterized fluorescence probes designed for this purpose, or an electrophysiology approach where a high capacity, plate based, automated patch-clamp technique is employed. We had interrogated a 200,000 member small molecule library against Kv2.1 for a diabetes project, and in that way identified the first potent and selective, small molecule Kv2.1 inhibitors, using Ion Works Quattro instrumentation (Herrington et al., 2011), but that effort cost ca. $250,000 in supplies, even when doing the initial screen with mixtures of 10 compounds and subsequently deconvoluting positives to identify active molecules. Thus, interrogating a library of >1 million compounds against the BK channel target using automated electrophysiology was not cost effective at the time and is only slowly becoming more feasible as an approach with development of higher density patch-clamp plate formats. So we used a biochemical approach initially to configure an HTS assay for processing large sample numbers in search of channel agonists (Ponte et al., 2012).

First, a BK channel representative of the smooth muscle complex (alpha/beta-1) was expressed heterologously in CHO cells; CHO cells were employed to limit density of BK channel expression ([^{125}I]IbTX binding was used to quantitate channel levels) and to have cells that do not express a high level of background K currents. Resting membrane potential in these cells can be monitored using either a commercially available FRET dye pair originally developed by Roger Tsien or a single dye-based assay which was commercialized by molecular devices. These cells sit rather depolarized because intracellular Ca levels are not high enough to activate expressed BK channels, so when high K is added to depolarize their plasma membrane, there is only a modest signal change detected using either fluorescent dye system. However, if BK channels are activated by exposure to an agonist, cells hyperpolarize to potassium equilibrium potential (−80 mV), and then when high K is added, there is a large signal generated, signifying a correspondingly large depolarization event (Ponte et al., 2012). This is a very simple and facile way to screen for BK channel agonists in a large sample deck, and the assay can be prosecuted in 96-, 384-, 1536-, or 3456-well plate formats. BK agonists detected by causing CHO cell hyperpolarization can further be profiled using any number of automated electrophysiology platforms, like Ion Works Quattro, Sophion Q patch, Nanion Patchliner, or Patch Express. From experience, a biochemical screening strategy is the most cost-effective way of screening large chemical libraries in a timely, efficient, robust, and precise fashion.

In order to prosecute a BK channel agonist HTS properly, each test plate should be spiked with a positive standard to maintain quality control. At the time, we had not identified an appropriate positive control to use in the screen through our own efforts, nor were any of BK channel agonists described in the literature deemed sufficient because of their various limitations. However, we borrowed from another approach that was being conducted in parallel with our screening exploration for novel channel agonist templates. As is typical in Pharma, there is avid monitoring of patent literature to follow what new ideas and chemical space competitors are working on; this exercise can be valuable if it affords ideas for fragment-based drug design, or if it provides for a more straightforward and traditional "patent busting" effort. In this way, a core structure of a BK channel agonist had been identified from a 2005 patent publication (Lee, Marino, & Zhao, 2005). Substructure search of the Merck sample collection afforded a number of compounds with chemical similarity and one, compound (**VI**; Fig. 3),

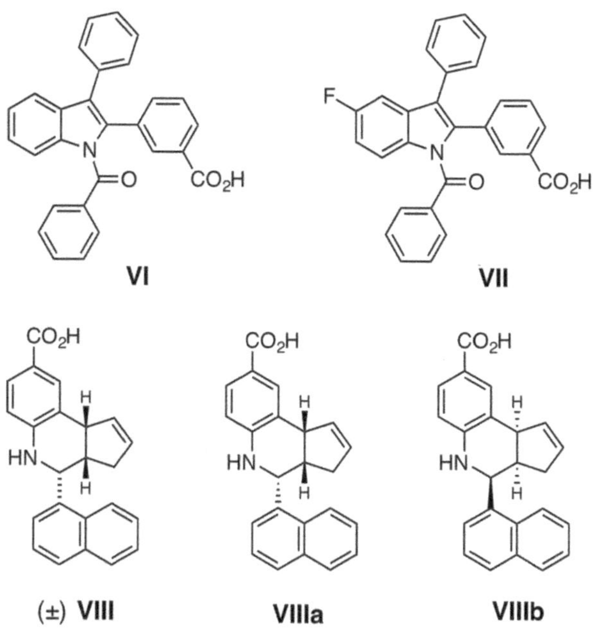

Fig. 3 Structures of synthetic, small molecule BK channel agonists.

had BK agonist activity in functional assays (Ponte et al., 2012). This finding stimulated a small exploratory Medicinal Chemistry effort that produced only a few active compounds, such as one typified by a fluoro-substituted analog (**VII**; Fig. 3). Each of these compounds increased BK channel activity modestly in patch-clamp recordings by 5- and 14-fold at 10 μM, respectively, but they did give a significant, paxilline-sensitive signal in the proposed membrane potential-based HTS assay (EC_{50} values of 1.32 and 0.63 μM, respectively) and, therefore, would be useful as a positive control. However, it was clear from profiling this series that many analogs demonstrated significant ancillary pharmacology; sometimes, they possessed Cav1.2 inhibitory activity (eg, IC_{50} of (**VI**) ~10 to 30 μM, and (**VII**) <10 μM), and, more importantly, they inhibited different isozymes of the PDE family (ie, PDE1–6), often in an unpredictable, and sometimes very potent manner. It became a major Medicinal Chemistry effort to increase BK channel agonist activity, while improving selectivity at the same time, so this approach was abandoned in favor of prosecuting an HTS to identify new potent and selective BK channel agonists. Such is a typical example of what one finds when mining the patent literature for BK channel agonists.

To interrogate that portion of Merck sample collection which was available for screening at the time (>1.6 million compounds), the assay described earlier was adapted from 384- to 1536-well format, a single fluorescence dye system was used for ease of operation and to eliminate wash steps, and the assay was optimized to maximize a high Z' factor in final HTS format; entire primary screening campaign required 6 weeks to execute and the assay consistently performed in a robust fashion (Ponte et al., 2012). Compounds were tested at a final concentration of 6 μM; 5300 compounds were identified in the first pass that increased fluorescence signal in cells by ≥50% relative to 1.8 μM (**VII**) and were thus selected for confirmation. Of these, 1400 compounds repeated where the corresponding signal response was blocked by paxilline. These compounds were then triaged by testing for activity in automated electrophysiology and determining potency in the fluorescence membrane potential assay. Compounds with EC_{50}'s <10 μM were further subjected to selectivity profiling by assaying against Cav1.2, Nav1.5, Kv11.1 (hERG), and a panel of PDE enzymes. Final active candidates were prioritized based on potency, selectivity, and chemical tractability, which resulted in ca. 6 chemical series, of which one was a family of tetrahydroquinolines that were selected for an exploratory Medicinal Chemistry effort. From this SAR work, a more potent and selective tetrahydroisoquinoline analog (**VIII**; Fig. 3) was identified.

Compound (**VIII**) is a racemic mixture (Ponte et al., 2012). In the FRET fluorescence-based CHO cell assay, (**VIII**) demonstrates an EC_{50} of 270 nM, and hyperpolarization response was completely blocked by indole diterpene BK channel inhibitors. Compound (**VIII**) has little inhibitory activity against Nav1.5 or Kv11.1, blocks Cav1.2 with an IC_{50} of 14 μM, and is a weak blocker of PDE1, PDE3, PDE4, PDE5, and PDE6 enzymes; it was not tested against PDE2 isoform. Overall (**VIII**) is more potent that (**VI**) or (**VII**) and also has better selectivity against Cav1.2 and PDE enzyme off target activities. Remarkably, when (**VIII**) is resolved into its two enantiomers, (**VIIIa**) and (**VIIIb**), there is ca. a 30-fold difference in potency between these two compounds in the FRET agonist assay; EC_{50}'s of 3100 and 110 nM, respectively, were determined, and importantly, all agonist responses were blocked by BK channel inhibitors. In a functional Cav1.2 assay, (**VIIIa**) and (**VIIIb**) inhibited with IC_{50}'s of 16 and 11 μM, respectively. To complete the selectivity profiling of (**VIIIb**), it was tested at two CRO's. In a PanLabs screen against 150 receptors, ion channels, and enzymes, (**VIIIb**) was found to exhibit no significant ancillary activity at <10 μM. In an EMD electrophysiological Cardiac Profiler assay, (**VIIIb**) had no significant inhibitory effect at concentrations of up to 10 μM against Kv4.3/KChIP2, Kv1.5, Kv7.1/mink, Kv11.1, HCN_4, or Kir2.1, but there was a 28% increase in Kv7.1/mink current at 10 μM, which increased to 50% at 30 μM test compound; a potential interaction between this BK channel agonist structural class and IKs, where a reproducible weak activation of the cardiac slowly activating delayed rectifier K channel was observed, deserves further study.

Both (**VIIIa**) and (**VIIIb**) increased BK channel activity in an automated electrophysiology study with CHO cells expressing BK channel alpha/beta-1, with EC_{50} values of 2 and 0.5 μM, respectively, and maximal increases in current amplitude of ca. six-fold were observed with each compound (Ponte et al., 2012). When experiments were repeated by manual patch with CHO cells expressing only BK channel alpha subunits, (**VIIIb**) increased maximal current by 11-fold with a slightly less potent EC_{50} of 2 μM. Therefore, activation of BK channels by tetrahydroquinoline analogs occurs via an interaction with channel pore-forming subunit. Interestingly, and completely independently, scientists at Amgen identified a series of similar tetrahydroquinolines as BK channel agonists, although profiling of their channel-activating properties and ancillary pharmacology was limited (Gore et al., 2010). A small number of these compounds were made and profiled in our group; some (eg, compound 31 in the reference) were found to have significant PDE inhibitory activity; this PDE activity precludes using such compounds as selective BK channel agonists in smooth muscle relaxation assays.

Next, pharmacology of tetrahydroquinolines was examined (Ponte et al., 2012). Compounds (**VIII**), (**VIIIa**), and (**VIIIb**) were first analyzed in in vitro smooth muscle relaxation assays of phenylephrine-contracted rat thoracic aorta rings; all compounds promoted relaxation in a dose-dependent fashion with a rank order of potency predicted based on ability of compounds to activate BK channels. IbTX reversed relaxation produced by these compounds when added at the end of the experiment, and pretreatment of aortic rings with IbTX prevented any relaxation from occurring in the first place. Neither apamin (KCa2.1 channel blocker), glibenclamide (K-ATP channel blocker), nor 4-AP (Kv channel blocker) had any effect on tetrahydroquinoline-induced smooth muscle relaxation, which gives an indication of specificity of action of these compounds. Importantly, when aortic rings were contracted by 80 mM K, neither (**VIIIa**) nor (**VIIIb**) had any relaxant effect, consistent with lack of significant Cav1.2 inhibitory activity for these compounds. In carbachol-induced stimulation of bladder detrusor smooth muscle strip contractility, (**VIIIa**) and (**VIIIb**) reversed spontaneous activity in a concentration-dependent fashion, with expected rank order of potency. IbTX reversed the effect of (**VIIIa**) and (**VIIIb**), while apamin and glibenclamide did not. These tissue bath experiments demonstrate prototypical behavior expected of a selective BK channel agonist, but absolute concentrations needed to elicit relaxation effects are shifted to the right approximately 10-fold because apparently these compounds bind to tissue, which limits their free effective concentrations.

Consistent with this notion is observation that when members of the tetrahydroquinoline class were profiled in rats to determine their in vivo pharmacokinetic properties, they had exceptionally long plasma half-lives when administered by either i.v. or p.o. dosing regimens. These compounds are highly protein bound (>95%) which accounts for their prolonged $t_{1/2}$ in vivo. Unfortunately, this significantly precludes characterization of in vivo pharmacology because of low free drug fraction; this property must be corrected by appropriate chemical modification of the structure in order for this lead class to be of any value in terms of drug development. Nevertheless, data presented above (Ponte et al., 2012) are the first demonstration that a selective BK channel agonist displays predicted in vitro pharmacology in terms of smooth muscle relaxation. The fact that compound (**VIII**) can be prepared in one synthetic step, easily separated by chiral HPLC into enantiomeric components (**VIIIa** and **VIIIb**) that demonstrate marked differences in potency for BK channel activation, and that these compounds display in vitro activity in smooth muscle contractility paradigms, makes identification of tetrahydroquinoline BK channel agonists a significant discovery. These molecules provide important new tool compounds for the field to use in characterizing roles of BK channels in integrated system biology, at least at the level of in vitro models.

There are literature reports that certain voltage-sensitive fluorescent dyes, such as $DiBAC_4$ (an oxanol derivative), are agonists of BK channels and that their activity depends on beta subunit composition of the BK channel complex under investigation (Morimoto et al., 2007; Scornik et al., 2013). For example, $DiBAC_4$ activates BK channels containing beta-1 or beta-4 subunits at submicromolar concentrations, is less potent when tested on alpha subunits alone, and apparently blocks BK channels containing beta-2 subunits. In our FRET BK channel agonist assay screen, we used 4 μM $DiSBAC_2$ as the membrane permeable oxanol probe to detect changes in membrane potential upon BK channel activation and, in separate experiments, found that this compound also reversibly activates alpha/beta-1 channels. To be noted is that this activity was never found to interfere with operation of our FRET-based assay, agonist (**VIII**) was discovered in an HTS that did not employ a FRET assay or an oxanol reagent, and functional activity of each putative BK channel agonist was always confirmed by electrophysiology; it cannot be ruled out, however, that use of an oxanol dye in a BK channel agonist screen might sensitize the BK channel, making it easier to find other compounds with agonist activity in this type of assay.

Regarding unexpected effects of dyes on BK channels, it is curious that the recently described GoSlo-SR family of BK channel agonists is derived from a dye molecule (Roy et al., 2012). Observation that the dye, reactive blue 2, is a weak (10 μM) activator of BK channels in bladder smooth muscle, has provided a lead for this effort; a structurally related dye analog, acid blue, is also a weak BK channel agonist. Reactive dye 2 is not selective for BK channels because it also blocks ATP-activated ion channels and P2Y purino-receptors, and to our knowledge has never been adequately profiled in terms of its ancillary pharmacology. The Medicinal Chemistry strategy chosen to explore this lead (Roy et al., 2014) has been to simplify parent dye structure by eliminating the triazole substituent and derivatizating the anthraquinone nucleus; ca. a 10- to 50-fold maximal increase in potency has been reported so far, but it is not clear how these changes affect molecular selectivity of resulting analogs. Further, the group prosecuting this lead has focused on the mechanism by which prototype compounds function to activate BK channels and where on transmembrane domains of the alpha subunit compounds bind, rather than on exploring compound specificity (Large et al., 2015). It appears that parent compound in this series activates BK channels without influence of beta subunits, but with potency increases, agonists require presence of beta-1 or beta-4 subunits to elicit maximal channel activation (Webb et al., 2015). While the described approach may be interesting in terms of testing potential BK channel gating mechanisms, it is our and others' long-term experience that starting with a weak and nonselective compound as a lead for an ion channel project severely hampers POS in terms of identifying a viable therapeutic agent. What often happens is that SAR during structural exploration is flat, potency increases track with increasing hydrophobicity of substitution, and extent of potency increase is capped (usually in the 100 nM range); the latter phenomenon is related to compound hydrophobicity causing local drug concentration in membrane to increase which artificially increases potency, but the magnitude achievable is limited. What one hopes for in development of a therapeutic agent is that increasing potency also leads to greater selectivity, but starting with a weak, nonselective compound suggests that the drug-channel protein binding site is not a well defined, high affinity one. Goal of successful drug discovery is to identify a high-affinity region on a protein target that is important for protein function and to optimize a molecule's interaction specifically at that site.

Many different types of compounds have been identified which will activate BK channels (Torres et al., 2014), and some of these target specific alpha/beta subunit channel complexes to elicit activity; fatty acids, arachidonic acid metabolites, ethanol, steroids like 17β-estradiol, dehydroepiandrosterone, and the structurally related lithocholic acid, sodium tungstate, NO, CO, and various membrane active dyes have all been described. However, it is hard to imagine how any of these agents could be a lead for a drug discovery program, or be of therapeutic importance in a traditional sense. Therefore, there is still an important need to identify viable BK channel agonist lead structures which can support a drug development program. Clearly, HTS strategies have been reported which can successfully identify active compounds after screening large chemical libraries. Interesting structural classes with agonist potencies <1 μM can be identified depending on quality of the collection interrogated. Procedures for triage to separate viable, chemically tractable actives from uninteresting hits which can be prosecuted by Medicinal Chemists are well described. And, critically, efficient support of Medicinal Chemistry SAR studies using current technologies to aid in identifying a development candidate is now rather facile. Success is largely dependent on quality of the lead discovered, degree of support from Medicinal Chemistry to progress the lead and enthusiasm of the therapeutic area in bringing the project to fruition. Despite lack of success so far in the small molecule BK channel agonist field, recent progress has demonstrated that such effort is not futile. Moreover, by selecting the most relevant, tissue-specific BK channel complex to target and establishing a screen with that channel form, POS may be increased. If small molecule discovery remains unsuccessful, an alternative strategy is to employ a molecular biology approach and introduce native or genetically modified BK channel beta subunits into tissues by gene therapy to create a situation where BK channel activity would be artificially elevated to achieve therapeutic benefit. In this case, a selective BK channel inhibitor delivered locally might be useful for further fine tuning physiology of that system.

5. IDENTIFICATION OF POTENT AND SELECTIVE SMALL MOLECULE BK CHANNEL INHIBITORS

It is becoming more evident that BK channels are important in controlling physiology in other tissue types besides smooth muscle and CNS, where they have traditionally been studied, and that BK channel inhibitors might have therapeutic utility in some pathophysiologies associated with expression of this channel in other cells. For example, in rheumatoid arthritis, fibroblast-like synoviocytes participate in joint inflammation and damage, and upregulation of BK channels appears critical to support invasiveness of these cells in both rat arthritis models and human patients (Tanner et al., 2015). Studies with IbTX and indole diterpene BK channel inhibitors reduced proliferative and invasiveness properties of these synoviocytes, and stopped disease progression in animal models. Based on these data, BK channels might be a new therapeutic target for treating rheumatoid arthritis. Similarly, in certain other types of cell proliferation and metastasis, like prostate, brain, mammary, skin, or pancreatic cancer, data suggest that BK channels play a critical role in tumor development; inhibiting BK or other K channels, therefore, might provide a novel chemotherapeutic approach for treating specific cancers (Huang & Jan, 2014). Recently, a role for BK channels in controlling breathing activity has been identified and a small molecule BK channel inhibitor, GAL-021, was shown to diminish opioid-induced respiratory depression by functioning at the level of the carotid body; hence, therapeutic roles in drug-induced respiratory depression or sleep apnea have been postulated for inhibitors of BK channels (Golder et al., 2015). Finally, it was reported in a patent from Allergan (Adorante, WoldeMussie, Ruiz, Kopper, & Moore, 1999) that BK channel inhibitors when injected intracamerally (peptides like ChTX and IbTX, or indole diterpenes) or applied topically (indole diterpenes) to rabbit eyes caused lowering of intraocular pressure (IOP). These data suggest that BK channel blockade would be a potential new mechanism for treating glaucoma.

In order to prosecute any of these medical needs, a chemically tractable, small molecule BK channel inhibitor is required as a lead to support that effort. Moreover, systemic exposure of the drug must be avoided so as to limit toxicity due to widespread inhibition of normal BK channel function. Identification of the first, small molecule, BK channel blocker was made possible as a follow up to discovery that structurally complex indole diterpenes (Fig. 2) were potent inhibitors of this channel. To ascertain whether any simple substituted indole might possess BK channel inhibitory activity, a small, structurally diverse collection of such compounds (ie, an indole "test" kit) was screened at 10 μM in the $[^{125}I]$ChTX sarcolemmal smooth muscle membrane binding assay. Not surprisingly, given that 100,000 small molecules representing compounds newly submitted to the Merck sample collection had previously been screened in that assay without scoring any hits, nothing was active, except for one rather peculiar molecule (**IX**; Fig. 4) which inhibited the binding reaction with an IC_{50} of 100 nM.

In the late 1960s, Merck developed the NSAID, indomethacin (a cyclooxygenase inhibitor that blocked prostaglandin production) which was registered as a drug to treat inflammation, fever, and pain. One adverse effect of this drug is gut irritation and, in an effort to deal with that issue, amides and esters of indomethacin were synthesized for use as potential prodrugs using a carbodiimide coupling reaction. When indomethacin is treated for an extended period with *N,N*′-dicyclohexylcarbodiimide (DCCD) before adding a second amine or alcohol containing reactant (to drive formation of the corresponding amide or ester derivative of indomethacin, respectively), the indomethacin–DCCD adduct can rearrange to form a stable acyl guanidine (**IX**) that precipitates out of solution because of its limited solubility; (**IX**) is a synthetic artifact that was submitted to the Merck sample collection in 1969 by a diligent chemist who synthesized it by accident and it had never previously tested positive in any assay. When (**IX**) was tested in electrophysiology experiments, it was found to inhibit BK channels and to be the most potent and selective small molecule inhibitor of any K channel

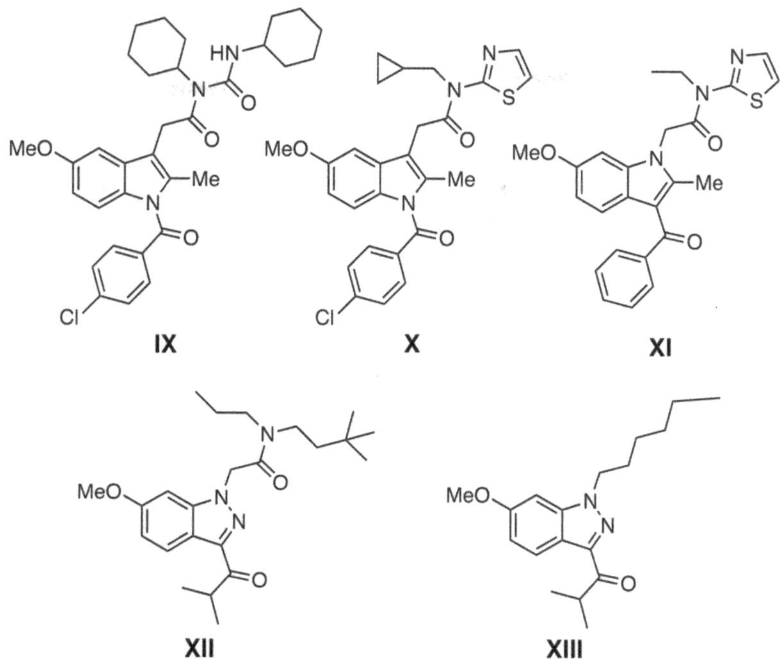

Fig. 4 Structures of synthetic, small molecule BK channel inhibitors.

known at the time of its discovery (Garcia, Kaczorowski, & McManus, 2003a); members of the class of BK channel inhibitors derived from this lead are similarly highly regarded today in terms of their potency and selectivity. Profiling of (**IX**) revealed that it inhibits BK channels (alpha, or alpha/beta-1 compositions) in patch clamp with an IC_{50} of 0.6 nM; (**IX**) has no effect at 1 μM on functional activity of Kv1.X, Kv2.X, Kv7.1/mink, Kv11.1, Nav1.5, or Cav1.2 channels, (**IX**) is completely free of ancillary pharmacology in a PanLabs screen of 150 targets when tested at 10 μM (with exception of blocking somatostatin type 4 receptor binding with an IC_{50} of 1.4 μM; no activity was observed in a follow-up functional receptor assay); i.v. dosing of (**IX**) elicited no effects on any cardiovascular parameter in rat or dog (highest plasma concentration achieved was 13 μM), and it was nontremorgenic in mice at 5 mpk p.o.; (**IX**) is >99% protein bound, so plasma free fraction of this compound is very low in in vivo safety profiling experiments. Unfortunately, (**IX**) was deemed chemically unattractive as a lead by Medicinal Chemistry because of its poor aqueous solubility, potential chemical reactivity of the acyl guanidine group, and lack of a suitable, nonreactive chemical isostere for that functionality.

To search for other indole-containing BK channel inhibitors, and to assuage concerns that an [^{125}I]ChTX binding assay might be missing compounds with functional activity, two biochemical cell-based BK channel screens were established and optimized for detecting channel blockers. In one (Doherty, Chen, Liu, Natarajan, & Tynebor, 2009), BK channels were over expressed in HEK cells (lines with ca. 10^6 channels per cell were identified using [^{125}I]IbTX binding) so that basal channel activity would set cell resting membrane potential at E_K, ca. −80 mV. Interrogating membrane potential with a FRET-based assay, after preincubation with test compound and addition of high K to cause cell depolarization, would produce a sizable fluorescence signal change, unless test article inhibited BK channels; that would cause prior cell depolarization itself before high K was added and no signal change would occur. The other assay employed monitoring Tl flux through BK channels into HEK cells using an intracellular Tl sensitive fluorescence dye as a probe, which is a means to monitor basal channel activity directly in this system; a similar procedure was successfully developed to support other K channel drug discovery programs, such as the ROMK project (Garcia & Kaczorowski, 2014). The latter assay has an advantage that there is excellent correlation for blocking BK channel activity, no matter whether this is measured by Tl flux or electrophysiology, whereas BK channel blocking potencies are shifted 10- to 20-fold to the right in the membrane potential-based assay because >95% of channels must be blocked to elicit HEK cell depolarization, and concentration dependence of inhibition is steep; for example, potency of (**IX**) in FRET-based assay is 11 nM, compared to 0.6 nM in manual patch-clamp measurements. Nonetheless, the FRET-based assay is facile, robust, high capacity, and dyes supporting this assay were commercially available many years before the Tl sensitive dyes were sold. Screening of ca. 1000 indole-containing molecules obtained by similarity searching of our sample collection with (**IX**) as a template afforded no hits in BK channel FRET-based assays, except for identifying the corresponding acyl guanidine to (**IX**) that is obtained after prolonged treatment of indomethacin with *N*,*N*′-diisopropylcarbodiimide and isolating the stable rearrangement product; this compound was ca. 40-fold less active as a BK channel blocker than (**IX**), indicating that channel inhibition occurs with a defined SAR (Garcia et al., 2003a). The FRET-based assay was also used to interrogate ca. 300,000 sample deck from the Merck sample collection that was chosen to reflect a cross section of chemical diversity; again no confirmed potent (<1 μM) hits were detected. Thus, conclusion at the time was that structurally interesting, small molecule inhibitors of

BK channels were extremely rare. Having spoken with other members of Pharma who carried out similar screening exercises to find BK channel inhibitors after Allergan disclosed their work on the link between BK channels and glaucoma, it was general consensus in the field that inhibitors with any marked potency or selectivity were nearly impossible to find.

To investigate Allergan observations concerning role of BK channels in controlling IOP further, distribution of BK channels was mapped in rabbit and human eye tissue in collaboration with scientists (Drs. H.-G. Knaus and M. Dichtl) at University of Innsbruck, Austria. Using receptor autoradiography and binding studies with [^{125}I]IbTX and Western blotting and immunohistochemistry with an antibody directed against the alpha BK channel subunit, highest expression of BK channels in the eye was found in retina, followed by nonpigmented epithelium of ciliary body; lower channel levels were also found in other eye tissue (eg, ciliary muscle, corneal endothelium, iris smooth muscle, trabecular meshwork, etc.), but channels are not present in lens. Similar results had been reported in the literature demonstrating functional BK channel activity in human nonpigmented ciliary epithelium (Barros, Lopez-Briones, Coca-Prados, & Belmonte, 1991), and in bovine (Stumpff, Strauss, Boxberger, & Wiederholt, 1997) and human (Thieme et al., 2001) trabecular meshwork. Therefore, a hypothesis was put forth that constitutively active BK channels in nonpigmented ciliary epithelia facilitate aqueous humor secretion by forming a pathway for K counter ion movement which allows Cl flux to occur, with ensuing water movement into the posterior chamber of the eye; blocking inflow of aqueous humor by inhibiting BK channel activity would be predicted to lower IOP in patients with open-angle glaucoma or ocular hypertension. In support of this hypothesis, when human ciliary epithelial cells were subjected to patch-clamp recording (Garcia et al., 2003a), BK channel currents were observed that were sensitive to reversible block by EGTA, TEA, paxilline, or (**IX**). IOP-lowering experiments in rabbit reported by Allergan with peptide and indole diterpene BK channel blockers (dosed by either intracameral injection or applied topically) were confirmed with a number of different BK channel blockers and extended by demonstrating that topically applied indole diterpenes also lowered IOP in a glaucomatous monkey model. Moreover, (**IX**) also lowered IOP in rabbit when applied topically using a peanut oil formulation, which was required because of poor solubility of that compound in other vehicles. Thus, based on these data, there was pressing need to identify a chemically tractable, small molecule BK channel inhibitor as a Medicinal Chemistry lead to progress the project.

Through use of a very small and focused exploratory chemistry effort at a CRO, Oxford Asymmetry (which is now part of Evotec), about 100 derivatives of (**IX**) were synthesized. It was shown that the acyl guanidine substituent of (**IX**) could be replaced by a simple disubstituted amide, (**X**; Fig. 4), where in vitro potency for inhibiting BK channels in electrophysiology measurements is maintained (IC_{50} of 0.1 nM), while improvement in physical and chemical properties of the series could be achieved (Boyd et al., 2008). These findings provided a strategy for an internal Medicinal Chemistry program which generated ca. 2500 compounds before identifying a development candidate, and a backup candidate that was a prodrug. During this process, a reverse indole template (Chen et al., 2003) was employed to address metabolic stability (**XI**, IC_{50} of 1.2 nM; Fig. 4), and also an indazole core structure replacement for the indole template (**XII**, IC_{50} of 0.06 nM; Fig. 4) was employed in this series (Chen, Doherty, Liu, Natarajan, & Tynebor, 2005). In parallel (Garcia, Kaczorowski, & McManus, 2003b), there was a Medicinal Chemistry effort that focused on an indole diterpene natural product core structure which was produced by fermentation and subjected to SAR analoging studies (Brnardic, Doherty, Dorsey, et al., 2009; Brnardic, Doherty, Ellwood, Fillmore, & Malaska, 2009). In all, there are more than 25 patents issued which demonstrate the progress and success of these efforts; chemistry strategies that were pursued for lead progression of BK channel inhibitors are well illustrated in these documents.

Although neither biophysical nor ancillary pharmacological properties of any of these small molecule BK channel inhibitors were formally reported in the literature, a competitor made a representative compound from a Merck patent and published its profile (Zeng et al., 2008). The molecule chosen was of average potency (**XIII**, IC_{50} of 2 nM; Fig. 4), but it was shown to be a highly selective BK channel inhibitor in that it had no effects on a number of other ion channels or on action potential waveforms in guinea pig or rabbit ventricular myocytes. This agent inhibited BK channels composed of either alpha subunit alone, alpha/beta-1 or alpha/beta-4, with similar potency, as well as, native BK channels in detrusor smooth muscle and in vagal neurons; BK channel inhibition was not voltage dependent, did not require channel opening for channel inhibition to occur, and inhibitor decreased channel probability of opening by increasing time constants of closed channel states; there were no effects observed on single channel conductance. We can confirm a number of these findings from our own unpublished data, but mechanism of channel inhibition by this series was never extensively investigated at Merck. However, there is similarity in the mechanism of action reported for these BK channel inhibitors with that observed for the indole diterpene BK channel inhibitor, paxilline (Zhou & Lingle, 2014) as might be expected if both structural classes of inhibitors interact at the same binding site on the BK channel. In addition, our data indicate that there is a very strict SAR for BK channel inhibition (eg, demethylation of the methoxy substituent on the core structure results in severe loss of inhibitory activity) and that the most potent inhibitors obtained had IC_{50}'s of ca. 10 pM. Clearly, these BK channel inhibitors have unique characteristics for small molecule ion channel modulators in terms of their potency and selectivity properties, especially when compared in general with other K channel inhibitors that have been disclosed. Undoubtedly, BK channels have unique features in terms of their drug binding sites distinct from those of other voltage-gated K channels.

The development candidate identified to take forward for studies in patients with glaucoma had promising characteristics. In addition to acceptable potency and lack of any significant ancillary pharmacology, it lowered IOP in animal models from two species, had a favorable drug metabolism profile, displayed no issues in multiday ocular tolerability studies in two species, and completed multiple safety assessment studies in two species without any issue. However, it encountered an unexpected issue in Phase 1b studies in patients when PK of topically applied drug was studied in subjects undergoing cataract surgery. It became clear that there was a 10^6-fold difference between concentration of drug applied externally to the eye in a topical formulation and resulting drug concentration present in aqueous humor. In other words, human cornea, that is composed of hydrophobic exterior and interior layers, with a hydrophilic core region sandwiched in between, presented a very significant barrier which prevented BK channel inhibitor from efficiently accessing its target in ciliary epithelium. This observation was confirmed in in vitro human corneal penetration studies. In follow-up PK experiments where radiolabeled drug was applied topically to either rabbit or monkey eyes, followed by corneal removal and analysis by autoradiography, results clearly demonstrated that very high levels of drug were bound to hydrophobic exterior and interior corneal surfaces of each species. At present, with development of nanopharmacology and new specialized formulations that deliver hydrophobic drugs efficiently across previously considered impermeant barriers, there might have been a positive outcome, but acceptable ophthalmic formulations available at the time did not provide a solution. It is not uncommon to experience formulation issues with topically applied glaucoma drugs; the prostaglandin prodrug, Xalatan®, took many years to develop because of the same issue that we encountered, but eventually that effort was successful. The BK channel drug candidate was designed to be delivered topically to treat glaucoma, not by intracameral injection, and physicochemical properties of the molecule combined with characteristics of the corneal barrier unfortunately prevented the BK channel inhibitor/IOP-lowering hypothesis from being adequately tested in human subjects. In retrospect, given corneal PK findings in animals used as the basis for glaucoma models, these data call into question IOP measurements that were used to make program decisions, as well as, veracity of the type of in vivo animal pharmacology that was used to support this glaucoma project.

Interestingly, it has been postulated that a BK channel agonist might also have utility in lowering IOP. During development of Rescula® (unoprostone isopropyl), a docosanoid that lowers IOP in patients with ocular hypertension or primary open-angle glaucoma, mechanistic studies demonstrated that the drug affects the contractility of trabecular meshwork and ciliary muscle to increase aqueous outflow (Thieme et al., 2001). This initial study also demonstrated that unoprostone increased an IbTX-sensitive outward K current in bovine and human trabecular meshwork cells. Subsequent work (Cuppoletti, Malinowska, Tewari, Chakrabarti, & Ueno, 2012) indicated that unoprostone and its primary metabolite are potent activators of BK channels in human trabecular meshwork and smooth muscle cells. However, it is not clear that BK channel activation has any significant physiological consequence in the eye since the IOP-lowering effect of unoprostone is lost in prostanoid FP receptor knockout mice (Ota, Aihara, Narumiya, & Araie, 2005). The physiological effect of unoprostone action may be more complex because the drug has also been reported to activate ClC-2 Cl channels in smooth muscle (Cuppoletti, Malinowska, Tewari, Chakrabarti, & Ueno, 2011). To clarify the involvement of BK channels in the mechanism by which unoprostone lowers IOP, it would be interesting to investigate drug-mediated effects on IOP in mice where the BK channel has been genetically ablated.

The BK channel inhibitor clinical candidate was designed to be metabolically labile for fast clearance from plasma if any systemic exposure should occur, as a way to restrict potential adverse effects from inhibition of normal BK channel function. That is why compound was dosed topically to place it in close proximity to its target, rather than by giving it orally where drug would be delivered during systemic exposure. This limits ability to repurpose the molecule for other therapeutic purposes. However, a situation occurs in hemorrhagic shock (Gutierrez, Reines, & Wulf-Gutierrez,

2004), a condition which is often associated with accidents where significant loss of blood takes place causing the circulatory system to fail to maintain adequate perfusion of vital organs, at which point such a BK channel inhibitor might be useful. One reason for vasoplegia is that blood vessels fail to constrict in response to pressor agents and massive vasodilation takes place. One mechanism triggering blood vessel relaxation, based on data from animal models, appears to be activation of BK channels in response to NO/cGMP regulation, as well as, to increases in intracellular Ca due to ATP depletion (Clapp & Orie, 2007; Zhao et al., 2007). Indeed, preliminary data obtained by investigators at University of Innsbruck (Drs. H.-G. Knaus and V. Wenzel) demonstrate that IbTX is cardioprotective in a pig model of hemorrhagic shock, although further experiments are necessary to quantitate the magnitude of this protection. Therefore, a small molecule BK channel blocker with limited plasma half-life might be therapeutically useful to treat vasoplegia and could be administered i.v. to accident victims by Emergency Services. Or perhaps even IbTX application by itself might be a useful therapeutic treatment in this regard. However, in order to register a drug for this purpose, outcomes of clinical trials must be assessed using a mortality index; developing a drug using this clinical strategy could be an issue for traditional Pharma, but, nonetheless, treatment of hemorrhagic shock remains a serious unmet medical need.

The BK channel inhibitor represented by (**X**) is an excellent template for further Medicinal Chemistry SAR studies. Molecules synthesized previously were for a specific application, and, therefore, their properties were optimized around goals of the glaucoma program. Different therapeutic applications will require identification of development candidates with different properties, but intrinsic potency and selectivity are likely to be retained during the process if this template is used as a starting point for such an effort. Given new nanopharmacology and creative synthetic strategy approaches, novel compounds from this series may find their way as successful treatments for other diseases where a BK channel inhibitor is required to treat a medical problem.

6. CONCLUSIONS

When considering our current understanding of ion channel molecular pharmacology, BK channels stand out as distinct from other members of this super family of proteins. For example, voltage-gated Na and Ca channels display rather promiscuous pharmacology; it is not uncommon when screening against targets from these specific ion channel families that the number of primary hits identified is overwhelming. However, it has been very difficult to develop molecular pharmacology of BK channels since so few potent and selective channel agonists or inhibitors have been identified, despite a fairly significant and sustained effort over the years. Nevertheless, work described in this review was able to move this endeavor forward significantly. The first potent and particularly selective peptidyl BK channel blocker (IbTX), which is used currently as a standard reagent by the field, was identified. The first potent and selective natural product BK channel agonist was identified (compound **I**) whose mechanism of action demonstrated for the first time that beta subunits in the BK channel complex (ie, beta-1) can influence the molecular pharmacology of the BK channel. A family of tremorgenic indole diterpene alkaloids was identified (Fig. 2) as highly potent and selective natural product BK channel inhibitors, members of which are also now commonly used as standard reagents by the field; in vivo tremorgenic activity of these alkaloids is due to BK channel inhibition in the CNS. The first potent and selective small molecule BK channel agonist, whose mechanism displays stereochemistry in drug action and whose in vitro pharmacological profile in smooth muscle relaxation experiments displays the prototypic phenotype expected for a BK channel opener, was identified (compound **VIII**). Discovery of a derivatized indomethacin analog (compound **IX**) with potent BK channel inhibitory activity, that may mimic indole diterpene channel blockers at the level of the channel, and which provided a lead for developing a series of extremely potent and selective small molecule BK channel inhibitors is a seminal discovery (even perhaps a small miracle); to our knowledge, these BK channel inhibitors are the most potent and selective K channel inhibitors to be disclosed to date, and may provide a structural template for further BK channel small molecule inhibitor development. Even with the paucity of potent and selective BK channel modulators identified so far, precedent has now been established and further progress in developing molecular pharmacology of this channel is expected; then, therapeutic agents will follow.

ACKNOWLEDGMENTS

The authors acknowledge valuable and innumerable contributions of Senior and Associate staff of the Merck Rahway Ion Channel Department, who supported 15 drug discovery/development programs on ion channel and membrane transporter targets, and mentored Postdoctoral Fellows, Visiting Scientists, and 63 Summer Interns, over a 29 year period from 1980 to 2009. The effort outlined in this review would not have been possible without dedication of Dr. Jim Doherty and his Medicinal Chemistry team, Drs. Michael

Goetz, Guy Harris, Tony Lam and Sheo Singh from Natural Products Chemistry, and Dr. Dennis Schmatz from Department of Animal Health. In addition, continued support from Profs. Hartmut Glossmann and Hans-Günther Knaus, Univ. of Innsbruck, Innsbruck, Austria, and Prof. Guilherme Suarez-Kurtz, National Cancer Institute, Rio de Janeiro, Brazil was invaluable in carrying out these studies. Finally, without the vision of Drs. Gene Cordes and P. Roy Vagelos, establishment and continued support of the Rahway Ion Channel group would never have occurred.

We thank Dr. Jed Fisher, Univ. Notre Dame, for his kind effort in helping to prepare the figures.

REFERENCES

Adorante, J. S., WoldeMussie, E., Ruiz, G., Kopper, K., & Moore, A. M. (1999). Method for reducing intraocular pressure in the mammalian eye by administration of potassium channel blockers. US5925342, Allergan.

Barros, F., Lopez-Briones, L. G., Coca-Prados, M., & Belmonte, C. (1991). Detection and characterization of Ca^{2+}-activated K^+ channels in transformed cells of human non-pigmented ciliary epithelium. *Current Eye Research*, *10*, 731–738.

Bentzen, B. H., Olesen, S.-P., Ronn, L. C. B., & Grunnet, M. (2014). BK channel activators and their therapeutic perspectives. *Frontiers in Physiology*, *5*, 389.

Boyd, E. A., Fisher, M. H., Garcia, M. L., Kaczorowski, G. J., Meinke, P. T., Parsons, W. H., et al. (2008). US7414067, Merck & Co.

Brenner, R., Chen, Q. H., Vilaythong, A., Toney, G. M., Noebels, J. L., & Aldrich, R. W. (2005). BK channel beta4 subunit reduces dentate gyrus excitability and protects against temporal lobe seizures. *Nature Neuroscience*, *8*, 1752–1759.

Brnardic, E., Doherty, J. B., Dorsey, J., Ellwood, C., Fillmore, M., Malaska, M., et al. (2009a). Preparation of indole diterpene alkaloids as Maxi-K channel blockers for the treatment of glaucoma. WO2009048559, Merck & Co.

Brnardic, E., Doherty, J. B., Ellwood, C., Fillmore, M., & Malaska, M. (2009b). Maxi-K channel blockers and methods of use. WO2009048558, Merck & Co.

Cheeke, P. R. (1995). Endogenous toxins and mycotoxins in forage grasses and their effects on livestock. *Journal of Animal Science*, *73*, 909–918.

Chen, M. H., Doherty, J. B., Liu, L., Natarajan, S., & Tynebor, R. M. (2005). Ophthalmic compositions for treating ocular hypertension. WO2005026128A1, Merck & Co.

Chen, M. H., Liu, L., Meinke, P. T., Natarajan, R., Parsons, W. H., Shen, D.-M., et al. (2003). Ophthalmic compositions for treating ocular hypertension. WO2003105847A1, Merck & Co.

Clapp, L. H., & Orie, N. N. (2007). Stoking up BK_{Ca} channels in hemorrhagic shock: Which channel subunit is really fueling the fire? *Circulation Research*, *101*, 436–438.

Contreras, G. F., Castillo, K., Enrique, N., Carrasquel-Ursualez, W., Castillo, J. P., Milesi, V., et al. (2013). A BK (*Slo1*) channel journey from molecule to physiology. *Channels*, *7*, 442–458.

Cuppoletti, J., Malinowska, D. H., Tewari, K. P., Chakrabarti, J., & Ueno, R. (2011). Unoprostone isopropyl and its primary metabolite M1 activate BK and ClC-2 channels and prevent ET-1 induced smooth muscle effects. *Investigative Ophthalmology & Visual Science*, *52*, 3104.

Cuppoletti, J., Malinowska, D. H., Tewari, K. P., Chakrabarti, J., & Ueno, R. (2012). Unoprostone isopropyl and metabolite M1 activate BK channels and prevent ET-1-induced $[Ca^{2+}]i$ increases in human trabecular meshwork and smooth muscle. *Investigative Ophthalmology & Visual Science*, *53*, 5178–5189.

Doherty, J. B., Chen, M. H., Liu, L., Natarajan, S. R., & Tynebor, R. M. (2009). Ophthalmic compositions for treating ocular hypertension. US7547720, Merck & Co.

Dopico, A. M., & Bukiya, A. N. (2014). Lipid regulation of BK channel function. *Frontiers in Physiology*, *22*, 312.

Du, W., Bautista, J. F., Yang, H., Diez-Sampedro, A., You, S.-A., Wang, L., et al. (2005). Calcium-sensitive potassium channelopathy in human epilepsy and paroxysmal movement disorder. *Nature Genetics*, *37*, 733–738.

Engler, M. B. (1992). Effects of omega-3, omega-6 and omega-9 fatty acids on vascular smooth muscle tone. *European Journal of Pharmacology*, *215*, 325–328.

Galvez, A., Gimenez-Gallego, G., Reuben, J. P., Roy-Contancin, L., Feigenbaum, P., Kaczorowski, G. J., et al. (1990). Purification and characterization of a unique, potent, peptidyl-inhibitor of the high-conductance calcium-activated potassium channel from venom of the scorpion *Buthus tamulus*. *Journal of Biological Chemistry*, *265*, 11083–11090.

Garcia, M. L., Garcia-Calvo, M., Hidalgo, P., Lee, A., & MacKinnon, R. (1994). Purification and characterization of three inhibitors of voltage-dependent K^+ channels from *Leiurus quinquestriatus* var. *hebraeus* venom. *Biochemistry*, *33*, 6834–6839.

Garcia, M. L., & Kaczorowski, G. J. (2014). Targeting the inward-rectifier potassium channel ROMK in cardiovascular disease. *Current Opinion in Pharmacology*, *15C*, 1–6.

Garcia, M. L., Kaczorowski, G. J., & McManus, O. B. (2003a). Ophthalmic compositions for treating ocular hypertension. US6545036, Merck & Co.

Garcia, M. L., Kaczorowski, G. J., & McManus, O. B. (2003b). Method for treating ocular hypertension. US6548535, Merck & Co.

Garcia, M. L., Shen, D.-M., & Kaczorowski, G. J. (2007). High-conductance calcium-activated potassium channels: Validated targets for smooth muscle relaxants? *Expert Opinion on Therapeutic Patents*, *17*, 831–842.

Garcia-Calvo, M., Knaus, H.-G., McManus, O. B., Giangiacomo, K. M., Kaczorowski, G. J., & Garcia, M. L. (1994). Purification and reconstitution of the high-conductance Ca^{2+}-activated K^+ channel from tracheal smooth muscle. *Journal of Biological Chemistry*, *269*, 682–767.

Giangiacomo, K. M., Kamassah, A., Harris, G., & McManus, O. B. (1998). Mechanism of maxi-K channel activation by dehydrosoyasaponin-I. *Journal of General Physiology*, *112*, 485–501.

Golder, F. J., Dax, S., Baby, S. M., Gruber, R., Hoshi, T., Ideo, C., et al. (2015). Identification and characterization of GAL-021 as a novel breathing control modulator. *Anesthesiology*, *123*, 1093–1104.

Gore, V. K., Ma, V. V., Yin, R., Ligutti, J., Immke, D., Doherty, E. M., et al. (2010). Structure-activity relationship (SAR) investigations of tetrahydroquinolines as BKCa agonists. *Bioorganic & Medicinal Chemistry Letters*, *20*, 3573–3578.

Gutierrez, G., Reines, H. D., & Wulf-Gutierrez, M. E. (2004). Clinical review: Hemorrhagic shock. *Critical Care*, *8*, 373–381.

Herrington, J., Solly, K., Ratliff, K. S., Li, N., Zhou, Y.-P., Howard, A., et al. (2011). Identification of novel and selective Kv2 channel inhibitors. *Molecular Pharmacology*, *80*, 959–964.

Holtzclaw, J. D., Grimm, P. R., & Sansom, S. C. (2011). Role of BK channels in hypertension and potassium secretion. *Current Opinion in Nephrology and Hypertension*, *20*, 512–517.

Huang, X., & Jan, L. Y. (2014). Targeting potassium channels in cancer. *Journal of Cell Biology*, *206*, 151–162.

Imlach, W. L., Finch, S. C., Dunlop, J., Meredith, A. L., Aldrich, R. W., & Dalziel, J. E. (2008). The molecular mechanism of "ryegrass staggers", a neurological disorder of K^+ channels. *Journal of Pharmacology and Experimental Therapeutics*, *327*, 657–664.

Jensen, B. S. (2002). BMS-204352: A potassium channel opener developed for the treatment of stroke. *CNS Drug Reviews*, *8*, 353–360.

Kaczorowski, G. J., McManus, O. B., Priest, B. T., & Garcia, M. L. (2008). Ion channels as drug targets: The next GPCRs. *Journal of General Physiology*, *131*, 399–405.

King, V. F., Garcia, M. L., Himmel, D., Reuben, J. P., Pan, J.-X., Lam, Y.-K., et al. (1988). Interaction of tetrandrine with slowly-inactivating calcium channels: Characterization of calcium channel modulation by an alkaloid of Chinese medicinal herb origin. *Journal of Biological Chemistry*, *263*, 2238–2244.

Knaus, H. G., McManus, O. B., Lee, S. H., Schmalhofer, W. A., Garcia-Calvo, M., Helms, L. M., et al. (1994). Tremorgenic indole alkaloids potently inhibit smooth muscle high-conductance Ca^{2+}-activated K^{+} channels. *Biochemistry*, *33*, 5819–5828.

Korsgaard, M. P. G., Hartz, B. P., Brown, W. D., Ahring, P. K., Strobaek, D., & Mirza, N. R. (2005). Anxiolytic effects of Maxipost (BMS-204352) and retigabine via activation of neuronal K_v7 channels. *Journal of Pharmacology and Experimental Therapeutics*, *314*, 282–292.

Large, R. J., Kshatri, A., Webb, T. I., Roy, S., Akande, A., Bradley, E., et al. (2015). Effects of the novel BK ($K_{Ca}1.1$) channel opener GoSlo-SR-5-130 are dependent on the presence of BK_β subunits. *British Journal of Pharmacology*, *172*, 2544–2556.

Lee, D., Marino, J. P., & Zhao, Y. (2005). Novel compounds. WO/2005/009993, Smithkline Beecham.

MacKinnon, R., & Miller, C. (1988). Mechanism of charybdotoxin block of the high-conductance, Ca^{2+}-activated K^{+} channel. *Journal of General Physiology*, *91*, 335–349.

McManus, O. B. (2014). HTS assays for developing the molecular pharmacology of ion channels. *Current Opinion in Pharmacology*, *15*, 91–96.

McManus, O. B., Harris, G. H., Giangiacomo, K. L., Feigenbaum, P., Reuben, J. P., Addy, M. E., et al. (1993). An activator of calcium-dependent potassium channels isolated from a medicinal herb. *Biochemistry*, *32*, 6128–6133.

McManus, O. B., Helms, L. M., Pallanck, L., Ganetzky, B., Swanson, R., & Leonard, R. J. (1995). Functional role of the beta subunit of high conductance calcium-activated potassium channels. *Neuron*, *14*, 645–650.

Miller, C., Moczydlowski, E., Latorre, R., & Phillips, M. (1985). Charybdotoxin, a protein inhibitor of single Ca^{2+}-activated K^{+} channels from mammalian skeletal muscle. *Nature*, *313*, 316–318.

Morimoto, T., Sakamoto, K., Sade, H., Ohya, S., Muraki, K., & Imaizumi, Y. (2007). Voltage-sensitive oxonol dyes are novel large-conductance Ca^{2+}-activated K^{+} channel activators selective for beta1 and beta4 but not for beta2 subunits. *Molecular Pharmacology*, *71*, 1075–1088.

N'Gouemo, P. (2011). Targeting BK (big potassium) channels in epilepsy. *Expert Opinion on Therapeutic Targets*, *15*, 1283–1295.

Nardi, A., & Olesen, S.-P. (2008). BK channel modulators: A comprehensive overview. *Current Medicinal Chemistry*, *15*, 1126–1146.

Norris, P. J., Smith, C. C., De Belleroche, J., Bradford, H. F., Mantle, P. G., Thomas, A. J., et al. (1980). Actions of tremorgenic fungal toxins on neurotransmitter release. *Journal of Neurochemistry*, *34*, 33–42.

Ota, T., Aihara, M., Narumiya, S., & Araie, M. (2005). The effects of prostaglandin analogues on IOP in prostanoid FP-receptor deficient mice. *Investigative Ophthalmology & Visual Science*, *46*, 4159–4163.

Pomposiello, S. I., Alva, M., Wilde, D. W., & Carretero, O. A. (1998). Linoleic acid induces relaxation and hyperpolarization of the pig coronary artery. *Hypertension*, *31*, 615–620.

Ponte, C. G., Mcmanus, O. B., Schmalhofer, W. A., Shen, D.-M., Dai, G., Stevenson, A., et al. (2012). Selective direct activation of high-conductance, calcium-activated potassium channels causes smooth muscle relaxation. *Molecular Pharmacology*, *81*, 567–577.

Roy, S., Akande, A. M., Large, R. J., Webb, T. I., Camarasu, C., Sergeant, G. P., et al. (2012). Structure-activity relationships of a novel group of large-conductance Ca^{2+}-activated K^+ (BK) channel modulators: The Go-Slo-SR family. *ChemMedChem*, *7*, 1763–1769.
Roy, S., Large, R. J., Akande, A. M., Kshatri, A., Webb, T. I., Domene, C., et al. (2014). Development of GoSlo-SR-5-69, a potent activator of large conductance Ca^{2+}-activated K^+ (BK) channels. *European Journal of Medicinal Chemistry*, *75*, 426–437.
Sanchez, M., & McManus, O. B. (1996). Paxilline inhibition of the alpha-subunit of the high-conductance calcium-activated potassium channel. *Neuropharmacology*, *35*, 963–968.
Scornik, F. S., Bucciero, R. S., Wu, Y., Selga, E., Calero, C. B., Brugada, R., et al. (2013). $DiBAC_4(3)$ hits a "sweet spot" for the activation of arterial large-conductance Ca^{2+}-activated potassium channels independently of the β_1-subunit. *American Journal of Physiology Heart and Circulatory Physiology*, *304*, H1471–H1482.
Singh, S. B., Goetz, M. A., Zink, D. L., Dombrowski, A. W., Polishook, J. D., Garcia, M. L., et al. (1994). Maxikdiol: A novel dihydroxyisoprimane as an agonist of maxi-K channels. *Journal of the Chemical Society, Perkin Transactions*, *1*, 3349–3352.
Singh, H., Lu, R., Bopassa, J. C., Meredith, A. L., Stefani, E., & Toro, L. (2013). MitoBK (Ca) is encoded by the *Kcnma1* gene, and a splicing sequence defines its mitochondrial location. *Proceedings of the National Academy of Sciences of the United States of America*, *110*, 10836–10841.
Stretton, D., Miura, M., Belvisi, M. G., & Barnes, P. J. (1992). Calcium-activated potassium channels mediate prejunctional inhibition of peripheral sensory nerves. *Proceedings of the National Academy of Sciences of the United States of America*, *89*, 1325–1329.
Stumpff, F., Strauss, O., Boxberger, M., & Wiederholt, M. (1997). Characterization of maxi-K-channels in bovine trabecular meshwork and their activation by cyclic guanosine monophosphate. *Investigative Ophthalmology & Visual Science*, *38*, 1883–1892.
Sugg, E. E., Garcia, M. L., Reuben, J. P., Patchett, A. A., & Kaczorowski, G. J. (1990). Synthesis and structural characterization of charybdotoxin, a potent peptidyl inhibitor of the high conductance calcium-activated potassium channel. *Journal of Biological Chemistry*, *265*, 18745–18748.
Tanner, M. R., Hu, X., Huq, R., Tajhya, R. B., Sun, L., Khan, F. S., et al. (2015). KCa1.1 inhibition attenuates fibroblast-like synoviocyte invasiveness and ameliorates disease in rat models of rheumatoid arthritis. *Arthritis & Rheumatology*, *67*, 96–106.
Thieme, H., Stumpff, F., Ottlecz, A., Percicot, C. L., Lambrou, G. N., & Wiederholt, M. (2001). Mechanisms of action of unoprostone on trabecular meshwork contractility. *Investigative Ophthalmology & Visual Science*, *42*, 3193–3201.
Torres, Y. P., Granados, S. T., & Latorre, R. (2014). Pharmacological consequences of the coexpression of BK channel α and auxiliary β subunits. *Frontiers in Physiology*, *5*, 1–15.
Tsantoulas, C., & McMahon, S. B. (2014). Opening paths to novel analgesics: The role of potassium channels in chronic pain. *Trends in Neurosciences*, *37*, 146–158.
Valverde, M. A., Cantero-Recasens, G., Garcia-Elias, A., Jung, C., Carreras-Sureda, A., & Vicente, R. (2011). Ion channels in asthma. *Journal of Biological Chemistry*, *286*, 32877–32882.
Webb, T. I., Kshatri, A. S., Large, R. J., Akande, A. M., Roy, S., Sergeant, G. P., et al. (2015). Molecular mechanisms underlying the effect of the novel BK channel opener GoSlo: Involvement of the S4/S5 linker and the S6 segment. *Proceedings of the National Academy of Sciences of the United States of America*, *112*, 2064–2069.
Xin, W., Cheng, Q., Soder, R. P., & Petkov, G. V. (2012). Inhibition of phosphodiesterases relaxes detrusor smooth muscle via activation of the large-conductance voltage- and Ca^{2+}-activated K^+ channel. *American Journal of Physiology Cell Physiology*. *302*, C1361–C1370.

Yang, H., & Cui, J. (2015). BK channels. In J. Zheng & M. C. Trudeau (Eds.), *Handbook of ion channels* (pp. 227–239). Boca Raton, FL: CRC Press.

Yao, Y., Peter, A. B., Baur, R., & Siegl, E. (1989). The tremorgen aflatrem is a positive allosteric modulator of the gamma-aminobutyric acid A receptor channel expressed in Xenopus oocytes. *Molecular Pharmacology*, *335*, 319–323.

Zeng, H., Gordon, E., Lin, Z., Lozinskaya, I. M., Willette, R. N., & Xu, X. (2008). 1-[1-Hexyl-6-(methyloxy)-1H-indazol-4-yl]-2-methyl-1-propanone, a potent and highly selective small molecule blocker of the large-conductance voltage-gated and calcium-dependent K^+ channel. *Journal of Pharmacology and Experimental Therapeutics*, *327*, 168–177.

Zhao, G., Zhao, Y., Pan, B., Liu, J., Huang, X., Zhang, X., et al. (2007). Hypersensitivity of BKCa to Ca^{2+} sparks underlies the hyporeactivity of arterial smooth muscle in shock. *Circulation Research*, *101*, 493–502.

Zhou, Y., & Lingle, C. J. (2014). Paxilline inhibits BK channels by an almost exclusively closed-channel block mechanism. *Journal of General Physiology*, *144*, 415–440.

INDEX

Note: Page numbers followed by "*f*" indicate figures, and "*t*" indicate tables.

A

N

CONTENTS OF RECENT VOLUMES

Volume 39

Volume 40

Volume 41

Volume 42

Volume 43

Volume 44

Volume 45

Volume 46

Volume 47

Volume 48

Volume 49

Volume 50

Volume 51

Volume 52

Volume 53

Volume 54

Volume 60

Volume 61

Volume 62

Volume 63

Volume 67

Volume 68

Volume 69

Volume 70

Volume 71

Volume 72

Volume 73

Volume 74

Volume 75

Volume 76

Volume 77

Volume 78

Volume 79

Volume 80

Volume 81

Volume 82

Volume 83

Volume 84

Volume 85

Volume 86

Volume 87

Volume 88

Volume 89

Volume 90

Volume 91

Volume 92

Volume 93

Volume 94

Volume 95

Volume 96

Volume 97

Volume 98

Volume 99

Volume 100

Volume 101

Volume 102

Volume 103

Volume 104

Volume 105

Volume 106

Volume 107

Volume 108

Volume 109

Volume 110

Volume 111

Volume 112

Volume 113

Volume 114

Volume 115

Volume 116

Volume 117

Volume 118

Volume 119

Volume 120

Volume 121

Volume 122

Volume 123

Volume 124

Volume 125

Volume 126

Volume 127

FSC
www.fsc.org
MIX
Papier | Fördert
gute Waldnutzung
FSC® C083411

Kolibri 360 GmbH
Ferdinand-Jühlke-Straße 7
99095 Erfurt, Deutschland
produktsicherheit@kolibri360.de